Foothold in the Heavens

The Seventies

Ben Evans

Foothold in the Heavens

The Seventies

 Springer

Published in association with
Praxis Publishing
Chichester, UK

Ben Evans
Space Writer
Atherstone
Warwickshire
UK

SPRINGER–PRAXIS BOOKS IN SPACE EXPLORATION
SUBJECT *ADVISORY EDITOR*: John Mason, M.B.E., B.Sc., M.Sc., Ph.D.

ISBN 978-1-4419-6341-3   e-ISBN 978-1-4419-6342-0
DOI 10.1007/978-1-4419-6342-0

Springer Dordrecht Heidelberg London New York

Library of Congress Control Number: 2010925487

Springer Science + Business Media, LLC, © 2010

Cover design: Jim Wilkie
Copy editor: David M. Harland
Typesetting: BookEns, Royston, Herts., UK

Printed on acid-free paper

Springer is a part of Springer Science + Business Media (www.springer.com)

Contents

Illustrations

Author's preface

When I set out to write a five-volume history of humanity's exploration of space, it seemed a big project, though fairly straightforward. Starting with Yuri Gagarin's pioneering voyage in April 1961, the journey through five dramatic decades promised to be an exciting one, with specific breakpoints between the volumes: the resumption of manned lunar landings in the Seventies, the arrival of the Shuttle in the Eighties, the development of the International Space Station in the Nineties and the increasing 'privatisation' of getting people into orbit and the first fee-paying 'tourists' in the Noughties. My intention was for something a little more detailed than a basic log of manned expeditions into space, but as time has rolled on, the project evolved into a far larger and more complex task than envisaged.

I must, therefore, ask the reader to forgive me for failing to strictly track an *entire* decade with each volume. The first volume, *Escaping the Bonds of Earth*, had to take into account some of the advancements of the Fifties as a prerequisite to focusing on 'its' decade, the Sixties. In this second volume, *Foothold in the Heavens*, I had to focus on particular episodes in considerable depth – the historic flight of Apollo 11 being a notable example – at the expense of covering an entire decade. Furthermore, I realised that spaceflight was not, and *is* not, a unique phenomenon, outside of public or political control. Rather, it has been an integral part of our social, economic and cultural fabric, and the lightning speed or snail's-pace slowness of its progress through the decades has been increasingly dictated by outside influences: the Bay of Pigs, the mythical 'missile gap' between the United States and the Soviet Union and the Cuban Crisis of 1962 were all instrumental in determining space policy. In the early Seventies, the progressive thawing of relations between the United States and the Soviet Union similarly impacted their respective space programmes. I feel it would be unconscionable to discuss our progress in space without paying tribute to *why* we were doing so, the obstacles we had to overcome in order to get there and the opinions, attitudes and feelings of the people whose dollars and roubles were paying for such endeavours.

By the last year of the Sixties, almost three dozen astronauts and cosmonauts had journeyed into space. They had circled the globe in their pressurised ships, they had ventured outside in bulky, life-sustaining suits and they had embarked on the first mission to the Moon. Culturally, socially and technologically, the world around them had changed in a thousand ways. Against the backdrop of a deepening crisis in Vietnam, floodlit by the problems of the United States at war with itself over issues of racial and social inequality and with an increasingly disaffected youth struggling to make its voice heard, the American programme lost much of the popular appeal that it had once enjoyed. In the Soviet Union, the increasing regression and repression of Leonid Brezhnev's regime led to widespread condemnation and mass emigrations from a country that considered itself the embodiment of the beauties of the communist state. Yet, ironically, it was actually the steady thaw of relations between these bitter foes which spelled the end of limitless budgets for human space exploration.

By the early Seventies, men had walked on the Moon and had occupied space stations in orbit around Earth and both the Soviets and the Americans *could* still take pride in their achievements. Astronauts, managers, scientists and even some politicians could see no reason why a manned expedition to Mars and a permanently-occupied lunar base should *not* be achieved before the end of the century. It might not be on the scale of Arthur C. Clarke's imaginings, but it was certainly more than a mere dream. However, for an increasingly apathetic public, the cost was excessive. The race to reach the lunar surface in the Sixties would be superseded by a considerably more frugal attack on the heavens in the Seventies. As Apollo wound down, America's efforts switched to developing a cheaper means of accessing space: the Space Shuttle, which would bring both tragedy and triumph and effectively confine the nation to low-Earth orbit for the next several decades. Similarly, the Soviet leadership lost interest: its practice of signing blank cheques for Chief Designers to score major propaganda victories against the capitalist West declined markedly. Its ambitious effort to put men on the Moon fell apart and it, too, adopted a more conservative, gradual system of mastering the new frontier of space.

My intention in writing this second volume has been to briefly explore some of the reasons why the political, social, cultural and economic climate changed so significantly for both superpowers in the final years of the Sixties and those first few pivotal years of the Seventies and how this impacted our aspirations in space. For as Westerners grew their hair and sideburns steadily longer, so the budgets for space exploration steadily shortened and steadily thinned.

Four decades later, we continue to live with the consequences of those frugal times, for our species remains chained in low-Earth orbit. Today, we are further away from having our representatives return to the Moon than John Kennedy was when he made his famous speech in May 1961. The establishment of Earth-circling space stations, although ever larger, more advanced, more sophisticated and more 'international', has done little to escape what Konstantin Tsiolkovski once called the cradle of humanity. In a sense, therefore, the early portion of the Seventies was juxtaposed by two ironic, opposing themes, for as the last Apollo crew left the Moon our species reached the zenith of its accomplishments – and abruptly *stopped*. Four decades later, though we have had, and continue to have, a secure foothold in the heavens, we are still waiting for the next team of intrepid explorers to carry us deeper into the Universe around us.

Ben Evans,
Atherstone, February 2010

Acknowledgements

This book would not have been possible without the support of a number of individuals to whom I am enormously indebted. I must firstly thank my wife, Michelle, for her constant love, support and encouragement throughout the time it has taken to plan, research and write this manuscript. As always, she has been uncomplaining during the weekends and holidays when I sat up late typing on the laptop or poring through piles of books, old newspaper cuttings, magazines, interview transcripts, press kits or websites. It is to her, with all my love, that I would like to dedicate this book. My thanks also go to Clive Horwood of Praxis for his enthusiastic support and to David Harland for reviewing the manuscript; I appreciate not only their advice, but also their patience in what has been an overdue project which proved more difficult to write than I had envisaged. I must once again thank Ed Hengeveld for supplying many of the illustrations. Others to whom I owe a debt of gratitude include Sandie Dearn, Ken, Alex and Jonathan Jackson and Malcolm and Helen Chawner. To those friends who have encouraged my fascination with all things 'space' over the years, many thanks: to Andy Salmon and Andy Rowlands, Dave Evetts and Mike Bryce and Rob and Jill Wood. Our two golden retrievers – the ever-hungry Rosie and the the attention-seeking Milly – have provided a ready source of light relief. Lastly, there are two others that I would like to mention as having played a great role in my life in the last year. Firstly, my friend Roy Lane, who tragically lost his battle with cancer in March 2009, and secondly, our beautiful nephew, Callum Jay Hooper, born on St George's Day, 23 April 2009, whose never-ending smile and chuckling laugh are a joy to behold.

1

New directions

A DIFFERENT PATH

As the Sixties entered their final year, humanity stood on the brink of an event which, in Richard Nixon's words, would come to be known as "the greatest since the Creation". By the end of 1969, not one, but four Americans would have trodden the Moon's dusty surface, fulfilling a bold promise made by President John Kennedy eight years earlier. The Soviet Union, whose own tally of celestial successes included the world's first artificial satellite, the first man in space and the first spacewalk, was forced to watch dumbstruck from the sidelines. To be fair, the writing had been on the wall for some time: that all hopes of beating the United States to the Moon and planting a flag bearing the Hammer and Sickle into the lunar soil had been irretrievably lost.

Still, the Soviet dream of reaching our nearest celestial neighbour would not evaporate for some years: in the early Seventies they would land a pair of automated rovers, the Lunokhods, on the Sea of Rains and in Le Monnier Crater, and would test, albeit unsuccessfully, their own enormous N-1 Moon rocket. As impressive as these feats were, in the public eye they represented little more than a diversion from what had actually been a breathtaking triumph for the United States and its ideals of liberal democracy and freedom of speech. The Lunokhods gathered a wealth of valuable scientific data, it is true, but the sight of astronauts – *real* human beings – bouncing around in one-sixth of terrestrial gravity garnered far more popular appeal. At least, that is, in the early days. Moreover, the N-1 itself, despite being more powerful than America's Saturn V, would fail repeatedly even to attain Earth orbit. The last of its kind vanished in a fireball shortly after launch one morning in November 1972.

It seemed obvious that, after denying for some years that they even *had* plans to put a man on the Moon, the Soviets needed to fundamentally shift the paradigm of their spacegoing philosophy in a quite different direction. By the spring of 1969, they finally had the means to do so. It came in the form of a three-part spacecraft known as 'Soyuz' ('Union') and, late the previous year, an ebullient cosmonaut named Georgi Beregovoi had nursed it through a four-day shakedown cruise. He had tested its ability to manoeuvre in space and rendezvous with an unmanned Soyuz, but failed to dock the pair together.

In time, Soyuz would become the most-used manned spacecraft in history. As of 2009, variants of its original design have ferried over a hundred crews of different nationalities into orbit and supported more than half a dozen space stations. The goal of establishing a long-term human presence in Earth orbit would supersede and effectively replace the Soviets' failed lunar dream. On 14 January 1969, the first steps of this 'new' dream would be taken, when two Soyuz rendezvoused and docked and a pair of cosmonauts spacewalked from one craft to the other. On the face of it, the exercise, though risky, was highly successful, but not for almost three decades would the outside world learn of its close brush with disaster.

Disaster and tragedy had become virtual bywords for Soyuz since its conception, particularly in the wake of its first manned flight. In April 1967, cosmonaut Vladimir Komarov was launched aboard Soyuz 1, in the expectation that he would rendezvous and dock with Soyuz 2, crewed by Valeri Bykovsky, Alexei Yeliseyev and Yevgeni Khrunov. During the joint flight, Yeliseyev and Khrunov would don pressurised suits and clamber from Soyuz 2 over to Soyuz 1. They would then return to Earth with Komarov, leaving Bykovsky to land alone in Soyuz 2.

Unfortunately, only hours after Soyuz 1 and Komarov reached orbit, the first problems reared their heads. A solar array failed to unfurl, knocking out more than half of the spacecraft's electrical supply; then glitches with orientation and other sensors rendered the planned mission impossible. Soyuz 2 was cancelled and the focus shifted to bringing Komarov home safely, but this failed when the parachute lines became entangled during descent. The spacecraft hit the ground with all the force of an unrestrained meteorite, killing its occupant instantly. Not until October 1968 and the flight of Beregovoi would the much-improved Soyuz finally show what it could do.

A rendezvous between Beregovoi's Soyuz 3 and the unmanned Soyuz 2 went well, but was dramatically upstaged by the American flight of Apollo 8 in December, during which astronauts Frank Borman, Jim Lovell and Bill Anders became the first men to orbit the Moon. Less than three weeks later, the next stage of the Soviet plan to establish an Earth-circling space station was ready: the ambitious docking and spacewalking extravaganza, thwarted in 1967, would finally become reality. It would not have the same impact as a lunar mission, admittedly, but during the joint flight cosmonauts Yeliseyev and Khrunov would seize a couple of impressive 'firsts'. They would become the world's first two-man spacewalking team and complete the first-ever transfer of crew members from one craft to another, landing in a different vehicle from the one in which they were launched. Ironically, for one of the other cosmonauts, the mission would also entail a hazardous re-entry and he would come within a whisker of losing his own life.

UNION IN SPACE

Soyuz was the brainchild of Sergei Korolev, the famous 'Chief Designer' of early Soviet spacecraft and rockets, with the original intention of undertaking both Earth-orbiting missions and an ambitious series of lunar ventures to rival the United States' Apollo effort. As early as 1964, the design and definition of Soyuz was well underway and technical documentation and a 'boilerplate' mockup revealed it to be a craft capable of lofting two, and later three, cosmonauts. In his seminal work *Challenge to Apollo*, Asif Siddiqi noted that when Korolev saw the mockup for the first time, he is said to have declared proudly that Soyuz was "the machine of the future".

With the exception of his vague moniker, not until long after his death on a hospital operating table would the outside world learn anything of substance concerning Sergei Pavlovich Korolev. Yet this man of outstanding engineering genius had masterminded some of the most remarkable triumphs in the exploration of space. It was he who designed the R-7 missile which launched first Sputnik and then Yuri Gagarin, it was his design bureau which assembled Vostok – the world's first piloted spacecraft – and it was he who oversaw and orchestrated the first three-man orbital mission, the first spacewalk and his nation's first faltering steps towards the Moon. His brilliance and unwavering devotion to a lifelong dream of exploring the heavens was balanced by an all-or-nothing obstinacy which often manifested itself in a violent temper, capable of exploding without warning. He had lived a hard, driven and thankless life of service to the Soviet state and it was this, ultimately, which consumed him.

Born in 1907 in the central Ukraine, his interest in aviation and rocketry emerged at a young age. Under Joseph Stalin's regime, with its ingrained fear of the power of the individual, there had been little opportunity for the *intelligentsia* to prosper and, as a highly regarded engineer, Korolev quickly found himself arrested and sentenced to ten years of hard labour in the notorious Kolyma gulag. The Nazi invasion of the Soviet Union in 1941 prompted his release in support of the war effort. Later, he set to work developing an arsenal of rockets and missiles, which he hoped could someday transport instruments into the high atmosphere and, finally, into space. His masterpiece, the R-7, though principally intended for the Soviet military as an intercontinental ballistic missile, would be used to put satellites and humans into orbit. By giving it a dual-purpose use, he displayed a trait of his canny character, keeping his military critics quiet by satisfying their needs in parallel with his own.

In the early Sixties, the regime of Stalin's successor, Nikita Khrushchev, was generally supportive of Korolev and his projects, which yielded a regular delivery of space 'firsts' that the feisty and erratic Soviet premier could use to enforce an ideological advantage over the United States. The boorish Khrushchev was far more interested in the glamour, political and military impact of spacegoing rocketry and, to an extent, that was fine with Korolev because it provided him with ready supplies of manpower and funding to pursue his space ambitions. It remains a pity, though, that he never truly received the recognition that he deserved in life. After Gagarin's triumphant orbit of the Earth in April 1961, the Chief Designer had been barred from publicly wearing his medals and even had to thumb a lift into Moscow when his car broke down. Efforts by the Nobel Prize Committee to establish an award for this unknown man fell on deaf ears.

It is bitterly ironic, therefore, that his untimely death during a routine stomach operation in January 1966 should have finally opened a chink in the Soviet armour and uncovered something of the mysterious Chief Designer as a real person for the first time. Within the cosmonaut corps, his death was immediately recognised for the calamity that it was: Yuri Gagarin, in a solemn eulogy, described Korolev as being "synonymous with one entire chapter in the history of mankind" and Khrushchev's successor, First Secretary Leonid Brezhnev, was one of the pallbearers who carried the ashes for interment in the Kremlin Wall.

Many cosmonauts felt that the men who followed Korolev – his deputy, Vasili Mishin, together with Georgi Babakin, Vladimir Chelomei and the famous rocket engine designer Valentin Glushko – exhibited entirely different personalities which damaged the Soviet Union's chances of beating the Americans to the lunar surface. Had Korolev lived just a few years longer, wrote cosmonaut Alexei Leonov, "we would have been the first to circumnavigate the Moon".

Korolev had always described Voskhod 2 – the March 1965 mission during which Leonov became the first man to walk in space – as his life's last great work. However, remarkable as that achievement was, it was perhaps the development of Soyuz which has had the most long-lasting impact on the world. Since its first manned flight, a year after Korolev's death, Soyuz and a heavily modified version of the original R-7 continue to be used operationally today; a fitting legacy to an enduring talent.

In his 1988 book, Phillip Clark traced its history back to a three-part 'Soyuz complex' – a manned craft, a 'dry' rocket block and a propellant-carrying tanker – which Korolev had envisaged in the early Sixties being assembled in low-Earth orbit to fly circumlunar missions. The first part, which Clark identified as 'Soyuz-A', but which the Soviets catalogued as 'Soyuz-7K', was closest in physical appearance to the spacecraft which actually flew and it was to this that Korolev committed himself and his Kaliningrad-based OKB-1 design bureau in March 1963. Measuring 7.7 m long, Soyuz-7K had three components: a cylindrical 'orbital module', a bell-shaped 'descent module' for the crew and a cylindrical 'instrument module' for manoeuvring equipment, propellant and electrical power. According to Korolev's earliest blueprints, it weighed around 6,450 kg, but, unlike the final design, was not equipped with solar panels.

Supporting Soyuz-7K were the 'dry' Soyuz-B rocket block and the Soyuz-V propellant tanker, known to the Soviets by the designations of '9K' and '11K', respectively. Clark hinted that a typical flight profile would have begun with the launch of a 9K, followed, at 24-hour intervals, by as many as four 11Ks, which would dock, transfer their propellant loads and then separate. When the 9K had been fully fuelled, a manned 7K would be despatched to dock with the rocket block. "Mastering rendezvous and docking operations in Earth orbit may have been one of the primary objectives of the Soyuz complex," wrote Asif Siddiqi, "but the incorporation of five consecutive dockings in Earth orbit to carry out a circumlunar mission was purely because of a lack of rocket-lifting power in the Soviet space programme." In fact, it was the sheer 'complexity' of the Soyuz complex which seems to have foreshadowed its restructuring sometime in 1964 and effected a delay of its maiden voyage until at least the spring of 1966.

By the end of the decade, seven manned Soyuz spacecraft would have rocketed into orbit. However, a key physical difference between these vehicles and the original 7K was that they employed a pair of rectangular solar panels, mounted on the instrument module, to generate electrical power. The total surface area of these wing-like appendages was 14 m^2, with each wing measuring 3.6 m long and 1.9 m wide. The rest of the craft's design was strikingly similar to the 7K: a spheroidal orbital module, 2.65 m long and 2.25 m wide, the bell-shaped descent module, itself 2.2 m long and 2.3 m wide at the base, and the instrument module, a cylinder 2.3 m long and 2.3 m wide.

This shape emerged at the end of almost a decade of planning, theoretical work and aerodynamic modelling. As early as 1958, Mikhail Tikhonravov and Konstantin Feoktistov, engineers at Korolev's bureau, envisaged a multi-purpose spacecraft capable of both Earth-circling and circumlunar missions. As has been written by Rex Hall and Dave Shayler, the shape of the descent module was decided at least partly by a desire to touch down on land, rather than in water, and several designs were sketched out. The first utilised aerodynamic surfaces, facilitating an aircraft-like return to a runway, whilst the second adopted a 'missile principle', entering space in a ballistic manner and descending beneath parachutes.

By 1961, concerns about mass and the need for adequate thermal protection during re-entry had effectively eliminated the winged design from consideration. The missile principle, though, needed further work in order to man-rate it: a ballistic re-entry would impose significant duress on both the vehicle and its occupants and Tikhonravov and Feoktistov moved instead toward the concept of a 'glancing' re-entry in order to reduce stress. If the new craft was ever to undertake lunar flights, its return trajectory from the Moon would produce correspondingly higher re-entry speeds of perhaps 40,000 km/h, prompting the engineers to design a 'double-dip' profile which, by reducing the velocity in stages, would lessen the G loads on the cosmonauts and thermal stress on the vehicle.

When consensus had been reached on the method of re-entry, a trio of designs were explored by OKB-1 engineers and researchers at the NII-1 and NII-88 aerodynamic institutes: one nicknamed the 'segmented sphere', another called the 'sphere with a needle' and a third dubbed the 'sliced sphere'. The segmented version emerged as the most promising design, with Vladimir Roshchin's group at OKB-1 promoting a descent module with a displaced centre of gravity as a means of generating aerodynamic lift. By 1962, this had evolved into a shape approximating a car's headlamp or a beehive, which aerodynamic simulations predicted would avoid the high deceleration and thermal loads of a ballistic descent and have sufficient lift to be able to steer towards a given landing site. A plethora of proposals also surrounded Soyuz' means of landing, with helicopter-like rotors, fan-jet or liquid-propelled engines, controlled parachutes, ejection seats and shock-absorbing inflatable balloons all being considered. By 1963, however, Korolev approved the design which remains in use today: a combination of braking parachutes and a soft-landing apparatus of solid-fuelled rockets.

Even with these important decisions made, the appearance and layout of the spacecraft remained somewhat fluid, with early designs for a space station ferry called 'Sever' and a lunargoing concept known as 'L-1' both adopting its concept of a descent module for the crew, attached to an instrument module for propulsion and power. Already, though, the design was expanding further to encompass a habitable orbital module and there was disagreement about where this should be located. In some initial drawings it appeared between the instrument module and the descent module and in others it was above the descent module. The idea of placing the orbital module *below* the descent module was soon rejected since it would necessitate cutting a hatch into the descent module's base, potentially compromising its heat shield. The final layout, with the descent module in the middle, was in place by the end of 1962. By this time, it had also received the name of 'Soyuz'.

In spite of Korolev's assertion that it was the machine of the future, Soyuz had been mired for some years in technical and bureaucratic problems, to such an extent that by 1964 its development was virtually paralysed by the Soviet drive for the Moon. Early plans called for it to carry one or two cosmonauts, but by December 1963 the basic design of the Earth-circling version, known as 'Soyuz-7K-OK' ('Orbitalny Korabl' or 'Orbital Ship'), had grown to accommodate a three-member crew. Its purpose would be to support automated rendezvous and docking, spacewalking, manoeuvring and scientific research, thereby fulfilling a number of key requirements for its eventual use as a space station ferry.

During 1964, Korolev directed a small group under Boris Chertok, one of the deputy chief designers at OKB-1, to explore alternative uses for the basic 7K-OK craft. One proposal called for docking two Soyuz together in orbit, thus demonstrating their rendezvous capabilities, and having a cosmonaut spacewalk from one ship to the other. Not only would this ambitious plan offer valuable engineering experience in its own right, but it would also support early ideas for a Soyuz-based Moon mission in which a cosmonaut would transfer from the command ship to the landing craft in lunar orbit by 'extravehicular activity' (EVA). In February 1965, Korolev presented this 'new' version of Soyuz, with an emphasis on near-Earth operations, to the Scientific-Technical Council of the State Committee for Defence Technology and received the approval he needed.

Beginning at its base, the instrument module, also known as the 'service module', carried chemical batteries and two large solar panels to charge them, together with a thermo-regulation radiator and an integrated propulsion and attitude-control system. The latter, designated 'KTDU-35', comprised a pair of engines, one primary and the other backup, sharing the same oxidiser and fuel supply. The primary engine had a thrust of 417 kg and was capable of a change in velocity of some 2,750 m/sec, equivalent to a specific impulse of around 280 seconds. On the basis of early reports which speculated that this engine could boost Soyuz to an altitude of 1,300 km, Phillip Clark suggested that the spacecraft must have a propellant capacity of 755 kg. The propellants took the form of unsymmetrical dimethyl hydrazine and an oxidiser of nitric acid, carried in spherical tanks inside the instrument module. Attitude

A Soyuz spacecraft is prepared for launch at Tyuratam. Note the spheroidal orbital module at the top and the beehive-shaped descent module beneath. Part of the cylindrical instrument module is also visible.

control was by a set of 22 primary and eight backup hydrogen peroxide thrusters. Guidance, rendezvous, communications and environmental equipment occupied the remainder of the cylindrical compartment.

The descent module, whose shape has variously been likened to a bell, a beehive or a car's headlamp, sat directly above the instrument module and housed the crew during ascent and re-entry. It had a habitable volume of some 2.5 m^3. The commander's seat was located in the centre, flanked by a flight engineer and a research or 'test' engineer. Many of the Soyuz' flight regimes would be pre-programmed from the ground. Consequently the main instrument panel presented the crew with readouts and visual displays of the performance of on-board systems, together with a monitor for the external television camera, an optical orientation viewfinder called 'Vzor' for attitude manoeuvres and the Globus device to show the spacecraft's position above Earth. In the event of a failure of the automatic systems, and to facilitate rendezvous and docking, it was expected that the commander could assume manual control. As a result, a pair of joystick-like hand controllers – one for changing the vehicle's velocity and the other for its attitude in space – were located directly beneath the instrument panel in front of the commander.

Rendezvous and docking were supported by the Vzor, together with a system of gyroscopes, attitude-control sensors and thrusters and the 'Igla' ('Needle') radar. The latter would automatically navigate the spacecraft to its target and draw to a halt at a range of 200–300 m, after which the crew would take charge and accomplish the final approach and docking. The systems needed to enable physical contact had undergone extensive development since 1962. At first, OKB-1 engineers Viktor Legostayev and Vladimir Syromiatnikov advocated a 'pin-cone' device to allow two vehicles (one in a 'passive' role, the other 'active') to dock. At this stage, however, there was no provision for the internal, 'shirt-sleeves' transfer of cosmonauts from one craft to the other and, sometime in 1965, Korolev's proposal to change this was rejected by Feoktistov on the basis that a significant amount of work had already been done and additional revisions would put the development further behind schedule. The docking system featured a pin on the active spacecraft, which would be captured by a cone-like funnel on the passive one, essentially cancelling any remaining velocity or angular displacement.

The descent module would be the only component capable of surviving the intense heat of atmospheric re-entry and bringing the cosmonauts back to Earth. Towards the end of a mission, the instrument and orbital modules would be jettisoned and the descent module would employ half a dozen hydrogen peroxide engines, each producing a thrust of 10 kg, to provide roll, pitch and yaw controllability prior to and during the early portions of re-entry. To protect its occupants, it was coated with a heat-resistant ablator, together with a thermal shield at its base that would detach shortly before touchdown to reveal the four solid-propellant landing rockets. A 14 m^2 drogue parachute would automatically deploy 9.5 km above the ground, followed by a main canopy. Seconds before hitting the ground, an altimeter would command the landing rockets to fire to cushion the impact.

Atop the descent module, the spheroidal orbital module contained a bunk, a cupboard for food and water supplies, life-support gear, controls for experiments, cameras and a variety of other equipment appropriate to each individual mission. It was, wrote Boris Mandrovsky in January 1970, "intended for rest, recreation and sleep, as well as for scientific research". For the first, and only, independent mission involving a spacewalk (Soyuz 4/5), the space suits would be stored in the orbital module. However, in view of the fact that its internal volume was barely 6.5 m^3, the difficulty of this task should not be underestimated. Indeed, in May 1968 cosmonaut Boris Volynov, clad in a training version of the suit, was brought face to face with these difficulties: it was hard – unsafe, even – to squeeze through the hatches between the descent and orbital modules and then get outside.

Difficulties aside, Soyuz promised to be one of the safest manned spacecraft ever built, possessing as it did the Soviets' first 'true' launch escape system. This consisted of a tower atop the R-7's payload shroud and a multiple-nozzle rocket engine. In the event of an emergency during the period from 20 minutes before launch until about 160 seconds into the ascent, the shroud would split at the base of the descent module and the escape tower's engine would lift the descent and orbital modules to safety. At the top of the arc, the descent module would be released to parachute back a couple of kilometres from the pad. Early predictions estimated that the crew could be exposed to gravitational loads as high as 10 G during such a scenario.

Launching Soyuz, which was considerably heavier and more complex than the earlier Vostok and Voskhod craft, demanded further improvement of Korolev's original R-7. The basic design of the missile remained physically the same: a two-stage behemoth, fed by liquid oxygen and a refined form of kerosene known as 'Rocket Propellant-1' (RP-1). Strapped around its lower stage were four tapering boosters, each 19.6 m long. The upper stage had an uprated engine which enhanced its thrust from 27,210 kg to 27,573 kg. This 'new' rocket, wrote Asif Siddiqi, was "an extremely rare case of a Soviet launch vehicle developed first for civilian goals". With the escape tower in place, the upgraded R-7 stood 49.3 m tall, produced 411,650 kg of thrust at liftoff – a 3 percent increase over the earlier Vostok version – and could insert a 6,900 kg payload into a 200 x 450 km orbit.

Like Vostok and Voskhod before it, the R-7 was rolled to the launch pad horizontally on a railcar. It is the method still used today. The spacecraft's propellants were loaded prior to its attachment to the rocket's upper stage, after which the shroud was installed and, following rollout, the entire stack was tilted upright. Four cradling arms, known as the 'tulip', supported the R-7 and a pair of towering gantries provided pre-launch access. Cosmonauts entered the Soyuz through a side hatch in the orbital module and dropped into their seats in the descent module. On 14 January 1969, Vladimir Alexandrovich Shatalov would become only the third cosmonaut to ride a Soyuz into orbit. Like his predecessors, he would launch alone. Unlike them, however, he would return to Earth with company.

MAN OF GRANITE

It was the dead of winter in Tyuratam – a time of bitter snow and hurricane-strength blizzards – when Shatalov, clad in fur boots, arrived at the launch pad on 13 January for his first flight into space. He would begin his momentous journey as all previous cosmonauts had done: from a bleak, featureless expanse of steppe, some 200 km east of the Aral Sea. In the local Kazakh tongue, the name of the place, 'Tyuratam', roughly translates as the gravesite of Tyura, beloved son of the great Mongol conqueror Genghis Khan, whose medieval empire spanned much of Asia. According to some sources, it began as an ancient cattle-rearing settlement on the north bank of the Syr Darya River, although at least one Soviet-era journalist has given it a more modern origin, hinting at its foundation as recently as 1901 as an outpost to replenish steam engines passing between Orenburg and Tashkent.

Tyuratam's importance over the past half a century, though, cannot be disputed. It was from this sparsely populated region, five decades ago, that the first steps of a journey far more audacious, much longer and considerably more difficult than any the Great Khan could have envisaged were taken. It was from this place that Yuri Gagarin, the first man in space, began his historic flight. It was from here that

Valentina Tereshkova became the first woman to travel into orbit and Alexei Leonov became the first person to 'walk' in space; and it is from here, indeed, that manned missions continue to be launched each year. It was to this remote corner of old Soviet Central Asia – a region swarming with scorpions, snakes and poisonous spiders, whose climate is characterised by vicious dust storms, soaring summertime highs of 50°C and plummeting wintertime lows of -25°C – that Lieutenant-Colonel Shatalov of the Soviet Air Force came to secure his unenviable place as Russia's 13th man in space.

Shatalov was born in Petropavlovsk in northern Kazakhstan on 8 December 1927. His father worked as a locomotive driver and the young boy grew up listening to the clang of carriages and the whistles of the brakemen. Yet it was not his father's stories of promotion from stoker to driver's assistant to driver and, finally, to office-based dispatcher that filled Shatalov with such pride. Rather, it was Alexander Shatalov's tales from the Great Patriotic War against the Nazis, when he drove trains to mend damaged communication lines and signalling equipment and helped lay the 'Road of Life' across frozen Lake Ladoga to link besieged Leningrad with the mainland. Even more exciting for the young Shatalov were his father's memories of seeing the very first Soviet pilots.

After secondary education, the boy studied and graduated from Kachinsk Military Pilot School and Monino Military Academy. He then worked as an instructor for some years and moved into the world of experimental test flight. One interesting anecdote from this time concerns Shatalov and his friend Valentin Mukhin, who approached the chief of the test pilot school to apply for admission ... and to recommend their comrade, Yevgeni Kukushev, too. The chief smiled and told them that Kukushev had come to see him a few days earlier with a similar request for all three of them! "I already have your names on my list," he concluded, and all three were eventually selected.

After several more years flying for a number of Soviet Air Force units and meeting and marrying his wife, Muza, Shatalov was chosen as a cosmonaut in January 1963. By this time, the requirements for these 'star sailors' had changed. When the first group, which included Yuri Gagarin, was selected in 1960, the focus was exclusively upon Soviet Air Force pilots who had proven themselves under exposure to hypoxia, high pressures and varying G loads and had undergone rigorous ejection-seat and parachute training. That requirement remained very much the same when Shatalov was picked, but with the exception that military engineers and navigators were also considered and university credentials were mandatory. Further, the age limit was lifted: instead of the 25–30 range in the first cosmonaut intake, the new class had to be 40 or younger. Aged 35 at the time of his selection, Shatalov made the cut. Consequently, unlike the cadre of relatively inexperienced fliers of the first group, Shatalov's class were far more accomplished pilots and also held advanced engineering degrees.

This second class of cosmonauts was interesting because Shatalov himself, then a senior instructor and pilot of the new Su-7 aircraft, had been approached by his commanding officer to select his five best fliers for consideration. When Shatalov saw the list of requirements, he realised that "the selection criteria apply to me".

Despite the commander's admonition that he was insane to apply – he had a comfortable position and rank, a bright future and probably would not make the final cut anyway – Shatalov persevered. (So too did one of his 'recommendees', a pilot named Anatoli Filipchenko.) At the time of selection, Shatalov had piloted 17 different types of aircraft, including helicopters, and had accrued 2,500 hours of flying time; ten times more than many of the 1960 cosmonauts.

Described by others as a man of impeccable reliability and a person capable of carrying heavy physical and moral burdens, the light-haired and powerfully built Shatalov would adopt as one of his radio callsigns the word 'Granit' ('Granite'), indicative of his steely and hard-as-nails disposition. That disposition, and the respect in which he was held, were recognised as early as June 1964, when he was recommended to lead one of the early Soyuz flights. Nikolai Kamanin, the Soviet Air Force general in command of the cosmonaut corps, considered but ultimately rejected Shatalov, preferring veterans to rookies on the first few missions. A few months later, following exams, Shatalov was rated as one of the most outstanding of his 15-strong group. By March 1965, Kamanin had named him as part of the team to fly Voskhod 3 and, following the cancellation of this mission, assigned him to Soyuz. When Georgi Beregovoi headed into orbit aboard Soyuz 3 in October 1968, Shatalov served as his backup.

A month before his own launch, on 17 December, he passed his final exams in support of the joint Soyuz 4/5 mission, once again scoring outstanding marks. The excitement of those final days, however, was tinged with disappointment. When Shatalov and the Soyuz 5 crew – cosmonauts Boris Volynov, Alexei Yeliseyev and Yevgeni Khrunov – arrived at Tyuratam on Christmas Eve, they received the grim news that Apollo 8 had just entered lunar orbit. Working until late that night, for Shatalov the pot boiled over when Kamanin told him that a 'recommendation' had been received, calling for Soyuz 4 and 5 to dock automatically, not manually. The four cosmonauts objected, arguing that, as pilots, they had the skills and training and ought to be permitted to execute a manual linkup. Finally, Shatalov exploded with fury. "Here we are debating this for the tenth time," he yelled, "while the Americans are orbiting the Moon!"

A DOCKING AND A TRANSFER

Despite the excitement of the impending mission, Apollo 8's success placed morale within the cosmonaut corps at a very low ebb. The question of whether to give cosmonauts active control of their machines had been hotly disputed since the early days of the Soviet space programme. Kamanin had frequently locked horns with Korolev over the issue and his memoirs, preserved in a series of quite remarkable diary entries, first published in 1995, reveal a tough, bitter military man who blamed his country's loss of the Moon race on Soviet engineers' unwillingness to yield control of a spacecraft to its crew. The argument over the role of the cosmonauts in the Soyuz 4/5 docking proved a perfect case in point.

Over the years, Kamanin's diaries have been interpreted and used to paint a portrait of a man who fought fiercely for 'his' cosmonauts and to illustrate the close relationship that he had with them during their time together on the isolated Kazakh steppe. Others have seen him quite differently: he has been described by space analyst Jim Oberg as "an authoritarian space tsar, a martinet" and by Soviet journalist Yaroslav Golovanov as "a malevolent person ... a complete Stalinist bastard". Still others, including Alexei Leonov, have proven more complimentary, seeing Kamanin as "very approachable" with a keen love of sports, especially tennis. Nonetheless, his brand of military leadership, in most cases, successfully prepared the first generation of space explorers for their ventures into the heavens.

Automation, though, remained a key operating principle. Under the cover names of Cosmos 186 and 188, a pair of unmanned Soyuz spacecraft had satisfactorily performed a rendezvous and docking exercise in October 1967. Although they did not achieve a 'hard' link-up – there remained an 8.5 cm gap between them, which prohibited full electrical connections – their mission underlined the reality that the Soviets now had a grasp of rendezvous and docking with exciting possibilities for the future. Unfortunately, these flights did not end well. Cosmos 186 suffered a failure of its solar-stellar sensor, which altered its descent trajectory into a purely ballistic fall from orbit. It landed hard, but in one piece, on Soviet soil. Cosmos 188, on the other hand, re-entered the atmosphere at too steep an angle; so steep, in fact, that its self-destruct package had to be remotely triggered, spraying debris close to the Soviet-Mongolian border.

Success finally came the following April, when two more Soyuz craft, this time under the cover names of Cosmos 212 and 213, rendezvoused automatically and successfully 'hard' docked. In the eyes of many cosmonauts and engineers, this cleared the way for a manned rendezvous, docking and spacewalking flight later in 1968, involving Georgi Beregovoi aboard the 'active' Soyuz 2 and Boris Volynov, Alexei Yeliseyev and Yevgeni Khrunov aboard the 'passive' Soyuz 3. Sadly, it was not to be.

Trials of the spacecraft's backup parachute were not deemed good enough to assign a human pilot and it was considered likely to rip during deployment with a crew of three and a total weight of up to 1,300 kg. Vasili Mishin, who had assumed Korolev's mantle as Chief Designer of OKB-1 in May 1966, proposed reducing the Soyuz 3 crew to just two men to circumvent this risk and suggested postponing the risky spacewalk to a later mission. Others, including Mstislav Keldysh, head of the Soviet Academy of Sciences, were even more cautious, refusing to endorse *any* manned flights until another automated test had been successfully performed. Their reluctance was understandable. The previous year, Soyuz 1 had been lost and its pilot Vladimir Komarov killed when both the primary and backup parachutes failed ...

By the end of May 1968, a compromise was suggested by Mishin: two Soyuz would dock in orbit, one of them unmanned, the other carrying a single cosmonaut. Assuming the success of that flight, the next crews would attempt the transfer mission, perhaps as early as September. Dmitri Ustinov, chair of the Military-Industrial Commission and *de facto* head of all Soviet missile and space projects during this period, demanded a wholly automated flight. This would slip the

Artist's concept of the Soyuz-to-Soyuz extravehicular transfer, performed by cosmonauts Yevgeni Khrunov and Alexei Yeliseyev in January 1969. Originally intended as a key stepping stone toward manned lunar flights, by the time it was eventually undertaken, it was two years overdue, virtually obsolete and far too late to beat the Americans to the Moon.

intended August date for the manned mission until October at the earliest. On 10 June, the Soyuz State Commission convened and decided to launch the automated flight in July, followed by the joint mission with Beregovoi in September and the full-scale docking and spacewalk transfer in November or December.

To this, Ustinov added a proviso that the spacewalk should transfer not one, but two cosmonauts between craft. Although this had been the original plan for Komarov and Bykovsky's joint flight back in April 1967, concerns over the backup parachute led to a call to fly a crew of two, with just one spacewalker. Ustinov's request was borne out by Boris Volynov, whose work in a training version of the bulky space suit had revealed a major obstacle: a single spacewalker risked getting stuck in the hatchway between the descent and orbital modules. Moreover, if he then experienced difficulties getting outside, there would be no one to help him. The commander, in the sealed-off descent module, would be unable to assist, making a *pair* of spacewalkers – capable of supporting each other – the only safe and practical option.

For a time, it had been thought prudent to adopt a so-called '2 + 2' profile, whereby only one of the spacewalkers would actually perform the external transfer and both missions would return to Earth with crews of two. This neatly avoided the risk of bringing a Soyuz back to Earth with three men and a potentially dangerous parachute situation. By the end of September 1968, however, it seemed that the parachute woes had been resolved and the original, pre-Komarov plan for both Yeliseyev and Khrunov to spacewalk over to the other craft was reinstated.

In the meantime, Cosmos 238 had been launched late in August (a month behind the schedule agreed on 10 June) and apparently conducted at least one major manoeuvre before touching down after a near-flawless four-day flight. Finally, on 25 October, the unmanned Soyuz 2 was launched, followed by Beregovoi aboard Soyuz 3 the next day. During his mission, the cosmonaut managed to rendezvous with his automated target, but did not physically dock with it; a peculiarity which perplexed western observers for many years. The Soviets explained that on their first manned flight after the Soyuz 1 disaster, they did not want to subject Beregovoi to any undue risk. However, in a 2002 interview, quoted by Hall and Shayler, Konstantin Feoktistov accused Beregovoi of committing "the grossest error" by failing to notice that Soyuz 2's orientation was mismatched with that of his own craft. This caused Soyuz 3 to inexplicably bank 180 degrees relative to the target, despite the cosmonaut's best efforts to counter it.

Suspicion that the Igla rendezvous device might have been to blame was vigorously denied by its designer, Armen Mnatsakanyan, and Feoktistov agreed that it was simply a classic case of pilot error. Beregovoi's failure to notice the orientation mismatch with Soyuz 2 caused him to waste "all the fuel intended for the ship docking" and forced managers to cancel the remainder of the rendezvous. Years later, Siddiqi postulated that if the cosmonaut had recognised the problem and managed to stabilise Soyuz 3 along a direct axis to the target, he might still have achieved a successful docking. To be fair, though, Beregovoi was a rookie, obliged to perform the tricky exercise very soon after launch, whilst out of direct contact with the ground and during orbital darkness. It has also been reported that he suffered from a bout of disorientating space sickness early in his flight.

Lessons clearly would need to be learned before the docking between Soyuz 4 and 5. Indeed, it was decided that when Shatalov guided his spacecraft to link up with that of Volynov, Yeliseyev and Khrunov, he would do so a full 48 hours after launch, to allow him time to fully adapt to the strange 'microgravity' environment of space. He would also complete his docking during orbital daylight and within range of Soviet ground stations. Furthermore, this plan gave the Soyuz 5 crew, upon whose shoulders lay the burden of the hazardous spacewalk, a full day to acclimatise themselves.

During the docking, Shatalov would rely heavily upon the Igla. This system would later be used to enable a docking between Soyuz and the first Salyut orbital station. Broadly, its goal was to control the relative motion and attitude of two vehicles, the 'passive' of which carried a radio beacon for use as a homing aid by its 'active' counterpart. Firstly, the passive craft (Soyuz 5 in this case) would transmit a continuous-wave beam signal, which the active craft (Soyuz 4) would use to orientate itself to acquire its 'target' in a similar manner to the Cosmos 186 and 188 rendezvous. Next, Soyuz 4 would start to transmit an 'interrogation' signal to Soyuz 5 through its narrow beam antenna. Finally, Soyuz 5 would switch off its continuous-wave beacon and retransmit the interrogation signals through its own narrow beam antenna to establish a secure 'lock' between the pair.

Bearing as he did the unenviable reputation of becoming the Soviet Union's 13th spacefarer, Shatalov could perhaps have anticipated a run of bad luck during the days preceding his mission. Matters were not aided by the fact that his home telephone number ended in '13' and the launch itself was set for 1:00 pm Moscow Time – 13:00 hours – on 13 January, which also happened to be a Monday, traditionally regarded by the Russians to be a most difficult day ...

As it happened, Shatalov's only real bad luck transpired shortly after clambering into Soyuz 4 on the morning of 13 January, when he fell victim to the first launch scrub in Soviet space history. Despite temperatures of -24°C and winds gusting at 8–10 m/sec, the fuelling of the R-7 proceeded normally and the cosmonaut settled into the spacecraft and began running through his pre-launch system checks. Minor irritations came in the form of voice communication dropouts whenever Soyuz 4's television camera was in use, prompting it to be switched off. Then, with nine minutes to go, a problem was detected within the R-7's gyroscopes, apparently related to the ambient temperature and humidity.

By the time this problem was resolved, the launch time had slipped to mid-afternoon and Shatalov had been lying on his back for over two hours. Moreover, with a mission whose planned duration was almost exactly three days, to launch at this time of the day would produce a landing in the half-light of a gloomy midwinter's afternoon. This was considered far from ideal on such a complex flight. Ultimately, mission rules decided the outcome: fuel temperatures could not fall below -2°C at night; otherwise the loss of specific impulse would reduce the R-7's thrust by more than 5 percent. The managers therefore opted to postpone the launch.

Shatalov concealed his disappointment well and as he was extracted from his couch he declared that he had just set a new record for the world's shortest space flight and the very first to return to its exact point of liftoff! Years later, he would admit to an interviewer that, despite the run of thirteens, he was not an overly superstitious man. Still, he said, the decision to scrub the launch "hit me like a sledgehammer ... I had been waiting for the day [for] six years; dreamed, worked and trained hard".

His quick wit, superstition and disappointment aside, a number of potentially serious challenges remained. Although Soyuz had been designed to touch down on solid ground and was capable of performing a water landing, unmanned experience had shown that it might not be totally waterproof and, indeed, could sink. The chances of either Soyuz 4 or 5 splashing down somewhere in the ice-covered Aral Sea were estimated at only 0.003 percent, but, erring on the side of caution, recovery forces despatched rescue helicopters and a trio of B-12 seaplanes in readiness for such an eventuality.

In addition, the debate continued about how to conduct the rendezvous, with Dmitri Ustinov and Space Minister Sergei Afanasyev pressing Vasili Mishin for an automated flight profile. Both were aware of how flawlessly this had been executed by Cosmos 186/188 and 212/213 and remained mindful of Beregovoi's difficulties the previous October. The matter was decided the day before launch by Mishin, who, though he normally favoured automated systems, ruled in favour of the cosmonauts. Nevertheless, on the evening before launch, Nikolai Kamanin took Shatalov aside and told him that if he encountered difficulties then he should revert immediately to the automatic systems.

Shatalov's mission finally got underway at 10:30 am Moscow Time on 14 January with a perfect launch and insertion into a 173 × 225 km orbit, inclined at 51.7 degrees to the equator. For a time it had looked as if another scrub was on the cards, when a fault was detected in the R-7 which could normally only be resolved by lowering the rocket into a horizontal position. According to Hall and Shayler, a young pad technician saved the day by volunteering to strip off most of his clothes and, in freezing temperatures, managed to squeeze through a narrow hatch into the rocket's bowels to correct the problem.

For Shatalov, a veteran, 41-year-old test pilot, the experience of rocket launch, the weightlessness of space and the mesmerising view of Earth, was profound. "When we look into the sky," he explained later, "it seems to us to be endless. We breathe without thinking about it, as is natural ... and then you sit in a spacecraft, you tear away from Earth and within ten minutes you have been carried straight through the layer of air and beyond there is nothing. The 'boundless' blue sky, the ocean which gives us breath and protects us from endless black ... is but an infinitesimally thin film." Shatalov was not the first spacefarer, nor would he be the last, to remark upon the fragility of the world from which he came. Weightlessness, he recounted at the post-flight press conference at Moscow State University, "took me about three to four hours to master". (The fact that he spoke of adaptation lent some credence to the rumours that Beregovoi had experienced difficulties.)

Back on Earth, watching the launch from Tyuratam's Site 17, were a trio of cosmonauts who undoubtedly wished that they could be in his place. Their turn would come the very next morning, 15 January, when they were destined to blast off aboard Soyuz 5 and adopt a passive role as Shatalov performed the world's first-ever link-up between two manned spacecraft. For Lieutenant-Colonel Boris Valentinovich Volynov, who had overcome a string of obstacles in his 34 years of life to reach the point of commanding a spacecraft, his first flight into orbit would almost become his last.

THE MAN WHO ROCKED THE BOAT

Volynov's journey to the launch pad had already been thwarted several years earlier, a fact which many observers have attributed, at least partly, to his Jewish heritage; for he would become the first Jew in space. Selected as one of the original cosmonauts in March 1960, Volynov served as Valeri Bykovsky's backup on the Vostok 5 mission and was widely expected to receive assignment to a later flight or perhaps command of the first Voskhod. By January 1964, however, Marshal Sergei Rudenko, the Soviet Air Force's deputy commander-in-chief, requested Volynov's transfer from the Voskhod to the Soyuz training group. This apparent 'downgrading' was vetoed by Nikolai Kamanin, who, in July of that year, named Volynov along with Vladimir Komarov to train for the Voskhod 1 command.

Despite receiving formal certification to fly in August and seeming to be a strong contender to lead a crew, Volynov was dropped from the mission only days before its launch. So too was fellow cosmonaut Georgi Katys, whose father had been executed in one of Stalin's purges and who had a pair of half-siblings living in Paris. In spite of the solid support of Mstislav Keldysh and the Soviet Academy of Sciences, Katys was cast aside, with Kamanin noting in his diary that an "unfavourable background ... spoils the candidate for flight".

When Sergei Korolev heard of the decision to drop Volynov purely on the basis of his Jewish heritage, he was reportedly furious, but was advised by Nikita Khrushchev not to "rock the boat ... it's not worth it!" Unperturbed, Volynov next trained to lead the long-duration Voskhod 3 mission and, after its cancellation, moved over to Soyuz. In May 1968, he was assigned to command the passive rendezvous flight and, despite passing all of his exams in September and being commended for his "mastery" of the systems, almost lost this assignment too. When the Central Committee of the Communist Party met on 20 December to discuss the crew selections, further 'unhappiness' was expressed over Volynov's background. On this occasion, however, good fortune smiled upon him and his position on Soyuz 5 was accepted.

The man who aroused all this debate was born in Irkutsk in southern Siberia on 18 December 1934. He received schooling in Prokopyevsk and developed a love for the wildness of the taiga which would remain with him throughout his life. He moved to Prokopyevsk before the Great Patriotic War, where his mother, a paediatrician, surgeon and traumatologist, raised him on her own. After graduation, Volynov applied to his local Komsomol Committee for a letter of recommendation to a pilots' school and, with this in hand, experienced a taste of military life: endless drilling, guard duty and kitchen detail. His first solo flight, it is said, gave him little satisfaction, but, thanks to his instructor, Veniamin Reshetov, and his air force squadron commander, Major Ivanov, eventually developed a love of aviation.

Completion of the military pilots' school in Novosibirsk in 1955 was followed by marriage to his childhood sweetheart, Tamara, and the birth of their first son, Alexei. Then, in the closing months of 1959, with a glowing reference from Major Ivanov, he was one of hundreds of young Soviet Air Force pilots to be interviewed about flying a quite different machine: the Vostok spacecraft. Acceptance in March of the following year marked the beginning of his long wait for a mission into space. It also marked the end of a long and difficult selection process.

The search for the world's first spacefarers had begun in earnest in May 1959, when representatives of the armed forces, the scientific community and the design bureaux met at the Soviet Academy of Sciences in Moscow, under the auspices of Mstislav Keldysh, to discuss methods of choosing the most suitable candidates for Earth-orbital missions. Aviators, rocketeers and even car racers were considered, but at length, bowing to the Soviet Air Force, Keldysh agreed to narrow the criteria to pilots from this branch of the military.

Despite an obvious vested interest in wanting to have 'its' fliers take the first manned spacecraft beyond the atmosphere, the logic was inescapable: as already noted, military pilots had proven themselves under exposure to hypoxia, high pressures and varying G loads and had undergone rigorous ejection-seat and parachute training. In addition to their flying experience, candidates would only be admissible if they could meet the height and weight requirements of the Vostok spacecraft: they needed to be no taller than 1.75 m and weigh no more than 72 kg. Moreover, in the expectation that they would be embarking on lengthy careers as 'cosmonauts', the age limit was firmly set at between 25–30 years old.

Throughout 1959, groups of physicians were sent to air bases in the western Soviet Union and by August the selection teams had the records of over 3,000 pilots available for inspection. Most of these were eliminated at a fairly early stage, on the basis of not meeting the height, weight, age or medical criteria – some, indeed, were dropped for bronchitis, angina, gastritis and colitis, renal and heptic colic and pathological cardiac shifts. The remainder were then systematically interviewed from early September, still unaware of exactly what the so-called 'special flights' project entailed. The list was soon reduced to a little over 200 candidates. They were despatched in groups of about 20 for further tests at the Central Scientific Research Aviation Hospital in Moscow. In addition to more interviews, the candidates were spun in stationary seats to assess their vestibular apparatus, placed in low-pressure barometric chambers and spun around on a centrifuge to evaluate their performance under high-G loads. Original plans, it seemed, called for seven or eight pilots, but Sergei Korolev insisted on tripling this number, for no other reason than because he wanted a larger team than the United States' seven-strong group for Project Mercury.

In January 1960, Marshal Konstantin Vershinin, commander-in-chief of the Soviet Air Force, formally signed plans to establish a centre for cosmonaut training in Moscow. Although it was nominally under the control of physicians, the Air Force General Staff eventually assigned Nikolai Kamanin command of cosmonaut affairs. It was he who approved a final shortlist of 20 Air Force candidates in late February: Ivan Anikeyev, Pavel Belyayev, Valentin Bondarenko, Valeri Bykovsky, Valentin Filatyev, Yuri Gagarin, Viktor Gorbatko, Anatoli Kartashov, Yevgeni Khrunov, Vladimir Komarov, Alexei Leonov, Grigori Nelyubov, Andrian Nikolayev, Pavel Popovich, Mars Rafikov, Georgi Shonin, Gherman Titov, Valentin Varlamov, Boris Volynov and Dmitri Zaikin. The age criteria was waived in a couple of instances out of respect for their exemplary performance during testing and ran from just 23 for Bondarenko to 34 for Belyayev. Some of these men would become the most famous names in the history of space flight, whilst others would remain anonymous ... and, in a few cases, fall into disgrace.

On 7 March, the cosmonauts were given their welcoming speech by Vershinin at the Central Scientific Research Aviation Hospital. A week later, after settling their affairs at their individual air bases, they began training with Vladimir Yazdovsky's first class in aerospace medicine. The following four months were consumed by a mixture of in-depth lectures and an intense physical fitness regime, the latter of which included two hours daily of intensive calisthenics at the Central Army Stadium in Moscow. Their parachute training was conducted in the Saratov region, near Engels. They jumped from a converted An-2 aircraft and within six weeks each man made between 40–50 jumps over water and land, from high and low altitudes and in daylight and darkness.

Almost a full decade later, on 14 January 1969, as he stood at Tyuratam and wistfully watched Vladimir Shatalov's rocket vanish from view, Boris Volynov – the man whose heritage and religious beliefs came close to barring him from space – had the satisfaction of knowing that the next day it would be his turn.

AMUR AND BAIKAL

A little more than six hours after launch, at 4:35 pm, Shatalov adjusted Soyuz 4's orbit to 207 × 237 km, showed television viewers his spacious descent module with two (tellingly) empty 'extra' seats, then retired into the orbital module for his first night's sleep in space. Next morning, at 3:00 am, an An-12 aircraft from Moscow touched down at Tyuratam with an unusual cargo: ten newspapers and a batch of letters to be delivered to Shatalov by the Soyuz 5 crew in another, somewhat dubious, 'first': the world's first space mail service. Several hours later, the crew took their places in the spacecraft, with Volynov assuming the centre seat, flanked by flight engineer Yeliseyev and research engineer Khrunov.

Shortly after 9:30 am, with barely 25 minutes to go, a piece of electrical equipment failed and, despite the fully fuelled state of the R-7 booster, was replaced by Engineer-Captain Viktor Alyeshin. He also noticed that the crew access hatch in the aerodynamic shroud was secured by only three bolts, instead of the required four. Nonetheless, at 10:04 am, Soyuz 5 roared aloft and a few minutes later was precisely inserted into a 200 × 230 km orbit, trailing Shatalov by some 1,200 km. Although Volynov executed a thruster firing later that day to further refine his orbital parameters to 211 × 253 km, his craft would remain essentially passive during the rendezvous. In addition to the joint programme with Soyuz 4, the crew had their own scientific agenda: Yeliseyev would be "concerned with geological/geographical phenomena" and Khrunov would be "occupied with medical and ionospheric radio-propagation experiments", according to *Flight International*. It was also revealed that Khrunov would play a key role in the final stages of the rendezvous, by operating the ship's onboard sextant.

However, according to Hall and Shayler, there had already been some discussion on the ground about the precise order in which to launch the two missions. The cosmonauts, it seemed, wanted to fly the passive spacecraft *ahead* of the active one, as this would provide Yeliseyev and Khrunov with additional time to adapt to the weightless environment before their spacewalk. Moreover, they reasoned, Georgi Beregovoi had followed – not preceded   his target into orbit. In true Soviet fashion, with its ridiculous emphasis on revealing absolutely nothing except successes, there was another advantage. If the second launch was cancelled, the 'joint' nature of the mission could be disguised from the outside world by saying that a three-man flight with a spacewalk was a logical step. In the end, Nikolai Kamanin overruled his cosmonauts, on the grounds that it would be too complicated to change the launch plans at such short notice.

By mid-morning on the 15th, therefore, Shatalov, Volynov, Yeliseyev and Khrunov were in orbit, equalling the United States' record of four men in space, set during the Gemini VII and VI-A missions in December 1965. As radio chatter crackled between the two Soyuz spacecraft and ground control, their callsigns were revealed as 'Amur' for Shatalov and 'Baikal' for Volynov's crew. This choice apparently had much to do with a revival of plans around this time to finish laying the broad-gauge Baikal-Amur railway. Construction of this strategic artery had begun in the Thirties, stalled after Stalin's death in 1953 and resumed in the wake of increasingly strained relations with China in the late Sixties.

The Baikal-Amur line, finally completed in 1991, runs approximately 690 km north of, and parallel with, its more famous Trans-Siberian cousin; the latter, however, traces the Chinese border along many sections of its route and as the Sixties drew to a close it was feared that any attack would effectively sever transportation links with the Russian Far East. It was perhaps in response to these newly revived plans to finish the project that the men in command of Soyuz 4 and 5 named their respective spacecraft in its honour.

The two ships apparently established mutual radio contact shortly after Soyuz 5 reached space. At 8:06 am on 16 January, Shatalov made his final orbital adjustment in preparation for the rendezvous. At that point, he was partway through his 34th orbit and Volynov, Yeliseyev and Khrunov were on their 18th circuit of the globe. At 10:37 am, high above the South Pacific, Shatalov switched on his Igla radar to begin the automated 'ballet' which ended over Africa at 11:05 am when the separation distance between the two craft was just 40 m.

Speaking after the mission, he recalled that his most important aids during this critical period were his instruments and his own eyes. At 40 m, he said, "Boris Volynov and I performed several manoeuvres, in the course of which we changed the relative position of the spacecraft … Further approach and docking were performed within the zone of direct TV contact with the ground stations. To avoid sharp contact with each other, the relative speed of approach was reduced to several centimetres per second." Contact itself came at 11:20 am as Soyuz 4 and 5 flew above the Yevpatoria control centre in the Crimea. The crews may have been intently focused on their instruments, but for a few seconds after docking it would seem that one of the Soyuz 5 cosmonauts was thinking of something quite different. As Soyuz 4's 'Shtir' docking probe penetrated their own craft's 'Konus' receptacle, one of the men – some sources say Volynov, others Khrunov – could not help but visualise the sexual connotations of the now-linked ships. Without checking himself, over the radio link and within clear earshot of ground controllers, he blurted out: "We've been raped! We've been raped!"

It was an unfortunate choice of words, reflecting, perhaps, the over-excitement of a young rookie, but the outstanding success of the docking – a *manual* docking, at that, with Shatalov firmly in control – was eagerly proclaimed by the state-run Tass news agency. It was the first 'new' undertaking in orbit since the spacewalk by Alexei Leonov and it sent a clear message to the west that Russia was back in the game. "There was a mutual mechanical coupling of the ships … and their electrical circuits were connected," Tass said. "The world's first experimental cosmic station with four compartments for the crew was assembled and began functioning." These 'four compartments', it was revealed, had an internal volume of 18 m^3 and weighed an impressive 12,924 kg.

A few days later, *Time* magazine wondered about future Soviet space plans: would this "four-compartment version", it asked, lead next to "a roomy orbiting laboratory"? Many observers, though, had a more fundamental question. As *Flight International* pointed out on 23 January: "It is not clear whether astronauts can transfer from one vehicle to another through a tunnel joining the two vehicles." This was the crux of the debate over whether Soyuz 4/5 represented a 'true' experimental space station. Left unsaid, conveniently, by Tass, was the reality that those 'four rooms' – the two orbital modules and two descent modules, though electrically and mechanically mated – did not permit internal transfer from one spacecraft to the other. Nevertheless, as would shortly be demonstrated by Lieutenant-Colonel Yevgeni Khrunov of the Soviet Air Force and civilian engineer Alexei Yeliseyev, Soyuz 4 and 5 were by no means 'inaccessible' to one another.

THE GALLANT PILOT AND THE MUSKETEER

The space suits to be worn by Yeliseyev and Khrunov were quite different from the one worn by their comrade Alexei Leonov during his spacewalk four years earlier. On that occasion, the 'Berkut' ('Golden Eagle') ensemble had proven stiff and had ballooned dangerously as Leonov tried to re-enter the airlock. By the time he returned inside the Voskhod 2 cabin he was drenched in sweat, breathing hard and exhausted. By contrast, the 'Yastreb' ('Hawk') suits of Yeliseyev and Khrunov were more flexible, benefitting from a complex array of lines and pulleys for dexterity and their 50 kg life-support and environmental control units could be worn on either their chests or shins in order to help them get through the relatively small hatch of the Soyuz spacecraft's orbital module.

The size of this hatch, in fact, had almost proven a show-stopper a few years earlier. In July 1966, Yastreb's designer, Gai Severin – the Soviet Union's foremost manufacturer of attire and ejection seats for both MiG fighter pilots and the early cosmonauts – advised Nikolai Kamanin that the OKB-1 bureau, now headed by Vasili Mishin, had restricted the diameter of the orbital module's hatch at just 660 mm. A fully suited cosmonaut with his bulky life-support gear, Severin pointed out, needed the opening to be at least 700 mm wide. Simulations on the ground and in conditions of temporary weightlessness aboard a modified Tu-104 aircraft under-lined the problem: when fully pressurised, the suit swelled to 650 mm, just a few millimetres less than the diameter of the hatch itself and the men simply could not get through the hatch without twisting and contorting their bodies in remarkable feats of gymnastics. Kamanin deemed the situation wholly unacceptable. At length, Mishin conceded: although the first few Soyuz spacecraft had already been built, subsequent orbital modules would have a 720 mm hatch.

Yevgeni Vasilyevich Khrunov, known to his friends as 'Zhenka', was at the centre of these troubles. Chosen in 1960 as one of the original cosmonauts, his career appeared to have been leading inexorably towards a spacewalk. Initially assigned to the Voskhod 2 training group, he had supported both Pavel Belyayev and Alexei Leonov, even donning his space suit with them on launch morning in

Spacesuited Alexei Yeliseyev (left) and Yevgeni Khrunov are pictured during a training session. Between them is Boris Volynov, with Vladimir Shatalov on the right.

March 1965. By the end of the following year, despite Mishin's favoritism of his civilian engineers – of whom Yeliseyev was one – Khrunov, having completed more than 50 flights in the Tu-104, was considered by far the best-qualified candidate for the Soyuz-to-Soyuz spacewalk. He narrowly missed the chance to perform it in April 1967, when Vladimir Komarov's mission went so tragically wrong, but remained the leading contender and continued training in the expectation of eventually flying.

Born in the village of Prudy, in the Tula region, south of Moscow, on 10 September 1933, Khrunov's childhood was disrupted repeatedly by war. Prudy was torched by the retreating Nazi forces and, following his father's death, Khrunov's older brother, now head of the family, sent him to receive his education in an agricultural secondary school. After finishing his studies in Kashira and completing practical work on the collective farm, Khrunov developed a keen understanding and expertise with machinery. It is said that he was able to dismantle and then reassemble from scratch all the units of a grain harvester. However, as with so many other youngsters in the war years, and in particular many future cosmonauts, he was ultimately drawn by the magnet of his life's true calling: aviation.

He duly applied for a military pilots' school and in 1952 was accepted for training in the Soviet Air Force. Shortly thereafter, he met and married his wife, Svetlana, and their son Valeri was born a few years later. In the autumn of 1959, like so many other pilots, he was summoned by his commanding officer before a mysterious recruiting panel which asked unusual questions about flying "aircraft of unheard-of models" and committing himself to work which would require him to "study and try hard". Early the following spring, Khrunov, having been declared an excellent pilot by his commanding officer, a gallant flier by his friends and a competent and intelligent aviator by his instructors, was picked as a cosmonaut candidate.

It would appear that he fit in reasonably well. Gherman Titov, who in August 1961 followed Yuri Gagarin into space, described Khrunov as "cheerful", and Alexei Leonov paid tribute to his friendship and wide range of interests and skills; in gymnastics, tennis, literature and the theatre. When a group of nine cosmonauts collected their diplomas from the prestigious Zhukovsky Engineering Academy in December 1968, Khrunov was the only one to receive the highest honours. Certainly, his love of physical exercise would prove exceptionally useful as Soyuz 5's chief spacewalker: during training runs in the Tu-104 and in the vacuum chamber in late 1968, fully suited, it was found that exertion levels in the Yastreb were in the order of 600–900 calories per hour ...

In some ways, the other spacewalker of the mission, Alexei Stanislavovich Yeliseyev, despite his excellence as an engineer, owed his position as a cosmonaut to Vasili Mishin. When the latter succeeded Sergei Korolev as Chief Designer of OKB-1 in May 1966, he had quickly asserted himself by insisting on populating the cosmonaut corps with civilian engineers from the bureau. This infuriated Nikolai Kamanin, who fumed in his diary that Mishin put no value in Kamanin's six years' worth of experience of training military cosmonauts to fly into space. Kamanin considered it absurd that Mishin wanted, in just a few months, to prepare civilians for Soyuz command positions, with no pilot training, no parachute experience, no medical screening and no practice in the centrifuge. Eventually, under pressure from Dmitri Ustinov, Mishin was forced in July 1966 to accept veteran military pilots for Soyuz command posts and OKB-1 engineers in support roles. It was the first of many standoffs which would significantly damage the professional working relationship between Mishin and Kamanin.

To be fair, Mishin's desire to fly civilians into orbit had been shared by Korolev and, intermittently in the early Sixties, a handful of OKB-1 engineers had passed preliminary screening. When eight military cosmonauts began training in September 1965 for early Soyuz flights, Korolev entrusted one of his engineers to explore the possibility of forming a parallel group of civilians. Eleven candidates passed initial testing at the Institute of Biomedical Problems in Moscow and on 23 May 1966 Mishin signed an official order to establish the first non-military cosmonaut team. Candidates Sergei Anokhin, Vladimir Bugrov, Gennadi Dolgopolov, Georgi Grechko, Valeri Kubasov, Oleg Makarov, Vladislav Volkov and Alexei Yeliseyev seemed to have little hope of actually reaching space and, indeed, the language used to describe them – 'cosmonaut-testers' – implied that they would be of limited use. With this in mind, it is ironic that Yeliseyev would go on to jointly become the first cosmonaut to record three space flights.

Despite his own doubts, Kamanin was finally appeased when Grechko, Kubasov and Volkov passed tests at the Central Scientific Research Aviation Hospital and arrived at the cosmonauts' training centre, 'Zvezdny Gorodok' ('Star Town'), near Moscow, on 5 September. Within two months, they were joined by Yeliseyev and Makarov. All five, wrote Asif Siddiqi, were accomplished engineers. Unfortunately, Anokhin, Bugrov and Dolgopolov did not pass Air Force screening and were never considered for any space missions.

For the others, a seat on a space flight seemed only months away. By late 1966, military pilot Vladimir Komarov was assigned to command Soyuz 1, owing to his expertise. Mishin, naturally, wanted two civilians amongst the three-man Soyuz 2 crew, but Kamanin opposed this move, feeling that the complexity of the early missions made it ill-advised. A compromise was reached, thanks to Ivan Serbin, the chief of the Communist Party's Defence Industries Department, who suggested flying an Air Force pilot (Khrunov) and an OKB-1 engineer (Yeliseyev) with a veteran military commander (Valeri Bykovsky). By the end of November, in a triumph for the civilians, Yeliseyev was officially assigned to the Soyuz 2 crew. Although Komarov's death in April 1967 put everything on hold, Yeliseyev remained, with Khrunov, a prime candidate for the spacewalk transfer.

Yeliseyev was born in the town of Zhizdra in the Kaluga district on 13 July 1934 with engineering and the sciences seemingly in his blood, his chemist mother having headed a laboratory in the Institute of Physical Chemistry at the Soviet Academy of Sciences. His birth surname, 'Kuraitis', came from his Lithuanian-born father, who had been arrested and sent to the gulag; Yeliseyev had adopted his mother's maiden name to erase connections with his disgraced patronym. With a plethora of hobbies including physics, solving mathematical problems, a keen love of chess and a mastery of fencing, Yeliseyev readily passed the entrance exams for the prestigious Bauman Polytechnical Institute in Moscow. He received his engineering degree in 1957, his Candidate of Technical Sciences credential ten years later and would earn a doctorate in 1973. Despite glowing praise from his professors, regret was expressed by his former fencing partners. "It's a pity Alexei stopped taking part in contests," one of them said. "He's such a promising sportsman, fast and full of energy; a real musketeer."

Following the launch of the world's first satellite, Sputnik, in October 1957, Yeliseyev sought a means of becoming part of the fledgling space effort and eventually joined the team of Boris Raushenbakh, a leader in rocket dynamics, which was ultimately subsumed into Sergei Korolev's OKB-1 bureau. Working under the Chief Designer, Yeliseyev's early tasks included supporting the design of controls for the Vostok spacecraft and, shortly after Yuri Gagarin's flight in April 1961, he approached one of his superiors to ask for an endorsement in applying to become a cosmonaut. Another civilian engineer, Vitali Sevastyanov, did the same and both men quickly found themselves at the Air Force's Central Scientific Research Aviation Hospital undergoing physical tests.

Although Yeliseyev performed well, his mistake was that he applied without Korolev's knowledge or permission. When the Chief Designer paid a routine visit to the hospital to discuss – ironically – changing the medical criteria for selecting civilian cosmonauts, he was told about a candidate named 'Yeliseyev' and was reportedly furious that one of his men had applied unofficially. Yeliseyev was promptly removed from consideration. He continued to work on instrumentation for Vostok and Soyuz until, in July 1965, a group of OKB-1 engineers passed the preliminary medical screening for cosmonaut selection. Korolev had always believed that the civilians were being unfairly assessed compared to military pilots and so he sent his candidates to the Institute of Biomedical Problems for physiological and psychological assessment.

"The majority of my colleagues were irritated by the psychologist," Yeliseyev mused later. "The interview lasted three to four hours. He asked us about our grandparents: did we remember our grandmothers and grandfathers and the most common topic of their arguments? He wrote in his notes much more than we actually spoke! Now a question about our other relatives ... and again he scribbled. He did not permit smoking although he was himself smoking. We were all very careful during this interview." This ingrained fear of the power of the 'shrink' was mirrored in the United States' astronaut corps, many of whom told their own tales of how much care they had to exercise in order not to talk themselves into a psychological hole. For many cosmonauts and astronauts during this period, the physical and physiological tests came as a blessed relief by comparison.

In spite of his reservations, Yeliseyev was accepted into the cosmonaut corps in May 1966. Yet the unknowns remained. In his book about the first Salyut space station, Grujica Ivanovich pointed out that Yeliseyev was tapped early as a candidate for the tricky Soyuz-to-Soyuz transfer, but noted an 'unusual' psychological technique for finding out how he would react during an emergency: interviewing his *mother* and his *wife*. It was suggested that, should the young engineer become hysterical during the spacewalk, he might only respond to the voice of a woman who was close to him! Unfortunately, this went nowhere. Yeliseyev's mother had no idea that her son was even a cosmonaut and his ongoing divorce from his wife, Valentina, made *her* participation in the process unlikely, to say the least ...

However, Yeliseyev, who knew a great deal about spacecraft design and had helped prepare for Alexei Leonov's spacewalk back in March 1965, soon came to be accepted and respected in the predominantly military ranks of the cosmonauts. He worked closely with Yuri Gagarin, who described him as "very demanding when it comes to work . . . he has become inseparable from our group" and admired his "vast design experience". It was with this experience and more than two years of preparation under their belts that, shortly after midday Moscow Time on 16 January 1969, Yeliseyev and Khrunov finished donning their space suits and began the greatest engineering challenge – and the biggest thrill – of their lives.

A WALK OUTSIDE AND A WILD RIDE HOME

Years later, Khrunov would recall his first impressions upon opening the hatch of Soyuz 5's orbital module: the astonishing view of Earth – a stunning blue jewel, speckled with flecks of intense white cloud and backdropped by a sky of the pitchest black. The feeling, he said, was like those final euphoric seconds before embarking on a parachute jump or the sensation felt by an adrenaline-charged athlete about to perform the stunt of a lifetime.

With the exception that the cosmonauts' stunt was carried out far higher and in a much more hazardous environment, the parallels drawn by Khrunov were apt: just like a parachutist or an athlete, the spacewalk required a considerable amount of self-awareness and demanded every ounce of energy and stamina that he and Yeliseyev could muster. Although Boris Volynov was on hand to assist them for a while – checking their life-support apparatus and communications gear and helping them to attach and secure their gloves – the time inevitably came for him to retire to his couch in Soyuz 5's descent module, seal the hatch and depressurise the orbital module. With all this preparation in mind, it is quite remarkable that Khrunov, the first man outside, swung open the hatch and floated into the void at around 12:43 pm, scarcely an hour after the docking with Shatalov.

His departure was not, however, entirely untroubled. One of his oxygen hoses became entangled and he accidentally closed the tumbler of the suit's ventilator. Although he succeeded in freeing it and solving the problem, the momentary concern and the need to help his partner distracted Yeliseyev, who forgot to install a movie camera outside the hatch. As a result, the world was denied the film of the spacewalk and had to make do with a poor-quality video transmission.

Not surprisingly, no evidence of any of these problems appeared in either cosmonaut's official recollections. In fact, Yeliseyev's failure to set up the movie camera was 'saved', it seems, by Khrunov, who assembled a still camera inside Soyuz 4's orbital module. It was this which yielded the very few grainy images from the joint mission. Still, both men would tell a press conference at Moscow State University on 24 January that their Yastreb suits had performed well; their ventilators and heat exchangers worked effectively and they experienced no 'fogging' of their visors from condensation. Khrunov recounted that, far from 'walking', the most efficient form of locomotion was a hand-over-hand progression using rails attached to both spacecraft. "Moving along the rails in this way," he added, "I approached the camera. Then, gripping the rail with one hand, I removed the camera with the other from the bracket and disconnected it from the onboard electric mains.

Then, 'walking on hands' in the same manner, I moved along the outer surface of the assembled space station and entered the compartment of Soyuz 4."

Khrunov's misleading reference to the combination as a 'space station' was endorsed by Yeliseyev, who went further by stressing that "the choice of the method of transfer, through open space rather than by means of a tunnel, was not an accidental one". Not everyone in the western world was fooled. *Time* may have pondered the Soviets' possible space station ambitions, but American astronaut Deke Slayton, for one, remained sceptical. In his autobiography, Slayton pointed out that the space station comparison was "sort of a stretch" and speculated that the Soviets were simply trying to upstage Apollo 9, an American rendezvous, docking and crew-transfer flight, scheduled for launch several weeks later. On the Apollo mission, however, crew members were to transfer *internally* between their two spacecraft in shirt-sleeves.

Yet the spacewalk was still an impressive feat. By 1:30 pm, both men were inside Soyuz 4's orbital module – the hatch of which had been automatically cranked open by Shatalov – and, shortly thereafter, had assumed their new places in the descent module. At 3:55 pm, four hours and 35 minutes after docking, the two spacecraft separated and Volynov fired his thrusters to pull away. A variety of scientific tasks were performed by both crews during their final hours aloft: Volynov operated the RSS-1 spectrograph for geophysical studies and Khrunov supervised a series of experiments to analyse the passage of radio waves through the ionosphere. Both vehicles had externally mounted aluminium plates to characterise the presence of helium-3 and titanium and the crews participated in observations of cometary tails, terrestrial cloud coverage, evidence of storm formation, snow and ice cover and the forms of glaciers.

Next morning, Shatalov initiated re-entry and he, Yeliseyev and Khrunov descended through a wintry blizzard and thumped onto the snowy Kazakh steppe at 9:53 am, some 48 km south-west of the coal-mining city of Karaganda. Shatalov, whose performance during the rendezvous and docking was later described as exemplary, became the first cosmonaut to maintain a running commentary during re-entry, using a VHF antenna embedded in the hatch of the descent module.

For all the doubts over the validity of the 'space station' claims, Soyuz 4/5 became the first manned flight to exchange crewmembers in orbit. At the instant of touchdown, Shatalov had spent a little less than three days in space, whilst Yeliseyev and Khrunov concluded missions of almost 48 hours apiece. Despite landing in a blizzard, with 60–80 cm of snow on the ground and temperatures of -37°C, all three men were safe and were picked up by a recovery helicopter within minutes. However, the perils of their wintry landing would pale in comparison to the trauma suffered by Boris Volynov during his return to Earth early the next day.

In fact, so harrowing was the tale of Volynov's return – and so close was his brush with death – that it would be almost three decades before the western world heard anything about it. Even those closest to the Soviet space programme, including Vasili Mishin, were caught totally unaware as the prospect of a re-entry disaster of Columbia-like proportions unfolded before their eyes. The euphoria surrounding the safe landing of Soyuz 4 had given way to a mistaken sense that Volynov's return to

Earth would be a walk in the park. Shortly after Mishin arrived in the Yevpatoria control room at around 8:00 am on 18 January, apparently still hungover from the night's festivities, he and everyone else was brought face to face with a harsh reality: that space flight was by no means routine.

The main worry that morning was anti-cyclonic conditions at the landing site, coupled with frigid temperatures hovering at close to -35°C. The plan called for Volynov to manually orient Soyuz 5 for retrofire and make his landing at 9:30 am Moscow Time. After rehearsing the steps for this procedure during his final orbit, he reported he could not do it within the allotted nine minutes. Nevertheless, he was told to try. Commands were also provided for a second, automatic retrofire, in the event that the manual effort failed. The intended retrofire time came at 8:48 am, but, eight minutes later, Volynov reported that he had been unable to complete the orientation manually and controllers prepared to uplink the commands for an automatic burn on the next orbit. It would seem that weather conditions on the ground were also instrumental in the one-orbit delay.

Re-entry finally got underway high above the Gulf of Guinea at 10:26 am, but, wrote Hall and Shayler, it soon became alarmingly clear "that the spacecraft was ... violently tumbling". Having already lost Vladimir Komarov during a bungled return to Earth two years earlier, it was obvious to the Yevpatoria staff that yet another cosmonaut might very soon fall victim to the hazards of space flight. What was not known to them at the time, however, was that as re-entry began Soyuz 5's instrument module was still securely attached to its descent module. For Volynov, the implications of this were catastrophic.

Under normal circumstances, shortly after retrofire, a series of pyrotechnics should have sheared the two apart, enabling the bell-shaped descent module to adopt its correct re-entry orientation, with the heavily protected base facing into the direction of travel to shield Volynov from the brunt of 5,000°C frictional heat. For this reason, the base was coated with a 15 cm thickness of ablative material, half of which was designed to char, melt and peel away during re-entry, safeguarding the descent module from the heat flux. Unfortunately, the final half-hour of Soyuz 5 was far from normal.

With the instrument module still in place, the base's thermal shield was covered, unable to fulfil its purpose, and worse still, the combined spacecraft was forced to adopt the most aerodynamically stable orientation – with the 'dome' of the heavy descent module and its thin hatch facing into the direction of travel and about to feel the full force of a searing hypersonic re-entry. Unlike the base, the top of the descent module was coated with just 2.5 cm of ablator – and the heat of re-entry would char away at least *three times* as much off the base, so a re-entry in this attitude could only end in catastrophe.

At 10:32 am, Stockholm radio analyst Sven Grahn and his colleague Chris Wood, based in Fiji, noted that shortwave communications signals from Soyuz 5 had abruptly stopped; an instant "normally assumed to be the time of separation of the instrument module, and in all probability it was the time when the separation pyros fired ... Probably the electrical connections, but not the mechanical connection between the re-entry vehicle and the instrument module, were severed". Aboard

Soyuz 5, Volynov heard the pyrotechnics fire, but was stunned when he glanced through his window to see the solar panels and whip antennas of the still-attached instrument module. According to Grahn on his website, www.svengrahn.pp.se, the cosmonaut reported what he saw "through some coded radio channel" to ground controllers. This was probably done on shortwave, since he was out of VHF range with the Soviet Union at the time.

When they realised what had happened, or more accurately what had *not* happened, many controllers buried their faces in their hands. Another space tragedy was unfolding before them. One officer removed his cap, dropped three roubles into it and passed it along the line; within minutes, it had filled with coins for Volynov's young family. Against such overwhelming odds – essentially a nose-first return to Earth, with the least-protected section of his craft exposed to the greatest thermal stress and the cosmonaut himself subjected to G forces more than nine times their terrestrial load – it seemed that Volynov's fate was sealed.

Not until 1996, almost three decades after the event, was he finally able to speak publicly about what had happened during that terrifying final half-hour. Rather than being pushed back into his couch, as would be expected in a normal, base-first re-entry, Volynov found himself 'pulled' outward against his harnesses. Yet he still managed to repeat "no panic, no panic" over and over. In what he assumed would be the final minutes of his life, he continued to report his status into an onboard voice recorder and even tore the last few pages from his rendezvous notebook, jamming them into his pockets, in the vain hope that they might somehow escape incineration. Like the American astronaut John Glenn, who had endured his own harrowing re-entry in February 1962, Volynov thought only of working through his procedures ... and, at the same time, maintaining "a deep-cutting and very clear desire to live on when there was no chance left".

From his couch, he could only watch helplessly as flickering tongues of flame licked at the descent module's windows and washed over the cabin. The thin hatch, directly in front of his eyes, visibly bulged inwards under the tremendous heat and pressure. All of Soyuz 5's hydrogen peroxide propellant had been expended shortly after the onset of re-entry, when the automated systems struggled fruitlessly to properly orient the descent module. Gradually, the intense heat – a heat which Volynov, clad only in a light flight garment, rather than a pressurised suit, could physically *feel* – began to melt the gaskets which sealed the hatch and the cabin started to fill with noxious fumes. He clearly heard a roar as the propellant tanks in the instrument module exploded, together with a prolonged and disturbing grinding sound as the stresses of deceleration took their toll on the unusual configuration.

"Through it all," wrote Asif Siddiqi in *Challenge to Apollo*, "there were terrifying moments. Once, there was a sharp clap, indicating that the propellant tanks ... had blown apart with such force that the crew hatch was forced inwards and then upwards like the bottom of a tin can ... "

At length, thankfully, the struts holding the instrument module severed, the two modules separated and the descent module's offset centre of mass caused it to assume a base-first orientation. It tumbled violently as it fell ballistically. The descent ended at 11:08 am with a touchdown close to Orenburg, hundreds of kilometres off-target in the snowy Ural Mountains. 'Touchdown', however, at least in this context, was something of a misnomer; for the element 'touch' implies a safe and soft landing. For Volynov, it was anything but soft and, he soon realised, placed him in yet more danger.

Despite having endured and survived one of the space programme's most terrifying re-entries, the cosmonaut's ordeal was not over. Heat damage and the tumbling had caused Soyuz 5's parachute lines to entangle and as a result their canopies only partially inflated. Moreover, one of the solid-fuelled soft-landing rockets in the module's base failed to fire, resulting in a particularly hard touchdown – so hard, in fact, that the hapless Volynov was torn from his couch and thrown across the cabin, breaking several teeth. As the incessant noise and vibration of the last half-hour was replaced by the absolute silence, ethereal stillness and bitter cold of a late winter's morning in the Urals, he was able to reflect on how lucky he was to be alive.

Alive, yes, but certainly not safe. The temperature outside was close to -40°C and the superheated metallic surfaces of the spacecraft now hissed in the snow. Volynov knew that he was far from his planned landing site and would have to wait several hours for rescue. On the other hand, spending hours in Soyuz 5 in sub-zero conditions would mean certain death. He clambered outside and, spitting blood and bits of teeth into the snow as he went, set off in the direction of a distant column of smoke until he reached a peasant's cottage, where he took refuge, knowing that the rescue party would find the spacecraft and then follow the 'tracks' of his bootprints and blood. Through a mouthful of broken teeth, the traumatised Volynov had just four words for them: "Is my hair grey?"

ASSASSINATION BID

For Leonid Ilyich Brezhnev, the month of January 1969 began with the grim news of the American flight to orbit the Moon and ended with an attempt on his own life. By this point, Brezhnev was in his early sixties, entering his fifth year as *de facto* master of the Soviet state ... and was already regarded internationally in a deeply unpopular light. In October 1964 he had been instrumental in the overthrow of Nikita Khrushchev, due to economic incompetence, erratic behaviour and humiliating withdrawal from a dramatic face-off with the United States in Cuba.

After this bloodless coup, Brezhnev was appointed as the new First Secretary of the Communist Party. In steadily consolidating his power he assumed the title of 'General Secretary' in April 1966 – the first time that this controversial moniker had been used since the days of Stalin – and adopted various regressive and repressive socioeconomic measures. The most visible consequence of this was the trial and imprisonment of Andrei Sinyavski and Yuli Daniel for writing 'anti-Soviet' satirical texts. Although Brezhnev's years in office were not marred by a return to the infamy of Stalin's purges, the much-feared KGB regained many of its former powers. Notable pro-communist foreigners, including the French writer Louis Aragon, saw the imprisonment of writers as dangerous. As he wrote in *Time*: "To

Leonid Brezhnev congratulates the joint crews of Soyuz 4/5 on their achievement. From the left are Brezhnev, Vladimir Shatalov, Alexei Yeliseyev, Boris Volynov and Yevgeni Khrunov.

make *opinion* a crime is something more harmful to the future of socialism than the works of these two writers could ever have been. It leaves a bit of fear in our hearts that one may think this type of trial is inherent in the nature of Communism."

Surprisingly, objection was also voiced within the Soviet Union, with many luminaries putting their names to letters petitioning Brezhnev to show lenience and not return to the dark days of Stalinism. The letters went unheeded, but the seeds of dissent were already being sown. The Sinyavski-Daniel trial has been seen by many historians as perhaps the single pivotal moment at which many in the Soviet bloc finally showed that repression was unacceptable. "Little did [Sinyavski and Daniel] realise," added historian Fred Coleman, "that they were starting a movement that would help end communist rule."

Nonetheless, for now, the repression continued. It was at Brezhnev's direction that an attempt in August 1968 by elements of Czechoslovakia's communist leadership to institute liberal reforms and improve human rights – the so-called 'Prague Spring', under Alexander Dubček – was crushed by 250,000 Warsaw Pact troops. By invoking their right to meddle in the internal politics of a fellow Pact signatory, through the provisions of the notorious 'Brezhnev Doctrine', the Soviet-led invasion left 72 Czechs and Slovaks dead, culminated in the arrest of Dubček and other would-be reformers and provoked a wave of emigrations from the country on a scale never seen before or since. There was widespread condemnation, most visibly in the self-immolation of student Jan Palach in Prague's Wencesas Square on 16 January 1969. Other students, including Jan Zajíc and Evžen Plocek, followed his example in February and April, bringing more international attention onto Brezhnev's despotic regime. His support of Russia's space programme, too, was a shadow of what it had been in earlier years. In *Challenge to Apollo*, Asif Siddiqi noted that Brezhnev was "considerably less sympathetic" towards space than Nikita Khrushchev had been. The latter, to be fair, had been chiefly fascinated in the glamour and propaganda impact that it imposed on the world, whereas his successor "supported space only if it brought political dividends".

It was under the harsh glare of this international spotlight that Brezhnev sought on 22 January 1969 to exploit one of the few success stories of his year so far, by organising a lavish ceremony to applaud the joint mission of Soyuz 4/5, which ended shortly after Palach set himself alight. Cosmonauts Shatalov, Yeliseyev and Khrunov – with the injured Boris Volynov conspicuously absent – were the event's star guests. "A motorcade was to bring them to the Kremlin through cheering crowds gathered around Red Square," wrote Alexei Leonov, who was present that day. "While the cosmonauts rode at the head of the motorcade in an open limousine, behind them, in a Zil-117, rode Leonid Brezhnev." Leonov was in another car, immediately behind that of Brezhnev. With him were Valentina Tereshkova, the first woman in space, her husband Andrian Nikolayev – himself a decorated cosmonaut – and Soyuz 3 veteran Georgi Beregovoi. Little did any of them know that their own lives would be placed in great peril that day.

"After crossing the River Moskva," Leonov wrote, "the limousine carrying the crews of Soyuz 4 and 5, together with the car in which Brezhnev ... [was] riding, abruptly veered off to the side to approach the Kremlin through a different gate. The rest of the motorcade, with our car now in front, proceeded towards Borovitskaya Gate." The diversion of Brezhnev's car had been ordered following the theft, the previous day, of a pair of Makarov semi-automatic pistols from a Leningrad barracks. Those pistols were now in the hands of a young army lieutenant named Viktor Ilyin, who, dressed in his uncle's police uniform, was poised close to Borovitskaya Gate. As the limousine drew alongside him, Ilyin sprang from the crowd and unleased eight shots at the car he thought was carrying Leonid Brezhnev.

Unfortunately for Ilyin, the theft of the pistols had set off alarm bells amongst the Moscow authorities and all main roads into and out of the Soviet capital were closed for fear of an imminent terrorist attack. That attack was now directed not at the man behind the invasion of Czechoslovakia, but instead at the limousine bearing Leonov, Tereshkova, Nikolayev and Beregovoi. It has been suggested that Ilyin 'knew' that Brezhnev always travelled in the second limousine of a motorcade and may, indeed, have mistaken the bushy-browed Beregovoi for his target. Whatever his thoughts and motivations, Ilyin fired indiscriminately in the direction of the car, with several shots hitting the driver and four missing Leonov by a whisker. "Had the car not veered away from the gunman," he wrote, "one of his shots would certainly have killed me."

It would later become apparent that Ilyin had fled his barracks with the pistols and arrived at Moscow's Sheremetyevo Airport intending to assassinate Brezhnev. He was described as a short-tempered young man whose quarrel with the repressive regime had come to a head when he joined the army; he had expressed disapproval of the invasion of Czechoslovakia and was dissatisfied with the Communist Party's monopoly in Russia. Subsequently, the KGB hinted that agents had been watching Ilyin, but did not expect to find him in a policeman's uniform, and two of his comrades were later given five-year prison terms for failing to turn him over to the military authorities.

The driver of the cosmonauts' limousine, Ilya Zharkov, was a substitute on his final day before retirement; he died from severe neck and head injuries. Beregovoi was hit in the face and a bullet grazed Nikolayev's back, although it was he who had the presence of mind to seize the wheel from Zharkov and continue to steer. In the middle of the chaos, Vasili Zatsipilin, one of the motorcycle escorts, aimed his machine directly at Ilyin and succeeded in bringing him down. Later that day, Yuri Andropov, head of the KGB and future master of the Soviet Union, quizzed Ilyin over his actions. The lieutenant said a man "should live, not exist" and confirmed that his intention was to slay Brezhnev and somehow establish a non-communist government.

Surprisingly, Ilyin escaped the death penalty; so absurd was his desire to kill Brezhnev, it seemed, that he was declared criminally insane and remanded for 20 years in a Kazan mental asylum. Following his release in 1990, he would express regret for the death of Zharkov. Yet the final and strangest irony in the story of Viktor Ilyin – the man whose disdain and dislike of the Soviet system could only be attributed to madness – is perhaps this: because he was never officially dismissed from military service, the soldier who came closest to slaying his own commander-in-chief was able to sue the army for two decades of unpaid salary ...

THE TROIKA FLIGHT

When one reads the Tass and *Pravda* accounts of Boris Volynov's recollections at a Moscow press conference on 24 January 1969, it is hard to believe that he had almost lost his life six days earlier. In recognition of his achievement, the cosmonaut was awarded two of Russia's most exalted accolades – the Hero of the Soviet Union and the Order of Lenin – before being quietly spirited away to hospital. So severe were the physical and psychological scars from the Soyuz 5 re-entry that he was withdrawn from active status for two years and told that he would never fly again. Ultimately, he did make another space mission, but nothing – absolutely nothing – of what he endured that day was ever revealed in the Soviet press. The lack of information from behind the Iron Curtain was illuminated in the stifled reports of western journalists. *Flight International*, for example, summarising the joint flight on

23 January, told its readers that Volynov had "re-entered and landed uneventfully" and that all four cosmonauts "were said to be in excellent health … "

For all the glamour and political headway that it afforded, there was little denying the reality that Yeliseyev and Khrunov's spacewalk was two years behind schedule and virtually obsolete in terms of the value that it offered for a 'real' space station. Nikolai Kamanin, in his ubiquitous diary, revealed a few days after the flight that this transfer method had already outlived its purpose and, indeed, Soyuz 4/5 would be the only time this was done. Many analysts in the west were quick to point out that an extravehicular transfer from a Soyuz to a space station simply would not do: in the interests of safety, to say nothing of practicality and sheer technical obstacles, cosmonauts needed to be able to transfer *internally*, in shirt-sleeves.

Of course, before any space stations could be launched or occupied, the Soyuz itself required further qualification. On 26 April 1969, the State Commission that investigated the failure of Soyuz 5's instrument module to separate presented its report. This analysed dozens of possible scenarios, eventually settling on the 'most likely' cause: a 'hang-up' of one of the 102 clamps between the two modules. Simulations had shown that these could separate cleanly with the usual 70 kg of force imparted by the pyrotechnics, but in some cases the force could be greater and this could cause a clamp to rebound and close again, preventing separation. An overhaul of the separation system, including its latches, clamps and pyrotechnics, was finished and tested by the end of September.

In spite of Soyuz 5's close shave with disaster, plans were well advanced for further missions, one of which even called for a duration of up to 30 days. By mid-February, however, Nikolai Kamanin was promoting a more realistic, week-long 'solo' flight by Soyuz 6 in April or May, followed in August or September by a docking between Soyuz 7 and 8 in which the two spacecraft would remain linked for up to three days. However, Dmitri Ustinov considered a solo flight too conservative and demanded a "more solid" mission. By the time the State Commission's report appeared, therefore, the plan was to fly *three* missions in August, the first of which would station-keep at a distance of some 50 m and film a rendezvous and docking between the other pair.

The timing, of course, was especially interesting. The United States was only days away from launching Apollo 10, its final mission before attempting a manned lunar landing in July 1969, and the direction of the Soviet space effort, by complete contrast, was in disarray. In his diary, Kamanin fumed at this lack of direction, blaming the "complete absurdity and record short-sightedness" of the Soviet leadership and bemoaning the fact that there was "not a single person in the country who can tell when the next space mission will take place".

Now, under pressure from Ustinov, the revised Soyuz 6 mission, in addition to its role in filming the rendezvous, would also conduct a wide range of scientific and technical studies. Its cosmonauts would test an experimental welding furnace, known as 'Vulkan' ('Volcano'), and perform observations of clouds and cyclones, examine geological and geographical targets, characterise terrestrial landmarks under differing light levels and analyse micrometeoroid erosion of the spacecraft's windows. Both it and the two flights that were to dock would be followed by the more ambitious Soyuz 9 and 10, which were tapped to fly for perhaps 16 days apiece and test a new rendezvous radar, originally for lunar flights, called 'Kontakt' ('Contact'). Flying for this length of time would soundly beat America's 14-day endurance record, set at the end of their Gemini VII mission in December 1965, and would once more cement Soviet credentials as a major spacegoing power. It would also provide renewed impetus for future near-Earth projects involving long-term orbital stations.

First, though, rendezvous and docking had to be duplicated and perfected. Each of the Soyuz 6, 7 and 8 missions of the so-called 'troika' ('threesome') was expected to last four or five days. Georgi Shonin and Valeri Kubasov would launch first, followed by Anatoli Filipchenko, Vladislav Volkov and Viktor Gorbatko and, finally, Andrian Nikolayev and Vitali Sevastyanov. By the end of August, yet more changes had occurred; this time with respect to the crewing. Nikolayev and Sevastyanov, who were to have flown the 'active' Soyuz 8 to dock with Filipchenko's 'passive' Soyuz 7, apparently performed so poorly in their preparatory exams in late July that they were dropped in mid-September in favour of their backups, Vladimir Shatalov and Alexei Yeliseyev. In spite of Nikolayev's protests – and, interestingly, also the complaints of his wife, Valentina Tereshkova – that he should be allowed to fly, Nikolai Kamanin remained firm in his conviction that Shatalov should command Soyuz 8.

The story of the replacement of Nikolayev's crew has been told by Grujica Ivanovich. Yeliseyev vacationed in Central Asia after his January mission and, upon returning to work, was assigned with Shatalov to backup *all three* of the troika flights, teamed with Pyotr Kolodin for Soyuz 7 duties. A summons to Vasili Mishin's office one day in the late summer of 1969 came as a great surprise to Yeliseyev. He knew the Soyuz 8 prime crew were "ill-prepared", but "I did not expect such a turn of events". Mishin told him, point-blank, that Nikolayev and Sevastyanov's exam scores were unacceptable and that both men worked "thoroughly badly".

Having executed his own docking with precision several months earlier, the confident and rendezvous-savvy Shatalov proposed a method for a more fuel-efficient link-up with Soyuz 7. Rather than the pilot assuming 'hands-on' control only for the final approach, he advocated totally uncoupling the automatic system and accomplishing both the 'terminal' rendezvous and the docking manually. Flight specialists agreed to review the suggestion and Kamanin endorsed it on the basis of its time and propellant-saving benefits, together with its potential applications for "intercepting enemy satellites", but Vasili Mishin was against it, saying that it would increase the chances of failure. Soyuz had been explicitly designed to perform automated rendezvous. A pilot would require indicators for range and range-rate readouts to attempt a manual approach, neither of which were provided.

There were other problems, too. Defects in all three Soyuz craft, which some sources said numbered into the dozens, required the replacement of several key components. So intensive was the workload that very little time was available for the cosmonauts to train in their assigned craft as August wore into September. It was said that the first time the men even saw some of the experimental hardware they were to use was at *Tyuratam* itself ... only days before launch! When they arrived on

22 September, they confidently expected their first launch on 5 October, but delays in the Politburo granting approval forced yet more slips. Nikolai Kamanin's frustration is evident in the pages of his diary. "The indecisiveness of our highest leaders," he wrote, "will delay the launch of the ships by at least a week." Yet the Politburo's caution is understandable: for *this* mission would be seen by the world as the Soviet Union's answer to Apollo 11. Even putting three craft into orbit at the same time would not be enough to capture the world's imagination and if it ended in catastrophe, the political ramifications were dire.

At length, on 10 October, the western world heard the first hint that the troika mission was imminent. The Soviets had already revealed that all three craft would manoeuvre in close proximity to one another, test common control mechanisms and perform scientific studies, with a clear emphasis on a future space station. In typically ambiguous Soviet style, the magazine supplement *Nedelya* observed that "man must build himself a home, wherever he goes: on the tundra, in the forests, in the mountains, on the bottom of the oceans and now in space". What actually transpired over those few days in mid-October 1969 was far from a home in space and came to be seen by many in the west as another Soviet effort to achieve a dubious space 'first'. Although Deke Slayton, watching from afar, admitted that there was some technical value in controlling three manned spacecraft simultaneously, he and others struggled to see any requirement for such a mission.

Nonetheless, the tracking support for the troika was enormous. "Transmissions," wrote Asif Siddiqi, "were normally limited to flight over the Soviet land mass or with a small flotilla of modest seafaring vessels under the control of the Department of Naval Expeditionary Work under the Academy of Sciences since 1967." Since then, however, the Soviets had begun building a series of larger ships, the first of which was named in honour of fallen cosmonaut Vladimir Komarov and was laden with huge arrays of communications and tracking antennas. The ship would serve as just one piece in a complex jigsaw, also involving the Soyuz themselves, the Molniya satellites in Earth orbit and the control centre in Yevpatoria.

The first stage of the troika got underway on 11 October, when Shonin and Kubasov were awakened to a rainy, overcast and whirlwind-whipped Tyuratam. Soyuz 6 blasted off at 2:10 pm. Moscow television, eager to capitalise on the new mission, especially in light of America's successful lunar landing the previous July, interrupted its afternoon programming to show the launch live. Shortly afterwards, the fuel tanks of the approach and orientation engines failed to automatically pressurise, threatening to deny the craft the ability to perform orbital manoeuvres. Fortunately, an override was transmitted from the ground and on the third orbit Shonin was able to manually pressurise the tanks using controls inside the orbital module. At 1:44 pm the following day, Filipchenko, Volkov and Gorbatko set off into light rain and low cloud cover aboard Soyuz 7. (Interestingly, four of these five men had backed-up the Soyuz 4 and 5 crews in January: Filipchenko for Shatalov and Shonin, Kubasov and Gorbatko for Volynov, Yeliseyev and Khrunov.) On the afternoon of 12 October, whilst waiting for the third element of the troika to launch, the Soyuz 7 crew used the 'Svinets' ('Lead') optical device to track ballistic missile plumes – clearly a military experiment – and transmitted live television pictures from

Nikolai Kamanin (left) plays chess with Soyuz 6 commander Georgi Shonin (right), whilst flight engineer Valeri Kubasov looks on.

their craft. The final launch came at 1:19 pm on 13 October, when Shatalov and Yeliseyev thundered aloft in Soyuz 8. It is interesting to note that, unlike Soyuz 4/5, on this occasion the 'passive' craft was launched ahead of its 'active' counterpart, just as the cosmonauts had previously advocated.

When placed into context, three months after Neil Armstrong and Buzz Aldrin had triumphantly trodden the lunar Sea of Tranquillity and a few weeks ahead of Apollo 12, the spectacle of seeing rapid-fire Soviet launches simply did not have the same impact as it had when pairs of Vostoks were launched in 1962 and 1963. The United States, indeed, had come a long way from sending its astronauts on relatively puny 15-minute suborbital 'hops' or orbital missions of just a few hours and the sense of humiliation in the Soviet Union was palpable. Much of what was now being done – multiple spacecraft in orbit, rendezvous, docking, spacewalking – had been done, repeatedly, by the Americans. As one young Muscovite glumly put it: "It's not much compared with the Moon, is it?"

Pessimism aside, the missions continued. Shonin and Kubasov's work with their own Svinets experiment reportedly went well. Also successful were their observations and photography of the Caspian Sea and Volga River deltas, the huge forested regions of central Russia, cloud movements and whirlwinds in Kazakhstan, typhoons and hurricanes battering Mexico and vast tropical storms in the Atlantic and Indian Oceans. In general, wrote Asif Siddiqi, all three Soyuz had quite different research programmes: Shonin and Kubasov's focus was upon biomedical experiments and Earth-resources photography, Filipchenko's crew undertook terrestrial and astrophysical studies using a battery of multispectral imagers and Shatalov and Yeliseyev's work included making observations of the polarisation of sunlight reflected by the atmosphere. One experiment of note was 'Fakel' ('Torch'), in which Shonin and Kubasov used the Svinets to visually discern the plumes of three R-16 missiles launched from Tyuratam on 12 October. However, "all the launches were at night," noted Siddiqi, "limiting the applicability of the experiment. It is unlikely that the Svinets instrument would have been capable of detecting launches during daytime."

Conditions aboard the three spacecraft were reportedly good, with the cosmonauts working and living in loose gym-style sweat suits and consuming four meals per day, with a caloric intake of 2,600 Kcal. Typically, a Soyuz breakfast comprised apricots, bread, chocolate and blackcurrant juice, followed by a lunch or supper of dried or smoked fish, pate, chicken, bread, cookies, prunes and candies, washed down with milk, coffee, cocoa or fruit juices. "It contains everything necessary for nutrition, including vitamins," wrote Boris Mandrovsky, "and at the same time it is sufficiently diversified." However, the appeal and tastiness of such culinary combinations can only be guessed at . . .

What proved not so successful was the rendezvous itself. The plan called for Shatalov and Yeliseyev to perform the link-up around 24 hours into their own flight. The orbital manoeuvres calculated by the ground gradually brought Soyuz 8 within a kilometre of Soyuz 7. At a distance of 100 m, Shatalov was to take manual control and perform the docking. Unfortunately, the Igla rendezvous device failed to achieve an automatic 'lock' onto Soyuz 7. This difficulty was compounded by onboard indicators which confirmed that it was working correctly. (Indeed, rumours over the years hinted that one of the craft might even have been inserted into the wrong orbit.) A stressed Shatalov "bravely", it is said, asked to complete the approach manually, without Igla support, but by the time Vasili Mishin gave his approval the two craft had drifted more than 3 km apart. Shatalov was advised that he could perform a manual rendezvous, so long as his range was less than 1,500 m, and a second attempt was set for the following day.

Western observers offered their own clues as to what was happening. Desmond King-Hele of the Royal Aircraft Establishment at Farnborough in Hampshire, England, an expert in satellite orbits, reported seeing two objects, apparently 12 km apart, at 8:43 pm Moscow Time on 14 October. Others, including Pierre Neirinck, saw a pair of objects some 3.5 degrees of arc apart – roughly corresponding to 30 km – with a third, presumably Soyuz 6, some 470 km distant. In view of the fact that Shonin should have closed to within 50 m of Filipchenko and Shatalov to film their rendezvous, it was clear that something had gone seriously wrong.

The second rendezvous attempt should have begun early on 15 October, but it was not until 12:40 pm that Soyuz 7 and 8 were close enough – at some 1,700 m – for Shatalov to commence a manual approach, aided by ground-based ballistics data. "If the Igla system had been working," wrote Hall and Shayler, "the main engines could have been used to close the distance; but without it, Shatalov could only use the smaller ... approach and orientation engines. The cosmonauts therefore had to rely on visual cues and there was little chance of success." The problems mounted. Yeliseyev, in Soyuz 8's orbital module, found it difficult to 'see' Soyuz 7 against the backdrop of Earth, and Shatalov himself, after performing four small manoeuvring burns, had no accurate idea of his range from the target. As a result, the docking attempt was called off and, according to a Tass news release, the two craft drifted past each other at a range of 500 m.

Later that same day, Shonin and Kubasov manoeuvred their own Soyuz to a location around 800 m from Soyuz 7, apparently as part of a dress rehearsal to rendezvous with "an uncooperative object" in space. Since Soyuz 6 was not equipped with an Igla, the cosmonauts had to use data supplied by ground controllers, together with an onboard sextant. Although Shonin benefitted from having a second control panel in the orbital module for his approach and orientation engines – a modification which Shatalov had suggested following his Soyuz 4 experience – the evidence suggests that Soyuz 6 was unable to achieve a 'close' approach to either of the other craft. However, according to the journal *Novosti Kosmonavtiki*, at one stage its crew got a brief glimpse of Soyuz 8 through their portholes.

Also able to see each other's craft were the flight engineers of Soyuz 7 and 8, Volkov, Gorbatko and Yeliseyev, but they had no reliable data to assist the rendezvous process. Reports hinted that their heart rates climbed as high as 100 beats per minute, which was taken as evidence of extreme nervousness. Hall and Shayler noted that Shatalov made two more close approaches on 16 October, "but ballistics experts introduced errors into the orbital computations and both attempts failed". All further docking attempts were called off and plans initiated to return all three crews to Earth.

CIA intelligence reports, declassified in 1997, pointed to a total of five failed attempts to dock Soyuz 7 and 8. According to the reports, the cosmonauts' efforts were hampered by a combination of Igla problems, excessive use of attitude-control propellant, poor timing, insufficient time to correct out-of-plane separation errors between the two vehicles and difficulties controlling lateral velocities during final approach. It was, wrote Siddiqi, "a complete mess". In his diary, Nikolai Kamanin criticised the Soyuz design, because an Igla failure left a crew with no means to take effective manual control.

Some success derived from Soyuz 6, which became the first mission to successfully demonstrate a welding tool in space ... albeit at the expense of nearly burning a hole in the wall of the orbital module! The Vulkan furnace had been provided by the Institute of Electrical Welding in Kiev. When Boris Paton, the director of that institute, proposed the development in 1963, his expectation was that an institute engineer – probably Vladimir Fartushny – would operate it, but the furnace weighed 50 kg and spacecraft limitations prevented flying a second flight engineer.

Vulkan required the internal hatch to be sealed and the orbital module depressurised, with the operation being performed automatically and Kubasov monitoring it from the descent module. The two-piece furnace was a squat green cylinder which resembled "a round refrigerator". It comprised command and power systems in a pressurised nitrogen atmosphere and three welding devices to be evaluated: an electron beam, a low-pressure compressed arc and a consumable electrode. An electron gun was provided to perform the welding, which used samples of titanium, aluminium alloys and stainless steel, with the samples mounted on a turntable. A protective shield covered the unit.

Although these welding techniques were considered simple, reliable and relatively easy to automate, Boris Mandrovsky noted that they also produced large quantities of steam and gases. "Consequently," he wrote, "it was necessary to find out how they would react under conditions of vacuum and weightlessness and to design special equipment which would be suitable for working in space."

As Soyuz 6 completed its 77th circuit of the globe, only a few hours before landing, Kubasov closed the hatch between the orbital and descent modules and flipped switches to begin the experiment. Samples of stainless steel and titanium were welded together, then cut, after which – according to Mandrovsky – Kubasov reopened the hatch to the orbital module and "carried out a hand-held welding operation using part of the Vulcan device while Shonin photographed this performance". The results were said to be "in no way inferior" to terrestrial welds and Boris Paton lauded the operation, declaring that "a stable processing of welding metals by methods of melting is feasible". Welding, added Mandrovsky, "will be needed for the building of large orbital stations as well as for repair of vehicles which have been in space for long periods of time".

When it became clear that the Vulkan had nearly caused a catastrophe for Shonin and Kubasov, considerable doubt was cast on what Paton referred to as its "reliability". In 1990 it was revealed that the experiment almost burned a hole in the side of the orbital module; the low-pressure compressed arc having inadvertently aimed a beam at its wall. When the cosmonauts entered the orbital module to recover the welding samples, they were shocked to discover the damage and, fearing a depressurisation, retreated back to the descent module. If Kubasov did any hand-welding at all, as Mandrovsky claimed, he must have done so with near-miraculous speed! Eventually, after assuring themselves that the orbital module (for now) remained structurally sound, they hurriedly retrieved their samples and shortly thereafter returned to Earth as planned. Kubasov later admitted that Vulkan was a strictly experimental device – not an 'operational' one – and said some of its elements could be used to perform repair work in emergencies! It is ironic, therefore, that the only 'emergency' on Soyuz 6 was very nearly caused by the Vulkan and that if any 'repairs' *were* needed, their very need would have come from damage caused by the furnace itself . . .

On 18 October, 48 hours after the Vulkan test, Shatalov and Yeliseyev became the last of the troika to return to Earth touching down safely at 12:10 pm Moscow Time. Despite the disappointment, there was some euphoria. Nikolai Kamanin wrote in his diary that his 61st birthday began with yells from Soyuz 8 . . . as the cosmonauts celebrated successfully closing the interior hatch! (Whilst tightening the wheel on the hatch lock before launch, Shatalov had accidentally cracked three of its spokes. Despite Vasili Mishin's assurance that the hatch would be fine, so long as pressure integrity was maintained, the prospect of decompression had worried both

Alexei Yeliseyev (left) and Vladimir Shatalov outside the Soyuz 8 descent module after touchdown.

cosmonauts throughout their flight.) A few hours later, Shatalov publicly declared his joy over the 'successful' mission, which, he said, had been "a flight into the future" and a key stepping stone towards orbital stations. Describing the landings of the three Soyuz, on successive days, he could only exult: "They landed like airplanes!"

Shonin and Kubasov were first to return, thumping onto the steppe, 180 km north-west of Karaganda, at 12:52 pm on 16 October. Their descent module landed upright and by the time recovery crews arrived, the cosmonauts had already extricated themselves from their couches, opened the hatch and were walking around. Soyuz 7 followed suit the next day, its only problem being a warning light in the cabin that advised Filipchenko that the automatic landing sequence had been activated. Kamanin told him that it was probably an electrical glitch, but it caused concern: controllers feared that the landing sequence might begin too soon, perhaps affecting the timing of the parachute deployment and the firing of the solid-fuelled landing rockets. The scare eventually led to a consensus that all was well and the "excited" cosmonauts landed safely at 12:26 pm.

Each mission had lasted barely an hour shy of five full days and the troika of perfect touchdowns seemed to have forever banished the ghosts of Vladimir Komarov's tragic re-entry and Boris Volynov's brush with disaster. In fact, wrote Boris Mandrovsky, the three returns "were so precise that the search and rescue helicopters were on hand at the time of the landing". Sadly, as time would tell, it did not bring down the final curtain on the Soviet record of disastrous returns from orbit. One more tragedy would occur in less than two years' time ... an *avoidable* tragedy in which a misguided sense of complacency brought the dangers of space exploration into stark relief.

MAGNIFICENT SEVEN

When the seven cosmonauts of the troika mission – Georgi Shonin, Valeri Kubasov, Anatoli Filipchenko, Vladislav Volkov, Viktor Gorbatko, Vladimir Shatalov and Alexei Yeliseyev – arrived in Moscow a few days later, they were feted as heroes. The men who had smashed the United States' record of placing the most people into orbit simultaneously were escorted into the Soviet capital by a squadron of six MiG-21 fighters. They were surrounded by journalists, invited to dinners, asked to deliver toasts, posed for photographs and obliged to attend meetings. Yet the sight of this Magnificent Seven, walking to receive their Hero of the Soviet Union and Order of Lenin awards, illustrated how different these cosmonauts were from the exclusively military brotherhoods which had previously dominated both the American and Russian spaceflying corps. Only four wore the uniform of a Soviet Air Force officer; the others, thanks to Vasili Mishin, were dark-suited, black-hatted civilian engineers from his OKB-1 design bureau.

When they were selected for cosmonaut training three years earlier, it seemed unlikely that such civilians would fly at all; now, Yeliseyev had flown twice in just nine months ... and, when teamed with Shatalov for another mission in April 1971, would jointly become the first Soviet cosmonaut to make three space voyages. The other two men, Valeri Nikolayevich Kubasov and Vladislav Nikolayevich Volkov, had been chosen together with Yeliseyev in the summer of 1966 and both would make their own impact on space history: the former as a member of the first US-Soviet joint mission, the latter as one of the first occupants of an Earth-circling space station ... and one of the last Russian cosmonauts to die during a space flight.

Kubasov had been born in the town of Vyazniki, in Vladimir Oblast, some 200 km east of Moscow, on 7 January 1935 and seemed destined to become a cosmonaut from his earliest days as an engineer at OKB-1. In May 1964, working for Sergei Korolev, he had been one of a handful of civilians who survived the preliminary medical screening to be considered for a seat on a Voskhod mission. Two years later, after some 'relaxation' of existing medical rules, he, Volkov and another engineer named Georgi Grechko were officially accepted into the newly established civilian cosmonaut team. When Soyuz 5 lifted off in January 1969, Kubasov was Alexei Yeliseyev's backup, having trained to perform the risky spacewalk transfer.

Vladislav Volkov, whose rugged good looks would later earn him a reputation as a pin-up idol for young Russian girls, had been born in Moscow on 23 November 1935. Among the early cosmonauts, he proved something of a 'Renaissance Man': in addition to his engineering expertise, he played football, ice hockey, tennis, handball and chess, was a skilled athlete, dabbled in boxing for a time and was a talented guitarist. His strength of character has been illustrated by Grujica Ivanovich in telling the tale of his first day at school, when a bully stole his breakfast. Volkov, despite being short and skinny, fought fiercely for it and won his classmates' respect.

His enthusiasm for aviation came from his parents. His father was an aeronautical engineer, his mother a worker in an aircraft factory, and their home was close to Tushino Airport, so the young Volkov grew up watching a variety of different machines taking to the skies. It was his uncle, a veteran combat pilot from the Second World War, who advised him to study aeronautical engineering and Volkov duly enrolled at the Moscow Air Force Institute, meeting and marrying his wife Ludmilla whilst a student.

Following graduation early in 1959, he began working as an electromechanical engineer for aircraft missiles and later, from April of that year, for Sergei Korolev's bureau. This immersed him for the first time in the design and development of the world's first manned spacecraft, Vostok, and its stop-gap successor, Voskhod, for which he became a deputy leading designer in February 1962. Volkov also indulged his love of flying, joining an aeroclub and receiving a diploma as a sports pilot, capable of flying solo in a Yak light aircraft.

When the crew selections for the first Voskhod mission were made early in 1964, it was revealed that – for the first time – a civilian engineer would be chosen for one of three cosmonaut positions and Volkov was among the 14 candidates to undergo physical and psychological screening. Along with Kubasov, he passed the tests, but his name did not appear on the final shortlist. However, Volkov approached Korolev himself to push his credentials and the Chief Designer is said to have responded: "You are young. There is still time."

That time finally arrived a couple of years later, when Volkov was selected as one of eight civilian trainees for Soyuz. Despite completing a gruelling course of parachute training, flights in MiG-15 jets, altitude-chamber runs and exposure to weightlessness in Tu-104 aircraft, not all of the eight would make the final cut. Nikolai Kamanin insisted that they also be screened by the Air Force and, by the end of 1966, only a handful – Kubasov, Grechko, Volkov and later Yeliseyev and Oleg Makarov – had passed.

In his diary, Kamanin recorded with some disdain that Volkov was one of the "invalid" civilians pushed by Vasili Mishin to fly an early Soyuz mission instead of a 'real' military cosmonaut. It was nothing personal, it seemed, and when Volkov, Kubasov and Grechko opined that they could complete cosmonaut training within two months, Kamanin firmly advised them that – despite their expansive knowledge of space systems – they would require at least one or two years of full-time, intensive training before receiving a seat on a crew.

Kamanin's advice seemed accurate, in Volkov's case at least, because until the end of 1968 he was training as a backup flight engineer for the 'passive' half of the Soyuz 4/5 joint mission. Then, when the decision was made to fly three Soyuz in October 1969, he was assigned to the prime crew of the passive craft for that rendezvous. It is recorded that his commander, Filipchenko, occasionally had to "restrain" the energetic Volkov, whose engineering training and knowledge of the systems made him eager to play a greater role in the simulator. Volkov's attitude, it seems, was more than just an expression of self-confidence. He overtly disdained many of the military cosmonauts, thinking himself intellectually superior to them; an outlook which would also haunt his next mission in 1971. Nonetheless, Volkov enjoyed his five days aboard Soyuz 7, despite the crushing disappointment of not being able to complete a docking. He carried among his personal effects a piece of Sevastopol soil, given to him by his young son, Vladimir, and even scored another Soviet 'first' by becoming the first accredited journalist in space, having penned a number of articles for the newspaper *Krasnaya Zvezda* (Red Star), under the name 'Vladimir Volkov'.

Clearly, it would seem that all three civilians performed well during the troika mission, with Kamanin noting that he played tennis with Volkov just two days after the engineer returned to Earth. No ill-effects of weightlessness were apparent. In fact, the only medical item of note was that all seven men lost weight, ranging from 1.5 to 3.5 kg, with the greatest loss being suffered by the Soyuz 7 commander, Lieutenant-Colonel Anatoli Vasilyevich Filipchenko.

Filipchenko came from the village of Davydovka in the Voronezh region, close to the border with Ukraine, where he was born on 26 February 1928. After finishing secondary education, he graduated from the Voronezh Air Force specialised school in 1947 and received early flying instruction. Three years later, the young Filipchenko completed the Kharkov Military Aviation School of Pilots and in 1961 graduated via correspondence from the Soviet Air Force Military Academy in Monino. By the time of his selection as a cosmonaut, in January 1963, he had built up an impressive resume: as deputy commander of a fighter squadron, a senior flying instructor and an accomplished parachutist. (In fact, Vladimir Shatalov, serving as an instructor, had actually *recommended* Filipchenko to the cosmonaut selection team.) By the mid Sixties, Filipchenko was one of a team preparing to fly the Soviet Union's winged orbital spaceplane, called 'Spiral'. He later transferred to the Soyuz training group, backed up Shatalov on Soyuz 4 and was made commander on Soyuz 7.

Aside from Shatalov, the other two military pilots aboard the troika were Lieutenant-Colonels Georgi Stepanovich Shonin and Viktor Vasilyevich Gorbatko, both chosen as cosmonaut trainees alongside Yuri Gagarin in March 1960. It is interesting, therefore, that Filipchenko, a member of the *second* group of cosmonauts, selected in January 1963, should have assumed the mantle of command on Soyuz 7, with Gorbatko, ostensibly more senior, as his 'research engineer'. (On the other hand, the second group were more experienced and, indeed, Gorbatko would go on to command two space station missions in the Seventies.) He was born in the village of Ventsy-Zarya in the Krasnodar region of southern Russia on 3 December 1934 and, aged 15, completed a course at a stud farm, propitiously named 'Voskhod'. Later, in 1953, he graduated from the Eighth Military Aviation School of Basic Pilot Training in the Ukraine and subsequently flew as a senior pilot in the Odessa Military District of Moldova and served as a parachute instructor.

Like Shonin, Gorbatko fulfilled the requirements of the 1960 cosmonaut intake: he was under 30 years old, less than 1.7 m tall and below 70 kg in weight. In fact, he

The combined crews of the Soyuz 6/7/8 mission. From left to right are Valeri Kubasov, Georgi Shonin, Vladislav Volkov, Anatoli Filipchenko, Viktor Gorbatko, Vladimir Shatalov and Alexei Yeliseyev.

barely met the criteria, standing almost 1.69 m and weighing 69.5 kg. By the time that Shatalov and Filipchenko were chosen three years later, the criteria had been opened to allow military navigators and engineers to be considered, as well as pilots, and the age limit had risen to 40. Moreover, the second group of cosmonauts needed to have academic credentials from either a military academy or a civilian university.

Gorbatko's early assignments were within the Vostok project and, as late as September 1963, he confidently expected to fly sometime in the following year. When plans for additional Vostoks were cancelled, he was redirected to train as a possible commander for the Voskhod 2 mission, featuring the world's first spacewalk. Subsequently, he served as Yevgeni Khrunov's backup on the original Soyuz 2 mission and later shadowed his spacewalk training for the Soyuz 4/5 transfer. In the final weeks before the January 1969 launch, however, Nikolai Kamanin noted that both Gorbatko and Georgi Shonin made some minor mistakes in their final exams. These mistakes did not seem to preclude their own chances of reaching space nine months later.

Shonin, in the same vein as many of the early cosmonauts, joined and worked his way up through the ranks of the Soviet Air Force, graduating from the Yeisk

Military Pilot School in 1957 and later from the Zhukovsky Engineering Academy in December 1968. Prior to selection as a cosmonaut, he flew for the Soviet Air Force's Red Banner Baltic Fleet. Shonin had been born on 3 August 1935 in the coal-mining city of Rovenky, in the Luhansk region of south-eastern Ukraine and, like Gorbatko, was considered for a seat on a late Vostok mission. By January 1964, he had begun work in the Soyuz training group and, in September of the following year, Kamanin felt that Shonin was one of the most outstanding members of the cosmonaut team. It was perhaps for this reason that he was assigned, with Boris Volynov, to fly the long-duration Voskhod 3 mission.

Bearing this glowing praise in mind, it is a pity that Shonin's performance in the wake of Soyuz 6 tainted his reputation. Only days after his return to Earth, it was unanimously declared that, with the exception of weight loss, no adverse effects of microgravity exposure were apparent. However, 18 months later, Kamanin and a team of military physicians would conclude otherwise. In March 1970, Shonin was disciplined over "many bad reports" about his behaviour; nevertheless, he remained on active flight status and was considered a likely contender to lead the first space station mission.

Then, early in February of the following year, the cosmonaut apparently turned up to a training session in a state of intoxication and went on to drink vodka in front of engineers and controllers. By the time Kamanin confronted him, Shonin was sober and the general reflected with astonishment how much this young man – whom he had known and trusted for more than a decade – had changed. Military physicians who examined Shonin speculated that he probably started drinking heavily after Soyuz 6 and in March 1971 he was sent to a sanitorium for rehabilitation. A month later, Vladimir Shatalov commanded Soyuz 10, the mission on which Shonin might otherwise have flown. He never flew into space again and quietly left the cosmonaut team in April 1979, "for medical reasons". It was a sad end for a man whose spaceflying career once seemed so full of promise.

MURMURS OF DISCONTENT

Soviet attempts to portray the equality of the communist state as a better and fairer alternative to the inequality of the capitalist west was, in reality, little more than skin deep and in many ways utterly hypocritical. Already, the unfair discrimination of Soviet Jews had reached the ranks of the cosmonaut corps, negatively influencing the career of Boris Volynov in the process. Anti-Semitic feelings had flared in the wake of Israel's victory over the Palestinians in the 1967 conflict: the Soviet Union broke off diplomatic ties with the Jewish state, religious observances were discouraged and even banned and, as a result, many thousands of Russian Zionists applied for permission to return to their homeland.

Such permissions were deliberately difficult to obtain, for the Brezhnev regime had no desire for the socialist state to be seen as one from which thousands of people were trying to escape. Common excuses for the Soviet refusal to issue exit visas included claims that applicants had been privy to classified national security information. The process was compounded still further by the requirement for a 'request letter' from a family member living in the country to which the applicant intended to emigrate, plus documents such as approval from the children's schools, a recommendation from an employer, evidence of financial status, approval from parents and even from divorced partners. Applications were slow, often taking six months or more to be processed. Moreover, in many cases, even requesting an exit visa was considered an act of social parasitism and an offence against the state. As a consequence, it is hardly surprising that only 4,000 exit visas were granted by the Soviet authorities during the Sixties.

The plight of the other would-be émigrés drew the international spotlight in May 1970, just a few weeks before cosmonauts Andrian Nikolayev and Vitali Sevastyanov were launched aboard Soyuz 9, when a group of 'refuseniks' – people who had been refused permission to leave the Soviet bloc – tried to hijack an airliner and fly it to the west. Led by dissident Eduard Kuznetsov and including amongst their number a military pilot with the rather unfortunate name of Mark Dymshits, the group bought up all the tickets for a Leningrad-to-Priozyorsk flight. The attempt failed and Kuznetsov and Dymshits were both given the death penalty for high treason, but a tide of international protest succeeded in reducing their sentences to 15 years of incarceration. Both men were freed in 1979 as part of a exchange deal for a pair of Soviet intelligence officers who had been arrested in New Jersey.

Predictably, the failure of the Kuznetsov-Dymshits hijacking led to a harsh crackdown, including a 'diploma tax', whereby individuals who had received higher education in the Soviet bloc had to pay many times their annual salary in a partially successful attempt to stem a 'brain drain' to the west. Notwithstanding these measures, continuing international pressure ultimately forced Brezhnev to increase the Soviet Union's emigration quota and by the end of the Seventies more than a quarter of a million people had left for Israel and the west. When placed alongside the many *illegal* defections and numerous attempts by would-be escapees to scale the Berlin Wall from east to west (but rarely vice versa), this proved hugely embarrassing for the hard-line communist regimes of eastern Europe. The reality that hundreds of thousands of people *wanted* to escape from a workers' paradise became ever harder for the dictators to explain away. Having said this, and despite an acute period of economic stagnation which forced increased attempts to foster détente with the west, the Soviet Union reached an unprecedented level of power and relative internal calm under Brezhnev. A Public Opinion Foundation poll conducted in 2006 found a 61 percent approval rating for his brand of leadership. Many people would have preferred to live during the Brezhnev years than at any other period of Russian history in the 20th century. Nevertheless, the effect of the Sinyavski-Daniel trial, the need to crack down on dissenters and the mass emigrations of the early Seventies would underline an equal reality that life under the communist fist was not quite as idyllic as it was portrayed. The fall of the Soviet Union was still two decades away … but in 1970, the first murmurings of discontent had already been heard and the first steps to bring about its fall had already begun.

RECORD-BREAKERS

When Vladimir Shatalov met the Dutch space journalist Peter Smolders in 1970, he toed the party line by declaring that no docking had ever been planned between Soyuz 7 and 8. The spacecraft, he told Smolders, "were not equipped for docking" and reinforced the point by reminding him that the Soviets "had already done this in January". Although it seemed peculiar, to say the least, to attempt such an intricate and close-range rendezvous with absolutely no intention to dock, this belief was accepted by many western observers for many years. Today, of course, in a new era of 'openness', and particularly since the fall of the Soviet bloc, we know otherwise: there was indeed a docking planned and its failure can be attributed to one critical piece of hardware: the Igla rendezvous system.

The purpose of Igla has already been mentioned and it actually reared its head as one possible reason for Georgi Beregovoi's inability to dock with Soyuz 2 in October 1968. Its designer, Armen Mnatsakanyan, rejected this allegation and Beregovoi's woes were eventually blamed almost exclusively on 'pilot error'. By the end of the following year, however, the failings of Mnatsakanyan's rendezvous system were clear. During ground tests, engineers used a 95 percent helium pressurisation mixture, which harmed Igla's radio components and thermostats. In the wake of the Soyuz 7/8 fiasco, a commission chaired by Boris Chertok concluded that the piezoelectrical quartz crystals used in its transmitter and receiver had been affected by the thermostat glitch. Since the crystals could not be maintained at the correct temperature, their frequencies had drifted. It was decreed that in future, the system would utilise a pressurisation mixture of either inert gases or a 5 percent helium solution.

Unlike Beregovoi, the two cosmonauts aboard Soyuz 8, Vladimir Shatalov and Alexei Yeliseyev, were recognised and respected as the best-qualified and most experienced in rendezvous and docking techniques. As Nikolai Kamanin mused, the fact that *they* could not complete a successful link-up clearly pointed to the inadequacies of the rendezvous hardware. With the emphasis on launching a space station within the following couple of years, it was obvious that further work on automated rendezvous and docking systems was acutely needed.

Against this backdrop, plans were afoot to attack a second strand of the plan to put a space station into orbit: the need to determine the biomedical effects on the human body of long-term exposure to weightlessness. By the close of the Sixties, the longest Soviet mission had been that of cosmonaut Valeri Bykovsky, aboard Vostok 5 in June 1963, who had spent an hour shy of five full days aloft. In fact, during their troika flight, the Shonin, Filipchenko and Shatalov crews had each fallen short of breaking his record by less than half an hour. Although Bykovsky's achievement was a 'record' for the Soviets, his feat had been almost tripled by the Americans ... *four years earlier*, when astronauts Frank Borman and Jim Lovell spent two weeks in space aboard Gemini VII. With some sense of urgency, therefore, particularly in the wake of Apollo 11 and the failure of the troika, the long-duration Soyuz 9 mission in the early summer of 1970 was to firmly re-establish the Soviet presence in the heavens.

Impressive night launch of Soyuz 9.

Its crew, Andrian Grigoryevich Nikolayev and Vitali Ivanovich Sevastyanov, had not had a very satisfying year as cosmonauts. They had been earmarked to fly Soyuz 8, but their poor knowledge of docking procedures and less than satisfactory performance in their final exams had eliminated them in favour of their backups, Shatalov and Yeliseyev. Then, in May 1970, just days before Soyuz 9 was due to blast off, they were 'caught' smoking – like a couple of naughty schoolboys, it seems – and an "aghast" Kamanin noted in his diary that only the close proximity to launch prevented him from dropping them again in favour of their backups, Anatoli Filipchenko and Georgi Grechko.

This poor opinion is surprising, particularly in the case of Nikolayev, whose reputation as a cosmonaut had burned with near-consistent brightness since his selection in March 1960. Along with Yuri Gagarin and a few others, he had been a prime candidate for the first man in space. When he was shortlisted as one of six finalists back in January 1961, Nikolayev was described by his examiners as "the quietest". By the time he launched into orbit aboard Vostok 3 in August of the following year, he had earned another nickname: 'Iron Man'. His astonishing stamina and ability to sit on his own in an isolation chamber, without stimulus or awareness of the passage of time, for four whole days, made him an ideal choice for long-duration space flight. Now a full colonel in the Soviet Air Force, he was about to make his second space flight. His Vostok 3 mission established a four-day record, but Soyuz 9 was to spend nearly 18 days in orbit. It would be a difficult slog that would leave even the Iron Man fatigued and physically weakened and would raise new questions over how well men could operate over long periods in space.

Born on 5 September 1929 on a collective farm in the village of Sorseli, in the forested Chuvash region of the Volga River valley, Nikolayev was one of four children. His love of aviation began when, aged eight, he visited a nearby airfield. One story from his formative years tells how he clambered into the branches of a tree and announced that he intended to fly from it; fortunately, local villagers persuaded him to come down.

After his father's death in 1944, his intention was to support his family, although this was opposed by his mother, who wanted her son to achieve a full education. Nikolayev entered medical school, then tried his hand at forestry, serving as a lumberjack and timber camp foreman for a time, before joining the Soviet Army. He initially trained as a radio operator and machine gunner, demonstrating "composure under stress" when he crashed a flamed-out jet into a field rather than bail out. Undoubtedly, this proved a contributory factor in his selection as a cosmonaut trainee in March 1960.

A bachelor at the time of his Vostok 3 mission, he is famously said to have kissed his girlfriend goodbye at the foot of the launch pad. That 'girlfriend' – former cotton mill worker Valentina Tereshkova – would not only achieve renown as the first woman to fly into space in June 1963, but also become his wife. Amid much pomp and circumstance, they were married in November 1963 and, in June 1964, had a daughter, Yelena. She became the first child whose parents had both flown into orbit. However, their marriage did not last and Nikolayev and Tereshkova were not even living together by the end of that year, although they did not divorce (and *could not* divorce, it seems, without Leonid Brezhnev's formal approval) until 1982. To this day, speculation continues as to whether their union represented a genuine love match or a cynical propaganda ploy engineered by then-Soviet premier Nikita Khrushchev. Whatever the truth, Tereshkova and Yelena were at Tyuratam for the launch of Soyuz 9.

The flight engineer for the long-duration mission, Vitali Sevastyanov, was another civilian from the OKB-1 bureau, having been picked as a 'supplementary' cosmonaut in January 1967. He was born on 8 July 1935 in the town of Krasnouralsk ('Red Ural') in the Yekaterinburg region, famed as the location of the massacre of Tsar Nicholas II and the last Russian royal family. Sevastyanov graduated from the Moscow Aviation Institute in 1959, receiving a candidate of technical sciences degree five years later and becoming an aeronautical engineer at OKB-1. His early work involved design studies on Vostok and Voskhod and he also lectured at the cosmonaut training centre before joining the ranks of the spacefarers himself, alongside fellow engineer Nikolai Rukavishnikov. Yet his desire to become a cosmonaut stretched back much further: not long after Yuri Gagarin became the first man in space, Sevastyanov and Alexei Yeliseyev had undergone initial medical screening at the Air Force's Central Scientific Research Aviation Hospital, apparently without Sergei Korolev's consent.

A keen chess player and, indeed, later a president of the Soviet Chess Federation, Sevastyanov seemed directed toward one of the early circumlunar missions as a flight engineer, before being formally paired with Nikolayev on 26 April 1969 to fly Soyuz 8. Although within months their performance led them to be substituted for their backups, Shatalov and Yeliseyev, they seem to have retained the support of Vasili Mishin.

When the Soviet leadership ordered, on 30 December, that a 'solo' Soyuz mission of 17–20 days' duration would be flown to coincide with the centenery of Lenin's birth on 22 April 1970, Mishin wanted Sevastyanov, one of 'his' civilians, and Nikolayev to fly it. Nikolai Kamanin disliked the idea, preferring to fly two other cosmonauts – military pilot Pyotr Kolodin and OKB-1 engineer Georgi Grechko – but Mishin stood firm. In his diary, Kamanin made no reference to exactly why he objected so strongly to Nikolayev and Sevastyanov, although their performance in training must have been a major factor.

For many years, Kamanin had attacked the 'knee-jerk' and whimsical attitudes of the Soviet leadership towards human space flight matters: on several occasions, there had been lengthy spells of inactivity and indecision, followed by a sudden desire to launch a 'spectacular' mission on an impossibly short timescale and, sometimes, on a politically motivated date. The late decision on Soyuz 9 was a clear example of this. Nonetheless, despite only being granted *two days* to prepare a training schedule for Nikolayev and Sevastyanov, Kamanin remained convinced that a launch was possible in time for the Lenin centenery.

Ironically, it was the State Commission itself which vetoed these plans and postponed the flight until May 1970. The various scientific panels and bureaux responsible for developing Soyuz 9's experiment hardware had already warned that they would not be ready for an April launch. The doubts intensified when it became clear that more work was needed on the environmental control system. The latter had been certified to support a crew for up to five days, but not for a mission of two and a half weeks. Still, Vasili Mishin directed that the maximum permissible level of carbon dioxide in the cabin be doubled and in the early months of 1970 Nikolayev and Sevastyanov successfully completed a full-duration, 18-day 'mission' in the simulator, with apparently few ill-effects. New biomedical sensors and an 'improved' waste-management system also performed well.

By mid-May, the launch was officially scheduled for midnight on the 31st, with 18 days of orbital operations planned. However, the mission was essentially 'open-ended', with options to bring the cosmonauts home early if their condition deteriorated or indeed to extend the flight to 20 days if all was going well. Kamanin favoured the 18-day limit, whilst Mishin wanted to go for an extension if possible. On the afternoon of launch day, the wily general took Nikolayev and Sevastyanov aside and explicitly forbade them to ask for an extended mission whilst in orbit. Any extension, Kamanin felt, would severely strain the capabilities of the Soyuz and might place the cosmonauts' lives in jeopardy.

Regardless of the flight's duration, the environmental control system, it seemed, would certainly have its work cut out. If they launched on time, Nikolayev and Sevastyanov would become the first crew to blast off in darkness – another Soviet 'first' ticked off the list. During their final days on Earth, they adjusted their sleep patterns to put them in 'synch' with the operating shifts at the Yevpatoria control centre.

A delay until 1 June was effected shortly after the cosmonauts' delegation arrived at Tyuratam, when no fewer than 15 defects were discovered in the spacecraft, of which several were classed as 'serious'. Some points in the electrical harnesses, which should have carried 38 volts, were measuring greater than 60 volts, requiring Soyuz 9 to be put through vacuum chamber testing again, then fuelling and finally integration with the R-7 booster. Incorrect mounting of Nikolayev and Sevastya-nov's headrests and unusable photographic equipment also demanded last-minute attention from technicians. There were other problems, too. An outbreak of dysentery in the Tyuratam garrison forced physicians to give the cosmonauts prophylactic measures to prevent their catching any bugs. The cosmonauts' health and diet would be strictly controlled before, during and after the mission: four meals daily, with 105 g of protein, 102 g of fat, 342 g of carbohydrates and 847 g of water. A bungee cord was also provided in Soyuz 9's orbital module for daily exercise workouts.

Finally, on 31 May, the go-ahead was given for launch. The rollout of the R-7 to the pad commenced in the pre-dawn hours of the following morning. Despite the apparent entente of an international space conference in Leningrad the previous week and the visit of Apollo 11 astronauts Neil Armstrong and Buzz Aldrin to the cosmonaut training centre at Star City, the Soviets' ridiculous insistence on secrecy continued. Nikolai Kamanin spent part of the day rehearsing with Nikolayev and Sevastyanov what code words they would use over the radio: 'good' or 'excellent' would mean just that, 'normal' would require the resolution of problems already known to ground controllers and 'satisfactory' would indicate problems whose severity demanded an early landing. Kamanin told them not to take any unnecessary risks and expressed his support of any decisions they might make in an emergency.

As darkness fell across the Kazakh steppe that evening, the State Commission met and offered its unanimous approval for the launch to proceed on the stroke of midnight, local time; Tyuratam being two hours 'ahead' of Moscow. At around 5:45

pm Moscow Time, Nikolayev and Sevastyanov reported to the medical office, where they were fitted with their biosensor harnesses. On arriving at the pad less than two hours later, they formally declared their readiness to fly before the State Commission and boarded the spacecraft. Despite the spectacular light show and inherent danger of the first-ever nocturnal launch of a manned mission, Soyuz 9's on-time climb to space was thankfully uneventful, with excellent radio communications and a perfect insertion into an initial 208 x 220 km orbit.

Supporting their mission was a new computer, known as the 'Spacecraft Analogue Machine', which permitted a 'rendezvous' to be performed with an imaginary target. "The computer was capable of locating targets at a range of 30–50 km," wrote Asif Siddiqi, "and providing input on subsequent manoeuvres." All details of the intention to break the endurance record had been kept from western observers, with only the most low-key and matter-of-fact pronouncements made about the flight. Still, Neil Armstrong, who was at Tyuratam on 1 June, was clearly surprised when his host, Georgi Beregovoi, switched on the television to view film of the Soyuz 9 launch and told the Moonwalker: "*This* is in your honour!" The old tricks of the Soviet space programme, it seemed, were still alive and well ...

Notwithstanding the lack of a docking mechanism and, indeed, no 'real' target for the rendezvous, Nikolayev and Sevastyanov's first day aloft included two manoeuvres into a 247 x 266 km orbit, high enough to prevent orbital decay for the remainder of the flight without additional thruster firings. As Siddiqi has noted, these burns may have been in support of a mock rendezvous with an imaginary target. By this time, however, the two cosmonauts had settled down to work on their extensive programme of biomedical and scientific research.

Naturally, studies of the crew themselves were of great importance, particularly with respect to Soviet plans to fly their first 'real' space station later in the year. However, within the first four days of the mission, ground controllers were brought face to face with the reality that the timelines of the early station crews would need more realistic planning. Mundane activities readily achieved on Earth required considerably longer in space. For example, Nikolayev and Sevastyanov typically spent 50 minutes on what in training had been a half-hour exercise session. Still, by the beginning of their second week, both men reported that they were feeling better, physically and psychologically, than they had in their first couple of days.

However, their reduced consumption of water and oxygen was clearly indicative of growing fatigue and after watching television of the cosmonauts on 12 June, Kamanin described their faces as "somewhat puffy ... and listlessness and irritability can be sensed in their actions". Both men exercised twice daily, using the bungee device in the orbital module, which required an exertion tension of 10 kg, and also wore special suits known as 'Pingvin' ('Penguin') to simulate some of the loads imposed by terrestrial gravity. They regularly checked and recorded their arterial pressures, pulses, respiration and the contrast sensitivities of their eyes and Nikolayev performed simulated commands on Soyuz 9's computer to test his mental awareness. Their food consumption rate of some 2,600 calories per day was close to pre-flight estimates. Not until after the cosmonauts returned to Earth would the true extent of their physical deterioration become alarmingly clear.

Cosmonauts Vitali Sevastyanov (left) and Andrian Nikolayev are shown on the large television projection screen during a communications session with ground controllers. The view gives some impression of how little volume was available to the men during their 18 days in orbit.

Conditions aboard the cramped spacecraft – whose habitable volume comprised only the descent and orbital modules – must have become decidedly unpleasant as the long mission wore on. A food heater, admittedly, gave them the chance to warm their meals and make a 'fresh' cup of coffee at the start of each workday, but personal hygiene was by way of wet and dry towels and changes of underwear were performed once a week. With the possible absence of beer cans and takeaway packets, life onboard must have been the equivalent of student accommodation at its very worst ...

Matters were also not helped, psychologically, by a lack of regular contact with loved ones. Notable exceptions included the birthday of Nikolayev's daughter, Yelena, when the six-year-old was brought into the mission control centre on 8 June to speak with her father over an audiovisual link. In years to come, direct contact with family members on a reasonably frequent basis would become an essential psychological crutch for crews on long-duration missions. Today, all crew members aboard the International Space Station have regular 'private' communications with loved ones and this is considered critical for maintaining their emotional wellbeing.

Seeing that both Nikolayev and Sevastyanov were becoming fatigued, the intensive workload of scientific experiments – which had grown to consume 14–16 hours of each day – was drastically reduced as the mission wore on. Vasili Mishin had his sights set on pushing the endurance record to almost three weeks. At an expanded meeting of the State Commission on 16 June, he casually asked ballistics experts what the spacecraft's orbital parameters would be on its 20th day of flight! Nikolai Kamanin opposed any such move and, indeed, Nikolayev's reports suggested that it would be difficult to stretch the food rations past 18 days and probably not worth the risk. Ultimately, the decision was taken to end the mission after 18 days. In his diary, Kamanin gloated that "Mishin and [State Commission chairman Kerim] Kerimov, having promised the high command in Moscow that they would carry the flight out to 20 days, will *now* have to concur with *our* decision". At one point, a clearly enraged Mishin had even turned to Yevgeni Vorobyov, the Ministry of Health representative responsible for Soyuz 9's dietary provision, and blatantly accused him of not supplying the cosmonauts with enough food for a 20-day mission!

Otherwise, the final portion of the flight ran quietly, with scientific tasks including photography of terrestrial targets, celestial navigation using the SMK-6 sextant, a range of astrophysical observations and confirmation that Soyuz could accommodate a crew for a lengthy period of time. The navigation experiments, in particular, utilised a new stellar sensor, "to calculate," wrote Asif Siddiqi, "the orbital parameters and geographical latitude of the point above which the ship was flying, relative to the position of a selected star above the horizon, Vega, in the Lyra constellation". Nikolayev and Sevastyanov carried out this complex task over two complete orbits, whilst out of direct contact with the ground, as they manually maintained their spacecraft's attitude and measured the drift of its gyroscopes. On 14 June, they even explored the possibility of checking Soyuz 9's orientation using stars such as Arcturus and Deneb, in conjunction with ground reference points in Africa and South America. "All these experiments," concluded Siddiqi, "led to precise determination of orbital elements to refine future rendezvous exercises."

The cosmonauts' photographic work yielded several thousand images, including many in support of a weather formations experiment which was part of an integrated exercise between a Meteor satellite at 600 km altitude, Soyuz 9 at 240 km and the Soviet hydro-meteorological research vessel *Akademik Shirshou* in the western Indian Ocean. At other times, Nikolayev and Sevastyanov observed a tropical storm developing in the Indian Ocean, forest fires near Lake Chad in central Africa and even acquired multispectral imagery of rock and soil cover, the moisture content of glaciers, the location of shoals of fish and timber reserves. Using the hand-held RSS-2 spectrograph, which Anatoli Filipchenko's crew had also carried on Soyuz 7, they examined aerosol particles in the upper atmosphere and made 200 spectro-photometric measurements of terrestrial formations. Other work focused on plant behaviour, the division of chlorella cells, propagations of bacterial cultures in various liquids and the development of insects in weightlessness.

Soyuz 9 offered its own share of glitches, too. Towards the end of its first day in orbit, there had been alarm when controllers noted that the solar arrays' storage batteries were exhibiting higher levels of charge than they theoretically should. It was a problem caused by circumstances outside of anyone's control. The orientation of the spacecraft's orbit around Earth relative to the Sun meant that the arrays were exposed to near-continuous sunlight. Matters were compounded still further by a malfunction in the control switch which should have regulated the supply of electricity from the arrays and fed it into the batteries. During the 47th orbit, two days into the mission, Sevastyanov reported that, although energised, one array was generating barely 26 amps, instead of 31 amps, and Nikolai Kamanin noted that if it were to drop below 23 amps this would constitute an emergency and demand a return to Earth within a couple of hours. The cosmonauts had already been told to cycle the switch on and off, engaging and disengaging the batteries manually, on at least a dozen occasions by this point. The spacecraft was designed for this to be done no more than 50 times before the batteries ran down. With barely two days of a planned 18-day mission completed, it became increasingly likely that Soyuz 9 would be back on the ground within a week unless the problem was solved. Later that same day, 4 June, Nikolayev and Sevastyanov were ordered to rotate their spacecraft at a rate of 0.5 degree per second around its long axis, in effect instituting a 'barbecue roll' that would periodically turn the arrays away from the Sun. However, this procedure also drained the batteries and the crew had to be awakened two hours before the end of each sleep period to reorient Soyuz 9 and recharge them. The roll was suspended on a few occasions, when it began to cause high temperatures on some of the spacecraft's storage tanks. Neither cosmonaut reported any physiological difficulties from the roll, but their poor condition after the mission would be at least partly blamed on its effects. Other concerns surrounded the worrisome environmental control system, whose performance led to a sharp increase in atmospheric pressure to 900 mm. They were told to lower the internal temperature from 21°C to 18°C, as a way to reduce the pressure to 870 mm.

As the mission wore on, the tiredness of the cosmonauts became increasingly evident. They played chess via radio with cosmonaut Viktor Gorbatko at the Yevpatoria control centre on a 'day off' on 10 June, but they were beginning to make mistakes. In one case, they accidentally put a television camera on the wrong setting and on 15 June mission controllers were unable to wake them, despite three minutes of increasingly frantic radio calls. The two cosmonauts apologised for oversleeping, but a groggy and disorientated Sevastyanov inadvertently switched on the button for the automatic landing display as he tried to turn on the cabin lights. Although it was not considered 'dangerous', his action removed the first lock on the system, which was then 'armed' and would be activated by a barometric switch during descent at an altitude of 11 km. (Anatoli Filipchenko had accidentally hit the same switch towards the end of the Soyuz 7 mission a few months earlier. In his diary, Nikolai Kamanin recounted that he had asked Mishin to put a lock on the switch to prevent such situations from occurring, but the Chief Designer had done nothing ...)

The final landing commission met on 18 June and, early the following morning, the cosmonauts began stowing their equipment. Conditions at the touchdown site, 75 km west of Karaganda, were described as 12–15 km visibility, with wind speeds no higher than 5–7 m/sec. Following a normal retrofire and an uneventful re-entry, the descent module was picked up on radar at an altitude of 83 km. As it descended, its parachute was spotted by two helicopters, which followed it toward a landing at 2:59 pm Moscow Time. After a mission of 17 days, 16 hours and 59 minutes – and 285 circuits of the globe – Nikolayev and Sevastyanov had soundly smashed the long-held American record from Gemini VII.

Naturally, the reports from Tass and *Pravda* hailed the flight, and rightfully so, but little attention was paid at the time to the hapless state of the two men. Although they had been provided with the means and some of the equipment for physical conditioning during the mission, their workload often prevented them from using it or completing the stipulated one hour of exercise per day. In fact, on a couple of occasions, they were scolded for not following their prescribed fitness regime. Within minutes of the descent module hitting the Kazakh steppe, General Goreglyad, who was in charge of the recovery forces, and Colonel Popov were on the scene and cosmonaut Alexei Leonov reported, officially, at least, that the crew was well.

Years later, Leonov told a quite different story. "One of the crew members," he wrote, "was so weak ... that he could not even carry his own helmet when he stepped out of his landing capsule." It seems that even 'stepping' from the capsule was difficult: the cosmonauts had to be lifted out and plans to fly them back to Moscow's Vnukovo Airport for a celebratory reception were cancelled. When Nikolai Kamanin saw them for the first time, he was stunned. "Sevastyanov was sitting on the sofa," he wrote, "while Nikolayev was at a small table. I knew they were having a hard time enduring the return to the ground, but I had not counted on seeing them in such a sorry state ... pale, puffy, apathetic, without the spark of vitality in their eyes ... completely emaciated ... "

When they finally arrived in Moscow on 20 June, the men had to be supported and 'walked' off the aircraft by fellow cosmonauts Vladimir Shatalov and Alexei Yeliseyev. Nonetheless, Nikolayev made a feeble attempt to deliver his prepared speech to the State Commission, advising in an unconvincing manner that he and Sevastyanov had completed their flight plan and "await further orders". Post-flight medical observation revealed that their transition from recumbent to sitting positions brought about circulation disorders. As the days wore on, they began to walk unsteadily and even maintained an 'erect' stature at rest, "on account of a significant elevation of their centre of gravity".

Normal terrestrial gravity felt like 4–5 G to them and not until the end of June, almost two weeks after landing, would they regain their pre-flight strength. Even then, they could only work for a few hours per day. Walking was difficult, with both men tiring rapidly. Their pulse rates, body temperatures and blood pressures fluctuated daily.

The medical recommendation not to attempt missions beyond 20–25 days in duration was at loggerheads with Vasili Mishin's hopes of flying crews for up to two months on the space station later that year. Nikolai Kamanin felt that a gradual progression – from 30 to 40 to 50 days, and so on – was more prudent.

Despite Nikolayev and Sevastyanov's inconsistent approach to exercise and the sheer length of their mission, apparently neither of these issues directly caused the profound physical deterioration. Rather, wrote Asif Siddiqi, "it was increasingly clear that part of the reason for the[ir] very poor shape ... was the slow spin of the spacecraft throughout the mission". This produced a weak field of artificial gravity, which also affected the clarity of several experiments. Therefore, the triumph of setting a new endurance record was countered by the reality that occupying an orbital station for a lengthy period of time would be more difficult than originally supposed. This challenge would be faced by the teams of cosmonauts who trained throughout 1970 for a series of ambitious missions to equal and surpass the achievements of Nikolayev and Sevastyanov ... aboard the world's first 'true' space station.

In spite of the problems experienced by Boris Volynov during re-entry, the virtual obsolescence of the need for Alexei Yeliseyev and Yevgeni Khrunov's ship-to-ship transfer, the failings of Igla and the terrible condition of the Soyuz 9 crew, the Soviet *modus operandi* in space had shifted remarkably in less than two years. Their lunar dream remained alive, but by the end of 1970 their space aspirations had a more solid direction. Soyuz, though imperfect, was operational and had delivered a clear answer to America's conquest of the Moon. The roles were now reversed. At the beginning of the Sixties, it was the Soviets whose space effort was governed by the need for a series of spectaculars. Now the United States could be shown as the superpower in pursuit of the celestial stunts: flying a series of headline-grabbing Moon landings, with only a handful of ill-defined plans thereafter, whereas the Soviets could portray themselves as following a more gradual and logical method of mastering near-Earth space. In essence, their programme was being presented to the world as the polar opposite of Apollo. Whilst the American effort was seen by many as a politically motivated stunt – conjured up in response to a series of military blunders and diplomatic fiascos – with the sole intention of flaunting capitalist technology and muscle, the Soviets could claim *they* were going into space in search of more peaceful goals, for the benefit of *all* humanity. It was a clever masquerade and one behind which they would cloak themselves in the coming years.

The Stars and Stripes might adorn the flat mare and undulating foothills of the Moon, but, to borrow from the title of Jim Oberg's book, the Soviets would soon establish their own, semi-permanent Red Star in orbit in the form of a space station called 'Salyut' (a 'salute' to the first man in space, Yuri Gagarin). In 1971, cosmonauts would not simply *fly* in space ... they would *live* there. Yet tragedy continued to chomp at the heels of each triumph. Barely a year after Nikolayev and Sevastyanov's mission, three cosmonauts would triumphantly occupy Salyut for more than three weeks, really *living* in space for the first time. Tragically, none of them would live to see home again.

2

Luna incognita

THE MAN WHO CHOSE THE 'FIRST MAN'

Six days into the most momentous year of the most momentous decade in the United States' history, Deke Slayton made the most momentous decision of his career. It was a decision that would make history. It was the decision of who would become the first man to set foot on the Moon. Less than ten years earlier, Slayton had been full of hope that *that* man might well be himself, when he was picked from thousands of fighter and test pilots to become one of America's first seven astronauts, joining the newly established National Aeronautics and Space Administration (NASA). Now, at the dawn of 1969 and in his mid-forties, the gruff, crew-cut Slayton was virtually impotent as an astronaut, with almost no chance of ever reaching space. *Among* the astronauts, on the other hand, he was viewed quite differently: as 'Father Slayton', their boss and mentor, the man who decided which of them would fly and ensured that they kept their noses firmly to the grindstone of training. The bitterest irony, therefore, was that Slayton's lack of experience as a 'real' spaceflyer was balanced by an unassailable omnipotence within the astronaut corps.

It should have been quite different. In the spring of 1962, he had been assigned to the United States' second manned mission into Earth orbit and had even chosen a name for his Mercury spacecraft – 'Delta 7', the fourth letter of the Greek alphabet – which he pointed out was "a nice engineering term that described the change in velocity". His own velocity, both in spacecraft and high-performance jets, would decline markedly, thanks to a minor, yet persistent, heart condition, known as 'idiopathic atrial fibrillation'. This was the occasional irregularity of a muscle at the top of his heart, caused by unknown factors and rare in extremely fit adults like Slayton. It had first arisen while Slayton was making a centrifuge run in August 1959, when physicians noted traces of 'sinus arrhythmia', which NASA flight surgeon Bill Douglas later wrote "wasn't uncommon in healthy young men and ... the kind of thing that often went away with exertion". After the run, however, it was still present, prompting Douglas and his team to undertake a clinical electrocardio-gram at the Philadelphia Naval Hospital.

They concluded that Slayton had a 'flutter' in his heartbeat, although, in 1959, the astronaut himself "had no idea how much of a problem it was". Further tests at the United States Air Force's School of Aviation Medicine in San Antonio, Texas, verified that the condition was of little consequence and should not impair his eligibility for a space flight. Douglas informed Bob Gilruth, head of the Space Task Group and its Project Mercury man-in-space effort, who briefed NASA Head-quarters on the issue late in 1959. The Air Force surgeon-general also advised that no further action was necessary. The 'Slayton File', for a time, lay dormant.

The problem resurfaced a couple of years later, when speculation arose that one of the astronauts – perhaps John Glenn – had a heart murmur. Apparently, the call to Bill Douglas came from Air Force physician George Knauf, attached to NASA Headquarters, and had originated from "a source higher than the Department of Defense". Douglas denied that Glenn had a problem, but effectively opened a can of worms. Knauf asked next if Glenn's backup, Scott Carpenter, had a heart murmur; again, the response was negative. Then Douglas, to reinforce the point that the matter was of little relevance, revealed that Slayton had long been known to have a minor heart condition. He expected this to be the end of the matter. It wasn't.

Back in 1959, flight surgeon Larry Lamb had examined Slayton at Brooks Air Force Base in San Antonio and had become convinced that heart fibrillation should disqualify him from the selection process. "He hadn't said so in 1959," wrote Slayton, "but he said so now. I don't think it was anything personal – this was just his medical opinion." Lamb's judgement was very much a voice in the wilderness, but unfortunately he also happened to be Vice-President Lyndon Johnson's cardiologist and in the spring of 1962 began to question the astronaut's suitability to fly. (Matters were not helped when the publicity-seeking Johnson had been refused access to John Glenn's home during the course of his space flight.) In early March, NASA Administrator Jim Webb reopened Slayton's medical file and the astronaut and Douglas were summoned to the office of the Air Force surgeon-general in Washington, DC. A panel of military physicians signed Slayton off as fit to fly, a decision endorsed by Air Force Chief of Staff Curtis LeMay.

For Webb, though, it was not enough. The Secretary of the Air Force, Eugene Zuckert, insisted that a panel of civilian physicians should also examine Slayton at NASA Headquarters. On 15 March 1962, less than two months before his scheduled launch, Slayton was poked and prodded and had his heart monitored by Proctor Harvey of Georgetown University, Thomas Mattingley of the Washington Hospital Center and Eugene Braunwell of the National Institutes of Health. As Slayton waited afterwards, NASA Deputy Administrator Hugh Dryden entered the room and told him, point-blank, that he could not fly. None of the physicians had found a specific medical reason to keep him off Delta 7, but their consensus was that if NASA had pilots 'without' his condition, *one of them* should fly the mission instead. Slayton, in his own words, was "devastated".

Years later, Slayton felt that the decision to ground him was a political one.

"NASA knew it would have to publicly disclose my heart condition prior to my flight," he wrote. "There would be medical monitors at tracking stations all over the world who wouldn't know how to react otherwise. Everybody expected this to be a big deal. NASA would be opening itself up to a lot of medical second-guessing." Bill Douglas felt that problems could arise if Slayton started fibrillating on the pad – "do you scrub the launch or go ahead?" – but he, Bob Gilruth and Project Mercury's operations director Walt Williams had confidence that he was the best person to fly. All three men were prepared to take the heat, but Jim Webb's fear that it could trigger adverse headlines for NASA – the agency had been created barely four years earlier and was still finding its feet – drew a line in the sand. "It didn't matter that a whole lot of doctors thought I didn't have a problem," Slayton wrote of Webb's actions. "He was only going to listen to the few who did."

In Webb's mind, a launch abort could subject the astronaut to loads as high as 21 G and conjured the very real possibility that Slayton, dehydrated and perhaps fibrillating, could die during descent. The impact on the agency in such a dire situation would be profound. Less than a year earlier, the newly elected President John Kennedy had boldly committed the nation to landing a man on the Moon and returning him safely to Earth, before the end of the decade, in order to score a major scientific, technological, political and ideological triumph over the Soviets. Any incident in which an astronaut died could stop this plan in its tracks.

The next day, 16 March 1962, the grounded and furious astronaut was forced to sit through a lengthy press conference, in which the minutiae of the case were examined. Hugh Dryden remarked that, despite the decision, Slayton might remain eligible for future flights. One journalist asked if the problem had been caused by stress, to which Slayton responded no, and further that he did not even know about it until he had been hooked up to the electrocardiogram in 1959. The most stressful part of a space mission, he explained caustically, was "the press conference after the flight". Bill Douglas' own departure from NASA within days of the announcement, to return to the Air Force, was leapt upon by some journalists as 'evidence' of his bitterness over Slayton's treatment, but he was already at the end of a three-year detachment to NASA and his return had been in the works since mid-1961.

Despite his grounding, Slayton did not give up on flying. "I made some changes in my lifestyle," he wrote, "gave up drinking, started working out more regularly – quit doing everything that was fun, I guess!" Bill Douglas also secured Slayton an examination by Dwight Eisenhower's cardiologist, Paul Dudley White, in June 1962. White advised him that two-thirds of people with his condition would die young, whilst the remainder would probably never know they had it and might never be affected. The verdict: "Young man, you're going to live a long time." (Slayton lived to the age of 69.) However, White's report, which highlighted that Slayton did not appear to have a problem, also advised that if astronauts were present without the condition, it would be preferable to assign one of them in his stead. As it became clear that he would not draw assignment to any of the remaining Mercury flights, Slayton turned his attention to the two-man Project Gemini, only to be told by Bob Gilruth that his ailment would make him a "hard sell" to senior management.

Shortly thereafter, the Air Force decreed that Slayton no longer met the qualifications for a Class I pilot's licence – which meant he could no longer fly solo – and, at the end of November 1963, he resigned from the service.

Although he would eventually get his ride into space, it would not be until July 1975, when he was 51 years old. A lesser man might have thrown in the towel and departed NASA for pastures new, but not Slayton. With no guarantee of a mission, he decided to stay and in the summer of 1962, as the agency prepared to expand its astronaut corps by picking nine new pilots, he was appointed as Co-ordinator of Astronaut Activities. Initial plans to bring in a manager from the outside to oversee the corps were quashed by the astronauts themselves. "What we wanted the least," wrote Wally Schirra, one of the first group of astronauts, "was somebody who would outrank us and issue orders in a military way. We wanted someone who knew us, who trained with us. Deke was the one and only choice." During the next few years, as America pushed for the Moon, Slayton would be all-powerful within the astronaut corps, deciding the career paths of the men who would someday walk on the lunar surface ... and those who would not.

The names of the first men to tread the dusty surface of the Moon were announced in Slayton's office on 6 January 1969. By that time, the United States' effort in space had advanced tremendously. Six Mercury flights had been followed by ten Gemini missions, during which teams of astronauts had spent up to 14 days aloft, carried out spacewalks in pressurised suits, rendezvoused and docked with unmanned target vehicles and flown as high as 1,300 km. Like the Soviets, though, triumph had been tempered by tragedy: two astronauts had lost their lives in an aircraft crash and another three had asphyxiated in a fire aboard the new Apollo spacecraft during launch pad test in January 1967. Following a lengthy period of criticism and self-doubt, a redesigned Apollo had undertaken its first manned shakedown flight in Earth orbit in October 1968. Eight weeks later, astronauts Frank Borman, Jim Lovell and Bill Anders triumphantly flew Apollo 8 around the Moon.

Sitting in Mission Control in Houston, Texas, during the Apollo 8 flight were a civilian test pilot named Neil Armstrong and an Air Force colonel, doctor of science and rendezvous expert called Edwin 'Buzz' Aldrin. Together with rookie astronaut Fred Haise, they were wrapping up their duties as backups to Borman, Lovell and Anders. In the forthcoming year, each of their careers would take a dramatic turn. Ten days after Apollo 8's command module splashed down in the Pacific Ocean, Deke Slayton called Armstrong, Aldrin and another astronaut, Mike Collins, to his Houston office. In just two words, he gave them the news every astronaut had trained for years to hear: "You're it!"

'It', of course, meant that all three were being tapped for the coveted lunar landing, tentatively pencilled-in for Apollo 11 in July 1969. Armstrong and Aldrin were already up to speed following their recent backup duty and Collins, originally assigned to fly Apollo 8 himself, had been dropped in the summer due to a spinal problem which required neck surgery. Having fought his way back to full health and back onto active flight status, Collins would serve as Apollo 11's command module pilot (CMP) – the man who would remain in lunar orbit, whilst Armstrong and

Aldrin descended to the Moon's surface in a spidery craft known as the lunar module.

Of course, we know today that Kennedy's goal was indeed met. But as 1969 dawned, a lunar landing on Apollo 11 was by no means set in stone. (Mike Collins would later estimate the chance of success, in his mind at least, as no better than 50-50.) Still to be proven was the lunar module itself, built by Grumman, which Apollo 9 astronauts Jim McDivitt and Rusty Schweickart intended to test in Earth orbit late in February, as their colleague Dave Scott in the command module practiced rendezvous and docking. The space suit which astronauts would one day use to walk on the Moon was to be tested by Schweickart during a dramatic EVA in which he would climb out of the lunar module's hatch, onto its porch and then clamber over to the command module. A couple of months later, in May, Apollo 10 would do a full dress-rehearsal of the Moon landing ... 370,000 km away, in lunar orbit. Astronauts Tom Stafford and Gene Cernan would guide their own lunar module to just 15 km above the surface, before boosting themselves back up to rendezvous with crewmate John Young. Only if both of these highly complex missions, the details of which remained to be hammered-out, succeeded would Apollo 11 stand a chance of flying in July.

MOONSHIP

Project Apollo, which enabled the famous steps of Armstrong and Aldrin on the Moon, was born very soon after NASA's own creation in October 1958. At that time, it was recognised that exploration of the Solar System would be one arena in which the abilities of *humans*, rather than machines, would be required. A fundamental obstacle, however, was the distinct absence of large boosters capable of fulfilling such roles and in mid-December of that year, NASA Administrator Keith Glennan listened as rocketry experts Wernher von Braun, Ernst Stuhlinger and Heinz Koelle summarised the capabilities of existing hardware and stressed the urgent need for a 'new' family of launch vehicles. Landing men on the Moon, for the first time, was explicitly discussed as a long-term objective and Koelle suggested a preliminary timeframe for achieving this as early as 1967.

Von Braun's vision for the new family of rockets was that, first and foremost, their engines should be arranged in a 'cluster', directly carrying an aviation concept into the field of spacegoing rocketry. The renowned missile designer – responsible for helping to develop Nazi Germany's infamous V-2 – also discussed propellants and the idea of employing different combinations for different stages ... then broached the subject of precisely how such enormous boosters could deliver a manned payload to the lunar surface. Von Braun had five methods in mind: one involving a 'direct ascent' from Earth to the Moon, the other four involving some sort of rendezvous and docking of vehicles in space. In whatever form such a mission took, the rocket would be enormous, comprising, he said, "a seven-stage vehicle" weighing "no less than 6.1 million kg". Alternatively, he suggested flying a number of smaller rockets to rendezvous in Earth orbit and assemble a 200,000 kg lunar vehicle, which could then depart for the Moon. Stuhlinger followed up by pointing out that quite aside from the immense practical problems of building and executing such a plan, the real unknowns were of how men and machines could operate in a weightless environment, with temperature extremes, radiation, micrometeorites and corrosion an ever-present hazard.

Glennan's focus at the time was, of course, the Project Mercury man-in-space effort, although in testimony before Congress early in 1959 he and Hugh Dryden admitted that there was "a good chance that within ten years" a circumnavigation of the Moon might be achieved, although not a landing, and that by then similar projects might be underway in connection with Venus or Mars. In support of NASA's long-term aims, Glennan sought funding to initiate development of the cornerstone for such epic ventures – the booster itself – and presented President Dwight Eisenhower with a report on four possibilities for a 'national space vehicle programme': Vega, Centaur, Saturn and Nova. Although Vega and Nova scarcely left the drawing board, the others received development funding and von Braun's team, which had championed a rocket known as the Juno V, was backed to develop it under the new name 'Saturn'.

In April 1959, Harry Goett, soon to be appointed the first director of NASA's Goddard Space Flight Center in Greenbelt, Maryland, was called upon to lead a research steering committee for manned space exploration. The major conclusions of his panel were that, after Project Mercury had sent a man into orbit, the agency's goals should encompass manoeuvring in space, establishing a long-term manned laboratory, conducting a lunar reconnaissance and landing and eventually missions to Mars and/or Venus. "A primary reason," remarked Goett of the choice of the Moon as a major target, "was the fact that it represented a truly end objective which was self-justifying and did not have to be supported on the basis that it led to a subsequent more useful end."

Elsewhere, efforts to begin developing the Saturn were gathering pace. Its challenges were both vast and staggering, with propellant weights alone for a direct-ascent rocket producing a vehicle of formidable scale; indeed, even the prospects for constructing a lunar spacecraft in Earth orbit using a less powerful rocket would require more than a dozen launches and the added complexity of rendezvous, docking and assembly. At this early stage, the problems of being able to store cryogenics for long periods in space, to have a throttleable lunar-landing engine and takeoff engine with storable propellants and the need to develop auxiliary power systems were first identified.

Unfortunately, midway through Dwight Eisenhower's second term in office, and with the emphasis of his administration on balancing the budget "come hell or high water", it proved impossible for Glennan to commit NASA to a manned space programme beyond Project Mercury. Instead, small groups at the agency's field centres sprang up, including one within the Space Task Group which considered a second-generation manned vehicle capable of re-entering the atmosphere at speeds almost as great as those needed to escape Earth's gravitational pull. "The group was clearly planning a lunar spacecraft," wrote Courtney Brooks, James Grimwood and Loyd Swenson in their seminal work *Chariots of Apollo* and by the autumn of 1959 sketches of a lenticular re-entry vehicle had advanced sufficiently for its designers to apply for it to be patented.

Early January of the following year finally brought approval from Eisenhower for NASA to accelerate development of von Braun's Saturn and the first hint of political support for manned space efforts beyond Project Mercury. Within weeks, Glennan's request to Abe Silverstein, director of the Office of Space Flight Programs at NASA Headquarters, to encourage advanced design teams at each of the agency's field centres and also within the aerospace industry began to bear fruit: von Braun's group proposed a Saturn-based lunar exploration design and J.R. Clark of Vought Astronautics offered a brochure entitled 'A Manned Modular Multi-Purpose Space Vehicle'. Project Mercury had yet to accomplish its first manned flight, but the proposals continued regardless of their limited chances of receiving presidential or congressional approval. Other efforts focused on exactly how the spacecraft and other hardware could be delivered to the lunar surface in the most economical way. In May 1960, NASA's Langley Research Center in Hampton, Virginia, home of the Space Task Group, sponsored a two-day conference on orbital rendezvous, with several techniques being discussed, although it was recognised that little practical progress could be made until the agency secured funding for a flight test programme.

It was at around this time that the decision was made over what to name the spacecraft which would bring about the most audacious engineering and scientific triumph in the history of mankind. The name 'Apollo', formally conferred upon the programme on 28 July 1960 by Hugh Dryden, would honour the Greek god of music, prophecy, medicine, light and – perhaps above all – progress. "I thought the image of the god Apollo riding his chariot across the Sun," wrote Abe Silverstein, who had come up with the name after consulting a book on mythology, "gave the best representation of the grand scale of the proposed programme."

The scope of Apollo, Bob Gilruth and others revealed to more than 1,300 government, scientific and industry attendees at a planning session in August, was for Earth-orbital and circumlunar expeditions as a prelude to the first manned landing on the Moon. The guidelines for the design of the spacecraft were that it must be compatible with the Saturn booster under development by von Braun's team, it had to be able to support a crew of three men for a period of up to a fortnight and it needed to encompass the lunar or Earth-orbital needs of the project, perhaps in conjunction with a long-term space station. By the end of October, teams led by Convair, General Electric and the Martin Company were each awarded a $250,000 contract for initial studies.

In spite of this apparent brightening of the lunar project's chances, Glennan himself remained unconvinced that Apollo was ready to move beyond the feasibility stage and felt a final decision would have to await President Kennedy taking up office in January 1961. By this point, Glennan was estimating Apollo to cost around $15 billion and felt that the new administration would have to spell out, clearly, and with no ambiguity, its reasons for pursuing the lunar goal, be they for national prestige or the advancement of science. At around this time, Hugh Dryden and NASA Associate Administrator Bob Seamans directed George Low to head a Manned Lunar Landing Task Group and draft plans for a Moon programme, based on either direct-ascent or rendezvous, within cost and schedule guidelines, for use in budget presentations before Congress. When Low submitted his report in early February, he assured Seamans that no major technological barriers stood in the way and that, assuming continued funding of both the Saturn and Apollo, a manned lunar landing should be achievable between 1968–70. Moreover, Low was considerably more optimistic than Glennan in terms of cost estimates: he believed that spending would peak around 1966 and total some $7 billion and reasoned that by that time, the Saturn and larger Nova-type boosters would have been built and an Earth-circling space station would probably be in existence. He stressed, however, that manned landings would require a launch vehicle capable of lifting between 27,200 and 36,300 kg of payload; the existing conceptual design, dubbed the 'Saturn C-2', could send no more than 8,000 kg on a lunar trajectory. Low's group advised either that several C-2s would need to be refuelled in space or an entirely new and much more powerful booster had to be designed. Both approaches seemed realistic, the committee concluded, with Earth-orbital rendezvous probably the quickest option, yet still requiring the technologies and techniques to refuel in space.

Of pivotal importance in the eventual direction of Apollo was the new president, John Fitzgerald Kennedy, who had already appointed a transition team to assess a perceived American-Soviet 'missile gap' and investigate ways in which the United States could pull ahead technologically. This group was headed by Jerome Wiesner of the Massachusetts Institute of Technology – later to become Kennedy's science advisor – and it advocated, among other points, that NASA's goals needed to be redefined and sharpened. Another key figure, and long-time ally of NASA, was Vice-President Lyndon Johnson, who now pushed for Glennan to be superseded by Jim Webb, a man with immense experience in government, industry and public service. On 30 January 1961, Webb's appointment was approved by Kennedy.

It was Webb who would guide NASA through the genesis of Apollo; indeed, his departure from the agency would come only days before the project's first manned launch in October 1968. His influence on America's space heritage and the respect in which he continues to be held will be recognised, just a few years from now, by the launch of the multi-billion-dollar James Webb Space Telescope, which will succeed Hubble as the agency's primary astronomical instrument. Yet Webb's background was hardly scientific or in any way related to space exploration: a lawyer by profession, he directed the Bureau of the Budget and served as Undersecretary of State for the Truman administration, but throughout the Sixties he would prove NASA's staunchest and most fierce champion.

Also championing the agency's corner was Kennedy himself, who, only weeks before Yuri Gagarin dashed America's hopes of being the first to send a man into space, raised its budget by $125 million above the $1.1 billion appropriations cap recommended by Eisenhower. Much of this increase was funnelled into the Saturn C-2 development and, specifically, its giant F-1 engine. Built by Rocketdyne, this engine – which burned a refined form of kerosene, known as 'Rocket Propellant-1' (RP-1), with liquid oxygen – remains to this day the most powerful single-nozzled liquid engine ever used in service. Although it experienced severe teething troubles during development, notably through 'combustion instability', it would go on to become the cornerstone for a lunar landing capability.

By this time, Convair, General Electric and the Martin Company had submitted their initial responses to NASA, none of which overly impressed the agency's evaluators; in fact, all three companies had stuck rigidly with the same shape as the Mercury capsule, but some theoreticians had predicted that a Mercury-type design would be unsuitable for Apollo's greater re-entry speeds. In response, Space Task Group chief design assistant Caldwell Johnson began to investigate the advantages of an alternative, conical, blunt-bodied command module.

Early in May 1961, after more adjustments and rework, the contractors offered their final proposals. Convair envisaged a three-component system which had its command module nestled within a larger 'mission module'. Notably, it would return to Earth by means of a glidesail parachute and develop techniques of rendezvous, docking, artificial gravity, manoeuvrability and eventual lunar landings. General Electric offered a semi-ballistic blunt-bodied re-entry vehicle, with an innovative cocoon-like wrapping which would provide secondary pressure protection in case of cabin leaks or micrometeoroid punctures. Martin proposed the most ambitious design of all. Its conical vehicle was remarkably similar to the design ultimately adopted, although it featured a pressurised shell of semi-monocoque aluminium alloy coated with a composite thermal shield of superalloy and a charring ablator. Its crew would sit two abreast and one behind, in a set of couches which could rotate to better absorb the G loads of re-entry and enable better egress. All three contractors spent significantly more than the $250,000 provided by NASA. Martin's study topped $3 million and required the work of 300 engineers and specialists over a period of six months. In *Chariots of Apollo*, Brooks, Grimwood and Swenson noted that, had times been less fortunate, NASA may have been obliged to spend months evaluating the contractors' reports before making a decision. However, Gagarin's flight prompted Kennedy to press Johnson to find out how America could beat the Russians in space. After consulting, Johnson recommended setting the goal of landing a man on the Moon within the decade. On 25 May 1961, before a joint session of Congress, Kennedy laid down the gauntlet.

In the wake of Kennedy's speech, one of the key areas into which the increased funding would be channelled was a new booster called Nova; this was considered crucial to achieving a lunar landing by the direct-ascent method. At this stage, although NASA was "studying" orbital rendezvous as an alternative to direct-ascent, Hugh Dryden explained that "we do not believe ... that we could rely on [it]". More money and increased urgency for Apollo was not necessarily a good thing: both Webb and Dryden felt that decisions over direct-ascent or orbital rendezvous and liquid or solid propellants would be better made two years further down the line.

Nevertheless, rendezvous as an option was coming to the fore, with a realisation that it could provide a more attractive alternative to enormous and unwieldy boosters, allowing NASA to mount a lunar mission using two or three advanced Saturns with engines that were already under development. Although Earth-orbital rendezvous was considered safer, a lunar-orbit option would require less propellant and could be achieved using just one of von Braun's uprated Saturn 'C-3' rockets.

The Apollo spacecraft which would fly these missions to the Moon was also taking shape. Max Faget, the lead designer of the Space Task Group, set the diameter of its base at 4.3 m and rounded its edges to fit the Saturn for a series of test flights. These rounded edges also simplified the design of the ablative heat shield which would be wrapped around the entire command module. Encapsulating the spacecraft in this way provided additional protection against space radiation, although on the downside it entailed a hefty weight penalty. Others, including George Low, saw merits in both blunt-bodied and lifting-body configurations and suggested that both should be developed in tandem. Most within the Space Task Group, however, felt that a blunt body was the best option.

In August 1961, NASA awarded its first Apollo contract to the Massachusetts Institute of Technology, directing it to develop a guidance, navigation and control system for the lunargoing spacecraft. Two months later, five aerospace giants vied to be Apollo's prime contractor, with the Martin Company ranked highest in terms of technical approach and a very close second in technical qualification and business management. In second place was North American Aviation, which the NASA selection board recommended as the most desirable alternative. On 29 November 1961, word quickly leaked out to Martin that its scores had won the contest, but this proved premature; the following day, it was announced by Webb, Dryden and Seamans that North American would be the prime contractor, as a result of its long association with NASA and experience in developing the X-15 rocket-plane. The choice of North American, whose fees were also 30 percent lower than Martin, would return to haunt NASA in years to come. Of course, rumour quickly abounded that it was politics, and not technical competency, which had won North American the mammoth contract. Astronaut Wally Schirra, later to command the first Apollo mission, would recount that he felt the decision was made because companies in California had yet to receive their fair share of the space business. Others pointed to the company's lobbyist, Fred Black, who had a close relationship with Capitol Hill insider Bobby Gene Baker, a protégé of Vice-President Lyndon Johnson.

As North American and NASA hammered out their contractual details, the nature of Apollo's launch vehicle remained unclear, as was its means of reaching the Moon. It seemed likely that the production of large boosters capable of accomplishing a direct-ascent mission would take far longer than the development of smaller vehicles. The attractions of rendezvous were also becoming evident as a means of meeting Kennedy's end-of-the-decade deadline. At around this time, Bob Gilruth wrote that "rendezvous schemes may be used as a crutch to achieve early planned dates for launch vehicle availability and to avoid the difficulty of developing a reliable Nova-class launch vehicle".

As the debate continued, it was recognised that, in whatever form it took, the launch vehicle would be enormous and would require ground facilities of unprecedented scale. Under consideration were Merritt Island, just to the north of Cape Canaveral in Florida, together with Mayaguana in the Bahamas, Christmas

The largest rocket ever brought to operational status, the Saturn V departs the world's largest single-story structure, the Vehicle Assembly Building. The sheer enormity of both is amply illustrated by the technicians at the centre-right of the image.

Island, Hawaii, White Sands in New Mexico and others. Of these, only White Sands and Merritt Island proved sufficiently economically competitive, flexible and safe to justify further study and the final choice was a 323 km^2 tract of land on Merritt Island for a site which (after the assassination of John Kennedy) would be named the Kennedy Space Center. In addition, Cape Canaveral changed its name to Cape Kennedy in 1964. One of the most iconic structures to be built there in the mid-Sixties, and associated forever in the public mind with the lunar effort, was the gigantic Vehicle (originally 'Vertical') Assembly Building (VAB) in which the Saturns were 'stacked' on mobile platforms. Some 160 m tall, 218 m long and 158 m wide, the VAB had a floor area of 32,400 m^2 and to this day remains the world's largest single-story scientific building.

Elsewhere, a site near Michoud in Louisiana was picked for the Chrysler Corporation and Boeing to assemble the first stages of the Saturn C-1 and subsequent variants. In October 1961, NASA purchased 54 km^2 in south-west Mississippi and obtained easement rights over another 518 km^2 in Mississippi and Louisiana for static test-firing stands for the large booster, forcing around a hundred families, including the entire community of Gainsville, to sell up and relocate. It was around the same time that the decision to move the old Space Task Group – now the Manned Spacecraft Center (MSC) – from Virginia to Houston, Texas, was made.

Shortly after 10:06 am on 27 October 1961, the maiden mission in support of Apollo got underway with the test of the Saturn I (originally C-1) rocket from Pad 34 at Cape Canaveral. Although the vehicle was laden with dummy upper stages, filled with water, its performance was satisfactory and it proved a successful demonstration of the clustering of large rocket engines. By the time the tenth Saturn 1 was launched in July 1965, it had carried a 'boilerplate' Apollo into Earth orbit. However, the 590,000 kg of thrust of the Saturn I was insufficient to send Apollo to the Moon. Most engineers envisaged that the first stage of the lunargoing Saturn would need at least four, or perhaps even five, F-1 engines, each more powerful than the entire Saturn I cluster. Despite continuing interest in direct-ascent using a Nova with a first stage equipped with as many as eight F-1s, the decision was taken on 21 December to proceed with a rocket known as the Saturn C-5 (later the Saturn V), whose five F-1s would be capable of dispatching an Apollo spacecraft to the Moon's surface using either the Earth-orbital or lunar-orbital rendezvous modes.

Although direct ascent was still considered by many to be the safest and most natural means of travelling to the Moon, sidestepping the dangers of docking with other vehicles in space, procedures for how a lander might be brought onto the lunar surface remained sketchy, with some suggesting that the craft should touch down vertically on legs that would deploy from its the base and others suggesting it land horizontally on skids. An Air Force-funded study called 'Lunex', begun in 1958, had already addressed a direct-ascent method of reaching the Moon, but von Braun doubted it was possible to build a rocket large enough to accomplish such a mission and favoured rendezvous with smaller vehicles. Before von Braun's team was transferred from the US Army to NASA, it had proposed a mission known as 'Project Horizon', which called for making a lunar base for military, political and, lastly, scientific purposes. He felt that only Saturn was powerful enough to undertake such a mission and one of his conditions upon joining NASA was that its development should continue.

Against this backdrop came the lunar-orbital rendezvous plan, whereby a specialised craft would descend to the Moon's surface and, after completing its mission, return to dock with a 'mother ship'. The landing crew would then transfer to the mother ship and return to Earth. Since 1959, in fact, this had been recognised as the best way to reduce the total weight of the spacecraft and represented a crucial factor in determining the power of the launch vehicle. Many in NASA, though, were terrified by the prospect of attempting rendezvous so far from home. Proponents, on the other hand, considered it relatively simple, with no concerns about weather or air friction, lower fuel requirements and no need for a monster Nova rocket. It was NASA engineer John Houbolt who finally convinced Bob Seamans to place it on an equal footing with direct ascent and Earth-orbital rendezvous when a decision came to be made. The clincher, noted historian James Hansen, was that lunar-orbital rendezvous was the only way America could hope to achieve Kennedy's end-of-the-decade goal. By July 1962, lunar-orbital rendezvous had been adopted as the Apollo mission mode. The first steps to design the specialised lander got underway, with early plans ranging from short-stay missions involving one man for a few hours on the surface to seven-day expeditions with crews of two. One minimalist design took the form of an open, Buck Rogers-like 'scooter' with landing legs, which a fully suited astronaut would manoeuvre onto the Moon. As these plans crystallised, the paucity of knowledge of the lunar surface material and how it would be affected by exhaust gases made it imperative that the lander be able to 'hover' in order to select an appropriate landing spot.

North American, which had already been awarded the contract to build the Apollo command and service modules, strongly opposed the lunar-orbital rendezvous mode, partly because it wanted its own spacecraft to perform the landing. (Indeed, in August 1962, cartoons adorned its factory walls, depicting a somewhat disgruntled Man in the Moon looking suspiciously at an orbiting Apollo and declaring "Don't bug me, man!") With this in mind, North American made a strong bid to build the lander, which NASA rejected on the basis that the company already had its hands full developing the mother ship and the second stage of the Saturn V. By September 1962, eleven companies had submitted proposals for the lander and in November the Grumman Aircraft Engineering Corporation of Bethpage, New York, was awarded the $388 million contract. Although each bidder was judged technically and managerially capable, Grumman had design and manufacturing space available, together with clean-room facilities to assemble and test the lander.

The decision to proceed with lunar-orbital rendezvous eliminated the requirement for the Apollo command module to land on the Moon, but created a new problem: the need for a docking apparatus by which it could link up with Grumman's lander. The need was also quickly identified for a series of Earth-orbital missions to demonstrate and qualify the command module's systems before committing them to lunar sorties; the result was the Apollo Block 1 and 2 variants, the second of which provided the docking hardware and deep-space navigation and communications

Grumman's spidery lander is prepared for its first flight to the Moon: Apollo 10 in May 1969. Note the prominent descent engine and the craft's four legs, shown here folded into their launch configuration.

systems required to fly to the Moon. By mid-1963, North American had begun work on an extendable probe atop the command module, which would fit into a dish-shaped drogue on the lander.

As the design of the command module moved through Block 1 and 2 variants, so the lunar module itself was assuming its final form as a two-part, spider-like 'bug' which would deliver astronauts to the Moon's surface and back into orbit. The descent stage was to be equipped with the world's first large throttleable rocket engine, whilst the ascent stage, housing the pressurised cabin, would use a fixed-thrust engine to boost the crew away from the Moon. The appearance of the vehicle produced something which Brooks, Grimwood and Swenson described as "embodying no concessions to aesthetic appeal ... ungainly looking, if not downright ugly". Since operating within Earth's atmosphere would be unnecessary, aerodynamics was not an issue, but there were other factors to concern the Grumman designers. For example, when the time came for the ascent stage to lift off, the blast of its exhaust in the confined space of the inter-stage structures – referred to as 'fire-in-the-hole' – might tip the vehicle over. Clearly, many problems remained to be solved.

Shape-wise, the ascent stage was originally spherical, with four large windows for the crew to see forward and 'down' in the style of a helicopter. Weight, though, remained a critical constraint and this design was ultimately discarded when it became clear that the windows would need extremely thick panes and strengthening of the surrounding structure. The spherical cabin was replaced by a cylindrical one with a flat forward bulkhead cut away at various planar angles and two small, flat, triangular windows were installed, canted 'downwards' so that the crew would have the best possible view of the landing site. The change from a spherical to a cylindrical cabin meant that Grumman could not easily weld the structure. By May 1964, it had been decided to weld areas of critical structural loads, but to use rivets where this was impractical. The interior of the 4,930 kg ascent stage cabin had a volume of 6.7 m^3, creating the largest American spacecraft yet built, and NASA urged Grumman to make its instruments as similar to those in the command module as possible. As it evolved, the astronauts became an integral part of this, with veteran spaceflyer Pete Conrad working on the design perhaps more than anyone else. When he started, the control displays were drawings on plywood! He was instrumental in implementing electroluminescent lighting inside the lunar module, as well as the command module, which reduced weight and power demands.

Another crucial change was the removal of seats, which were too heavy and restrictive in view of the fact that the astronauts would be clad in bulky space suits. Bar stools and metal cage-like structures were considered, but the brevity of the lunar module's flight and the moderate G loads it would impose rendered such aids unnecessary. Moreover, *standing* astronauts would have a better view through the windows and the eliminated worry about knee room meant that the cabin could be reduced in size. Instead of seats, bungee-like restraints would be added to hold the astronauts in place and prevent them from being jostled around during manoeuvres.

It was initially envisaged that when the ascent stage returned to the mother ship, the docking would be performed using the same hatch as the astronauts used to gain access to the lunar surface, but when it was decided to install a small window above the commander's head and dock using the hatch in the roof, the forward hatch was changed from circular to square. This made it easier for men wearing pressurised suits and bulky backpacks to pass through.

At the base of the 10,334 kg descent stage were five legs (later reduced to four as part of a weight-versus-strength trade-off) each with a 91 cm diameter bowl-shaped footpad below which extended a frangible probe to detect contact with the lunar surface. It was essential to minimise the lander's weight and NASA paid Grumman $20,000 for every kilogram they could shave off. In fact, even the weight of the astronauts themselves was a factor in determining which men would fly the lunar module and which would not.

Astronaut Gene Cernan, who flew the ungainly machine on its second manned mission, summed up its incredible fragility and hidden strength. "What little remained looked downright scary," he wrote. "A dropped screwdriver could punch cleanly through it. The two glass windows were shaved so thin that they bulged out when the spacecraft was pressurised. Looks are deceiving. History would show the fragile, bug-like lander would do everything ever asked of it."

At launch, the lunar module would be nestled inside a conical section atop the third stage of the Saturn V with its legs folded against the structure of the descent stage; they were extended in space. In addition to its ascent and descent engines, the lunar module possessed 16 small attitude-control thrusters, clustered in quads around the ascent stage, pointing upwards, downwards and sideways for the best possible manoeuvrability. The ascent engine, built by Bell Aerospace, simply *had* to work in order to get the astronauts off the lunar surface; as a result, it was made as simple as possible, with a pressure-fed fuel system employing hypergolic propellants. The descent engine was more challenging, since it had to be throttleable, and it used flow-control valves and a variable area injector to regulate pressure, flow rate and fuel mixture in the combustion chamber.

The control centre for Apollo was the integrated command and service modules. This was far more complex than Mercury or Gemini. The 'command module', firstly, was a conical structure, 3.2 m high and 3.9 m across its ablative base, and provided its three-man crew with 5.95 m^3 of living and work space whilst aloft. Brimming with reaction controls, parachutes, propellant and water tanks, batteries, cabling and instrumentation, it would truly live up to its name as the command centre for future voyages to the Moon. The 'service module' consisted of an unpressurised cylinder, 7.5 m long and 3.9 m wide, housing propellant tanks, cryogenic oxygen and hydrogen for the fuel cells, water tanks, an S-band communications antenna, the giant Service Propulsion System (SPS) engine and four 'quads' of manoeuvring thrusters. The SPS was some 3.8 m long and burned a mixture of hydrazine and unsymmetrical dimethyl hydrazine with nitrogen tetroxide. It was to provide the impulse for inserting Apollo into, and removing it from, lunar orbit. It was a critical component without a backup and, like the ascent engine of the lunar module, it was made as simple as possible. In particular, there was no need for either fuel pumps or an igniter, because the propellants were pushed into the combustion chamber by helium and were 'hypergolic', meaning that they would burn on contact. In effect, the propellants would flow so long as the valves were held open and the solenoids of those valves were designed to be extremely reliable. The service module would be

Impressive image of the command and service module in lunar orbit during the Apollo 15 mission in the summer of 1971. Note the docking probe atop the command module's nose, the thruster quads on the surface of the service module, the S-band antenna and the large Service Propulsion System (SPS) engine.

jettisoned at the end of the mission, leaving the command module alone to make atmospheric entry.

By the spring of 1969, the command and service modules had been extensively tested during two unmanned and two manned Apollo missions, one of which made humanity's first voyage to the Moon. What had not yet been achieved, however, was an 'all-up' manned test of the entire Apollo combo, including the lunar module. The first flight-ready version of this spacecraft, designated LM-1, had been delivered to Cape Kennedy in June 1967 and was launched by a Saturn IB booster in late January of the following year. Its environmental control system was incomplete and it was not fitted with landing gear, but the inaugural unmanned flight was highly successful: its TRW-built descent engine was test-fired, as was the ascent engine, and the spidery vehicle's guidance system proved its worth. The flight was so successful that NASA cancelled a rerun with LM-2 and declared that the lander's next mission would be carried out with men aboard.

However, it still had many problems. The instability of its Bell-built ascent engine, in particular, was a cause for concern throughout 1967 and much of 1968. Although most high-level NASA managers felt that Bell had a good chance of solving the engine's fuel-injector woes, the agency nevertheless hired Rocketdyne to develop an alternate device. Following difficulties in both cases, Rocketdyne was ultimately chosen to outfit the lunar module's fuel injector. Other problems with the lander included windows blown out and fractured during high-temperature tests, broken wiring and stress corrosion in its aluminium structural members; the latter led to the formation of a team to identify the cause and to order corrective actions. Grumman analysed more than 1,400 components and heavier alloys were employed for newer sections of the lunar module. Weight, too, posed an issue. In 1965, more than 1,100 kg had been shaved from the lunar module and NASA offered incentives to Grumman to remove yet more. Despite a major effort by all concerned, it was clear that the first manned mission would not be ready until the end of 1968 at the earliest.

THE 'D' FLIGHT

The men who would put LM-3 through its paces and perform the first demonstration of the entire Apollo spacecraft in Earth orbit would do so on one of the most complex and critical missions ever attempted. To understand where the Apollo 9 flight 'fitted' in terms of the lunar landing goal, it is important to recap a number of fundamental changes to the plan in 1968. In September 1967, MSC defined a series of steps, labelled 'A' through to 'G', for building up to the first manned lunar landing. First came unmanned tests of the Saturn V carrying the command and service modules, flying as Apollo 4 in November 1967 and Apollo 6 in April 1968. The 'B' mission, completed by Apollo 5 in January 1968, was an unmanned checkout of the lunar module. A manned 'C' flight, involving the command and service modules in Earth orbit, was achieved in October 1968 by Wally Schirra, Donn Eisele and Walt Cunningham on Apollo 7. The final strides toward the Moon focused on four increasingly difficult missions: 'D' (a manned demonstration of the whole Apollo system in Earth orbit), 'E' (a repeat of D but in a high elliptical orbit with an apogee of 6,400 km), 'F' (a full dress-rehearsal in orbit around the Moon) and 'G' (the landing itself).

During the course of 1966, crews had been assembled to support the first of these flights. Assigned to the D mission was a pair of veteran astronauts – Jim McDivitt in command and Dave Scott as the command module pilot (CMP) – teamed with rookie spacefarer Rusty Schweickart as the lunar module pilot (LMP). The plan called for their command and service modules to be launched atop a Saturn IB rocket and a few days later a second IB would place the unmanned lunar module into orbit. McDivitt's crew would rendezvous, dock and carry out a series of joint exercises, including a test of the lunar EVA suit on a spacewalk by Schweickart. Years later, Deke Slayton would note that McDivitt had been intimately involved in the lunar module's development for some time and since Slayton required the command module pilot to have prior experience of rendezvous, Dave Scott was the obvious choice. (Both men already had one Gemini flight apiece under their belts and Scott had accompanied Neil Armstrong in making the first docking in space.) The E crew, meanwhile, would comprise astronauts Frank Borman, Mike Collins and Bill Anders, who, on this plan, would become the first humans to ride the enormous Saturn V lunar rocket.

The Apollo programme abruptly stalled in January 1967, when astronauts Virgil 'Gus' Grissom, Ed White and Roger Chaffee – assigned to fly Apollo's first manned mission – were killed in a flash fire which swept through their command module during a launch pad test. More than a year later, after extensive modifications and safety upgrades, the McDivitt and Borman crews seemed in line to fly the D and E missions as Apollo 8 and Apollo 9, sometime in late 1968 or early 1969. Following the successful first unmanned test of the Saturn V, a wave of optimism swept NASA that a lunar landing could indeed be accomplished before the end of the decade; to such an extent that by August 1968 – two months before Apollo 7 was due to fly – some managers were talking of expanding Borman's mission from high Earth orbit either to lunar distance or possibly even for a circumnavigation of the Moon. This audacious plan had been under consideration since April 1968, primarily as a result of discussions between George Low, now head of the Apollo Spacecraft Program Office, and Chris Kraft, one of NASA's first flight directors and the man now in charge of the agency's flight operations division. It was attractive because it would eliminate one of the steps in the A-G plan and achieve a landing much sooner. Low and Kraft wanted to change the E mission into something called 'E-prime', moving it from a 6,400 km apogee to the vicinity of the Moon, but this quickly became untenable when it materialised that LM-3 would not be ready in time. Shortly thereafter, Low came up with an even more radical idea: a 'C-prime' mission which would send the command and service modules – but not a lunar module – *around* the Moon in December 1968. Low knew from Rocco Petrone, director of launch operations at Cape Kennedy, that LM-3 would not be ready, but a meeting in Bob Gilruth's office at MSC on 9 August concluded that the navigation and trajectory teams, together with the astronauts and their training staff, could be ready for such a flight. The added risk of actually entering lunar *orbit* – a proposal tabled by Bill Tindall and John Mayer – would be beneficial in that it would provide empirical data on the thermal regime and allow engineers to better model the Moon's gravitational field.

The electrifying nature of that meeting and the personalities involved – Gilruth, Low, Kraft, Deke Slayton, Apollo director Sam Phillips, Wernher von Braun and others – is related in depth by Kraft in his autobiography. Simply following Kraft's account of the words which were, after all, spoken a mere 18 months after Grissom, White and Chaffee died, is enough to produce goosebumps in the reader. "The excitement," he described, "was flush". Von Braun was excited, Slayton was "almost bouncing with anticipation" and the stern General Phillips stood up to congratulate them all and express his pride at their gutsiness. "There was a sense in the room," Kraft noted, "that something big was about to happen."

At that time, NASA Administrator Jim Webb and his Associate Administrator for the Office of Space Flight, George Mueller, were at a conference in Vienna, and it was left to Deputy Administrator Tom Paine at NASA Headquarters in Washington to conditionally approve the plan. 'Conditionally', that is, because Paine harboured his own concerns that the performance of the giant Saturn V on the unmanned Apollo 6 mission was not good enough for the vehicle to be entrusted with a human crew. Nevertheless, like Phillips, he was impressed by the boldness of his subordinates. Mueller, though, was furious when he heard the news. According to Kraft, "he was convinced that we'd waited until he was out of the country, then replanned the Apollo programme behind his back". However, the next day Mueller agreed, albeit with some reluctance. For his part, Webb vehemently opposed the idea. He had been particularly lambasted by Congress after the Apollo fire and did not wish to be hauled over the coals if the Moon shot failed. On the other hand, with a new administration – of whatever party – due to take office in January 1969, Webb knew that his days at the helm of NASA were probably numbered and, indeed, he planned to retire anyway. With this in mind, on 16 August, he was finally won over, with reservations, to accept C-prime.

Three days later, the plan was formally set in motion by Sam Phillips, although some managers remained cautious about making such a bold move before a manned mission in Earth orbit had even been flown. Officially, until Apollo 7's shakedown of the command and service modules had succeeded, the 'new' Apollo 8 would be promoted as little more than "an expansion" of its predecessor, albeit with the telling addendum that its "exact content ... had not been decided". Despite the tantalising possibilities implicit in such a statement, Chris Kraft, for one, was surprised when the press did not latch onto it. "The sharpest space reporters in the world missed one," he wrote later, "and we avoided the questions, interviews and press conferences that we'd thought were inevitable."

The content of the mission may not have been decided, publicly at least, but the crew certainly had. On 10 August, Deke Slayton told Jim McDivitt that the flight order was being switched: that his D mission with the lunar module would now become Apollo 9, preceded by Borman's C-prime expedition around the Moon. "Over the years," McDivitt recounted in Slayton's autobiography, "this story has grown to the point where people think I was offered the flight around the Moon, but turned it down. Not quite. I believe that if I'd thrown myself on the floor and begged to fly the C-prime mission, Deke would have let us have it. But it was never really offered." Offered or not, McDivitt, Scott and Schweickart had been training for so long on the lunar module and its systems that they were the best-prepared and wanted to fly its maiden mission.

Privately, Frank Borman was pleased to receive command of C-prime. McDivitt's D mission "was a test-piloting bonanza," wrote Andrew Chaikin, "and Borman would have gladly traded places." Borman's own E flight, originally, was little more than a repeat of D, albeit in a higher orbit, so the change to C-prime and the chance to command the first lunar expedition certainly appealed to his thirst for challenge. In the summer of 1968, he was at North American's Downey plant in California, working on tests of Spacecraft 104 – the command module for the E mission – when he was summoned to take a call from Slayton. Shortly afterwards, he was back in Houston, in Slayton's office, hearing about C-prime, together with disturbing CIA reports that the Soviets might be only weeks away from staging their own manned loop around the Moon. When Slayton asked Borman if he would fly into lunar orbit, it was a question with only one answer.

A NEW START?

"Vietnam," wrote astronaut Gene Cernan, "did in President Johnson." Since taking office after the assassination of John Kennedy in November 1963, Lyndon Baines Johnson had steered the United States on a course which had won him friends, but a far larger number of enemies. As the decade drew to its close, by the provisions of the 22nd Amendment, he could have run for an another presidential term ... but, in March 1968, he announced that he would not seek, and would not accept, his party's nomination for the upcoming election. His decision to commit more than half a million American troops to an ongoing and seemingly unwinnable conflict in Vietnam and to deal with the inevitable backlash as casualties mounted, month after month, had made him one of the most unpopular presidents of the modern era. His Secret Service aides had refused to let him speak on college campuses or appear at 1968's Democratic National Convention in Chicago, for fear that the very sight of him might cause riots.

Johnson's claim that the war was being won was not borne out by reality and was certainly not the view of the press or the majority of the public. At the end of January 1968, a hammer blow struck the misguided sense of complacency that the Vietcong – a pro-communist organisation, operating against the South Vietnamese puppet government and its international allies – were little more than snipers. The so-called 'Tet Offensive', which ran in three devastating waves until September, was intended to strike military and civilian command centres throughout South Vietnam in the hope of sparking popular uprisings. Although it actually proved disastrous, militarily, for the Vietcong, it was so vast (countrywide) and so well-organised (involving more than 80,000 troops) that it shocked the Johnson administration and America's war-weary public.

Prior to Johnson's announcement that he did not wish to run for re-election, it had been widely expected that he would have little difficulty winning the Democratic nomination, since no one – not even Senator Bobby Kennedy, brother of the murdered president and a man known to be deeply critical of some of Johnson's policies – was prepared to run against him. Only Senator Eugene McCarthy presented an open challenge and narrowly missed winning the first presidential primary in New Hampshire, losing to Johnson by just 7 percent of the vote. In the weeks that followed Johnson's withdrawal, Kennedy played his hand by announcing his candidacy and an increasingly embittered series of state primaries were fought. Although Kennedy ultimately won many of the primaries, McCarthy's firm anti-war stance and the support of a gaggle of prominent Hollywood stars sustained the momentum of his campaign. Shortly after Johnson withdrew, Vice-President Hubert Humphrey also declared his candidacy. The assassination of Bobby Kennedy in June 1968 led many to debate whether he might have won the nomination had he lived. His death left Humphrey as an apparent front-runner for the Democrats. Despite Humphrey's association with the Johnson administration and his unpopularity amongst anti-war campaigners, he held the support of the party's traditional power bloc. Some of Kennedy's supporters went over to McCarthy, but others, remembering the bitterness of the primaries, refused. The Democratic National Convention that August brought some of the ugliest scenes of anti-war protest directly into the living rooms of America: of demonstrators hurling bottles and rocks and Chicago police replying with clubs and tear gas. Some Democrats supported the demonstrators, whilst others backed the police, but Humphrey won the nomination and then picked Edmund Muskey as his running mate for the vice-presidency.

The anti-war riots, however, ultimately crippled the Humphrey-Muskey campaign and by the election in November it was the Republican team of Richard Milhous Nixon and Spiro Theodore Agnew who were declared the victors ... although the final result was incredibly close. "The key states," noted the Wikipedia website, "proved to be California, Ohio and Illinois, all of which Nixon won by 3 percentage points or less. Had Humphrey carried all three of these states, he would have won the election ... Nixon won the popular vote with a plurality of 512,000 votes or a victory margin of about one percentage point." Nixon's victory in the electoral college was larger, with 32 states and 301 electoral votes, compared to Humphrey's 13 states and 191 votes, but when the loser left him a gracious message of congratulation, the man who would be in control of America for the era of the first Moon landings was particularly touched; for Nixon, the 55-year-old veteran senator, had himself narrowly missed out on the presidency eight years earlier, opposing Democrat John Kennedy. "I know how it feels," he said later, "to lose a close one." It is interesting to speculate that if Nixon had beaten Kennedy in the 1960 election, the Moon landing of Neil Armstrong and Buzz Aldrin might never have transpired.

In his campaign, Nixon was instrumental in the development of a 'Southern strategy' to appeal to conservative whites angered by the Johnson administration's relentless crusade for civil rights. As Wikipedia pointed out, adoption of the strategy "would prove more effective in subsequent elections and would become a staple of

Republican presidential campaigns". The restoration of law and order, following several years of angry protest and rioting in American cities, was also high on Nixon's agenda, although his opponents argued that he was actually appealing to white racial prejudice; after all, he had *opposed* desegregation bussing (the practice of trying to integrate schools by assigning students based on their locality, rather than upon their race) and had also *opposed* Chief Justice Earl Warren's attempt to promote more liberal civil rights policies.

Nixon's promise to end military conscription was seen by some as a clever attempt to undermine the anti-war movement by deleting the *requirement* for college students to fight and thus also removing the *need* for them to protest. At length, Hubert Humphrey, who had been reluctant to resist Johnson's Vietnam policies, lest this damage his own election campaign, began to steadily distance himself from the conflict and even called for an end to the bombing of North Vietnam. When Johnson himself announced this in late October 1968, it seemed to reinvigorate part of Humphrey's campaign. Then, just days before the election, Nixon's team set out to sabotage the peace negotiations that were underway in Paris by reassuring the South Vietnamese military leadership that a Republican administration would offer a better deal than a Democratic one. "The tactic worked," observed Wikipedia, "in that the South Vietnamese junta withdrew from the talks on the eve of the election, thereby destroying the peace initiative on which the Democrats had based their campaign."

However, accusations of 'treason' for sabotaging the talks did not prevent what was a narrow victory for Nixon. In January 1969, at his inaugural address, he pointed out that the United States had "suffered from a fever of words" which had promised more than could be delivered, degenerated into hatred and had failed to satisfactorily persuade the electorate. "We cannot learn from one another," he said, "until we stop shouting at one another, until we speak quietly enough so that our words can be heard as well as our voices." Much of the 'noise' in Vietnam would be quietened through rapid reductions in troop numbers, with 25,000 military personnel returning home in June 1969. A month later, the 'Nixon Doctrine' was announced: the United States would continue to provide military aid and supplies to South Vietnam, but that nation would be expected to supply the manpower for its own defence. As far as reducing troop numbers were concerned, Nixon seemed to be making good on his promise to curb the war, but his insistence on allowing incursions into Laos and Cambodia to interrupt the Ho Chi Minh Trail and hit North Vietnamese 'sanctuaries' prompted criticism and student protests in the spring of 1970. Efforts to end conscription were also set in motion through the Gates Commission, which confirmed that adequate military strength could be maintained without the draft, and in June 1973 it was finally ended.

Rises in payments for government benefits, such as Social Security and Medicare, and increases in food aid and other assistance were balanced by decreases in military and space spending. Between 1969 and 1973, the former dropped by 3 percent of the gross national product and it is ironic that Nixon was in office to witness NASA achieve the historic first manned landing on the Moon ... and then systematically slashed its budget from 2.1 percent of the federal budget in 1969 to 1.1 percent in

1973. However, his first administration would see the Space Shuttle rise from a paper project to the beginnings of a real system which would define and dominate the United States' space aspirations for the next three decades. On the down side, manned exploration of the Moon and hopes of seeing Americans on Mars faded, then vanished. Tom Paine, promoted to the agency's Administrator, was already working on such plans, with a goal to establish a permanent lunar base by the close of the Seventies and perhaps a mission to the Red Planet as early as 1981. Vice-President Agnew also supported Paine's Moon-Mars plan, but Nixon did not. Consequently, the belief that the Shuttle offered a significantly cheaper means of getting humans into space won the day and 'exploration' would henceforth become the exclusive preserve of robots.

One of Nixon's most disappointing comments about human space exploration is related in the pages of Andrew Chaikin's book, *A Man on the Moon*. In December 1972, as the final Apollo crew prepared to leave lunar orbit, a statement from the president was read to them from Mission Control: "This may be the last time in this century that men will walk on the Moon, but space exploration will continue." For astronaut Jack Schmitt, the first, and so far only, professional geologist to walk on the lunar surface, the statement was met with outrage. "He hated the words," Chaikin wrote, "hated them for their lack of vision. These words, from the leader of the nation! Even if Nixon really believed them, he didn't have to say so in a public statement, taking away the hopes of a generation of young people."

The short-sightedness of national leadership towards human space exploration was, in a sense, also mirrored in the Soviet bloc. Over the years, *both* superpowers have used space as little more than a means of satisfying their own needs. Arguably one of the main reasons why humanity has not yet – even after five decades – successfully established a permanent base on another celestial body is that the political will has never really been there ... or if it *has* been there, its primary focus has been to fulfil a political or military need. In the words of Moonwalker John Young, who would lead NASA's astronaut corps throughout most of the Seventies and command the first Shuttle mission: "Politicians are a strange bunch of critters!"

FIRST AMONG EQUALS

The political will for space should have been at its strongest in the days and weeks surrounding the first manned lunar landing in July 1969, barely six months into Richard Nixon's first term. The new president would pay all due reverence to the Moon voyagers – making a "historic telephone call" from the Oval Office as Armstrong and Aldrin stood on the lunar surface and watching their splashdown from the deck of the recovery ship *Hornet* – but in general he saw little political, scientific or technological value in human space flight. He liked heroes, it is true, and the Apollo crews would be feted as such, but his interests lay elsewhere. It is ironic, therefore, that in the opening months of 1969, the biggest political, scientific *and* technological triumph in the United States' history would be realised.

With the changes to the A-G lunar landing plan in the summer of the previous year and the decision to swap Apollo 9 and 8, disappointment had fallen onto the shoulders of astronaut Dave Scott. As command module pilot of the D mission, he had nursed Spacecraft 103 ('his' original ship) through months of testing and preparation at Downey and was now assigned another vehicle, Spacecraft 104, which, to him, represented a great unknown. Still, the flight would remain full of challenge: by this time, the original aim to launch its pieces separately on two Saturn IBs had evolved into a new plan to send the entire spacecraft, including LM-3, into orbit atop the giant Saturn V, as would be done on lunar missions. Scott's role as CMP would encompass flying the ship 'solo', whilst McDivitt and Schweickart tested the lunar module. If the spidery craft were crippled, he would also be responsible for rendezvous and docking with them, a tricky and complex affair. On later missions, CMPs would spend up to three days, alone, in orbit around the Moon. Theirs was one of the most responsible, respected and coveted posts on an Apollo crew ... but its very nature also meant that its incumbent would not walk on the lunar surface.

David Randolph Scott, along with a handful of others, would have an almost perfect spaceflying career in terms of his three missions: in addition to his CMP duties on Apollo 9, he would later command Apollo 15, during which he became the seventh man to place his footprints on the Moon. However, his first space flight, aboard Gemini VIII in March 1966, had been laced with disappointment and came

The Apollo 9 crew during training in the command module simulator. Front to back are Jim McDivitt, Dave Scott and Rusty Schweickart.

close to tragedy. Paired with Neil Armstrong, Scott had trained to perform the first docking with an unmanned Agena target vehicle, as part of an effort to perfect rendezvous techniques prior to Apollo. The docking itself was successful, but a faulty thruster caused a near-uncontrollable spin, the recovery from which forced the astronauts to abandon their three-day mission and return to Earth after just ten hours. In doing so, Scott missed out on performing a two-hour EVA. His performance on the flight, however, quickly assured him a seat on the backup crew for Apollo 1 in order to gain early experience of training with the new spacecraft, prior to assignment to the D mission.

Scott's selection as an astronaut in October 1963 had come following a strict, frugal, military life which had instilled in him the virtues of personal discipline and devotion to setting and achieving ambitious goals. Born on 6 June 1932 at Randolph Air Force Base in San Antonio, Texas – hence his middle name – he was the son of an Army Air Corps officer and would later recall watching biplanes soaring overhead as a three-year-old boy. He was fascinated, too, by the knowledge that aboard one of them was his father. From that tender age, the young Scott set his sights on someday becoming a pilot.

With a military breadwinner, the family moved many times during his childhood, from Texas to Indiana, abroad to the Philippines and, in 1939, back to the United States. Scott's father was posted overseas after Pearl Harbour to support the war effort and the young boy developed a keen interest in model aircraft. When his father returned at the end of the conflict, Scott received his first flying lesson.

Despite a desire to attend West Point, in 1949 Scott won a swimming scholarship to read mechanical engineering at the University of Michigan. After a year in Michigan, he was summoned for a physical at West Point, passed and headed to upstate New York to begin preparations for a military career. In his autobiography, Scott would credit his four years at West Point – plus his own upbringing – as "the most valuable and formative ... of my life". In 1954, he graduated fifth in his class and opted to join the Air Force. After initial flight training at Marana Air Force Base in Tucson, Arizona, he moved to Texas to begin working on high-performance jets. This was followed by gunnery preparation and assignment to a fighter squadron in Utrecht, Holland. Whilst in Europe, Scott flew F-86 Sabre and F-100 Super Sabre jets under a variety of weather conditions and, in October 1956, when Soviet tanks rolled into Budapest, his squadron was placed on high alert for the first time.

Three years later, as the Mercury Seven were introduced to the world's media, Scott watched from afar with scepticism; why, he wondered, were they abandoning such promising military careers? His own focus was upon gaining an advanced degree in aeronautics from Massachusetts Institute of Technology (MIT) and achieving admission to test pilot school. His work at MIT, he later wrote, "was like trying to drink water from a high-pressure fire hydrant ... Compared to the hard grind of MIT, the five or six years I had spent flying fighter jets felt like playing." Meanwhile, in 1961 his wife, Lurton, gave birth to their daughter, Tracy.

As part of his master's degree, Scott was introduced to the new field of 'astronautics' – "my first exposure to space" – and his dissertation focused upon the mathematical application of guidance techniques and celestial navigation. This undoubtedly proved beneficial to Projects Gemini and Apollo, both of which required rendezvous. In 1962, shortly after John Glenn completed America's first orbital mission, Scott passed his final exams, hoping for reassignment to test pilot school ... only to be detailed instead as a professor of aeronautics and astronautics at the Air Force Academy. Fortunately, a conversation with a sympathetic superior officer led to a change of orders and Scott reported to the Experimental Test Pilot School at Edwards Air Force Base in California.

Graduation was followed, in mid-1963, by the lengthy application process to join NASA's astronaut corps. In his autobiography, Scott recalled undergoing cardiograms, running on treadmills and enduring hypoxia evaluations, in which he was starved of oxygen to assess his physiological response. The psychologists were especially difficult to please. When asked about his MIT days and how he had liked Boston, Scott replied that he found New Englanders cold, aloof and "a little hard to get to know" ... only to discover that the stone-faced psychologist was a born-and-raised Bostoner! It obviously had little adverse impact on his application and in October 1963 he and thirteen other candidates – Edwin 'Buzz' Aldrin, Bill Anders, Charlie Bassett, Al Bean, Gene Cernan, Roger Chaffee, Mike Collins, Walt Cunningham, Donn Eisele, Ted Freeman, Dick Gordon, Rusty Schweickart and Clifton 'C.C.' Williams – were accepted by NASA. Of these, eight were test pilots, whom Slayton intended to use "for the more immediately difficult work" as solo CMPs for the Apollo missions, whilst the others represented a mixture of operational military fliers, engineers or researchers who "would get their chance, too, but on the development end of things".

All these men possessed their own share of the 'right stuff' which had epitomised NASA's two previous astronaut intakes, but the selection criteria for the 1963 class changed slightly: the necessity for test piloting credentials was dropped and applications were accepted from fliers with 'operational' backgrounds or holders of advanced degrees in engineering or the physical or biological sciences. Dave Scott, though, still fitted the mould of the classic Astronaut and historians have seen it as no accident that he was the first of his class to fly, the first of his class to serve as a solo CMP and the first of his class to command a mission. "In some circles," noted Andrew Chaikin, "there was a joke that if NASA ever came out with an astronaut recruiting poster, Dave Scott should be on it." Further, even astronauts who did not get along with Scott placed him at the top. On Apollo 9, this 'first among equals' would be presented with his toughest test-flying trial so far.

FLYING THE 'GIANT SPRING'

Since his assignment to the D mission, Scott had inherently known that, aside from the complexities of the flight itself, with its tricky rendezvous and docking commitment between two *manned* spacecraft, there were also a host of contingencies for which he needed to train. One of the most dire focused on what might happen if Jim McDivitt and Rusty Schweickart, after undocking, were somehow unable to control the lunar module. Depending on the severity of the problem, Scott might have to rendezvous and dock with them manually. Left largely unspoken was another, darker possibility – the harrowing possibility that every CMP would face, whether in Earth orbit or someday in orbit around the Moon: the possibility of an emergency so severe that he might be forced to abandon his comrades and return home alone.

"Bringing Apollo home as a one-man show," Scott wrote, "involved my mastering many aspects of all three jobs performed by the crew, Jim's as commander, Rusty's as systems engineer, my own as navigator. The sheer logistics of operating in all three positions, let alone learning the complex procedures this would require, was challenging, to say the least." Nor, indeed, was it a relatively straightforward matter of 'just' learning procedures by heart; in addition to his already daunting workload, Scott had to be ready to recover from any problem or fault or emergency or error message or master alarm that the command module threw at him. More than two years of training in the simulator and in the spacecraft itself at Downey had honed his skills. Eventually, he had developed a routine: first, checking Apollo's electrical, communications and environmental systems from Schweickart's right-hand seat, then moving to the centre couch in order to fulfil his own duties by setting up the relevant programs on the computer – be they for rendezvous or for re-entry – and finally shifting into McDivitt's position to perform the manoeuvre itself. In his autobiography, he noted that, despite practicing this effort numerous times, "it was pretty exacting", although he admitted that, in space, weightlessness would aid his movements. Equally exacting, of course, was the possibility that he might have to rescue McDivitt and Schweickart if the lunar module developed problems. For example, if LM-3 was unable to initiate its correct rendezvous manoeuvre within a minute of when it was planned, Scott would need to rescue them. If the two craft successfully redocked, but the pressurised tunnel between them was inaccessible, or if the hatches failed to open, McDivitt and Schweickart would need to leave the lunar module in their pressurised suits and spacewalk back over to the command module's side hatch.

Like the experience of Yevgeni Khrunov and Alexei Yeliseyev, this was both complex and difficult and hampered by the fact that McDivitt – who was not scheduled to make an EVA on the mission – would have been totally reliant upon Schweickart's emergency oxygen supply. "If he didn't make the EVA transfer within 45 minutes," Scott wrote, "he would die." For months, the astronauts and their backups – Charles 'Pete' Conrad, Dick Gordon and Al Bean – methodically rehearsed the complicated steps; steps which were further compounded as technicians and engineers tested them by dreaming up failures, systems faults and other emergencies. "Our launch was almost postponed," added Scott, "because we could not get enough training, especially for the rendezvous profile. After every sim was finished, we were debriefed and often had to explain why we had failed to deal with a particular situation. It was all pretty intense."

Apollo 9 was indeed postponed for a few days, though not for lack of training. Originally scheduled to fly at 11:00 am Eastern Standard Time (EST) on the last day of February 1969, the crew arrived at Cape Kennedy three weeks prior to launch to finish their training, preparation and quarantine. Living conditions had

Despite the constant worry of every CMP about bringing the spacecraft home alone, these few hours of Apollo 9 represented Dave Scott's only solo time. Here the command and service module Gumdrop is shown during rendezvous operations with the lunar module Spider.

changed quite markedly from the spartan crew quarters of Mercury and Gemini days: they each had individual bedrooms, a shared bathroom, a large and comfortable living area, a miniature gym in which they played handball at night and a superb chef, Lew Hartzell. "He was great," Scott related, "a former cook on a tugboat, who served us with marvellous meals every time we sat down." Then, in the final days before launch, Jim McDivitt's white blood-cell count was found to be low, hinting that he might be coming down with a cold. By this time, the giant Saturn V booster had been sitting on swamp-fringed Pad 39A for almost two months and its 28-hour launch countdown had begun. Thirty minutes into a planned three-hour hold at T-16 hours, managers decided to recycle the clock to T-42 hours and give McDivitt time to recover his strength. In his autobiography, Deke Slayton related that "all three of the crew came down with colds", a suggestion which Scott and Schweickart attempted to dismiss by donning jogging clothes and running around the perimeter of the launch complex. *Flight International* quoted a medical report from the astronauts' physician, Chuck Berry, which mentioned "sore throats and nose congestion". Whatever the truth, it certainly illustrated, in Slayton's mind, that the crews were being worked hard and were increasingly susceptible to "opportunistic infections". Further, the complexity of the first few days of Apollo 9 made it imperative that no chances be taken with the health of the crew.

The brief postponement did not detract from what promised to be a spectacular, though challenging, year for NASA: the year in which John Kennedy's very public promise to land a man on the Moon would be realised. Up to five missions were planned, each one building on – and *dependent* on – the success of its predecessor. If the D mission was a success, the stage would be set for veteran astronauts Tom Stafford, John Young and Gene Cernan to ride Apollo 10 to the Moon in mid-May for the F mission that would involve a full dress-rehearsal of the landing, including undocking their lunar module, descending to within 15 km of the surface and then returning to dock with the command and service modules. Only if Stafford's flight verified the performance of the lander's descent and ascent engines and consolidated knowledge of the Moon's gravitational field, could Apollo 11 – the long-awaited G mission, with Neil Armstrong in command – stand any chance of touching down on the lunar soil sometime in July. Despite the confidence, some doubts within NASA remained. If Apollo 11 did not succeed, the task of making the first landing would pass to Apollo 12 in September and if that failed there would be time for Apollo 13 in November to achieve Kennedy's deadline.

Even with the benefit of hindsight, four decades later, it remains remarkable that such enormous steps and astounding technical and human challenges were met and the *next steps*, of even *greater* challenge, set in motion within weeks. Some 4 percent of the federal budget (during the mid-Sixties, at least) had much to do with the hectic pace of Apollo's journey to the Moon, yet the ability of the United States to achieve a lunar landing on this timescale – barely eight years – certainly puts current projects to return humans to the lunar surface to shame. At the end of the first decade of the 21st century, we live in quite different times: war and ideological conflict remain, but the threatening presence of a competing superpower is now effectively gone from the United States' radar. In an ironic twist, closer and more cordial working ties with the

Soviet Union, cemented by the Strategic Arms Limitation Talks from 1969 onwards, eliminated the need for such a vigorous, all-or-nothing human space programme. During the mid to late Sixties, on the other hand, the race to the Moon was seen by NASA, quite literally, as 'peaceful warfare' with the Soviets and the astronauts were its frontline warriors. "We *were* at war," wrote Gene Cernan. "If they reached the Moon first, it would be Sputnik and Gagarin all over again, but much worse, and we would be the losers." Unlike Sputnik and Gagarin, failure to achieve the lunar landing would not simply be embarrassing, it would represent an abject failure to deliver on Kennedy's promise. As February wore into March 1969, the Apollo 9 crew – colds or no colds – could not launch soon enough.

Despite McDivitt's "mild respiratory viral infection", the processing flow for the mission had been exceptionally smooth. Rollout to Pad 39A had been performed on 3 January despite the need to replace nine aluminium brackets on the lunar module, lest they crack from stress corrosion. By 7:30 am on 1 March, when the clock resumed, the main concerns surrounded a low-pressure disturbance in the Gulf of Mexico, just south-west of the Cape, which was causing overcast weather conditions. As the countdown entered its final stretch, dull, storm-bringing stratocumulus clouds covered 70 percent of the sky, with darker blotches of altostratus high above them. Nonetheless, preparations continued. On the evening of 2 March, Scott and Schweickart drove out to the pad to watch as their Saturn V, floodlit by searchlights, underwent its final checkouts.

Next morning, clad in their pure white space suits and clutching their air-conditioning units like briefcases, the three astronauts arrived at the pad to a very different scene. The scores of engineers and technicians swarming up and down the elevators were now gone and it was an eerily silent place, the sense of stillness punctuated only by the hissing of the Saturn itself, which offered its own subtle hint that it would very soon come fully to life. As technicians helped Jim McDivitt into the left-hand seat of the command module and Rusty Schweickart into the seat over on the right, Scott waited patiently outside on the gantry and had a few precious minutes to gaze out across the marshy expanse of the Cape and watch the Sun peek over the horizon. "I could see it was a pretty morning," he wrote, "a beautiful scene."

Precisely on the stroke of 11:00 am, the five F-1 engines in the Saturn's first stage roared to life, raising a question in everyone's mind: had the rocket *risen* or had Florida *sunk*? This was the fourth orbital voyage of one of the largest and most powerful machines ever built; a machine which delivered in spadefuls the means and the mettle to send men to the Moon. Witnesses to Saturn V launches have employed various adjectives to describe its impact on them – its naked power, its noise, its light, its intense vibration – but they all agree that it was one of the most awesome experiences imaginable. On its maiden flight in November 1967, carrying the unmanned Apollo 4 spacecraft into orbit, the vibrations nearly shook Walter Cronkite's CBS News trailer to pieces ... more than *five kilometres* from the pad! Astronauts Tom Stafford and Mike Collins described the initial sensation as something one could *feel* through one's feet, rather than *hear* through one's ears ...

The merest mention of the name 'Saturn V' implies power. From a height, weight and payload-to-orbit standpoint, it remains the largest and most powerful rocket ever brought to operational status, although the Soviet Union's short-lived Energia booster had slightly more thrust at liftoff. It evolved from a series of rockets, originally dubbed Saturn 'C-1' through 'C-5', with NASA announcing in January 1962 its intention to build the latter. Soon thereafter, it was renamed the Saturn V. It was a three-stage vehicle with five F-1 engines on its first stage, five J-2 engines on its second stage and a single J-2 on its third stage. Both types of engine were built by Rocketdyne. These engines, when tested, had shattered the windows of nearby houses. The Saturn V was capable of delivering up to 118,000 kg into low-Earth orbit or placing 41,000 kg onto a lunar trajectory.

When a mockup of the rocket was rolled out to Pad 39A at Cape Kennedy on 25 May 1966, it amply demonstrated its colossal proportions. It stood 110.6 m tall – only a few centimetres shorter than St Paul's Cathedral in London – and 10 m wide across its first stage. It comprised an S-IC first stage, an S-II second stage and was topped by the S-IVB which would complete insertion into parking orbit, then be restarted to boost the Apollo spacecraft to the Moon with the six-and-a-half-minute translunar injection (TLI) burn. All three stages used liquid oxygen as the oxidiser. The fuel for the first stage was the refined kerosene known as RP-1, while the S-II and S-IVB utilised high-performance liquid hydrogen. It took 89 truckloads of liquid oxygen and 28 of liquid hydrogen, together with 27 railcars filled with RP-1, to fill up a single Saturn V.

The S-IC, built by Boeing, was 42 m tall and its five F-1 engines, arranged in a cross pattern, produced over 3.4 million kg of thrust to lift the vehicle to an altitude of 61 km. The centre engine was fixed, but the four 'outboard' engines could be gimballed to steer in flight. The S-II, built by North American, was 25 m tall and the same diameter as the first stage. It would make history as the largest cryogenic-fuelled rocket stage ever built. The S-IVB, built by the Douglas Aircraft Company, was narrower and only 17.85 m tall. On Apollo 9, for the first time, it provided a kind of 'garage' to house the folded-up lunar module. Above the S-IVB were the command and service modules, topped-off by the massive escape tower that would pull the command module clear in the event of a launch accident.

Today, more than three decades since its last flight, the Saturn V is renowned as one of the safest and most successful manned rockets; in 13 launches between November 1967 and May 1973, it never once failed to fulfil its mission, although on two occasions its performance was less than optimum. In the first case, Apollo 6 in April 1968, the S-IC suffered severe thrust fluctuations which caused the entire rocket to 'bounce' violently like a pogo stick. Then two of the S-II's J-2 engines shut down suddenly, requiring the others to burn for longer to compensate for the power loss. Finally, after achieving low orbit, the S-IVB refused to restart. It was because of these and other problems that Tom Paine, George Mueller and Jim Webb were initially so reluctant to approve Apollo 8's manned lunar shot. By the time Borman and his crew set off, however, all the problems had been overcome and their flight was nominal. Ten weeks later, McDivitt, Scott and Schweickart's ride to orbit was virtually trouble-free. "There were some oscillations in the S-II," admitted Deke Slayton, "but nothing serious."

Among the dignitaries in the viewing area of Cape Kennedy's firing room was Spiro 'Ted' Agnew, the newly inaugurated Vice-President, who was on hand to witness the Nixon administration's first manned launch. In time, both he and his boss would gain notoriety in United States political history – indeed, Agnew's proneness to making ill-considered gaffes had already caused embarrassment during the Republicans' election campaign – but the true extent of his criminality and the unsavoury exploits of 'Tricky Dicky' would not be revealed for several years. One of Agnew's key responsibilities in 1969 was to chair the Space Task Group, a panel charged with charting a direction for space exploration after Apollo. Sadly, his lack of real clout in the White House meant that only one of his four recommendations – the Space Shuttle – was actually endorsed by Nixon. Plans for lunar bases, expeditions to Mars and an expanded space station fell on deaf ears for a president who liked the 'heroic' image of astronauts, but whose actual enthusiasm and excitement for space travel was low. Unlike Kennedy (and, to a lesser extent, Johnson), Nixon saw little political, scientific or technological value in exploring the heavens. Still, during his ill-starred tenure in the vice-presidency, Ted Agnew was generally supportive of NASA and fostered close friendships with many astronauts, including Gene Cernan, the man who would gain the moniker of the last Apollo man to stand on the Moon.

As Agnew and others watched the Saturn V thunder into the clear morning sky, the perspective of the Apollo 9 crew was quite different. From their vantage point, 'inside' the very mouth of the behemoth, they could cast aside any notion of which of them had flown into space before and which had not: for on that morning of 3 March 1969, *all three* were rookies as far as riding the Saturn V was concerned. In his autobiography, Scott recounted that it was a fundamentally different experience from his Gemini VIII launch on a Titan II rocket – the forces and vibrations were significantly higher and the men were alternately thrown forward against their restraining straps and smashed back into their seats as one stage shut down and the next ignited. The entire episode, wrote Scott, was "like being compressed at the top of a giant spring".

At one point, McDivitt told fellow astronaut Stu Roosa, the capsule communicator (or 'capcom') at MSC in Houston, that the smoothness of the first stage was "an old lady's ride". That changed slightly, seven minutes into the ascent, when the first indications of pogo appeared in the S-II, somewhat greater than Frank Borman's crew had experienced, but neither McDivitt nor his two comrades expressed any concern. The S-II finally shut down at 11:08:57 am and the S-IVB took over, inserting Apollo 9 crisply into an orbit of 189 x 192 km, less than 15 minutes after lifting off.

However, the three men at the top of the giant spring had little time to 'unwind', for a full plate of work and a packed ten-day schedule lay ahead. First and foremost, less than three hours into the mission, Dave Scott – the astronaut whose chance to prove himself had been snatched away prematurely during the ill-fated Gemini VIII – would face his first major task. He had a date with a spider which he would soon come to describe as "the biggest, friendliest, funniest-looking" arachnid he had ever seen.

MAKE OR BREAK

Scott's task would effectively make or break the mission. Two hours and 41 minutes after launch, he separated the command and service modules from the spent S-IVB, whose adaptor panels were automatically jettisoned, exposing the chrysalis-like lunar module. Using the thruster quads on the side of the service module, Scott turned his ship 180 degrees and prepared to dock with the spidery lander. At this point, the first problem arose when he found that his translational thrusters were not functioning properly. "If we couldn't pull the lunar module out from its storage pod," he wrote, "we didn't have a mission. We had this short moment when it seemed it wasn't going to work." As he held the ship's position steady, McDivitt and Schweickart scrambled to identify the cause of the problem. Eventually, it was traced to several of the attitude engine indicators, whose propellant valves were showing up as 'closed'.

"After all the pre-flight testing of the valves," Scott continued, "this was certainly not supposed to happen. At one point ground control thought one of us must have bumped several of the switches closed as we were jostled around during launch, but we were strapped into our seats so tightly this was impossible. Later analysis concluded that the valves had flicked shut as a result of the shock caused by staging." The problem was fairly straightforward to rectify: McDivitt recycled the switches to 'open' and Scott was able to execute a perfect nose-to-nose docking at 2:02 pm, a little over three hours into the mission. The intricate procedure, known as 'transposition and docking', was critical: on missions to the Moon, it would be done *after* the S-IVB had performed the translunar injection burn; "and that," wrote astronaut Mike Collins, "was *no* time to discover that the scheme didn't work for some reason".

When one considers the sheer complexity of this first 'all-up' flight of the Apollo combo in Earth orbit, the chances of something going amiss were high. In the press kit, NASA identified no fewer than *seven* 'alternate' flight plans which McDivitt, Scott and Schweickart might pursue in the event of problems. Firstly, Alternate 'A' presupposed a problem with the extraction of the lunar module from the S-IVB and this would have enforced a full-duration, ten-day solo mission with the command and service modules. The other alternates applied to problems with Spider itself – covering everything from life-support system failures to a troublesome engine and from a faulty rendezvous radar to an unsafe descent stage – and in each case the crew would have performed as much as they could with the available set of options. Whatever happened, if possible, they would at least attempt station-keeping with the lander, Schweickart's EVA and a long burn of the descent and ascent engines.

In completing the first transposition and docking, the Apollo 9 crew had achieved the first American link-up between two pressurised, habitable spacecraft. The next order of business was to confirm that the docking probe and tunnel between the command and service modules and the lunar module was functional. The probe, wrote Deke Slayton, "was a complicated little mechanism that allowed you to latch two spacecraft together, then move from one to the other without doing EVA."

The lunar module Spider, housed inside the final stage of the Saturn V, is shown during transposition and docking with Gumdrop.

Unlike Khrunov and Yeliseyev's spacewalk in January, McDivitt and Schweickart would transfer between ships *internally*, in shirt-sleeves. "The probe and drogue, as we called it," continued Slayton, "was a tube that extended out from the nose of the command module and poked into a hole in the appropriate hatch on the lunar module. Once the probe was inserted, it would retract, pulling the two vehicles together in the proper alignment. The ring of the lunar module fit snugly inside the ring of the command module, which had a series of latches that clicked shut. Once you were sure there were no pressure leaks, you could disassemble the probe mechanism, store it inside the command module and swim through to the lunar module."

After pressurising the 1.2 m-long tunnel between the two craft, verifying the integrity of the 12 latches on the docking ring and connecting umbilicals to provide the lunar module with electrical power, it was time to separate from the S-IVB for good. Four hours into the mission, at 3:08 pm, Scott threw a switch which fired pyrotechnics to release the lunar module from the S-IVB and again pulsed the thruster quads to draw the complete Apollo craft free. Forty minutes after separation, the S-IVB's restartable J-2 engine was lit again to enter an 'intermediate coasting' orbit of 3,095 x 196 km, preparatory for a third and final burn. This final firing, which began some six hours into the mission and burned until the propellant supply was expended, allowed the S-IVB to escape Earth's gravitational influence and established it in a solar orbit with an aphelion of almost 150 million km and a perihelion of 130 million km. Major objectives of the repeat firings were to demonstrate the J-2's ability to restart and to test its performance under conditions beyond those for which it was designed.

In the meantime, McDivitt, Scott and Schweickart were preoccupied for the remainder of their first day in orbit with checks to ensure that the combined command, service and lunar modules were spaceworthy. Among their initial tasks were a series of firings of the Service Propulsion System (SPS) engine, one of whose primary roles on missions to the Moon would be to perform the lunar-orbit-insertion and transearth injection burns. The first SPS firing, conducted late on 3 March and lasting five seconds, verified that the combined spacecraft could withstand the stress and gauged their "oscillatory response". It performed as advertised, but the mass of the combined spacecraft – which formed the largest manned vehicle yet placed into orbit, weighing an impressive 37,085 kg – was evident; McDivitt would later recount that it took the entire five seconds to add 11 m/sec to their velocity. Next day, three more burns were performed, together with systems checks and pitch, roll and yaw exercises. The first two burns established Apollo 9 in an orbit of 509 x 207 km, while the third – which was initiated at 3:24:41 pm and lasted for 27.87 seconds – was a 'phasing' manoeuvre to shift the spacecraft into a better position for lighting, braking and docking later in the mission.

With this work complete, the time finally came for what Slayton called the mission's "big deal" – the checkout of the lunar module itself. After Schweickart's dramatic spacewalk, the lunar module was to undock for a series of joint rendezvous exercises. During his planned two-hour excursion, Schweickart would put the EVA suit with its bulky backpack, designed to enable astronauts to work on the lunar surface, through its paces. He was to exit through the small, square hatch and onto the lunar module's front 'porch'. As Scott had spent more than two years perfecting his knowledge of the command and service modules, so McDivitt and Schweickart had devoted themselves to understanding every inch of 'their' ship'. Each man had prepared exhaustively for his respective tasks. What they had not prepared for, or bargained on, was Schweickart's adverse reaction to the space environment, which almost stalled the mission in its tracks.

SPIDERMEN

When McDivitt and Schweickart removed the probe and drogue and floated through the connecting tunnel into the lunar module early on the morning of 5 March, their actions to prepare it for its first manned flight were honed to perfection by hundreds of hours of training. At 6:15 am, Schweickart entered its cramped cabin – about the size of a broom cupboard, dominated by the large, cylindrical ascent engine cover in the middle of its 'floor' – and was followed by McDivitt less than an hour later. Both men agreed that the whirring systems, particularly the environmental control unit, were noisy. By 8:00 am, the first major step in preparing their lunar module for independent flight was completed when its four spidery legs were swung away from the body. Time, however, was not on their side, and shortly thereafter McDivitt was forced to admit to Mission Control that they were behind schedule. The reason: Schweickart, two days into his first space flight, was sick.

'Space sickness' – today properly termed 'Space Adaptation Syndrome' (SAS) – is now known to affect around half of all astronauts and cosmonauts. Research over the last five decades has generally concluded that it is a kind of nauseous malaise, somewhat akin to motion sickness, which typically lasts no more than two or three days of a mission. It was first noted by Soviet cosmonaut Gherman Titov during his flight in August 1961 and usually manifests itself in sensations of disorientation and discomfort, coupled with dizziness and recurrent headaches. Even today, explanations and countermeasures for the condition remain imprecise. It appears to be aggravated by a subject's ability to move around freely in weightlessness and seems more prevalent in 'larger' spacecraft, as indicated by the fact that 60 percent of Shuttle astronauts report the complaint. Modern thinking postulates that the influence of weightlessness on the human vestibular system – the workings of the inner ear, which control balance – could offer a possible root cause. This disorientation arises, it is theorised, when sensations from the eyes and other areas of the body conflict with those from the vestibular apparatus and with information in the brain, derived from a lifetime of 'normality' in terrestrial gravity. A 'repatterning' of the central memory network occurs over the first few days, such that unfamiliar sensations from eyes and ears begin to be correctly interpreted. Today, motion sickness medicines have been shown to help, but are rarely used, because most space fliers prefer to adapt naturally in orbit, rather than risk starting their missions in a drowsy state.

Original plans called for Rusty Schweickart to make his way from the hatch of the lunar module Spider to the command module Gumdrop. These plans changed somewhat during the mission, when Schweickart suffered a severe bout of space sickness.

At the time of Schweickart's flight, the sickness was virtually unknown and the bulk of military fighter and test pilots in the astronaut corps, imbued as they were with seemingly limitless stores of testosterone and 'the right stuff', tended not to report experiencing it, lest their susceptibility impair their chances of being assigned further missions. "It had been accepted," wrote Gene Cernan, "that everyone felt woozy on getting up there and ... maybe you might even toss your cookies a couple of times, but you sure as hell didn't tell anyone and neither did your crewmates." Frank Borman only reluctantly admitted to throwing up all the way to the Moon during Apollo 8, swearing Jim Lovell and Bill Anders to secrecy. McDivitt's crew was determined to avoid such problems and, upon reaching orbit, tried not to make sudden head movements and took the anti-nausea drug Dramamine. "These precautions helped," wrote fellow astronaut Buzz Aldrin, "but they still felt dizzy and nauseous as they moved about the spacecraft."

The 'waves' of dizziness, Aldrin related, plagued the crew during their first couple of days in orbit, but space sickness really hit Schweickart during the effort to ready the lunar module for its flight on the morning of 5 March. He suffered a bout of nausea as soon as he awoke and then felt increasingly queasy as he donned his space suit in the weightless cabin, finding a vomit bag just in time. As Apollo 9's lunar module pilot, Schweickart performed his initial duties inside the lander, flipping switches to begin powering up its systems, but his condition steadily worsened. As Aldrin explained, "Rusty ... experienced brief vertigo as he floated *up* through the tunnel into the LM, and ended up staring *down* at the lander's flight deck." When McDivitt – who had also suffered from episodes of dizziness – joined his crewmate, Schweickart vomited again.

Unlike Borman, it was impossible, and indeed unsafe, to conceal Schweickart's condition from ground controllers. McDivitt knew that the timeline called for Schweickart to make his spacewalk on 6 March. "Throwing up inside a pressure suit," explained Deke Slayton, "would not only be unpleasant as hell, it might be fatal." Another problem lurked in the shadows, too. The very act of admitting that one of them was sick, in Gene Cernan's words, "was to admit a weakness, not only to the public and the other astros, but also to the doctors, which would give them reason to stick more pins in us". In the hyper-competitive fraternity of NASA's astronaut corps, racing against the decade to beat the Soviets to the Moon, weakness of any description was not tolerated. Although Cernan would later admit that Schweickart's sickness opened the door to closer medical exploration of the condition, ultimately, this civilian research scientist who became an astronaut "paid the price for us all. Nothing was ever said in public against him, but he never flew another mission".

Nonetheless, Schweickart's efforts to find a root cause for the ailment in the wake of Apollo 9 – even at the expense of missing out on the chance to fly again – were highly commendable. "After our mission was over," wrote Dave Scott, "Rusty volunteered to undergo research into it at the US military's special centre on airborne sickness at Pensacola Naval Air Station in Florida. It was a brave undertaking." For many months, Schweickart was subjected, on at least a weekly basis, to the torment of rotating chairs, tilting rooms and balance beams to induce sickness. No reliable indicators or predictors were found, with some astronauts sensitive to the malaise in tests and fine in space and others who were resistant on the ground and, as Cernan put it, "tossing their cookies" in flight. One technique used to some effect by Schweickart as a physical countermeasure on Apollo 9 was trying to adapt to the strange microgravity environment as rapidly as possible. His strategy for doing this involved repeatedly moving his head to stimulate the symptoms, then stopping close to the brink of sickness and holding himself still. "What he did not know," continued Scott, "was that this simply delayed adaptation until he was really needed on the third day of our mission." Many of the mysteries and inconsistencies of Space Adaptation Syndrome remain present today.

One of the greatest ironies, as ever more astronauts ventured into space, was that they were steadily becoming anonymous. Gone were the early days in which the likes of Al Shepard and John Glenn were recognisable anywhere. By the spring of 1969, six groups had been chosen, including two batches of scientists, and NASA's astronaut corps had more than 50 members. In his autobiography, Gene Cernan noted that areas of Houston known to be populated by astronaut families – Timber Cove, El Lago, Nassau Bay and Clear Lake – were frequently visited by tour buses and camera-clicking sightseers. When asked by the bus drivers if there were any astronauts living nearby, Cernan would scratch his head, point down the road and respond, "I think a couple of 'em live over yonder somewhere". Jim McDivitt and Rusty Schweickart, the two men in charge of putting the lunar module through its paces and playing a crucial role in making the first manned Moon landing possible, did exactly the same. "If we weren't in space suits," Cernan wrote, "no one recognised us."

Certainly, appearance-wise, 33-year-old Russell Louis Schweickart had the irreverent air of a scientist, rather than the bravado of a test pilot, and his most noticeable physical trait was his mop of wavy red hair. Born in Neptune, New Jersey, on 25 October 1935, his education and the professional career which guided him into the hallowed ranks of NASA's astronaut corps was quite different from his military brethren: after finishing Manasquan High School in his home state, he earned a bachelor's degree in aeronautical engineering in 1956 and a master's in aeronautics and astronautics in 1963, both from MIT. After finishing his undergraduate degree, he served for seven years as an Air Force pilot and a member of the Air National Guard, accumulating 4,200 hours of flight time, including 3,500 hours in high-performance jets. He also worked as a researcher in MIT's experimental astronomy laboratory, undertaking work in upper atmospheric physics, star tracking and the stabilisation of stellar images. Indeed, his master's thesis, completed the very year that NASA picked him as one of its third class of astronauts, focused on the experimental validation of theoretical models of stratospheric radiance.

Despite belonging to the same astronaut group, Dave Scott and Schweickart were polar opposites; the former an Air Force test pilot, the latter a man with a military background, but primarily a scientist and researcher. When NASA announced its intention to select at least a dozen new fliers in June 1963, the press release explicitly stated that the agency was looking for both military and civilian candidates. Nonetheless, each successful applicant needed at least 1,000 flying hours in his logbook and a bachelor's credential in engineering or physical or biological sciences; furthermore, they could be no more than 35 years of age, no more than 1.83 m tall and, naturally, had to be in excellent physical condition. On 18 October, Scott and Schweickart's names were published on NASA's list, though they can scarcely have expected to someday fly together.

By the spring of 1965, Schweickart had already been given long-term duties for Project Apollo, focusing on future experiments, and in March of the following year was assigned with McDivitt and Scott to the backup crew for its maiden voyage. When a team of other astronauts took over this role later in 1966, Schweickart, McDivitt and Scott found themselves pointed at the D mission for the first time. Opinions of the pilot-cum-scientist were mixed. Dave Scott described him as "a highly cultured man, very liberal, widely read" (and even admitted that his daughter, Tracy, took a shine to one of Schweickart's twin sons), whereas fellow 1963 selectee Gene Cernan referred to him, somewhat disparagingly, as a "red-headed kid with a sharp needle, looking for a balloon to pop".

Clearly, the intense competition within the astronaut corps in those heady days of the mid to late Sixties dictated both actions and attitudes and feelings ran deep. Cernan's judgement of Schweickart (and also of fellow astronauts Walt Cunningham and Buzz Aldrin, both of whom also combined fighter-pilot experience with advanced scientific or engineering credentials) seem to have been rooted almost entirely in the perception of them being 'scientists'. When Deke Slayton asked the astronauts to peer-review each other's competencies and capabilities, Cernan put Dave Scott at the top of his list and Rusty Schweickart near the bottom. The hapless scientist's recurrent bouts of space sickness on Apollo 9 must surely have reinforced in the military fliers' minds the old stereotype: that, unlike the steely test pilots, the civilians were inferior and remained several steps behind.

At the opposite extreme was James Alton McDivitt, veteran astronaut, colonel in the Air Force, outstanding test pilot and to this day one of only a handful of American fliers to have commanded every one of his space missions. (Indeed, his desire *not* to fly right-seat to anyone later cost him the chance to walk on the Moon.) On his first flight, Gemini IV in June 1965, he had been the 'Inside Man', running the spacecraft whilst Ed White performed the United States' first spacewalk. It was a role he would reprise on Apollo 9, handling the lunar module whilst Schweickart ventured outside. Described by *Time* as "whippet-lean", McDivitt's 1.8 m frame made him one of NASA's tallest astronauts.

Born in Chicago, Illinois, on 10 June 1929, the son of an electrical engineer and the first Roman Catholic astronaut, his background did not immediately mark him out as an obvious holder of the 'right stuff' for a budding spacefarer. After graduating from high school in Kalamazoo, Michigan, he worked for a year as a furnace repairman, then drifted into college in 1948, vaguely describing his ambitions for the future as either a novelist or an explorer. Two years later, he completed his education and opted to enter the Air Force, discovering a love of aviation whilst flying 145 combat missions over Korea. His achievements there were rewarded with three Distinguished Flying Crosses and five Air Medals. In 1957, McDivitt was sent by his parent service to read for an aeronautical engineering degree at the University of Michigan, where he proved himself to be a straight-A student, graduating first in his 607-strong class. He was selected for the Experimental Test Pilot School at Edwards Air Force Base in California and seemed a likely candidate to fly the X-15 rocket-plane, but applied instead for the 1962 astronaut class. Although *Time* labelled him "a superb pilot and a first-class engineer", McDivitt approached NASA from a purely practical and technical standpoint. "There's no magnet drawing me to the stars," he has been quoted as saying. "I look on this whole project as a real difficult technical problem – one that will require a lot of answers that must be acquired logically and in a step-by-step manner."

In the months after flying Gemini IV, he was assigned technical duties to oversee the development of the lunar module, a spacecraft with which he would become infinitely familiar by the time he flew it for real in March 1969. "Jim would look things over and very carefully make a decision," Deke Slayton wrote of him; and, indeed, McDivitt's response to the Apollo 9/8 switch is well known. Some, including Gene Cernan, believe that he *was* given first refusal to fly the circumlunar Apollo 8, but turned it down. Cernan thought McDivitt mad to refuse such an offer, although Dave Scott felt that his commander's decision was the right one: exciting as the circumlunar flight was, "it was not going to offer the same challenges of EVA, rendezvous and docking that we had been training so hard for". Jim McDivitt himself, of course, has asserted repeatedly over the years that no such 'offer' was ever tabled by Slayton. It will undoubtedly pass into the annals of space legend, but highlights two key facets of McDivitt's personality: firstly that he was a *test pilot* and secondly that his devotion to The Mission was unshakable.

These two men, the Test Pilot and the Scientist, an unlikely pairing, perhaps, were now tasked with putting the lunar module through its paces. In doing so, they quite literally earned themselves the moniker 'Spidermen'; the appearance of the lander having been likened to a bug or some sort of oversized arachnid. Yet there was a serious side. On Apollo 9, for the first time in the United States' space programme, *two* manned craft would rendezvous, dock and transfer crews, which demanded individual names. There were, admittedly, formal international designations – the main spacecraft was labelled '1969-018A' after reaching orbit, the S-IVB became '1969-018B', the crew-carrying ascent stage of the lunar module became '1969-018C' and its discarded descent stage became '1969-018D' ... but those, surely, would not do. *Names* for the two manned craft were definitely needed.

Frivolous or 'sensitive' names were frowned upon by NASA. Many managers had been unhappy when Gus Grissom wanted to call his Gemini 3 craft 'Molly Brown' and had also disliked Gordo Cooper's suggestion for Gemini V of 'Lady Bird' – a pejorative jab at then-President Lyndon Johnson's wife. Three years later, Wally Schirra's proposal to nickname his ship 'The Phoenix', in respectful recognition of the three astronauts who died in the Apollo 1 fire, was similarly vetoed. Yet for Apollo 9, names were necessary. The ungainly appearance of the lunar module and the conical profile of the command module made the choice of names an easy one for McDivitt, Scott and Schweickart: the former would be called 'Spider' and the latter 'Gumdrop'. Two possible reasons have come to light for the Gumdrop name: firstly, and quite obviously, the command module closely resembled the cone-shaped, sugar-encrusted sweet, and secondly, it may have been on account of the blue-coloured cellophane wrapping in which that craft was delivered to Cape Kennedy. Although introduced as nicknames, extensive use by the astronauts and their training teams led to their becoming official radio callsigns. This precedent would open up the way to a pair of even more frivolous names on Apollo 10, before some semblance of dignified, patriotic order returned in time for the historic landing mission in July.

ADVENTURE OF 'THE RED ROVER'

Aboard Spider, concern mounted about how to tackle Schweickart's forthcoming EVA. In a bid to preserve his comrade's privacy, McDivitt requested a closed-loop medical consultation over a discreet radio channel to Mission Control – causing the hundreds of gathered journalists to fly into a frenzy, making up all kinds of stories as to what might be going on – and it was decided that a planned spacewalk from Spider over to Gumdrop was too risky. The chances of Schweickart suffocating if he threw up again, *inside* his suit, did not bear thinking about. The plan called for him to spend two hours and 15 minutes outside, exiting the lunar module and working his way by handrails over to the open hatch of Gumdrop, where a fully suited Dave Scott would conduct his own 'stand-up' EVA to observe. Schweickart would then return to the lunar module. The purpose of this test, in addition to evaluating the suit, was to show that a returning lunar crew could spacewalk over to the command module in the event of being unable to pass through the docking tunnel.

Instead, a comparatively straightforward opening of the hatch for three-quarters of an hour during orbital daylight was advocated. By thus exposing themselves to vacuum, but remaining *inside* Spider, McDivitt and Schweickart could conduct at least some tests of the suits in the required conditions. As managers hurriedly rescheduled and reprioritised the EVA, the lingering question of how much information to release to the media had worked its way up NASA's chain of command to newly appointed Administrator Tom Paine himself. Deke Slayton, naturally, defended 'his' astronauts and declared that the tape of Schweickart's private medical consultation should not be divulged. Bob Gilruth, head of the Manned Spacecraft Center, disagreed, but finally acquiesced that the tape itself should be kept back and a paraphrased account issued to the journalists. Many astronauts were furious, with one declaring that he would "never tell the ground a goddamn thing from up there". Eventually, Paine concurred that the confidentiality of the Schweickart tape should be respected. By this time, sadly, some of the gathered media had convinced themselves that NASA was suppressing vital information, which can hardly have placed Paine – who had been sworn into office by President Nixon several weeks earlier – in a positive light.

If the spacewalk happened – and early on 5 March it seemed unlikely – the astronauts' third day in space was far too busy for them to worry about it. Their packed schedule of engineering and other objectives got underway at 9:28 am, with a five-minute televised transmission from inside Spider, showing its instruments and displays, various internal features and the faces of McDivitt and Schweickart themselves. Managers were suitably impressed by the quality of the images, but not by the sound, which they considered less than satisfactory. Three hours later, at 12:41 pm, McDivitt executed the first firing of the descent engine. In addition to evaluating the handling characteristics of the combo, the six-minute burn demonstrated the effectiveness of the lunar module's digital autopilot and how the descent engine behaved at full throttle. He was more than happy with the descent engine's performance. Only seconds after starting the burn, he yelled "Look at that [attitude] ball; my God, we hardly have any errors". Twenty-six seconds into the firing, those errors remained virtually non-existent and the commander even took a few seconds for a bite to eat.

Shortly after shutting down the lander's systems, the two men transferred back to the command module to perform another SPS firing – the mission's fifth in total – which circularised their orbit at 240 km in readiness for the undocking and rendezvous with Spider.

Schweickart awoke on 6 March, apparently much recovered, no longer nauseous or pale, and McDivitt, notwithstanding the reservations of ground controllers, decided to press ahead with his scheduled EVA onto the lunar module's porch. The cabin was duly depressurised at 11:45 am, although McDivitt found that he had to exert more force than expected to turn the handle and swing the waist-high hatch inwards. Clad in a bulky suit virtually identical to the one which astronauts would use on the Moon and anchored by means of a 7.6 m tether, Schweickart began moving onto the porch a little over 14 minutes later.

In essence, he was now a miniature spacecraft in his own right. The suit, designated 'A7L', was actually the seventh Apollo ensemble built by ILC Dover in Delaware. It incorporated significant improvements over earlier designs, including integrated thermal and micrometeoroid layering, and following the January 1967 launch pad accident had also been upgraded to be fireproof. Outwardly, it consisted of a single-piece, five-layer 'torso-limb' structure with convoluted joints of synthetic rubber at its shoulders, elbows, wrists, hips, ankles and knees and a system of 'link-net' mesh to prevent it ballooning under pressure. Metallic rings at its neck and forearms provided for the attachment of helmet and gloves. Astronauts entered the suit by means of a vertical zip which ran from the shoulder blades down the spine to the crotch. For the early missions – up to and including the third Moon landing in the spring of 1971 – the entire ensemble weighed 85 kg when fully charged with oxygen and other consumables. Later crews, who would spend up to three days on the lunar surface, would benefit from a more advanced design.

The torso-limb assembly worn by Schweickart featured six life-support connectors in two parallel columns on the chest, four supplying oxygen and the others providing hook-ups for electrical headsets, biomedical sensors and water coolant. Outside this was the 13-layered integrated thermal micrometeoroid garment, which protected the suit from abrasion, thermal solar radiation and micrometeoroid impacts; the content of its layers ranged from rubber and nylon to aluminised Mylar and non-woven Dacron and finally topped-off with fire-retardant Teflon-coated beta filament cloth. It also included a patch of Chromel-R woven steel to prevent abrasion caused by the suit's bulky backpack, the Portable Life Support System (PLSS).

Beneath all this bulk, Schweickart wore a form-fitting outfit which looked for all the world like a pair of long johns, threaded with water-carrying tubes of polyvinyl chloride. Composed of nylon and Spandex, this 'liquid-cooling and ventilation garment' utilised the tubing, together with a heat exchanger and sublimator in the PLSS, to keep its wearer's body temperature at a comfortable level. It minimised sweating, kept the head cool (indeed, some astronauts felt it made them *too* cool) and by preventing the build up of humidity in the helmet stopped the visor from fogging. In precis, it circulated cool water through the tubes, in direct contact with the skin, drawing heat away from the body to lower its core temperature. The water was returned to the PLSS to be cooled by the heat exchanger ready for recirculation. The PLSS, meanwhile, provided the astronaut with oxygen, communications gear, batteries and lithium hydroxide and charcoal filters and scrubbers. More importantly, it provided him with protection from the thermal and radiation extremes of the lethal space environment. Its pair of high-pressure oxygen tanks could support the wearer for some five hours, whilst an additional 'oxygen purge system', mounted atop the PLSS, directly behind the astronaut's helmet, provided an emergency 30-minute bailout in case the main supply failed. "What was important about this EVA," wrote Deke Slayton, "was that the lunar pressure suit was completely self-contained. All the suits used on the Gemini EVAs had relied on the

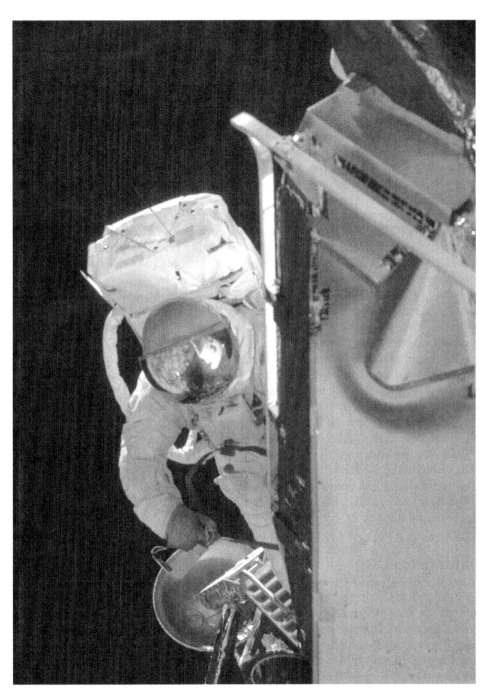

Rusty Schweickart, clad in a lunar-style EVA suit with the massive backpack, emerges onto Spider's porch. Note the circular footpad, visible just below his right hand.

spacecraft to provide oxygen and communications. The consumables and communication equipment for the lunar suit ... were all in [the] backpack."

With the satisfying gurgle of water coolant and a stable pressure indicator, Schweickart lost no time. As soon as he was outside, he secured his feet in a pair of so-called 'golden slippers' – boot restraints attached to Spider's porch, painted gold for thermal reasons – and gained his bearings before embarking on his first task: to observe, photograph and retrieve thermal material samples from the exterior of the lunar module. By now, Dave Scott, clad in a slightly different suit and dependent on Gumdrop's systems for life support, had opened *his* hatch for a 'stand-up' EVA. In some ways, his work mirrored that of Schweickart, albeit that *he* was responsible for collecting thermal samples from the outside of the command and service modules.

Next, at 12:39 pm, Schweickart – picking fun at his own hair colour by aptly naming himself 'The Red Rover' – began his first abbreviated attempt to evaluate his ability to move and control his body in the lunar suit. His planned transfer to Gumdrop and back had been cancelled, but he was able to obtain 16 mm Maurer motion-picture and 70 mm Hasselblad still photography of Scott's activities and imagery of the exteriors of both spacecraft. By 12:52 pm, he was back inside Spider. Scott retreated into Gumdrop and repressurised its cabin ten minutes later. Both excursions were successful, with a great improvement in the pictures and sound quality, compared to the previous day.

One sound that the astronauts apparently did *not* hear, was the repeated calling from Mission Control; indeed, they were so preoccupied and talkative that the duty capcom was forced to radio ten times – "Red Rover, do you read? Gumdrop, do you read? Hey, does *anyone* up there read me?" – before he received a response. The four-way radio chatter between Schweickart, Spider, Gumdrop and Houston was all loud-and-clear. The main discovery was that leaving the lander in a pressurised suit was more straightforward than expected and the depressurisation and repressurisation of the lunar module's cabin ran without incident. Another key hurdle on the road to the Moon had been overcome. Said NASA's public affairs officer Jack Riley: "You heard it here, live, first-hand – the adventures of Red Rover and his friends, Spider and Gumdrop!"

ORBITAL BALLET

Still to be tested were Spider's systems when flying independently of the command and service modules, including the rendezvous radar, descent and ascent engines, guidance computers and the docking mechanism. Not long after 3:00 am on 7 March, keen to get an early start and move ahead on their timeline, McDivitt and Schweickart shimmied down the tunnel into the lander and sealed the hatches between themselves and Gumdrop. At the appointed time, Scott flipped a switch to release Spider ... and nothing happened. The latches, it seemed, had 'hung-up'. Following several more tries, in which the CMP repeatedly flipped the switch back and forth, he was successful and the two craft separated cleanly at 7:39 am. Then, after performing a brief fly-around to ensure that all was well, Scott pulsed his thrusters to back away.

From his station on the left-hand side of the lunar module, McDivitt performed a 90-degree pitch manoeuvre, then yawed 360 degrees, to enable Scott to verify that its four legs were properly extended. Forty-five minutes later, McDivitt fired the descent engine for 19 seconds to insert Spider into a circular orbit some 20 km 'higher' than Gumdrop. As he throttled the engine for the first time, it ran smoothly until it achieved 10 percent thrust, but when advanced to 20 percent both astronauts noted a peculiar chugging noise. McDivitt paused, then resumed, and was delighted that the strange groaning did not recur, even when he throttled up to 40 percent.

Next, the pyrotechnics were fired to jettison the descent stage – McDivitt called the sensation "sort of a kick in the fanny ... but it went alright" – and the ascent engine blazed silently to life to place Spider into an orbit 16 km 'beneath' and 120 km 'behind' Gumdrop. The kick affected more than just the crew: as the two sections of the vehicle separated, a small cloud of debris hit the ascent stage and knocked out its strobe tracking beacon. In accordance with the laws of celestial mechanics, the lander, being in a lower orbit, began to gain on its quarry. Despite the apparent grace of this orbital ballet, the dangers were still present. Spider had no heat shield and could not return the crew safely to Earth, which meant that if McDivitt and Schweickart encountered insurmountable difficulties, Scott would have to rescue them. Even from a distance of 160 km or more, he could 'see' the lunar module's shiny surfaces using Gumdrop's sextant, although on one occasion, in orbital darkness, he briefly lost sight of them.

A few hours later, at 2:02 pm, the first all-up demonstration of the lunar module in flight was completed in spectacular fashion, with Scott lining up and McDivitt executing an almost-perfect docking. *Almost* perfect, that is, because lighting conditions were less than ideal: the Sun was shining directly through the small rendezvous window above McDivitt's head. "Using my alignment device," Scott wrote, "I talked him through the manoeuvres and he was right on target, though there were more than a few tense moments." Retracting the docking probe produced the welcoming 'ripple-bang' of capture latches engaging. McDivitt quipped that "that wasn't a docking ... *that* was an eye test!" During their six-and-a-half-hour independent flight, the two men aboard Spider had ventured as far as 183 km from Gumdrop and had cleared another obstacle on the road to the Moon.

Two hours after docking, McDivitt and Schweickart rejoined Scott inside Gumdrop for what would be a fairly relaxed final few days in orbit. At 4:22 pm, the now-unneeded ascent stage was jettisoned and one of the astronauts wistfully remarked that they hoped they had not left anything important behind. When the capcom asked if they had left the LMP – Schweickart – aboard by mistake, McDivitt replied that, no, "I didn't forget him ... I left him there on purpose!" Humour aside, a few important (and expensive) items *were* left in Spider's cabin, simply because there was no way to safely store them in the command module for re-entry. These included the Hasselblad and the Maurer, together with their television camera, whose cost amounted to somewhere in the region of $450,000 ...

Spider's ascent stage, viewed by Dave Scott from the windows of the command module Gumdrop. Clearly visible is the ascent engine and the square hatch.

The two parts of Spider would ultimately meet the same fiery fate, burning up in the dense layers of Earth's atmosphere, although at very different times. The descent stage lasted barely a couple of weeks, re-entering on 23 March, its remnants showering into the Indian Ocean just off the coast of eastern Africa. For the ascent stage, on the other hand, more than a decade would pass before it finally took its destructive plunge. Shortly after being jettisoned by McDivitt's crew, its engine was fired for just over six minutes to put it into an eccentric orbit with an apogee a couple of thousand kilometres high. Trajectory specialists predicted that it might burn up in the atmosphere sometime in the mid Seventies, but not until 23 October 1981 – this author's fifth birthday – did the last relic of what had been the Spider finally return to Earth in a firestorm of glowing debris.

As for the Apollo 9 crew, with the jettisoning of the ascent stage more than 97 percent of the mission was over and their final days were devoted to catching up on scientific and navigation experiments, Earth observations ... and some well-earned rest. A brief SPS firing, originally scheduled for 12:48 pm on 8 March, was postponed by two hours when the crew discovered that the reaction control translation needed for 'propellant settling' prior to ignition was improperly programmed. When it did occur, the 1.43-second burn lowered Gumdrop's perigee in order to provide better propellant margins for a contingency re-entry.

In his autobiography, Dave Scott recalled McDivitt turning to him on their final night and telling him that Apollo 9 would be his last mission. The exhaustion caused by the tremendous responsibility and the burden of command had taken its toll. "It was easy to burn out on missions," Scott wrote, "get permanently tired and not want to fly again. The great NASA team made them look too easy. They were really, really hard." McDivitt's next steps within the space programme would take him into senior management.

Scott himself, however, could not have felt more differently. Following his truncated ten hours in orbit and the loss of his spacewalk on Gemini VIII, the opportunity to spend ten *days* aloft on Apollo 9 and conduct an EVA, albeit only a stand-up one, gave him the chance to really relish being in orbit. "I really appreciated during this second time in space the astonishing beauty of the stars and the planet Earth," he wrote. "At one point, we dimmed the lights in the spacecraft so that we could get the best possible view of the amazing vista that lay before and below us. Most striking of all was the clarity with which it was possible to see streaks of lightning piercing the clouds during thunderstorms in the Earth's atmosphere, both along weather fronts and in swirling tropical air masses. Between flashes we could also see subtle streaks in the sky below us. They were so subtle that we weren't sure at first what they were, or if the others had noticed them, too."

It was Rusty Schweickart, the mission's rookie, who first mentioned these streaks, which turned out to be meteors burning up as they entered the atmosphere. He, too, was astounded by the experience. His work in MIT's experimental astronomy laboratory before joining NASA led him to have many interesting discussions with Scott as they beheld the enormity of the heavens, spread out around and 'above' them. Yet one of the most amazing observations for Schweickart was the Earth ... not just its astounding beauty, but the speed at which Apollo 9 was passing over it all. Writing in a 1977 article entitled 'No Frames, No Boundaries', he related with wonderful clarity the experience of circling the world in just 90 minutes: awakening over North Africa, breakfasting over the Mediterranean and the Middle East – recognising that, laid out, map-like, were all the millennia of human history – before sweeping serenely across the Indian subcontinent, past Burma and south-east Asia and into the deep-blue panoramic vastness of the Pacific. Despite the sense of separation, Schweickart wrote, the world felt 'friendly' to him and he felt part of it all.

"You look down there," he explained, "and you can't imagine how many borders and boundaries you cross, again and again and again, and you don't even see them. There you are – hundreds of people in the Middle East killing each other over some imaginary line that you're not even aware of, that you can't see. And from where you see it, the thing is a whole, the Earth is a whole, and it's so beautiful. You wish you could take a person in each hand, one from each side in the various conflicts, and say: 'Look. Look at it from *this* perspective. Look at that. *What's* important?'" Forty years on, Schweickart's words resound with a haunting echo, for our generation continues to see hundreds of people killed daily – and not just in the Middle East – over boundaries, religious faiths, ideas, ideologies and intolerances. Hatred, it seems, is just as prevalent a force in human nature as kindness; fear and oppression as powerful and influential as peace and prosperity. Many astronauts and cosmonauts have shared Schweickart's sentiment: the fragility of Earth should not – and *cannot* – be taken for granted and that our impact on it will play a part in deciding our species' ultimate fate. Perhaps the 'opening up' of space travel in the coming years, through private enterprise and fee-paying flights, will have an ever increasing impact in opening more eyes and minds to the unimportance of humanity's petty squabbles.

At the beginning of March 1969, as today, such squabbles dominated the headlines; in 2010, we concern ourselves daily with Iraq, Afghanistan and a handful of 'rogue states' – Iran and North Korea being the most prominent – that threaten Western perceptions of world peace and security, whereas back *then* the main military and geopolitical issues were focused on the Middle East, the Soviet bloc and, of course, the bloody conflict in Vietnam. As one looks back, with an unashamedly cynical eye, over half a century of human stewardship in space, it makes one wonder *what*, if *anything*, we have learned about our responsibility to what Jim Lovell once called this "grand oasis in the vastness of space", because our quarrels, our xenophobic fears and, above all, our wars, continue.

At the opposite extreme, another sentiment also came across in Schweickart's words: that on this blue, green and brown marble, parts of which he could literally cover at will behind his thumb, lived and breathed *everything* that he held dear: his family, his friends and, above all, his sense of *home*. "You think about what you're experiencing, and why," he wrote. "Do you deserve this? Have you earned this in some way? Are you separated out to be touched by God, to have some special experience that others cannot have? And you know the answer to that is No. There's *nothing* you've done that deserves this experience. It's not a special thing just for you. You know very well at that moment, for it comes through to you so powerfully, that you are the 'sensing element' for all of humanity: you, as an individual, are experiencing this for everyone. It's a feeling that says you have a *responsibility* ... not for yourself."

Providing a musical backdrop for such philosophical musings, at least on the day before landing, was Schweickart's cassette of the cantata *Hodie* – Vaughan Williams' last major choral-orchestral composition – which Dave Scott had hidden in one of his pockets for fun. "He never forgave me for that," Scott wrote.

Also fun, it seemed, and quite beautiful in a strange way, were the mission's urine dumps ... the periodic release of the crew's waste through a vent valve in the side of the command module. "It's really spectacular," Schweickart admitted to *Time* in January 1978, explaining that as soon as the urine escaped into the frigid cold of space, it froze into a shower of tiny ice crystals, forming a 'hemisphere' and spraying in all directions. Spectacular as the dumps may have been, the process of ridding oneself of one's wastes proved somewhat less entertaining. The Apollo urine-collection device looked innocuous enough, taking the form of a hose and condom-like fitting ... but the system in place for the men's 'other' toiletry need was anything but discreet and anything but pleasant. The 'defection bags' required 45–60 minutes to use and their very nature prompted many astronauts to postpone their 'need to go' for as long as possible. Bill Anders, who flew Apollo 8 to the Moon, would later tell Andrew Chaikin that, since nothing 'falls' in the microgravity of space, it was necessary to "flypaper this thing to your rear end and then reach in there with your finger – and suddenly you were wishing you'd never left home!" Adding insult to injury, the blue germicidal tablets, which prevented bacterial and gas formation, had to be *kneaded* into the contents of the defecation bag to ensure a good mix. The used bags were then sealed and stored in empty food containers in Gumdrop's lower equipment bay, but they still produced unpleasant odours. With more than a hint of humour, many astronauts who followed would pay tribute to the difficulties that these pioneers of the Sixties and Seventies – these 'real men' – had to endure.

Amidst all this clutter and grime and discomfort, the astronauts of Apollo 9 spent their last five days sharing their lives with a battery of photographic and scientific equipment in the command module. They performed a wide range of Earth resources observations, in particular multispectral images of the southern United States, together with Brazil, Mexico and parts of Africa. Simultaneous photographs were taken using four different combinations of films and filters with outstanding results, thanks to the long period of time available to wait for cloud cover to pass and Gumdrop's orbital inclination of 33.6 degrees, which allowed vertical and near-vertical coverage of never-before-imaged areas. The command module's windows, too, provided an almost crystal clear platform for these photographic sessions. Of particular note were the infrared images, whose quality was sufficient to distinguish locations of diseased vegetation in areas of healthy growth, and it was remarked at the time that such results could lay the foundations for a future Earth resources satellite.

Other work included the successful tracking of the Pegasus III micrometeroid-detection satellite on 11 March and using the planet Jupiter as a point of

Recovery of the command module Gumdrop on 13 March 1969.

navigational reference for their inertial measurement unit. Navigation, indeed, was a primary focus in those closing days, with the astronauts obtaining fixes on islands, capes and various other landmarks to establish Apollo 9's exact position in orbit. Tiredness, though, was setting in and having completed each and every one of their objectives, Mission Control finally cut McDivitt, Scott and Schweickart some slack and gave them some opportunity for well-deserved rest. On Sunday 9 March, one NASA official told the press that "the big events of today are the sleep cycle and the wake-up period!" The following morning, when the crew failed to call Houston at the scheduled time, flight controllers allowed them to sleep for another two hours ...

Despite the success of the mission, Apollo 9's return to Earth came an orbit later than planned, due to unacceptable weather in the prime recovery zone, near Bermuda in the Atlantic. In fact, whilst carrying out one of their photography tasks, the men had noticed whitecaps brewing in the area and it was with some reluctance that the capcom admitted to McDivitt that wind speeds and waves were definitely outside NASA's mandated safety limits. By the middle of the week, a day or so before landing, the decision was made to postpone re-entry by one revolution and bring the command module into the ocean in the calmer waters near Grand Turk Island in the Bahamas. The SPS engine was duly fired for the eighth and final time at 11:31 am on 13 March, a little more than ten days since Apollo 9 left Earth, to bring McDivitt, Scott and Schweickart home. Five minutes later, the service module was jettisoned and the command module continued on alone to plunge into the atmosphere. Splashdown came at 12:01 pm, within sight of the helicopter carrier

Guadalcanal, some 290 km east of the Bahamas. "TV cameras on the deck," *Time* reported a few days later, "zoomed in to show [the] astronauts ... tumbling into inflated rubber rafts – a surprisingly awkward operation after the precise manoeuvres and sophisticated procedure of space flight."

Ocean swells, whipped up by the rotors of the rescue helicopters, caused the cage-like sling meant to haul the crew to safety to miss its target several times. "When Scott was finally able to hitch a ride after ten misses," *Time* continued, "the cage swung widely back and forth in stomach-churning arcs as it was lifted to the helicopter." Schweickart, who had restricted his dietary intake to fruit and liquids during his final days in orbit, was similarly thrown around and McDivitt had to seek refuge on the flotation collar when the wind whipped across his raft. He earned himself a thorough soaking and a dizzying spin, before clambering to safety in the waiting chopper.

In the wake of a mission which Deke Slayton would later say had placed America "one giant step closer to going to the Moon", the spectacular success of Apollo 9 provoked an enormous wave of optimism about the upcoming lunar landing. In some circles within NASA, it was suggested that the landing – tentatively scheduled for mid-July – might be brought forward, perhaps to June. Officially, the next flight, Apollo 10, called for a full dress-rehearsal in orbit around the Moon, with Tom Stafford and Gene Cernan piloting their lunar module to an altitude of just 15 km above the surface. Mutterings abounded in the days after the return of McDivitt's crew that Stafford's team should attempt to land. For various reasons (discussed below) this would have been both unwise and impractical, but it certainly illustrated the electrifying mood within NASA at the time. This civilian agency, still barely ten years old, was finally drawing close to making good on John Kennedy's challenge to place American boots on the Moon within the decade and so accomplish the greatest scientific and engineering feat ever attempted in human history. "Whatever the decision," wrote *Time*, "there is now more confidence than ever that US astronauts will be walking on the surface of the Moon this summer."

It was not only the United States which was galvanised by the triumph of Apollo 9; so too were the Soviets, whose own lunar programme was faltering and who now seemed to be shifting the nature of their effort towards long-duration missions aboard Earth-orbital stations. Their lunar dream was far from dead, but it was clear that the Stars and Stripes would be planted in lunar soil long before a Hammer and Sickle. In his diary, Nikolai Kamanin, head of the cosmonaut team, lamented on 10 March that the Soviets had effectively lost the Moon race. Just two weeks later, allaying a plethora of rumours, NASA announced that it would maintain its current schedule: Apollo 10 would fly the dress-rehearsal in lunar orbit in mid-May and the first landing would be attempted by Apollo 11 in July.

HAMBURGER HILL

Despite Richard Nixon's drive to reduce troop numbers in Vietnam, the war continued to drag on with yet more casualties, yet more embarrassment and yet more angry calls for its immediate end. In the spring of 1969, the massacre of hundreds of unarmed civilians in the South Vietnamese hamlet of My Lai had reached the ears of United States lawmakers. Ron Ridenhour, a former member of Charlie Company, the unit primarily responsible for the act, had written to a number of congressmen, the Joint Chiefs of Staff, contacts within the Pentagon and President Nixon himself to detail an episode which has since become infamous in American military history. The last vestiges of public support for the war vanished in the autumn of 1969, when one of the soldiers was convicted of premeditated murder and two dozen other officers were charged with related crimes.

Against this backdrop of mounting public fury, just a few days before Apollo 10 left Earth for the Moon, another shambolic episode occurred at a place forever known as 'Hamburger Hill'. The hill lay in the A Shau Valley of Vietnam's north-central coastal stretch, an area which had already been at the centre of numerous bitter battles, since it offered an infiltration point for pro-communist guerrillas into South Vietnam. A direct assault was ordered on 10 May against well-entrenched North Vietnamese defenders on the heavily fortified hill. The plan called for five companies to systematically destroy an enemy which left no electronic signature of its movements, communicated by wire and runner and travelled by night under the camouflaging cover of triple-canopied jungle. At Hamburger Hill, the Americans figuratively and literally embroiled themselves in an uphill struggle.

The first resistance was first met by Bravo Company late on 11 May and a series of Cobra helicopter gunships were despatched to offer support. Unfortunately, in the heavy jungle, the Cobras mistook Bravo for a North Vietnamese unit and opened fire, killing two American soldiers and wounding three dozen others. Conditions did not improve over the following days, with treacherous terrain and the virtual invisibility of the well-camouflaged enemy constantly hampering the assault. Another company suffered losses in attempting a flanking manoeuvre and efforts to suppress anti-aircraft and small-arms fire from the ground were rendered almost impossible. By 16 May, the media had heard of the ongoing mission. Associated Press correspondent Jay Sharbutt questioned senior commanders as to why infantry was being used instead of aerial firepower against the hill. Two days later, a two-pronged assault was ordered, with Delta Company getting within 75 m of the summit, but at the cost of severe casualties. Efforts to co-ordinate the attacks from the air were rendered fruitless when a thunderstorm reduced visibility to zero, but by the afternoon of 20 May the Americans successfully secured the hill ... and the last remaining North Vietnamese defenders quietly vanished over the border into neighbouring Laos. A total of ten days' worth of fighting had left 72 American troops dead and 372 wounded; and furthermore, 1,800 men had been committed to the battle, together with ten artillery units, 272 aerial support missions and thousands of kilograms of bombs and napalm. The irony was that the hill itself – popularly dubbed 'Hamburger Hill' since it became 'chewed-up' just like a hamburger – was strategically insignificant and was actually abandoned just two weeks later.

The debacle prompted Congress to seriously question the United States' command in Vietnam. The June 1969 issue of *Life*, which contained photographs of 241 troops killed in a single week, was seen by many as the 'watershed' turning point in the war. As the Wikipedia website pointed out, "While only five of these were casualties on Hamburger Hill, many Americans had the perception that all the dead were victims of the battle." Shortly thereafter, President Nixon authorised the withdrawal of 25,000 troops from Vietnam. By the end of the year, when the My Lai revelations reached the ears of the public, demands for the war to end had intensified. Nixon's actions, wrote astronaut Gene Cernan, had sent him soaring in the public opinion polls. Sadly, the popularity would not last.

THE WORRY AND THE REWARD

When a spate of colds delayed the launch of Apollo 9 in late February 1969, astronaut Gene Cernan could scarcely have imagined that the annoyance would actually 'save' part of his own mission, scheduled to fly just two months later. Cernan, nearly 35 years old at the time, a decorated naval officer and veteran of one previous space flight, had flown to Cape Kennedy in Florida with his wife, Barbara, and daughter Tracy to watch the launch. After its postponement, he was invited to spend a few days sunning himself in St Lucia, whereupon he fell into conversation with Dick Iverson, vice-president of the Ryan Aeronautical Company of San Diego, which had built the lunar module's digital landing radar.

"At a certain point in space," Cernan wrote in his autobiography, "with the spacecraft oriented toward the Moon at a specific angle and altitude, the radar was to start scanning the surface and find the precise landing area. When we discussed the exact numbers, Dick's mouth fell open in surprise and he stammered, 'But it's not designed to do that, Gene!' I stared back blankly. By sheer accident, our shop talk uncovered the fact that something important had slipped through the cracks. The computer software designers of the landing radar apparently had been working from an early version of the flight plan and when the trajectory for our mission was refined, the changes somehow had not found their way back to Ryan." In effect, but for their conversation, it is quite possible that he and Tom Stafford could have experienced major difficulties in ascertaining their precise rate of descent towards the lunar surface.

Eugene Andrew Cernan was already something of a rising star within the hyper-competitive fraternity of NASA's astronaut corps. Almost three years earlier, as backup pilot for Gemini IX, he had been catapulted onto the prime crew when Elliot See and Charlie Bassett were killed in an aircraft crash. His commander for that flight was Tom Stafford. During three days in orbit, they rendezvoused with an unmanned target craft and Cernan performed the longest EVA to date; an EVA which showed the world the true hazards of venturing into the vacuum in a pressurised space suit. For more than two hours, Cernan struggled with his suit's mobility, weathered a fogged-up visor which left his visibility almost nil and encountered grave difficulties in moving his body in a controlled way. Nonetheless, his performance impressed managers and he was considered for another slot on the final Gemini mission. By early 1967, Cernan was immersed in training for one of the early Apollo missions, eventually filling backup duties on Wally Schirra's flight and on 12 November 1968 receiving the word that every astronaut of that era had waited for years to hear: He was going to the Moon.

In time, the man who described himself as a "second-generation American of Czech and Slovak descent" would also command Apollo's final mission to the Moon and would become the last person – to date – to have trodden lunar soil. Cernan's parents had migrated to America shortly before the outbreak of the First World War; on his mother's side, they were Czechs from a Bohemian town south of Prague, while on his father's side they were Slovak peasantry from a place close to the Polish border. Their children, Rose Cihlar and Andrew Cernan, would produce the boy who, by the spring of 1969, had already gazed down on Earth through the faceplate of a space suit and who would miss out on becoming one of the first men to land on the Moon by merely 15 km.

Gene Cernan was born in Chicago, Illinois, on 14 March 1934. As a young boy, he learned from his father how machinery worked, how to plant tomatoes, how to hammer a nail straight into a piece of wood and how to repair a toilet; all of which instilled in him the ethos "to always do my best at whatever I put my hand to". In high school, that ethos led him to play basketball, baseball and football, for which he was even offered scholarships, but eventually he headed to Purdue University in 1952 to read for a degree in electrical engineering.

Four years later, Cernan graduated and was commissioned as a naval reservist, reporting for duty aboard the aircraft carrier *Saipan*. After initial flight training, he received his wings of gold as a naval aviator in November 1957 and gained his first experience in jets by flying the F-9F Panther. He was then attached to Miramar Naval Air Station in San Diego and assigned to Attack Squadron 126. While flying an A-4 Skyhawk, he performed his first carrier landing. Then in November 1958, he flew Skyhawks from the *Shangri-La* in the western Pacific and frequently encountered Chinese MiG fighters in the Straits of Formosa.

Shortly after the Mercury Seven were selected in April 1959, Cernan looked into what it took to become an 'astronaut'. In his autobiography, he noted that he met just two of NASA's requirements – age and degree relevance – but had little of their experience and no test-piloting credentials. "By the time I earned those kind of credentials," he wrote, "the pioneering in space would be over." Still, the germ of a new interest, to someday become a test pilot and fly rockets, implanted itself in the young aviator's brain.

In the summer of 1961, now married to Barbara Atchley, and approaching the end of his five-year commitment to the Navy, Cernan was offered the chance to attend the service's postgraduate school for a master's degree in aeronautical engineering. It would provide him with a route into test pilot school. When NASA selected its second group of astronauts in September 1962, Cernan knew that, although he had the right education, becoming a test pilot was still years away. Ultimately, the decision was made for him when one of his superiors recommended him to NASA for the third astronaut class.

As 1963 drew to a close, and by now the father of a baby daughter, Tracy, whose initials he would someday scrape into the lunar dust at the valley of Taurus-Littrow,

Cernan was repeatedly summoned to a remorseless cycle of physical and psychological evaluations and interviews by the space agency. Like so many others before him, he checked into Houston's Rice Hotel under the false name of 'Max Peck' and sat, "like a prisoner before the parole board" at an interview with such famous men as Al Shepard, Wally Schirra and Deke Slayton. The questions were awkward. In his autobiography, he recalled, "Someone asked how many times I had flown over 50,000 feet. Hell, for an attack pilot like me, who had spent his life *below* 500 feet, that was halfway to space!" How to turn the question to his advantage? He flipped it around, telling them that he had flown very low and "if you're going to land on the Moon, you gotta get close sometime".

He was also getting close to actual selection, as friends began calling up to find out why FBI agents had visited them with questions about Cernan's character, his background, his military record, his education, his parking tickets and his disciplinary records. At the same time, he was close to completing his master's thesis, focusing on the use of hydrogen as a propulsion system for high-energy rockets. Then, just a few weeks before President John Kennedy's assassination, he received the telephone call from Deke Slayton that would totally change his life. Little did he know that one of his Navy buddies, Ron Evans, rejected by NASA on this occasion, would be hired in 1966 and the two of them would eventually travel to the Moon together.

Cernan's first two years as an astronaut were spent on technical assignments. During the early Gemini flights, he occupied the 'Tanks' console in Mission Control, overseeing pressurisation and other data for the Titan II's propellant tanks. Then, one day towards the end of 1965, a technician tapped on his office door and told Cernan that Slayton wanted him to get fitted out for a space suit. The reason was inescapable: a flight assignment must be imminent. It was a backup assignment on Gemini IX, but he moved up to prime crew after Elliot See and Charlie Bassett lost their lives at the end of February 1966.

Less than three years later, when Cernan, Stafford and John Young were formally named as the crew of Apollo 10, their expectations were high. Wally Schirra's mission had triumphantly tested the Apollo command and service modules in Earth orbit, clearing the way for Frank Borman's team to orbit the Moon on Christmas Eve 1968. Officially, the schedule then required Jim McDivitt's all-up demonstration of the complete Apollo spacecraft in Earth orbit to fulfil the D mission, after which Stafford, Young and Cernan could fly to the Moon for a full dress-rehearsal as the F mission.

Schedule or no schedule, many astronauts had watched with interest as elements of the programme were expedited or deleted entirely to achieve the landing sooner. Already, in August 1968, the E mission – which would have tested the re-entry systems of the command module from a high Earth orbit – had been unofficially scrapped. In his autobiography, Cernan would recall the exciting possibility that as 1969 headed towards its second half, the F mission might be transformed into a full-blown landing: he and Stafford *could* become the first humans to walk on another world! Stafford had the same thoughts, but unlike Cernan was decidedly unhappy about them. Early that year, he had been approached by George Mueller, who hinted strongly that, assuming the success of Jim McDivitt's mission, Apollo 10 might make the first manned touchdown on lunar soil. Stafford had his doubts. "Tom was not so adamant about being first on the Moon," recalled Cernan. "He never looked at it that way. He wanted to do what was the best thing to do to have a co-ordinated, planned programme." In fact, Stafford told Mueller in no uncertain terms that if Apollo 10 *was* rescheduled to make a landing, "the flight crew won't be on it. There was just so much to do". The main problem was that 'his' lunar module, designated LM-4 and shortly to be renamed 'Snoopy', was overweight; it was only by a few kilograms, but still too much to satisfy the safety margins for a successful lunar liftoff. Grumman engineers had long known that LM-4 was earmarked for an Earth-orbital or lunar-orbital test flight, rather than a lunar landing, and had not subjected it to their Super Weight Improvement Program (SWIP). Lunar Module No. 5, on the other hand, was slated for the first shot at the G mission and was significantly lighter. "The option, then, was to postpone Apollo 10 for a couple of months until [LM-5] was ready," wrote Deke Slayton. "When you added up what we would gain as opposed to what we would lose, the decision was pretty clear." On 24 March 1969, shortly after McDivitt's crew splashed down in the Atlantic, Apollo manager Sam Phillips made that decision: the F mission would go ahead as planned.

Over the years, it has been speculated that, had Stafford's mission taken place closer to the end of 1969, 'adjustments' could have been made to enable LM-4 to achieve a landing. A little offloading of fuel to shed some weight could have allayed some of the managers' concerns, but in truth other difficulties remained which simply *had* to be ironed out. One of these was the phenomenon of 'mass concentrations', or 'mascons'. The experience of Apollo 8 had highlighted the need for a clearer understanding of their nature. Mascons are regions that have a large positive gravitational anomaly; specifically, an excess distribution of mass on or just beneath the surface. Such concentrations exist on Earth in Hawaii and, on the Moon, in the enormous impact basins of Imbrium, Serenitatis, Crisium and Orientale. Gravitational modelling has suggested that such anomalies indicate some form of 'positive density' within the lunar crust or upper mantle which is supported by the lithosphere. Possible explanations include dense basaltic lavas on the lunar 'seas' (mare), some of which may be several kilometres thick, but there are cases which do not support this line of thought and other contributory factors probably also play a part. The effect of lunar mascons on a satellite can be profound: in certain regions, local gravitational conditions might so alter a satellite's orbit as to cause it to crash.

The discovery of mascons was reported in the journal *Nature* in August 1968 by Paul Muller and William Sjogren of NASA's Jet Propulsion Laboratory in Pasadena, California, after studying radio tracking data from the Lunar Orbiters. They regularly diverged from their predicted positions, sometimes by up to ten times greater than expected. From the perspective of trying to land at a specific point on the Moon, it was essential to understand how and where the mascons were located and how strong they were. Throughout the remainder of 1968, NASA operated a 'tiger team' to address the question. Three of the Lunar Orbiters had flown in

Apollo 10's Saturn V leaves the Vehicle Assembly Building, bound for Pad 39B.

equatorial orbits and the other two in polar orbits, so it proved possible to compile a gravity map of the near side of the Moon. Since the primary task of the F mission was to rehearse approaching the site assigned to the first landing, it was crucial to investigate how a spacecraft would be perturbed by the gravitational anomalies along the ground track. The F mission was the only opportunity for engineers to observe how the lunar module's guidance and navigation systems performed in this situation.

Shortly after Stafford, Young and Cernan were assigned as the prime crew for Apollo 10 in November 1968, NASA press releases described their mission as encompassing a range of options "from Earth orbital operations to a lunar orbit flight". When it became clear that Frank Borman's crew would indeed journey to lunar orbit, Apollo 10 began to firm up as a mission to either circle or land on the Moon. One of the key trajectory analysts involved in the planning was Bill Tindall, described by many as "the architect for all of the techniques that we used to go down to the surface of the Moon". During a planning session for Apollo 10 in December 1968, he advocated that instead of simply undocking the lunar module from the command and service modules, opening the range a little and then rendezvousing, it should descend close to the surface, thereby rehearsing the approach to the point at which a landing mission would initiate the powered descent. In doing so, a full test of the descent engine and the landing and rendezvous radars could be conducted. Tindall even wanted a 'fire in the hole' ignition of the ascent engine at low altitude to simulate an aborted descent. On this last point, however, he was overruled, on the grounds that the mission was already full of tasks. It would certainly be busy and would set the stage for the most ambitious test-flying challenge ever attempted in human history: the first piloted landing on the Moon.

FIRECRACKERS, FOOTBALLING AND THE F-86

It is more than a little ironic that bald-headed Thomas Patten Stafford Jr, the man saddled with the burden of command on Apollo 10, had been deemed "too green" to have a realistic chance in the Project Mercury selection in April 1959. At that time, as America's first group of astronauts were introduced, Stafford was barely 28 years old and about to graduate from test pilot school. Moreover, his height would have rendered him ineligible and, had it not been for NASA's decision to increase the height limit for its larger Gemini spacecraft, he would not have been picked at all. "The most important change for me," Stafford wrote in his autobiography, "was that they had raised the height limit from five feet 11 inches to six feet even."

In time, he would become one of NASA's most accomplished astronauts, flying four times into space, including the F mission to the Moon and the joint Apollo-Soyuz venture with the Soviets. His habit of trying to speak faster than he could think led fellow fliers to nickname him 'Mumbles' and he would take charge of the Astronaut Office in the summer of 1969 as its chief. He was born on 17 September 1930 in Weatherford, Oklahoma, the son of a dentist father and a teacher mother. He became an avid reader and an enthusiastic watcher of the silvery DC-3 airliners which soared above his childhood home. After the Pearl Harbour bombings, he took a paper round in order to buy parts and build his own balsa wood model aeroplanes and in 1944 took his first flight in a two-seater Piper Cub. That flight alone, he wrote, "made me eager to become a fighter pilot and help win the war". He would not engage in combat during the Second World War, but his dream to fly would eventually come true. In high school, he excelled in football, becoming team captain, although he recounted in his autobiography that he was far from perfect: shooting out streetlights with a BB gun, throwing a firecracker into the police station and attempting, together with his friends, to disrupt English lessons with a cleverly orchestrated symphony of coughing. "The neighbours could always tell when I had been caught," he wrote. "I would be out front painting the fence as a punishment, like Tom Sawyer."

Stafford's footballing abilities eventually drew the attention of the University of Oklahoma's coach, although he had applied for, and would receive, a full scholarship from the Navy to study there. He had already undertaken military training in 1947 as part of the Oklahoman National Guard and was called to temporary duty when the small town of Leedey was hit by a tornado. He also worked on manoeuvres to plot howitzer targets and his calculations contributed to his battery receiving an award for the most outstanding artillery unit. The following year, 1948, brought both success and tragedy: acceptance into the Naval Academy and the death of his father from cancer. During four years at Annapolis, he was assigned to the battleship *Missouri*, where he met another midshipman named John Young. "We would have laughed," he wrote, "at the suggestion that someday we would become astronauts flying in space and circling the Moon together." After graduation in 1952 with a bachelor's degree, his eagerness to fly the F-86 Sabre jet – "the hottest thing in the sky" – led him to join the Air Force, rather than the Navy. He achieved his coveted silver wings from Connally Air Force Base in Waco, Texas, late the following year. By now married to Faye Shoemaker, he received advanced training in the F-86 and the T-33 Shooting Star. He was then assigned to an interceptor squadron, based in South Dakota, and later moved to Hahn Air Base in Germany as a flight leader and maintenance officer for the Sabre.

A paucity of opportunities for promotion almost prompted Stafford to resign from the Air Force in 1957 and he even drafted application letters to numerous airlines ... before deciding to continue in the service when he first saw the F-100 Super Sabre jet and the forthcoming F-104 Starfighter. "If I went to an airline," he wrote, "I'd be flying the equivalent of cargo planes and could say goodbye to high-performance fighters." He tore up the letters and was promoted to captain the next year, then selected for the Air Force Experimental Test Pilot School at Edwards Air Force Base in California, where he had to work harder than ever before. "Each morning's flight generated a pile of data from handwritten notes, recording cameras, oscilloscopes and other instruments. We had to reduce this data to a terse report that we submitted to the instructors and we had a test every Friday." From such schools, pilot astronauts were, are and continue to be drawn.

Stafford graduated first in his class in May 1959, stayed on at Edwards as an instructor and, over the next couple of years, supervised a number of test pilot candidates, including astronauts-to-be Jim McDivitt, Frank Borman and Mike Collins. He also met a visiting aviator from the Navy's test pilot school, named Pete Conrad. Additionally, he co-authored two flight test manuals: the *Pilot's Handbook for Performance Flight Testing* and the *Aerodynamics Handbook for Performance Flight Testing*. By the spring of 1962, he was due for a permanent change of station and confidently expected to study for an advanced master's degree in a technical field, but was picked to attend Harvard Business School. This suited him just fine, as a business administration credential would benefit both his military career and any subsequent plans he had. Nevertheless, when he learned in April that NASA was recruiting its second class of astronauts, he sent his application. In July, after passing the Air Force's initial screening board, Stafford was summoned to Brooks Air Force Base "where we had the expected blood tests and EKG stuff but no centrifuge testing ... we were all pulling Gs on a regular basis in high-performance aircraft. You didn't need to be some kind of physical superman to fly in space."

Some tests, though, were pointless. Stafford recalled looking into an ocular device for long enough to see a sudden flash of light, part of an evaluation of how an astronaut's eyes might respond to a thermonuclear explosion. "It wasn't enough to damage your eye – at least, I don't think it was – but you couldn't see for several minutes after the test," he wrote. The Brooks tests were followed by hour-long technical interviews in Houston in August and, although Stafford felt confident he had a good chance of being selected, his attention was primarily focused on the impending start of classes at Harvard. In fact, his interview in Houston came partway through his family's move eastwards from California. Arriving in Boston, Massachusetts in early September, they unpacked enough belongings to live on temporarily and Stafford worked through three days of inaugural classes. Returning home on 14 September, he was greeted by his next-door neighbour and the news that someone at NASA by the name of Deke Slayton had called. The gruff Slayton told Stafford that, if he was still interested in the astronaut programme, he had been officially selected.

Three days later, on his 32nd birthday, Stafford sat alongside fellow selectees Neil Armstrong, Frank Borman, Pete Conrad, Jim Lovell, Jim McDivitt, Elliot See, Ed White and John Young for their first press conference at Ellington Air Force Base, near Houston. Little could he have imagined how closely entwined their professional lives would become during the next few years, as they would all earn their own places in history.

DRESS REHEARSAL

The 'mark' to be made by the Apollo 10 astronauts also extended to their official, shield-shaped crew patch, which displayed a large Roman 'X' on the lunar surface. Yet Tom Stafford and Gene Cernan would come no closer than 15 km from humanity's exalted goal, then return to dock with John Young in the command and service modules. Since their assignment to Apollo 10, they had tried to stay out of the spotlight. Despite his disappointment at missing out on being one of the first men to walk on another world, Cernan was philosophical and accepting of the reality that *their* flight would be a technical triumph in itself. "Our pathfinder role," he wrote, "meant launching aboard a Saturn V, flying a quarter of a million miles in space and leaving John in Moon orbit while Tom and I took the lander on a sweeping flight near the lunar surface. No one had ever done that before."

As a result of the photographic reconnaissance by the Lunar Orbiter missions, the Apollo Site Selection Board had settled on five 'candidate' areas which were deemed suitable for the first Apollo landing. Each site was chosen on its merits, taking into account the smoothness of the terrain to support a lunar module, the absence of large hills, cliffs or deep craters which might compound a landing attempt and the need to satisfy propellant and tracking constraints. They were all on the equator.

Oblique view of south-western Mare Tranquillitatis, viewed from Apollo 10. Sabine is the nearer of the two large craters in the middle distance and Apollo landing site 2 is indicated by the cross.

The first two sites were located in the eastern hemisphere on the relatively flat Mare Tranquillitatis (Sea of Tranquillity), one was near the meridian in Sinus Medii (Central Bay) and the other two were in the western hemisphere in Oceanus Procellarum (Ocean of Storms). After being inspected by Apollo 8, the easternmost site had been rejected. Various factors led to Apollo Landing Site 2, some 100 km east of the rim of the crater Sabine, becoming the preferred choice for the first landing. If the launch was delayed, then it would attempt a target further west in accordance with the migration of the line of the terminator, so that the landing would occur just after local sunrise. In making their low altitude pass, Stafford and Cernan were to visually inspect Site 2 as a final check of its suitability.

Some wondered why Apollo 10 could not have simply taken a chance: after travelling all the way to the Moon, and with all the hardware in place, why *not* take that historic final step? Others acknowledged that there were still too many unknowns about the lunar environment and the spacecraft itself. Indeed, Cernan's conversation with Dick Iverson while waiting for Apollo 9 to launch had uncovered a potentially serious problem with the landing radar. Moreover, the software and procedures needed as the lunar module descended in a precise, sweeping arc to the surface under the impulse of its throttleable descent engine had yet to be finalised. Two years earlier, many managers decreed that half a dozen different docking modes *had* to be demonstrated before a landing could be contemplated. "So far," wrote Deke Slayton, "we had demonstrated exactly *one*." The drive to reach the Moon was already pushing ahead at break-neck pace and had *already* been significantly and daringly accelerated by Apollo 8. To skip everything and attempt a landing on Apollo 10 was simply too rash to risk.

During the nine weeks between the return of Jim McDivitt's crew and their own launch, the astronauts of Apollo 10 spent virtually every waking hour in the Cape Kennedy simulators, rehearsing each step of their planned eight-day mission. In fact, throughout their entire six-month training regime, each man spent five hours practicing every *hour* of his 192 hours aloft. This included a handful of 'three-way' training scenarios between Stafford and Cernan in the lunar module simulator, Young alone in the command module simulator and flight controllers at their consoles in Mission Control.

One lesson carried over from Apollo 9 was a desire – in fact, a *need* – to impose individual callsigns on both the lander and the command ship. If McDivitt, Scott and Schweickart's choice of Gumdrop and Spider had gone down like a lead balloon in some NASA circles, then the choice made by Stafford, Young and Cernan would raise yet more eyebrows. Borrowing from the hugely popular *Peanuts* cartoon strip, they decided that *their* lander would become known as 'Snoopy' and the command ship would become 'Charlie Brown'. This was not just a bit of fun. For some years, NASA had awarded 'Snoopy pins' to staff for outstanding work. "The choice of Snoopy was a way of acknowledging the contributions of the hundreds of thousands of people who got us there," wrote Tom Stafford. "Once you had Snoopy, Charlie Brown couldn't be far away."

In his autobiography, Stafford recalled that he was vocal in his desire that the American public – who were, of course, picking up the tab for Apollo, through their taxes – *should* be made privy to the wonders of the historic journey and its findings. It was a sentiment shared by Max Faget and Chris Kraft, who could scarcely believe their ears when some NASA engineers and scientists expressed the opinion that television was an unnecessary frivolity. "You're willing to exclude the people of Earth from witnessing man's first steps on the Moon?" Kraft had yelled increduously. "I don't believe it, and if you think about it, I don't think you'll believe it either." With this in mind, during the final weeks before launch, Stafford pushed vigorously to carry a colour television camera. Working with a team of NASA engineers, he helped design a camera which was small enough and light enough to be carried inside the command module. "The colour imaging system," he wrote, "[used] a spinning wheel of blue, red and yellow colours to encode signals, which would then be decoded by a synchronised wheel in Mission Control. We obtained two low-light-level TV camera tubes, originally developed for use by the military in Vietnam, two lenses from a French camera and an actuating motor from a Minuteman missile and put the new unit together." Ten days before Apollo 10's scheduled liftoff, the camera was ready for flight.

By this time, the mission had already experienced several weeks' worth of delays. As early as January, the liftoff had slipped from 1 May to the 17th, in order to best fit the lunar launch window and permit additional training for the astronauts. Then, only a few days after McDivitt's crew splashed down, at the suggestion of Jack Schmitt, a scientist-astronaut involved in planning the lunar surface activities, Sam Phillips postponed the launch by an additional day so that when Stafford and Cernan made their low pass over Site 2 the illumination would better match that which would face Apollo 11 in making its landing attempt. When their launch day did finally arrive, it marked the Saturn V's first – and only – liftoff from Pad 39B. Situated 2.6 km north of its sister pad, 39B has carved out its own niche in the annals of space history: weathering three Saturn IB launches to Skylab and more than 50 Shuttle flights.

Roughly octagonal in shape, both the 39A and 39B complexes were originally expected to be part of a much wider network of five pads spread along the marshy coastline of Florida's Merritt Island. As early as 1963, blueprints called for the construction of 39A, 39B and 39C, running sequentially from 'north' to 'south', with sites for 39D and 39E available for possible future use. However, 39A was never built and in 1963 the 'original' 39C was renamed '39A'. Although Tom Stafford's flight would be the only lunar mission ever despatched from 39B, the pad was kept in reserve, lest a disaster wipe out 39A. However, in the weeks leading up to Apollo 10, its 'newness' was apparent. On 11 March 1969, as Stafford watched his Saturn V roll out of the Vehicle Assembly Building, he could not help but think of the many workers putting finishing touches to the new complex. As the gigantic rocket arrived on the pad, in fact, the *paint* was barely dry on the umbilical tower …

Despite having a brand-new launch pad and a Saturn V whose reliability had now been confirmed by two manned flights, other preparations for Apollo 10 suffered setbacks. During routine maintenance to Cape Kennedy's launch control centre, the electrical power was temporarily cut off. The Saturn's pneumatic systems sensed this, opened some valves – following the 'normal' protocol for such a failure – and dumped 20,000 litres of highly toxic RP-1 fuel onto the pad surface. In addition to this loss, one of the fuel tank bulkheads buckled. Technicians worked feverishly to remove all but a few 'wrinkles' from the tank and an inspector verified that there had been no internal damage. It is quite remarkable that the impact on Apollo 10's launch date was ... absolutely nil.

The training of the Apollo 10 crew and their backups, Gordo Cooper, Donn Eisele and Ed Mitchell, was hectic. Not until he was placed into quarantine, a few days before launch, could Stafford finally appreciate the enormity of the mission that he was about to undertake. On the evening of 16 May, the crews shared dinner with Vice-President Ted Agnew and James McDonnell, founder of the aerospace giant which designed and built NASA's Mercury and Gemini spacecraft. Late on the following afternoon, driving a little too fast back to the Cape Kennedy crew quarters after seeing his family, Cernan was pulled over by a deputy sheriff. A lack of papers in his globebox, an iffy-looking military driving licence which, apparently, *never* expired and an unusual, unlikely name aroused the officer's suspicions. Only the timely intervention of launch pad leader Guenter Wendt saved the day. In his thick German accent, Wendt finally satisfied the disbelieving cop that, no, 'Mr Kurnin' could *not* accompany him to the station because – motioning towards the distant Saturn V – he had *somewhere* important to go tomorrow ...

"NOTHIN' TO WORRY 'BOUT"

'Launch time' on 18 May also happened to be 'lunchtime' – for Apollo 10's liftoff was precisely timed at 12:49 pm – which pleased Tom Stafford, who had been forced to get up at ungodly hours for his two previous missions. In fact, Gene Cernan was awake first, attending a private mass with family friend Father Cargill and the astronauts' nurse, Dee O'Hara, before joining the others for breakfast. The edge was taken off the intensity of the occasion by the presence of Snoopy everywhere. The little black-and-white dog, clad in a red scarf and an astronaut's bubble helmet, had literally become the Apollo 10 mascot; by that time, indeed, Cape Kennedy was alive with sweatshirts, stickers, posters, dolls and buttons bearing his image. Years later, Cernan would note uncanny parallels between Apollo 10 and Jules Verne's famous 1865 novel of a cannon-borne journey to the Moon: both launched from Florida, both carried three men, both would land in the Pacific Ocean ... and in Verne's story, the crew also included a canine passenger, Satellite.

Snoopy was even present as the three astronauts headed down the hall of the Operations and Checkout Building, bound for the transfer van and the launch pad. Standing off to one side was Jamie Flowers, one of the crew secretaries, holding an enormous stuffed Snoopy. Stafford patted it on the head, Young swiped at it and Cernan playfully tried to grab it and take it with him.

Despite the jubilant atmosphere, the crew's arrival at the vast scaffold of Pad 39B brought a renewed sense of calm to the proceedings. "The elevator door rattled closed as we rose up," wrote Cernan, "higher and higher, and we could see clearly through the wide openings of the safety door. Every inch of the way the rocket beside

George Low, Sam Phillips and Deke Slayton (all standing) monitor the countdown of Apollo 10 from the Firing Room at Cape Kennedy.

us hummed and vibrated. Glass-like chunks of ice slid away as her cryogenic lifeblood, liquid oxygen and liquid hydrogen, boiled and bubbled in her guts. *She's alive!* The elevator jerked to a sudden halt at the 320-foot level and we stepped out." An engineer welcomed them to the 'Twelve-Forty-Nine Express'. In the distance were hundreds of cars and trucks, parked bumper-to-bumper, interspersed with thousands of spectators lining the Cape's beaches.

A few minutes after ten that morning, Tom Stafford became the first to enter the command module, scrunching himself down into the commander's seat on the left side of the cabin. Next came Cernan, assuming the right-hand LMP's couch, and finally Young took his place in the centre as the CMP. Pad leader Guenter Wendt, doubtless thankful that Cernan was finally *aboard* and not in a prison cell, wished them luck, tapped their helmets and gave final a thumbs-up before Charlie Brown's hatch was closed and sealed. Wendt would be their last human contact for the next eight days.

Yet launch itself was still more than two hours away. Alone now, Stafford, Young and Cernan broke out their checklists and began the laborious process of reading data to flight controllers and implementing computer updates. The stabilisation and control system was checked, telemetry and radio frequencies were verified, pyrotechnics were armed, internal batteries were inspected, the altimeter was updated and Charlie Brown's reaction controls were pressurised. Each and every step was checked, double-checked and *triple-checked* to ensure that everything was ready. In Cernan's words, there was "no time to think, just time to do". As they tackled each step on the checklist, their movements, their flowing fingers, their fleeting eyes, washed over the instruments which two years of training and preparation had taught them to *know*, instinctively, like a family member. They would not be the first, nor the last, Apollo crew to relate that by this stage, they were a *part* – literally an *extension* – of their ship.

As the minutes and seconds ticked away, the tension both inside the cabin and outside increased: external power sources were removed as the spacecraft was switched to its internal fuel cells, Stafford's hand controllers were activated and the Saturn V's internal guidance system assumed control. With a little under nine seconds to go, Stafford, Young and Cernan felt, then heard, the fuel valves on the S-IC first stage open and the five mighty F-1 engines, engorged with propellant, roar to

King Baudouin and Queen Fabiola of Belgium (behind the man in hat) watch the launch of Apollo 10 on 18 May 1969.

life. Five kilometres away, the public affairs officer blared the final seconds of the countdown to the waiting spectators: "Ignition sequence start ... engines on ... five, four, three, two ... all engines running ... " and, finally, as the Saturn ponderously rose, "Launch Commit ... Liftoff ... we have a liftoff at forty-nine minutes past the hour!"

In the VIP bleachers, Queen Fabiola of Belgium grabbed the arm of her husband, King Baudouin, in surprise as the shockwave and "unearthly howl" of one of the most powerful rockets in existence rolled over them; even King Hussein of Jordan, who had seen many launches, flinched at the spectacle. Anything living in the vicinity of the ascending behemoth was incinerated and even beach sand close to the launch pad was glazed into glass by the intense heat. Inside the shuddering command module, the astronauts were buffeted by a vibration which rifled vertically through the booster. It was something that even steely, no-nonsense Cernan, one of a relatively rare breed of astronaut who also possessed the talent of a wordsmith, could only describe as "absolutely *scary*".

That scariness was counterbalanced by a sheer, adrenaline-fed thrill as the Saturn commenced its climb for the heavens. Unlike their Gemini missions, on which the Titan II rockets left the pad quite quickly, the rise of Apollo 10 was slower, more stately and required a full 11 seconds merely to clear Pad 39B's tower. All three men would recall the first stage as little more than a smooth, guttural roar, which pitched and rolled them, as planned, out over the Atlantic Ocean for the first two minutes of the flight. Next came the ignition of the S-II second stage, which Jim McDivitt had already warned Stafford would generate a sharp jolt. "He was right," wrote Stafford. "We went from nearly five Gs to zero in a fraction of a second, flying toward the control panel."

At this stage, the effects of 'pogo' – the longitudinal oscillations which had plagued several missions, most notably Apollo 6 – returned with a vengeance. "The Marshall [Space Flight Center] engineers had shaved twenty thousand pounds of metal out of [the S-IC]," wrote Stafford, "making the booster walls more flexible and more prone to pogo. Also, there was a ground stabilisation bar inside the cockpit that connected our crew couches to the rear bulkhead. It was supposed to be removed before launch, but somebody forgot. The bar magnified the pogo." With seemingly instantaneous effect, the S-II came alive, its five J-2 engines slamming the men back into their seats with a noticeable *wham*. "But the pogo stayed with us," remembered Cernan, "worse than ever, as another million pounds of liquid hydrogen and liquid oxygen ... burned hot and hard for seven minutes, and we accelerated with breathtaking speed." As it accelerated, there were disturbing sounds from the rocket – low moans and creaking groans pointed to the straining of metal under excessive pogo forces – which the men could 'feel', 20 storeys beneath them. They could see nothing; the command module's only uncovered window was directly in front of Stafford's face. "We were, in effect, blind," Cernan wrote.

Their blindness ended abruptly when John Young pushed switches to jettison the Saturn's escape tower and launch shroud. These blew away with a tremendous roar and instantly filled the cabin with sunlight. It offered a final surprise for

Cernan. Although he had anticipated the instant of jettison, and knew to the second when it would occur, its violence left him wondering if the spacecraft had been torn loose from the rocket! The hair-raising ride to orbit was, however, rewarded shortly afterwards by the view through the window: directly beneath him, Cernan beheld the familiar vista of Africa's western coast, the intense bright blues of the ocean lapping at its shores and the unmistakable browns and tans of the desert inland. The astronauts had crossed the entire breadth of the Atlantic in less than 12 minutes ...

Yet, still, they were not quite in orbit. "We got another stomp when the third stage kicked in," Cernan wrote. Then, the ride on the S-IVB turned from an intense rocking and rolling, as it began steering, into something so smooth that he could only compare it to being borne along with the grace and style of a Cadillac. At last, Apollo 10 was established in a circular 189 km parking orbit, ready for its venture to the Moon. The onset of weightlessness – expected, but profound – flooded over them. Alas, the tumult of the previous minutes had concerned them so much that the sense of euphoria was delayed as they pondered whether the violent pogo might have damaged their spacecraft. Was their lunar goal in jeopardy?

The astronauts' worries were shared by Flight Director Glynn Lunney and his Black Team of flight controllers in Houston, although none of their telemetry showed any indication of a problem with either Charlie Brown or, nestled in the S-IVB beneath it, Snoopy. "The only thing they knew was that *something* seemed amiss," wrote Cernan, "but the computer numbers provided no clue. Lunney voted with the data, which said nothing was wrong, and when we picked up communication with the Carnarvon station in Australia, Flight gave us a Go for the burn that would get us to the Moon."

That six-minute burn for translunar injection (TLI) began at 3:19 pm, a little over four hours into the mission, when the single J-2 engine of the S-IVB came to life for the second time that day. For the first three minutes, things went well. The lights of Sydney, a couple of hundred kilometres 'below' them, vanished instantaneously as Apollo 10 was accelerated towards the velocity needed to depart Earth's gravitational well and set course for the Moon. "Unlike on Gemini," explained Cernan, "where dawn arrived at a relatively slowpoke pace, we now crashed into daylight, climbing away from Earth at incredible speed." Unfortunately, coupled with that incredible speed were some pretty incredible vibrations throughout the vehicle, which, at their peak, were so harsh that the astronauts could hardly read their instruments.

From the commander's seat, Tom Stafford kept his gloved hand tightly closed around the controller which would allow him to shut down the S-IVB and scrub the mission if necessary. Calmly, he radioed Houston: "We're getting a little bit of high-frequency vibrations in the cabin ... " and then, in his Oklahoman drawl, added "nothin' to worry 'bout". It felt similar to 'flutter', a piloting term for the aeroelastic phenomenon in which aerodynamic forces on an object, together with an aircraft's natural mode of vibration, produce rapid periodic motions. In particularly severe instances, flutter had been known to shake aircraft to pieces and the likelihood of its occurrence in the vacuum of space was both perplexing and mystifying to Stafford's test-flying brain. As he struggled to understand what was happening, he seriously felt the mission was over, but could not bring himself to twist the controller to initiate an abort. "An abort," he wrote, "would leave us in a giant, looping orbit. There would be no visit to the Moon, no test of the LM, just a two-day wait for re-entry. I told myself: If she's going to blow, she's going to blow."

On the opposite side of Charlie Brown's cabin, Gene Cernan was thinking of the abort procedures which the crew might need to implement in a few seconds' time. However, he shared his commander's sentiment and would not give up on the mission. Full stop. Stafford repeatedly whispered "C'mon, baby!" The S-IVB shut down precisely on time. Apollo 10's velocity was right on the money and had reached the 38,500 km/h needed to begin the three-day flight to the Moon. This scare illustrated a stark reality: that on the United States' second manned lunar mission, *absolutely nothing* could be taken for granted. Years of flying the most advanced aircraft on the planet, pushing them to their structural limits and finding answers to pre-determined questions counted for nothing. "Out here," Cernan wrote, "confronting a foreign and hostile environment where there was no horizon, no up or down, and where speed and time take on new meaning, we not only didn't know the answers – we didn't even know the *questions*."

STARK PLACE

Notwithstanding the reality that Apollo was a pioneering effort, the relief that Stafford, Young and Cernan were finally on their way was palpable. It would later become clear that the vibration was caused by the helium valves, used to pressurise the S-IVB's fuel. Thankfully, these vibrations were well *within* the design tolerance of both the spacecraft and the booster. As engineers on Earth worked to identify this problem, the crew pressed on with their mission.

Shortly before four in the afternoon, John Young pulled Charlie Brown away from the now-spent final stage and perfectly executed a transposition and docking to collect Snoopy. At 4:29 pm, the docked Apollo spacecraft – more than 17 m long and weighing in excess of 45,000 kg – separated from the S-IVB, which receded eerily into the blackness of the cislunar void. So too did the Earth. Before the TLI burn, the home planet resembled a gigantic map unfolded 'beneath' them. Now, its apparent size as Apollo 10 headed for the Moon had shrunk noticeably from filling Charlie Brown's windows to something akin to a basketball. By the time the astronauts reached lunar orbit, it was little bigger than a large marble. Even Stafford was astounded. "For the first and only time in my space flights, I felt strange," he wrote. They were a *long* way from home.

All this – human beings leaving their world for another – was captured, *live*, for posterity and for a billion-strong audience on Earth, by the colour television camera that Stafford had lobbied so hard to carry. As Gene Cernan pointed the camera back toward home, he was able to identify the unmistakable line of the Rockies and the long, finger-like sliver of Baja California and even claimed to have glimpsed the Los Angeles freeways. Further north, Alaska was "socked in" with thick cloud cover and, far to the east, a low-pressure weather system lurked over New England.

For Cernan, a man born and raised in the Catholic faith, yet by his own admission "not an overly religious person", it redefined everything he thought he knew; out here, the insignificance of minor problems back home and the sheer *smallness* of Earth, its continents and even its vast ocean trenches were dwarfed by the true infinity of the Universe around him. This beautiful, perfect, limitless expanse of nothingness *must*, he reasoned, prove the reality of some form of Creator, but to comprehend further went beyond his mortal understanding. "Someone, some being, some power, placed our little world, our Sun and our Moon where they are in the dark void," Cernan pondered, "and the scheme defies *any* attempt at logic."

These thoughts were undoubtedly with all three men at quiet times throughout their voyage, but such were the demands of a lunar expedition that no one had the opportunity to dwell upon them. Notions of infinity were, however, brought figuratively back to Earth by the grind of daily life aboard ship. The end of the decade was seven months away and achieving President Kennedy's challenge was on everyone's mind. None of the astronauts wanted to screw up, get sick or miss a step in the timeline. Sickness was a major concern, especially following Rusty Schweickart's experience on Apollo 9. All three men had experienced stuffy heads upon arriving in space, although the sensations cleared within a few hours for both Stafford and Young. For Cernan, it lingered a little longer and he tried making 'cardinal head movements', prescribed by NASA physician Chuck Berry, to counteract it. By 20 May, he felt fine and after the mission would actually blame the *head movements* themselves for exacerbating the problem.

It was more than a little ironic that Apollo 10 was the first American flight in which bread – *real* bread – officially became part of the crew's pantry. 'Officially', that is, because some years earlier one member of Stafford's crew had been lightly reprimanded for 'unofficially' taking a corned-beef sandwich into space. On the Gemini 3 mission, John Young had arranged for the treat to be sneaked aboard as a surprise for his crewmate, Gus Grissom. Unfortunately, after taking a bite, Grissom had been obliged to put it away when it started to crumble and bits began to float around the cabin. This problem was solved by the time Apollo 10 launched: slices of white and rye bread were flushed with nitrogen, which kept them fresh for up to a fortnight and prevented them from drying out and crumbling into fragments.

Drinking, on the other hand, gave Tom Stafford a rather unpleasant surprise when he forgot to open a valve to the ship's water tank – and was rewarded with an evil-tasting dose of highly chlorinated water. There were other problems, too. The drinking water was a byproduct of the fuel cells, which generated Charlie Brown's electricity, and on previous missions astronauts had complained about the presence of hydrogen bubbles in it. A new drinking bag was created, with a handle which allowed the astronauts to whirl it around and separate the gas from the water. Unfortunately, it did not work and caused the hydrogen bubbles to settle at the bottom, then remix with the water when a crewman took a sip. All three astronauts suffered what NASA euphemistically referred to as "gas pains", but an outbreak of diarrhoea was thankfully avoided.

It was perhaps due to the quality of the drinking water that the men's appetites remained low throughout the mission. Indeed, on more than one occasion, they skipped a meal. To be fair, the food itself was by no means *haute cuisine*: even Don Arabian, head of the Apollo Test Division, who once described himself as "a human garbage can", struggled to find anything appealing in the tasteless sausage patties and minuscule chicken bits. Early in May, he had volunteered to try Apollo 10's fare for four days ... but after three days of chewing foodstuffs which tasted like granulated rubber, he understandably lost the will to live! Some foods were better than others, of course, and some could even be eaten quite 'normally' with a spoon; but the dehydrated dishes needed reconstituting with water and *that* meant injecting an uncomfortable amount of hydrogen gas into their meals. Not surprisingly, the men ate little during their mission to the Moon.

Still, with Snoopy on Charlie Brown's nose, Apollo 10 provided a relatively large space in which to live and work. For Tom Stafford, whose two previous Gemini missions had been like sitting in the front seat of a Volkswagen Beetle for days on end, it felt almost akin to having an attic or an extra apartment. The job of opening up that apartment fell to the lunar module pilot, Gene Cernan, who floated through the tunnel early on 19 May ... to be greeted by a snowstorm of floating fibreglass crumbs! It turned out that a Mylar cover on the command module's tunnel wall had torn loose, releasing the cloud of snowy particles, which itched like hell, took hours to vacuum up, stuck to hair, eyebrows and lashes and left Cernan looking "like a hound dog who'd been in a chicken coop".

By the following morning, Apollo 10 was more than 240,000 km from Earth and its velocity had slowed to a relatively puny 4,000 km/h as the gravitational influence of the home planet waned. Shortly thereafter, it entered the Moon's sphere of gravitational influence and began to accelerate as it 'fell' toward its objective. "Our trajectory," wrote Stafford, "had been so accurate that three of our four mid-course correction burns had been cancelled." The *only* mid-course burn of Charlie Brown's SPS engine occurred on the afternoon of 19 May and it changed their velocity by barely 54 km/h. It was so accurate that the last two burns were subsequently cancelled. This also served to calibrate the engine for the forthcoming entry into lunar orbit.

Fourteen thousand kilometres from the Moon, they made another television transmission, giving their audience a spectacular view of Earth, which by this point had diminished to somewhere between a grapefruit and an orange. Such views gave Stafford a chance to jab at the British Flat Earth Society that "the Earth *is* round". Perhaps the use of the word 'round', rather than 'spherical', pre-empted the society president's defiant response: "Colonel Stafford, it may be *round*, but it's still *flat*, like a disk!" Yet the television camera was a marvel and gave the eager audience an unprecedented sense of 'being there'. By the end of the flight, Apollo 10 had made no fewer than 19 telecasts, spanning almost six hours of air time and providing such a 'different' and novel dimension to what was happening in space that Stafford, Young and Cernan were awarded a special Emmy. The resolution was so good that when they filmed the transposition and docking with Snoopy, viewers could actually count the tiny metal *rivets* on the lunar module's skin. "Finally," wrote Cernan, "the taxpayer would get a look at where their money was going."

If the taxpayer knew where their *money* was going, it was not until Apollo 10 passed around the limb of the Moon late on the afternoon of 21 May that the astronauts finally saw where *they* were going. Until then, their lunar goal had been virtually invisible. "During the entire mission," wrote Stafford, "we had been facing its nighttime side, which was almost totally black. Peering through his navigation equipment, John Young had been able to find a place in the sky where the stars were occluded, so we were pretty sure the Moon was out there." To this day, Cernan maintains that his first glimpse of the lunar surface was not grey or white ... but *blue*. This mistaken impression lasted barely a second, before its true greyish-brown hue became apparent. The trajectory planners and mathematicians had guided them to the Moon with pinpoint accuracy and *there it was*, the lunar surface, just 95 km from them; so close, it seemed, that they could almost touch it ...

A few minutes before five in the evening, Eastern Standard Time (EST), the SPS engine roared silently into the void to slow Apollo 10 by 5,900 km/h and insert it into an elliptical orbit, with a perilune of 110 km on the farside and an apolune of 315 km on the nearside. "I pitched the spacecraft over," wrote Tom Stafford, "so we could get a good view of the surface. We were looking at the so-called farside of the Moon, the tide-locked side facing away from Earth." Visible in sharp relief were forbidding mountains, pockmarked ridges and furrows and thousands of craters – including the gigantic Tsiolkovski basin, named after the humble Russian schoolmaster today revered as the father of theoretical cosmonautics. Indeed, the lunar farside looked so tortured that it reminded Stafford of a plaster-of-Paris cast.

On the nearside, the dark, basalt-rich Mare Crisium (Sea of Crises) was easy to spot; a near-circular, flat-floored, wrinkle-edged blob, clearly visible to the astronauts in the wonderful, eerie clarity of the early lunar morning. It really stood out, said Young. Stafford added that the ridges running across its floor went "straight down just like the Canyon Diablo in New Mexico". Originally given the Latin name 'seas' by early astronomers, who mistook their darkness for being open stretches of water, the lunar mare were actually formed by ancient volcanic eruptions, many of which (as a result of the samples collected by astronauts on the Moon) have been radiometrically dated to between 3.1 and 4.2 billion years old. Their intrinsic darkness comes from their iron-richness and, in total, some two dozen maria on both the near and far sides cover 16 percent of the surface.

Two orbits after their arrival, a second SPS burn roughly circularised Apollo 10's path around the Moon at 109 x 113 km. As the astronauts gawped through Charlie Brown's windows, their eyes adapted to distinguish finer gradations of colour in this lifeless world: for it was now early morning, lunar time, and the surface exuded a vivid spectrum from white to black and a myriad of greys, tans, sickly pale yellows and even hints of red in some craters and on a handful of mountaintops. The spectacle was completed by the awe-inspiring sight of their first Earthrise on the lunar horizon; even at this distance – some 370,000 km from home – they could still pick out the ice caps, the vast bulk of Antarctica, the southward-projecting finger of Baja California, the intense flecks of white cloud and the iridescent blues of the oceans.

A terraced wall crater on the lunar limb, seen from Apollo 10.

Moving into their third orbit around the Moon, Stafford, Young and Cernan again broke out the camera and treated their audience to the first-ever televised images of Earth's closest celestial neighbour ... *in colour*. The opening glimpse was of Mare Smythii, a patchwork amongst bright mountains, named for the 19th century astronomer William Henry Smyth. Although these early images were somewhat 'washed-out', thanks to the height of the Sun in the sky, they improved as Apollo 10 headed westwards, where the illumination was oblique and the terrain was brought into sharper relief. Capcom Joe Engle in Mission Control described the vast expanse of Mare Fecunditatis (Sea of Fertility) as "absolutely unbelievable". Other controllers were dumbstruck by the Langrenus impact crater, its walls 3.2 km high in places, its central, cone-like peak rising a thousand metres from an irregular, boulder-strewn floor that Apollo 8 astronaut Jim Lovell, the previous December, had described simply as "huge".

Stafford keyed his mike: "Houston, tell the world we have *arrived!*"

THE ASTRONAUT

Next morning, after a surprisingly fitful night's sleep, considering it was their first in lunar orbit, Stafford and Cernan shimmied through the short tunnel into Snoopy. "I was disorientated," wrote Cernan, "because the floor was now above my head, so I rolled into a weightless ball, flipped and let my eyes adjust to the new environment until my equilibrium returned." The interior of the cabin was about the size of a broom cupboard, dominated by the cylindrical ascent engine cover poking out of the floor and surrounded on three sides by rows of lockers, circuit breakers, plumbing and bundles of wiring. The controls took the form of a square instrument panel and, at waist height on either side of the small, square hatch, were a pair of hand controllers, one on the left for Stafford and one on the right for Cernan. Above the commander's head was the tiny rendezvous window and in front of each position was a triangular window. Someday, in the not too distant future, astronauts would gaze through these windows onto the alien lunar landscape.

Like Jim McDivitt and Rusty Schweickart before them, Stafford and Cernan knew the interior and systems of the spidery lander like the back of their hands. Still, they had to be ready for unexpected glitches. The first chore was to clean up the remaining fragments of Mylar, as they would otherwise pose a real risk to the delicate instruments. Then came an irritating problem with a radar gauge, then a communications difficulty and later an error with Snoopy's gyroscopic platform. Among the more serious, however, was discovering that the lander had 'twisted' by three and a half degrees at its junction with Charlie Brown. This had imparted stress on the docking mechanism and risked damaging or even shearing off several of the latching pins. If that happened, it might make redocking impossible ... leaving Stafford and Cernan no option but to make an *extravehicular* transfer in their space suits from Snoopy back to the command module. The problem probably arose when Young forgot to turn off the service module's roll thrusters, but after some analysis ground controllers assured the crew that they could proceed so long as the misalignment did not exceed six degrees.

Precisely on schedule, Apollo 10 disappeared behind the Moon on its 12th orbit and, when the radio blackout ended 40 minutes later, Stafford jubilantly announced that Charlie Brown and Snoopy had successfully parted company. After undocking at 2:00:57 pm, Young used his thrusters to withdraw from the lander.

"You'll never know how big this thing gets when there ain't nobody in here but one guy," Young drawled.

"*You'll* never know how small it looks when you're as far away as we are!" countered Cernan.

"Yeah," continued Stafford. "Don't get lonesome out there, John."

"And don't accept any TEI updates," added Cernan. It was a spot of gallows humour – none of them *wanted* Young to fire the SPS for the transearth injection (TEI) burn and return home without them, but this was a grim reality that all command module pilots were trained to face. If an emergency arose in the lander which Stafford and Cernan could not rectify, it was Young's job to lower his orbit and rescue them ... and if *that* proved impossible, he would have no choice but to fire the SPS and come home by himself, leaving his crewmates to their fate. Doing so was an intense, highly emotional and largely unspoken moral fear which many CMPs pondered in the wake of their flights and was summed up best by Apollo 11 astronaut Mike Collins in his autobiography, *Carrying the Fire*. "If they fail to rise from the surface," Collins wrote, "I am *not* going to commit suicide; I am coming home, forthwith, but I will be a marked man for life and I know it … "

As the range opened, it was Young's task to activate a homing receiver for the lander's rendezvous radar and he had to toggle a switch several times to make it work properly. Then, a glitch with the orientation of Snoopy's antennas affected communications with Charlie Brown. Next, the link between Charlie Brown and Houston fell silent. "A quick check of the system," wrote Stafford, "showed that a breakdown had occurred in the line between Houston and the tracking station in Goldstone, California." At length, the problems were ironed out. For the next eight hours, Young – who would later walk on the Moon himself, command the first Space Shuttle mission and become the longest-serving chief of NASA's astronaut corps – would score a new record: the first human being ever to fly solo in lunar orbit. It was just one more addition to a long tally of achievements for the man who continues to be known today as 'The Astronauts' Astronaut'.

Over the years, John Watts Young Jr has been regarded as something of an enigma, with his misleading "aw-shucks" demeanour and country-boy drawl cleverly concealing a sharp, analytical and talented engineering mind. He was born in San Francisco on 24 September 1930 and when he was three years old his family moved to Cartersville, Georgia, before settling permanently in Orlando, Florida. As he related in an interview, it was around this time that he began to construct model aircraft. It was a hobby that would remain with him throughout high school, together with rockets, which he chose for a speech to his classmates in the 11th grade.

In 1952 Young earned his degree in aeronautical engineering from Georgia Institute of Technology with highest honours, receiving coveted membership of the institute's prestigious Anak Society. Among his earliest assignments after joining the Navy in June of that year, he served as fire control officer aboard the destroyer *Laws*. During this time, he completed a tour in Korea and a former shipmate would remember his coolness under duress. Joseph LaMantia (quoted on www.johnwyoung.com) recalls: "Though only an ensign at the time, he was the most respected officer on the ship. When we sustained counter-battery fire and enemy rounds were striking the ship, it was John Young's leadership which kept us all cool and focused on returning that enemy fire … which won the day." On returning home, he entered flight school at Naval Basic Air Training Command in Pensacola, Florida, where he learned to fly props, jets and helicopters. Later, he undertook a six-month course at the Navy's Advanced Training School in Corpus Christi, Texas. With receipt of his wings came four years' service as a pilot in Fighter Squadron 103, flying F-9 Cougars from the *Coral Sea* aircraft carrier and F-8 Crusaders from the *Forrestal* supercarrier. During these years, colleagues would describe him as "the epitome of swashbuckling aviators … he exuded confidence coupled with uncommon ability".

This ability, indeed, would ultimately guide him into the hallowed ranks of

NASA's spacefaring corps. But not yet. The selection process for the Mercury Seven began early in 1959, at which time Young was just starting Naval Test Pilot School at Patuxent River, Maryland, and test-flying credentials were a prerequisite for astronaut training. After graduation, he served as a project test pilot and programme manager for the F-4H weapons system at the Naval Air Test Center in Maryland, evaluating armaments, radar and bombing fire controls for both the Crusader and the F-4B Phantom fighters. During one air-to-air missile test, he and another pilot approached each other's aircraft at closing speeds of more than three times the speed

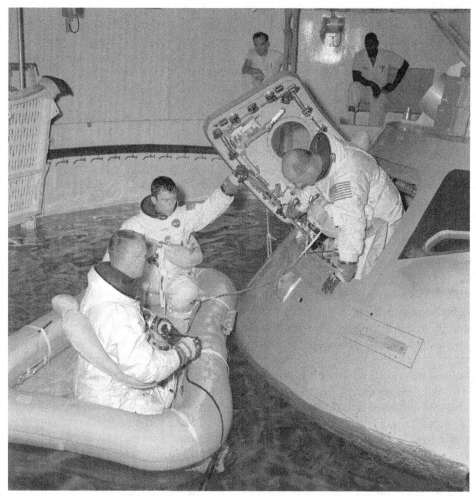

The most experienced spacegoing crew yet assembled: on the eve of launch, Apollo 10's astronauts had five previous flights between them. Here they perform an emergency water egress exercise from a command module simulator. Tom Stafford crawls into the life raft, whilst John Young holds the hatch. Meanwhile, Gene Cernan (back to camera) looks on.

of sound. "I got a telegram from the chief of naval operations," Young later quipped, "asking me *not* to do this anymore!"

In early 1962, he set two time-to-climb world records. By now a lieutenant-commander, Young's experience with the 'Phabulous' Phantom had made him the obvious choice to set the records as part of Project High Jump. The first, on 21 February, saw him climb 3,000 m above Naval Air Station Brunswick in Maine in 34.5 seconds; and six weeks later he made another attempt from Point Mugu in California and achieved 25,000 m in 230.4 seconds. In September of that year, after leaving active naval duties as a maintenance officer in Phantom Fighter Squadron 143, he got a call from Deke Slayton which marked the start of his astronaut career. The training was arduous. "You had to learn a lot of stuff," he said later. "You probably only needed to know one percent of all the stuff you had to learn ... but you didn't know which one percent it was!"

The relationship between Young and Tom Stafford is a perfect example of the complexities of the crew-selection process in the early Sixties. The latter was originally assigned to pilot the first manned Gemini in March 1965, but was shifted to a later flight when his commander, Al Shepard, was grounded by an inner ear disorder. Young ultimately replaced Stafford on that mission, which flew as Gemini 3, gaining some of the limelight as the first member of the 1962 astronaut class to fly. A little more than a year later, Young commanded his own mission, Gemini X, using one docked Agena as a 'shunt engine' to rendezvous with another one in a higher orbit.

Stafford, on the other hand, had already come to be recognised as one of the Astronaut Office's experts on rendezvous and had participated in two complex rendezvous flights: Gemini VI-A in December 1965 and Gemini IX-A in mid-1966. The first of these missions marked the first-ever rendezvous between two manned craft in orbit. Teamed with Gene Cernan, who had flown as Stafford's pilot on Gemini IX-A, the Apollo 10 trio were the most experienced crew ever assembled. Between them, they had spent more than ten days in space, had five missions under their collective belts and, thanks to Cernan, even had a measure of EVA experience thrown in for good measure. They would need every ounce of that knowledge as they prepared for the most difficult engineering challenge to date.

"DOWN AMONG 'EM!"

Somewhere behind the Moon, during the first of Snoopy's four independent orbits, Stafford fired the descent engine for the first time to reduce its velocity and begin dropping toward the lunar surface. Mindful, perhaps, of the chugging sound reported by Jim McDivitt in March, he started the engine at its minimum level – first 10 percent, to allow its gimbals to align the thrust through the lander's centre of mass, then opened the throttle to 40 percent – and, from his vantage point, John Young reported that they were moving noticeably away from him at more than 20 m/sec. For the men aboard the lander, on the other hand, the ride seemed relatively slow and pleasant and established Snoopy in an elliptical orbit with a perilune of just

15.4 km above the surface. The two astronauts, broadcasting on 'hot-mike' to the whole world, were exultant.

"We is down among 'em, Charlie!" radioed Cernan.

"Rog, I hear you're weaving your way up the freeway," replied capcom Charlie Duke from Mission Control.

Twenty kilometres above the Moon, as intended and precisely on cue, the radar detected the looming surface and began feeding altitude data into Snoopy's computer. The lunar mountains, rushing past below, seemed almost close enough to touch, their appearance and texture resembling wet clay. As they approached Mare Tranquillitatis, running along an imaginary 'lane' of physiographic features memorised from months of studying maps and charts, Stafford was astounded by the sheer barrenness of the forbidding terrain. "There are enough boulders around here," he breathed, "to fill up Galveston Bay."

Since their assignment to the mission the previous November, Stafford and Cernan had spent hundreds of hours poring over maps and photographs from the unmanned Lunar Orbiters of two of the candidate landing sites for Apollo 11. Both lay in the relatively flat Tranquillitatis region and the men had even tried to simulate part of their trajectory aboard a T-38 aircraft back on Earth. When the time came to overfly Site 2 for real, they knew the craters, mountains, rilles, bumps, hollows and furrows so well that they literally formed a familiar 'road', helping to guide them towards the landing zone. To their trained eyes, it was a virtual lunar highway and for this reason they had nicknamed it 'US 1'. Along the way, a range of low mountains was called the 'Oklahoma Hills', a rille which split into a pair of craters was dubbed 'Diamondback' and 'Sidewinder' and one ridge was even named in honour of Stafford's then-wife, Faye.

Attempting to shoot a photograph every three seconds as Snoopy passed over the landing area, Stafford was annoyed when the "goddamn" Hasselblad issued an ominous puff of smoke and jammed. (In his autobiography, he would relate apologising to Victor Hasselblad upon his return to Earth.) It was the first of a number of unfortunate outbursts from Snoopy's cabin. Deprived of his camera, Stafford resorted to relaying verbal reports to Mission Control, comparing the inhospitable appearance of the site to the high desert near Edwards Air Force Base in California. If Neil Armstrong and Buzz Aldrin were to find themselves heading for the 'near' end of the target ellipse, then they would have a smooth landing; but Stafford advised that a landing at the 'far' end would demand additional manoeuvring fuel to find a spot free of boulders.

Just beyond Tranquillitatis, and an hour after the first burn, Stafford again fired Snoopy's descent engine, this time using full throttle to accelerate them by almost 200 km/h and enter an eccentric orbit with a 350 km apolune. The high apolune would set up the timing so that, on the next pass over Site 2, the ascent engine could be fired to simulate a rise from the surface.

When the time came to jettison the descent stage and return to redock with Young, Stafford oriented the vehicle correctly but noted a slight yaw rate on his attitude indicators. "Telemetry suggested we might have an electrical anomaly," he wrote, "so I started to troubleshoot the problem." Shortly thereafter, Cernan,

Gene Cernan (left) and Tom Stafford rehearse their procedures in a mockup of the lunar module cabin. The lander's main instrument panel is visible to the far right of this image.

thumbing through his checklist, switched control from the Primary Guidance, Navigation and Control System (PGNS) to the Abort Guidance System (AGS). The former provided an exact navigational reading, whereas the latter, in his words, would "get us the hell out of there if unexpected trouble cropped up". In case Armstrong and Aldrin needed to abort in a hurry – punching their ascent stage away if their descent went wrong – a test of the AGS in lunar orbit was critical. Stafford and Cernan had rehearsed it a hundred times in the simulator. Clad in their bulky suits, however, high above the Moon, both men found it difficult to hit the right switches and an instant after Cernan set the control mode of the AGS to 'attitude hold', Stafford reached across and inadvertently changed it to 'auto'.

Seconds later, they were ready to blow the four explosive bolts to separate the ascent stage from the descent stage and begin to trek back to Charlie Brown. Suddenly, and without warning, Snoopy went berserk, lurching wildly in both pitch and yaw axes. "Gimbal Lock!" shouted Stafford urgently, believing the lander's gyroscopes to have frozen. Then, Cernan, over hot-mike and with the whole world listening in, yelled with unfortunately crystal clarity: "Son-of-a-*bitch*! What the *hell* happened?" As the menacing lunar terrain, the black sky and the grim line of the horizon alternately flashed past Snoopy's windows, both men knew they had just seconds to resolve whatever was wrong.

By activating the AGS and mistakenly setting it to 'auto', Stafford had in effect instructed Snoopy's radar to begin searching for Charlie Brown and the abort guidance system was causing the lunar module to flip wildly around its centre of mass. Quickly, he pushed the button to jettison the descent stage and steadily regained manual control of the gyrating ascent stage. Fearing that the inertial measurement unit was close to gimbal lock, Stafford executed a pitch manoeuvre, started working the attitude switches and finally calmed the ascent stage down. From Cernan's initial shout to Stafford's final report to Houston that Snoopy was back under control, three minutes elapsed.

Those minutes were unnerving not only for the lunar module crew, but also for John Young, listening helplessly aboard Charlie Brown. "I don't know what you guys are doing," he drawled in his typically understated manner, "but knock it off. You're scaring me!"

Years later, Cernan and Stafford would both say that it *was* a classic piloting error. "Neither Tom or I can be sure," Cernan related, "but when it came time to stage ... there was some switch that had to be changed and I changed it. And I'd be willing to bet that ... I put the switch in the new position and Tom went ahead and moved it back to the old one. His action was to move the switch. I'd already done it for him. But he didn't know that and, when he moved the switch, he just moved it back to where it was. In effect, *we* created the problem!"

RETURN TO EARTH

Although the cause of the glitch, a simple mistaken switch setting, could be easily rectified in time for Apollo 11 simply by refining crew procedures, the episode amply highlighted that nothing could be taken for granted on a lunar expedition. Charlie Duke, helpless to assist, had warned them from his data that they were close to gimbal lock, but things were moving far too quickly for him or anyone else in Mission Control to do anything. By contrast, the return to Charlie Brown was charmed; Stafford found Snoopy's ascent stage a little difficult to hold steady, but John Young had no problem docking and all three men were greeted by the welcome 'ripple-bang' of a dozen capture latches snapping shut. "Snoopy and Charlie Brown are hugging each other," chortled Stafford. A few minutes later, at 11:25 pm, back in the sanctuary of the command module, Young was given a couple of hugs, too.

Perhaps Stafford might not have been quite so welcoming had he realised that Young had used some of his relatively private time, alone in Charlie Brown, performing his *very first* 'bowel movement' in almost five days aboard Apollo 10. Consequently, when Snoopy's ascent stage was jettisoned a few hours later, into permanent solar orbit, it carried with it a United Nations flag, a small flag from each state of the Union, the command module's now-unneeded docking probe, a pile of empty food packets ... and Young's bag of faecal goodies. "We joked that Snoopy would have food, water, oxygen, organic material, all the ingredients for the creation of life," Stafford recounted with glee. "Maybe a few billion years from now, some kind of Snoopy monster, distantly related to John Young, will emerge from somewhere in the Solar System."

At 6:25 am on 24 May, sixty-one hours since entering lunar orbit, a perfect TEI burn sent Charlie Brown hurtling home. The 55-hour return journey gave them more than enough time to relax and, on the day before splashdown, to actually *shave*. Previously, astronauts had been forbidden to carry bladed razors and a $5,000 electric device, designed to suck up trimmed whiskers, failed miserably. Most Gemini and early Apollo fliers had returned grungy and heavily bearded. But Stafford's crew lathered their faces with thick gel, shaved using a safety razor and wiped away the residue with a wet cloth. They even brushed their teeth. During one of their final telecasts, the audience must have been astonished to see its latest batch of hardened heroes returning home with baby-soft skin and glistening teeth ...

The return to Earth on 26 May was totally different to what any of the three had experienced in their Gemini days. Re-entering the atmosphere at a lunar-return velocity close to 40,000 km/h, together with the different shapes of the vehicles, altered the effect in a hundred different ways. In his autobiography, Cernan remembered an enormous white and violet 'ball' of flame, literally sweeping up the conical command module like a glove. "It grew in intensity," he wrote, "and flew out behind us like the train of a bride's gown, stretching a hundred yards, then a thousand, then for miles ... and the whole time we were being savagely slammed around inside the cabin."

To some people, the fire and brimstone nature of re-entry illustrated God's wrath with this foul-mouthed team of space fliers. One man who was *not* happy, was an over-zealous Christian minister named Larry Poland. He was vocal in complaining that Stafford's crew had taken "the language of the street" with them to the Moon, and urged them to publicly apologise for their "profanity, vulgarity and blasphemy". However, recognising the reality that its men had been placed in a life-or-death situation, three hundred and seventy thousand kilometres from home, NASA managers stood by the crew, saying they had "acted like human beings". After splashdown, indeed, they would be greeted with tongue-in-cheek notices, which read: 'The Flight of Apollo 10: For Adult Audiences Only'. MSC director Bob Gilruth had burst out laughing when he heard the swearing and Chris Kraft summed up the general thinking in his autobiography: "They were out there doing man's work at the Moon. If a cuss word or three slipped out, well, who the hell cared anyway?"

Larry Poland cared greatly and for many others, too, laughing it off was not good enough. In the days and weeks after the mission, hundreds of letters and telegrams flooded into NASA's Washington headquarters, some tolerating the language, others condemning it. Years later, Cernan would recount that *he*, as the man responsible for the ungodly language, made his apologies. Young was not involved and Stafford, famously nicknamed 'Mumbles', had uttered *his* profanities under his breath – calling the Censorinus crater "bigger than shit" and referring to the malfunctioning Hasselblad as "goddamn". Cernan, though, was well and truly in Reverend Poland's firing line. *His* words could neither be misconstrued or explained away. So, at a news conference, Cernan apologised to the people he had offended and thanked those who understood how he came to say what he did. Privately, he was furious. It made no difference, he wrote, that Poland accepted his apology and forgave him, for Cernan "*never* got around to forgiving *that* self-righteous prig!"

"TH' INCONSTANT MOON"

When Charlie Brown plopped into the Pacific Ocean, 640 km east of American Samoa and a few kilometres from the recovery ship *Princeton*, at 12:52 pm on 26 May 1969, the final barrier on the road to the lunar surface was lifted. "It's all downhill from here," admitted Milt Windler, one of the four flight directors assigned to Stafford's mission. "I see nothing to constrain the launch of Apollo 11." Nothing, indeed, for the Saturn V rocket destined to loft Neil Armstrong, Mike Collins and Buzz Aldrin had been rolled out to Pad 39A while Apollo 10 was in flight and on the morning of splashdown there were banners hanging in Mission Control emblazoned with the legend: '51 Days To Launch'.

As those days ticked away, the Moon was indifferent to the historic events in which it was to play a part. As it had for countless millions of years, it travelled around Earth once every 27.3 days, turning on its axis at such a rate that it was *tidally locked* to its parent planet. With an equatorial diameter of 3,700 km, the Moon is distinctive by being the largest natural satellite with respect to its parent planet in the Solar System. It has been, and to this day continues to be, a source of mystery, inspiration and wonder for us. On the Shakespearean stage, Juliet famously implored Romeo to "swear not by the Moon / th' *inconstant* Moon". She was right: for throughout history, and even today, the Moon has been an inconstant companion and our perception of it has been ever-changing. We have viewed it as a divinity to be worshipped, as a cause of lunacy and the bringer of the shapeshifter and the werewolf, have seen the Man in the Moon in its nightly orb, have measured time and even the female menstrual cycle by it and, more recently, have begun to see it as more than just an eerie, shimmering lamp in the dark sky – as a *world* in its own right, a world with mountains and valleys to be explored. *That* realisation, in turn, inspired the first murmurings of science fiction, a growing knowledge of lunar geology and the first tentative efforts to actually reach it.

To trace one of the Moon's earliest representations on Earth, one must go to the Boyne Valley in Ireland, to the Neolithic passage grave of Knowth, largest of the Brú na Bóinne burial complex. This remarkable site consists of a huge mound and more than a dozen 'satellite' tombs, on the interior walls of which are numerous artistic renditions – including a rock carving, perhaps as much as 5,000 years old, which *may* represent the Moon. Neolithic opinion, of course, remains open to conjecture, but certainly later cultures supposed it to be a deity or some kind of supernatural phenomena. Such lunar deities have tended to be male – the Mesopotamian Nanna and Sin, the Germanic Mani, the Egyptian Thoth and the Aztec Tecciztecatl, for instance – but *female* variants, including the Greek Artemis and the Roman Diana, are perhaps better known.

Early *scientific* interpretation came from the Greek philosopher Anaxagoras in the fifth century BC, who theorised that both the Sun and the Moon were in fact giant rocks, the latter reflecting the glow of the former. Unfortunately, despite being remarkably close to the truth in his opinion of the Moon, Anaxagoras' atheistic views ultimately led to imprisonment and exile. Elsewhere, the Han Chinese also saw the light of the Moon as just a reflection of the Sun. By the 17th century, and the invention of the telescope, there came new recognition that the Moon was not only a *sphere*, but was also *not smooth*: its dark and light blotches were mountains, valleys and relatively flat expanses of terrain. Still, the nomenclature given to surface features – 'maria' (seas) for the dark bits, 'terrae' (continents) for the lighter ones – represented a belief that the Moon was quite Earth-like, with bodies of land existing in oceans of water.

These ideas, including the notion that the lunar surface harboured vegetation, animals and even sentient beings – nicknamed 'Selenites' – continued into the 19th century, with the Great Moon Hoax of 1835 being a particularly notorious attempt to fool a gullible public. Within a few years, scientists had concluded that the Moon had no bodies of water and probably no appreciable atmosphere. By this time, however, it was firmly in the public consciousness as a target of science fiction. One of the earliest such tales is attributed to the Greek-speaking Syrian writer Lucian of Samosata in the second century AD; his fantastic story encompassed a journey beyond the Pillars of Heracles (the Straits of Gibraltar) and included an account of a full-scale war between the King of the Moon and the King of the Sun. In reality, it was a satire on the wars waged by the Greeks. Although its precise position within the science fiction genre remains open to debate, Lucian's work nonetheless tabled the idea of extraterrestrial life, space travel and the colonisation of other worlds, many hundreds of years before they would enter mainstream thought.

Perhaps the most famous tale of lunar exploration is Jules Verne's 1865 story *De la Terre à la Lune* (*From the Earth to the Moon*), which presented a more sensible idea of launching three men from an enormous cannon, known as the 'Columbiad', from a site near Tampa in Florida. The practicalities of Verne's ideas have been criticised over the years: Konstantin Tsiolkovski calculated that his cannon would impose fatal acceleration loads in excess of 1,000 G on the Columbiad's crew, whilst Gerald Bull showed that such a projectile could only achieve 32 percent of the velocity needed to escape Earth's gravity. Leaving such problems aside, Verne's tale has been applauded by many for its closeness to reality. Indeed, the name of the cannon closely mirrored that of Apollo 11's command module – 'Columbia' – and other similarities such as the three-man crew, the Florida launch site and a Pacific splashdown at mission's end are particularly notable.

Others, from H.G. Wells to Arthur C. Clarke and Isaac Asimov to Robert Heinlein, have penned their own ideas of lunar expeditions. Still more have used the theme of travelling to our closest celestial neighbour to explore their own dreams of building fantastic flying machines and rockets, in allegorical expressions of their opinions of warfare and the morality of colonisation and – in the case of Johannes Kepler – to cleverly present controversial or potentially blasphemous theories about the heavens in the guise of fiction. Yet the lure and mystery of the Moon remains.

Many cultures continue to see a 'man' in the Moon – an imaginary face and head, sometimes even an entire body – composed of the dark and light areas of the lunar surface. Some have seen his eyes in the prominent Mare Imbrium and Mare Serenitatis, his nose in Sinus Aestuum (Bay of Billows) and his mouth in Mare Nubium and Mare Cognitum. Others have seen a *whole* man, carrying a burden on his back or accompanied by a small dog. Still others, in Elizabethan England, regarded the figure as that of a witch or an old man clutching a lantern. Even today, in supposedly more enlightened times, some Shia Muslims believe the name of the Prophet Muhammad's son-in-law to be etched in lunar dust and, during his solo flight around the Moon on Apollo 11, Mike Collins was advised to watch for a beautiful Chinese girl who had been banished there along with an enormous rabbit after stealing the pill of immortality from her husband ...

If all goes according to plan, humans will return to the Moon around half a century after the last Apollo explorers departed. Much has been learned about the Moon before, during and in the wake of those expeditions. Yet debate has raged over the years as to its precise origin. More than a century ago, George Darwin, son of the *Origin of Species* author, speculated that the young Earth, which he presumed to be in a fluid state, was rotating so rapidly that it expelled a fragment of itself. Others later speculated that the Pacific Ocean represents the 'scar' of this tumultuous event, but we now know that the oceanic crust in that area is at most 200 million years old. In any case, mathematical considerations have rendered Darwin's 'fission' hypothesis untenable. Still others have theorised that the Moon was gravitationally captured, fully formed, by Earth, but again there are mathematical issues and such a close encounter would probably have led to a collision or, at the very least, a highly elliptical orbit. Some have speculated that both formed together as a 'double system' as the Solar System accreted from a primordial solar disk of gas and dust, but even here there are problems, such as why the average densities of the two bodies are so different. As Apollo 11 loomed, there was no scientific consensus concerning the origin of the Moon, but it was hoped that an analysis of lunar material would settle the debate. What came as a shock was that the data appeared to rule out all of the options, causing a scramble to come up with something new.

Today, many lunar scientists favour the so-called 'giant impact hypothesis', independently formulated in the mid-Seventies by two research teams led by William Hartmann and Alastair Cameron. This argues that a collision occurred around 4.5 billion years ago between the young Earth and another celestial body with roughly the mass of Mars. A vast quantity of material was thrown into orbit and gradually coalesced to form the Moon. Lunar geologist Paul Spudis and others have seen this as an explanation for the Moon's dearth of volatile chemical elements, the evolution of its orbit, its relatively small iron core and the rapid rotation rate of Earth itself. "Moreover," Spudis wrote, "it makes the uniqueness of the Earth-Moon system seem more plausible" because events of such cataclysmic magnitude *must* have occurred only rarely. However, Spudis cautioned that the giant impact hypothesis' popularity only *remains* popular because *too little* is known to rule it out and that fundamental questions still linger.

The conjectural Mars-sized object, labelled 'Theia' by giant-impact theorists, supposedly struck Earth at an oblique angle, its iron core sinking into that of our own world and ejecting a substantial portion of both planets' mantles into orbit. It is quite possible that this material coalesced into the Moon in as little as a few years – and certainly no more than a century – and samples returned by Apollo astronauts have indirectly shown oxygen isotopes almost the same as Earth. Additionally, the largely anorthositic composition of the ancient lunar crust implies that the Moon was molten in its early youth. It has been calculated that as the debris ejected by a giant impact was accreted to create the Moon, the heat could easily have produced an ocean of magma at least 500 km deep. It is not clear whether Earth ever possessed such a magma ocean of its own. The magma ocean of the newly formed Moon, on the other hand, would have steadily cooled and crystallised. By around 4.3 billion years ago, a solid crust had formed, but the bombardment continued, mixing upper crustal materials, creating vast, multi-ringed basins and splashing pristine samples

Neil Armstrong (right) and Buzz Aldrin practice humanity's first Moonwalk in the 'sand box' at Houston's Manned Spacecraft Center.

from the interior across the surface. About 3.8 billion years ago, the events which created the Imbrium and Orientale basins are believed to represent the last major impacts on the Moon ... and half a billion years after *that*, wrote Paul Spudis, the rate of cratering had steadily declined to little more than the occasional formation of a large basin and a random 'peppering' by meteorites. "For all practical purposes," he concluded, "the Moon is now geologically dead."

As the space age dawned and Apollo prepared to bring bootprints as well as the pads of robotic landers to the lunar surface, this long epoch of geological inertness offered a tantalising opportunity to examine what is literally a near-pristine fossil from the early Solar System. Where to land the first team of astronauts, of course, could not be overwhelmed by scientific need; it had to take into account operational demands, such as fuel, tracking requirements, lighting and providing a safe spot in which a lunar module could be set down.

A landing close to the equator was highly desirable, since it could be reached with a minimum expenditure of fuel, but the site also needed to be at least 45 degrees 'west' of the Moon's eastern limb to permit Mission Control a few minutes of tracking data in the descent orbit and enable the lunar module's computer to be more precisely updated before initiating the powered descent.

By the time of Apollo 11, the prime site was in south-western Tranquillitatis and the backup was in Sinus Medii. Scientific opinion about the evolution and composition of the mare material was divided. On the one hand, the so-called 'cold-Mooners', led by Nobel Prize-winning chemist Harold Urey, maintained that the Moon was a frigid, relatively unchanged relic from the Solar System's creation, whereas the 'hot-Mooners', led by Gerard Kuiper, argued that ancient volcanism was responsible for most lunar features. Urey admitted it *was* likely that the surface had suffered episodic impacts of sufficient violence to create pools of molten rock ... but that this did not constitute volcanism.

When Apollo 11 returned to Earth with a few kilograms of basalt – a *volcanic* rock – from Mare Tranquillitatis, the dispute would finally be settled in the hot-Mooners' favour. Yet the very fact that the debate continued right up to the first manned landing – and did so with vigour, for Andrew Chaikin has likened it to "an almost religious fervour" – shows that our knowledge of our closest celestial neighbour was at that time, and *remains today*, far from complete.

MOON TEAM ONE

Neil Alden Armstrong, the man who would be first to pick up lumps of lunar rock, had lived a life of movement. Born on 5 August 1930 in Wapakoneta, Ohio, of Scots-Irish and German descent, his father was a government worker and the family moved around the state for many years: from Warren to Jefferson to Moulton to St Mary's, finally settling permanently in Wapakoneta in 1944. By this time, Armstrong was an active member of his local Boy Scout group and his mind was filled with dreams of flying. "I began to focus on aviation probably at age eight or nine," he told NASA's oral history project in 2001, "and [was] inspired by what I'd read and seen. My intention was to be an aircraft designer. I later went into piloting because I thought a good designer ought to know the operational aspects of an airplane."

When Armstrong enrolled at Purdue University in 1947 to begin an aeronautical engineering degree, he became only the second member of the family to undergo higher study and famously had learned to fly before he could drive. (Years later, he recalled first flying solo aged just 16. Alas, his early logbook entries were lost in a fire at his Houston home in 1964.) Under the provisions of the post-war Holloway Plan, Armstrong committed himself to four years of paid education in return for three years of naval service and a final two years at university. He was summoned to active military duty in January 1949, reporting to Naval Air Station Pensacola in Florida for flight training. Over the next 18 months, he qualified to land aboard the aircraft carriers *Cabot* and *Wright* and, several days after his 20th birthday, became a naval aviator.

His initial assignments were to Naval Air Station San Diego, then to Fighter Squadron 51, where he flew the F-9F Panther – "a very solid airplane" – and later landed his first jet on an aircraft carrier. By the late summer in 1951, Armstrong had been detailed to the Korean theatre and would fly 78 combat missions in total. His first taste of action came only days after arrival, whilst serving as an escort for a photographic reconnaissance aircraft over Songjin. Shortly thereafter, whilst making a low-altitude bombing run, his Panther encountered heavy gunfire and snagged an anti-aircraft cable. "If you're going fast," he said later, "a cable will make a very good knife." Despite his aircraft having a sheared-off wing and a lost aileron, Armstrong somehow managed to fly back over friendly territory, but the damage was so severe that he could not make a safe landing and had to eject. Instead of a water rescue, high winds forced his ejection seat over land, close to Pohang Airport, and he was picked up by a jeep driven by an old flight school roommate. By the time Armstrong left naval service in August 1952, he had been awarded the Air Medal, a Gold Star and the Korean Service Medal and Engagement Star. For the next eight years, he remained a junior lieutenant in the Naval Reserve.

After Korea and his departure from the regular Navy, Armstrong completed his degree at Purdue in 1955, was admitted to the Phi Delta Theta and Kappa Kappa Psi fraternities and met his future wife, home economics student Janet Shearon. They were married in January 1956 and their union would endure for more than three decades, producing three children, one of whom – a daughter, Karen – tragically died in infancy.

Armstrong's aviation career expanded into experimental piloting when he joined the National Advisory Committee for Aeronautics (NACA), forerunner of NASA, and was initially based at the Lewis Flight Propulsion Laboratory in Cleveland, Ohio. Whilst there, he participated in the evaluation of new anti-icing aircraft systems and high-Mach-number heat-transfer measurements, before moving to the High Speed Flight Station at Edwards Air Force Base in California to fly chase on drops of experimental aircraft. There, in an F-100 Super Sabre, he flew supersonically for the first time. On one occasion, flying with Stan Butchart in a B-29 Superfortress, Armstrong was directed to airdrop a Douglas-built Skyrocket supersonic research vehicle. However, upon reaching the release altitude, one of the

B-29's four engines shut down and its propellor began to windmill in the airstream. Immediately after dropping the Skyrocket, the propellor disintegrated, its debris effectively disabling two further engines. Nevertheless, Butchart and Armstrong were able to land the behemoth using its sole remaining engine.

His first flight in a rocket-propelled aircraft came in August 1957 aboard the Bell X-1B, reaching an altitude of 18.3 km, and three years later he completed the first of seven missions aboard North American Aviation's famous X-15 to the very edge of space. On one of these flights, in April 1962, just a few months before joining NASA's astronaut corps, he reached an altitude of 63 km. During descent, he held up the X-15's nose for too long and the aircraft literally 'bounced' off the atmosphere and overshot the landing site by some 70 km, but he returned and safely touched down. Although he was not one of the handful of X-15 pilots who actually reached space, exceeding 80 km, Armstrong's abilities in the rocket aircraft have been widely praised. The late NASA research flier Milt Thompson labelled him "the most technically capable" X-15 pilot.

In November 1960, by now flying for NASA as a civilian research pilot, Armstrong was chosen for the Dyna-Soar effort, ultimately leaving that project in the summer of 1962 as the selection process for the second group of astronauts got underway. At around the same time, he last flew the X-15, achieving a peak velocity of Mach 5.74. When his name was announced by NASA in September, he became one of only two civilian astronauts. Although Deke Slayton later wrote that nobody pressured him to hire civilians, fellow selectee Jim Lovell felt that Armstrong's extensive flying history within NACA and NASA made him an obvious choice for the final cut. In fact, Armstrong's application had arrived a week after the 1 June 1962 deadline, but, according to Flight Crew Operations assistant director Dick Day, "he was so far and away the best qualified ... [that] we wanted him in".

After admission into the so-called 'New Nine', Armstrong came to be regarded as by far the quietest and most thoughtful; indeed, his wife once told *Life* that 'silence' was an answer to a question and the word 'no' was an argument! As astronaut Frank Borman put it: "When he said something, it was worth listening to." His Apollo 11 crewmates Mike Collins and Buzz Aldrin would both characterise his nature as "reserved", and Dave Scott, the man who flew on Gemini VIII with him in March 1966, described Armstrong as "cool, calm and energised", who never operated in a frantic manner, but who could identify and resolve problems quickly, efficiently and smartly. All of these qualities proved vital on his first space flight, when the pair came close to losing their lives; and would prove yet more critical on his *second* mission, when he was tasked with the most exacting challenge of his aviation career.

A couple of days before Christmas 1968, as Apollo 8 was flying towards the Moon, Armstrong arrived in Mission Control as usual in his role as the mission's backup commander. Much of his time had been spent sitting at the capcom's mike or discussing experiments and making adjustments and recommendations to the flight plan. This day, however, would end quite differently. At some point in the afternoon, Deke Slayton pulled Armstrong aside for what James Hansen described as "a historic conversation". Slayton proposed that Armstrong should command Apollo 11, with Buzz Aldrin and Mike Collins for his crew.

By this time, Armstrong and Aldrin, who was also backing up Apollo 8, were coming up for reassignment and Slayton probably felt a pang of conscience for Collins, who had experienced a particularly difficult year as an astronaut. Collins was originally assigned to fly on Apollo 8, but was abruptly removed from the mission in the summer of 1968. One day, during a game of handball, he became aware that his legs did not seem to be functioning as they should, a problem which progressively worsened: as he walked down stairs, his knees would buckle and he felt a peculiar tingling and numbness. Eventually, with typical pilot's reluctance, he sought the flight surgeon's advice and was referred to a Houston neurologist. The diagnosis was that a small bony growth between his fifth and sixth cervical vertibrae was pushing against his spinal column and only surgery would relieve the pressure. Several days later, at the Air Force's Wilford Hall Hospital in San Antonio, Texas, he had an 'anterior cervical fusion' procedure in which the offending spur and some adjoining bone was removed and the two vertibrae fused together with a small dowel of bone drawn from his hip. Several months of convalescence followed, during which time his place on Apollo 8 was taken by his backup, Jim Lovell. Collins' rapid recovery from the surgery, together with an exemplary performance on his first space flight, Gemini X in July 1966, surely contributed to Slayton's decision to assign him to Armstrong's crew.

In fact, in his autobiography and in interviews, Collins has downplayed his resilience and skill, describing himself as "nothing special", "lazy" and "frequently ineffectual". Behind these words, however, was an obsessive hard-worker, a gourmet cook and a lover of French wine, whose gregarious nature led him to be called "smooth and articulate". By the time he applied to be an astronaut, the Air Force was sending its applicants to 'charm school' to teach social skills essential for spacefarers: wearing knee-length socks "that go on forever", abhorring hairy legs and needing to hold hands on hips in a particular way "because people you don't want to talk about hold 'em the other way!"

With a father, uncle and elder brother who would all rise through the ranks to become generals, it was obvious that Michael Collins would find a career in the military. He entered the world in Rome on 31 October 1930, becoming the first American astronaut born outside the United States, and during his childhood he was often on the move: from Italy to Oklahoma, to Governor's Island in Upper New York Bay, to Maryland, to Ohio, to Puerto Rico and to Virginia. Whilst in Puerto Rico, he took his first ride in a twin-engined Grumman Widgeon, but he would admit that as he neared graduation from West Point in 1952, his "love affair with the airplane had been neither all-consuming nor constant".

Nonetheless, he graduated from the Military Academy and his choice of the Air Force as his parent service was based on two factors. The first was sheer wonder over where aeronautical research would lead in years to come ... whilst the second was simply to avoid accusations of nepotism, "real or imagined", since his uncle happened to be the Army's chief of staff at the time! As a cadet, he completed initial flight training in Mississippi aboard T-6 Texans, before moving on to jets, flying the F-86 Sabre at Nellis Air Force Base in Nevada.

Nuclear-weapons-delivery training followed at George Air Force Base in

The amiable strangers of Apollo 11 check out the cabin of their command module, Columbia. From left to right are Neil Armstrong, Mike Collins and Buzz Aldrin.

California, as part of the 21st Fighter-Bomber Wing, and Collins transferred with the detachment to Chaumont-Semoutiers Air Base in France in 1954. Two years later, whilst participating in a NATO exercise, he ejected from his F-86 when a fire erupted behind his cockpit. He met Pat Finnegan in the officers' mess and, despite their differing religious beliefs – she being a staunch Roman Catholic, he a nominal Episcopalian – they were married in 1957.

Subsequent work as an aircraft maintenance officer, during which "dismal" time he trained mechanics, was followed by a successful application to join the Experimental Flight Test Pilot School at Edwards Air Force Base in California in August 1960. This involved flying on a totally new level. "Fighter pilots can be impetuous; test pilots can't," Collins recounted years later. "They have to be more mature, a little bit smarter ... more deliberate, better trained – and they're not as much fun as fighter pilots." By this time, he had accumulated over 1,500 hours in his logbook, the minimum requirement for a prospective student at the exalted school. (In fact, his class included future astronauts Frank Borman and Jim Irwin.)

Two years later, when John Glenn made America's first orbital space flight, Collins took notice and submitted his application for the 1962 astronaut intake. He underwent the full physical and psychological screening process, narrowly missing out on selection and, despite his disappointment, moved on to study the essence of space flight, flying the F-104 Starfighter to altitudes of 27 km and receiving a taste of weightlessness. He had barely returned to fighter operations when, in June 1963, NASA announced its intention to recruit another group of astronauts. Years later, Deke Slayton would write that the 1962 selection panel considered Collins a good candidate who was "held back to get another year of experience".

Initial instruction as part of the third group of 14 spacefarers, whom the press widely dubbed 'The Apollo Astronauts', included lunar geology, a subject for which Collins had no great enthusiasm or interest; ironic, perhaps, in view of where his career would eventually take him. Like Slayton, he felt that the New Nine was probably the best all-round astronaut group yet chosen, but admitted that the Fourteen were better educated: with average IQs of 132, an average of 5.6 years in college and one even having a doctorate from MIT. Completion of initial training led to an assignment to oversee the Gemini extravehicular suit and he would express annoyance at being left out of the loop in May 1965 when a closed-door decision was made to give Ed White a spacewalk on Gemini IV. In his autobiography, Collins described the suit and the astronaut's relationship with it as "kind of love-hate ... love because it is an intimate garment protecting him 24 hours a day, hate because it can be extremely uncomfortable and cumbersome". The suit, and the timeline for which astronauts were to get fitted for it, provided a never-ending source of rumour as to who would be next to be assigned to a mission. For Collins, recognition of his work came in June 1965 with a backup assignment to Gemini VII, along with Ed White. Despite falling ill with viral pneumonia soon thereafter, he recovered promptly and performed admirably, even taping a 'Home Sweet Home' card inside Jim Lovell's window on launch morning. His eventual assignment, with John Young, to Gemini X came in January 1966, by which time White had been named to the first Apollo mission. "I was overjoyed," wrote Collins. "I would miss Ed, but I liked John, and besides I would have flown by myself or with a kangaroo – I just wanted to fly." On a mission which included rendezvousing with *two* Agena target vehicles, Collins became the first man to make two spacewalks.

Subsequently named as Frank Borman's senior pilot on the original E mission, later redesignated C-prime and retasked as the first manned flight to the Moon, Collins' disappointment at not being able to fly that mission was bitter. As celebratory cigars were lit on 27 December 1968 and Apollo 8 triumphantly splashed down in the Pacific Ocean, Collins could only describe himself as "emotionally wrung-out". Ten days later, he, Neil Armstrong and Buzz Aldrin were called into Deke Slayton's office ... and into the history books.

However, there would be a slight shift of roles on Apollo 11. Following Jim Lovell's move to the prime crew of Apollo 8, Aldrin had been promoted to the backup 'senior pilot' role – later synonymous with the command module pilot and essentially a mission's second-in-command. However, in their discussion, both Armstrong and Slayton had more confidence in Collins to fill this role on Apollo 11. "I had a little difficulty putting Aldrin above Collins," Armstrong told James Hansen. "In talking with Deke, we decided, because the CMP had such significant responsibilities for flying the command module solo and being able to do rendezvous by himself and so forth, that Mike was best to be in that position." Thus, Aldrin missed out on being Apollo 11's second-in-command, but his 'demotion' had a silver lining: if the schedule ran as planned, he would become the second man to set foot on the Moon.

In many ways, Edwin Eugene Aldrin Jr stood out even amongst the over-achievers of the astronaut corps: he was a mathematical and engineering genius, the first astronaut to gain a doctorate prior to being recruited, an unquestioned expert in the field of orbital rendezvous ... and a constant worry to Deke Slayton. Aldrin had already raised managers' eyebrows during the Gemini IX-A mission in June 1966, when he suggested that spacewalker Gene Cernan use surgical scissors to cut a set of lanyards in order to free the jammed nose shroud on a docking target vehicle. Although it was not an overly outrageous suggestion, Aldrin's advice proved a little too adventurous. Senior managers were not the only ones with worries about him. In his autobiography, Cernan hinted strongly that Aldrin's intelligence was tempered by a tendency to fly off at tangents and drastically re-engineer everything, when NASA had little time to do so. Astronauts and their wives would roll their eyes whenever Aldrin cornered them, even over coffee, and spoke about the intricacies of celestial navigation and mechanics, giving rise to his nickname of 'Dr Rendezvous'. Coupled with reports of his performance in the simulator, he narrowly missed being dropped from the crew of Gemini XII. On that flight he made a total of three periods of extravehicular activity, both standing on his seat, poking his spacesuited head into the void, and actually leaving the craft to perform various tasks.

Born in Glen Ridge, New Jersey on 20 January 1930, the son of an Army Air Corps pilot and a mother with the propitious maiden name of Moon, Aldrin's development, even into adulthood, was very much controlled by his father. The senior Aldrin was a noted aviator in his own right and, it is said, had introduced Charles Lindbergh to rocket pioneer Robert Goddard, thereby indirectly causing the former to arrange a $50,000 Guggenheim research grant for the latter.

Naturally, in view of his father's military career, the man who would someday fly Gemini XII and become the second man to walk on the Moon was brought up with aviation in his blood. He first flew in an aircraft with his father in 1932, when he was barely two years old. (As a child, he earned the nickname 'Buzz' from his younger sister, who, unable to pronounce 'brother', called him her 'buzzer'.) Graduation was followed by enrolment in a military 'poop school' – aimed at preparing him for the Naval Academy at Annapolis – although Aldrin voiced an interest in the Military Academy at West Point. In spite of his father's preference for the Navy, which he felt "took care of its people better", his son persisted and eventually won his reluctant approval. But the pressure to excel continued. When Aldrin graduated third in his class from West Point in 1951, his father's reaction was a question: *who* had finished *first* and *second*? He was rejected for a coveted Rhodes postgraduate scholarship and instead entered the Air Force, earning his pilot's wings later that same year after initial training in Bryan, Texas. During the conflict in Korea, he was attached to the 51st Fighter Wing, flying F-86 Sabres, and by the time hostilities ended in the summer of 1953 he had no fewer than 66 combat missions in his logbook. A month before the end of the war, *Life* featured on its cover a gun-camera photograph from his aircraft showing a Russian pilot ejecting from his stricken MiG. In total, Aldrin returned from Korea having shot down three MiGs.

Back in the United States, Aldrin became a gunnery instructor at Nellis Air Force Base in Nevada and in 1955 was accepted into Squadron Officer School in Montgomery, Alabama. At around the same time, he met and married Joan Archer and shortly thereafter fathered a son, James. After a period serving as an aide to General Don Zimmerman, the dean of the new Air Force Academy, he was assigned to Bitburg in Germany in 1956 to fly the F-100 Super Sabre as part of the 36th Fighter-Day Wing. During this time, he became a father twice more: to Janice and Andrew.

Before pursuing his next ambition of test pilot school, Aldrin sought further education and was accepted into the Massachusetts Institute of Technology on military detachment for a doctorate of science degree in astronautics. His 259-page ScD thesis, completed in 1963, just months before he was selected as an astronaut, was entitled 'Line of Sight Guidance Techniques for Manned Orbital Rendezvous'. He chose this topic, he later wrote, because he felt it would have practical applications for the Air Force and aeronautics, although it also drew the attention of NASA, which was by now looking at lunar-orbital rendezvous for Apollo. His approach was so novel, it is said, that the doctoral committee accepted his thesis with 'reservations', but apparently a copy reached NASA, whose engineers and trajectory specialists began borrowing ideas from it even before Aldrin joined the astronaut corps. He had dedicated his thesis to future efforts in human exploration, wistfully remarking "if only I could join them in their exciting endeavours". In 1962 he had tried unsuccessfully to become an astronaut, seeking a waiver for his lack of test-piloting experience. When this requirement was dropped in 1963 he eagerly reapplied. A concern about his liver function, thanks to a bout of infectious hepatitis, did not prevent Aldrin from becoming one of the 14 astronauts named to the world that October.

Detailed to work on mission planning, Aldrin's early days saw him focusing his attention on rendezvous and re-entry techniques ... and, gradually, as each Gemini crew was named, he became increasingly frustrated at not receiving a flight assignment. At one stage, he even approached Deke Slayton to stress his confidence in his own abilities – that his qualifications and understanding of rendezvous far exceeded those of anyone else in the office – and was politely assured that his comments would be noted. Shortly thereafter, in early 1966, Aldrin and Jim Lovell were assigned as the backup team for Gemini X. His heart sank. Given Slayton's rotation system of first backing up a mission, skipping two and then flying the next, this was a 'dead-end' assignment because there would be no Gemini XIII to which Aldrin and Lovell could aspire. As Aldrin would later write: "Apparently, petitioning Deke – an arrogant gesture by 'Dr Rendezvous' – had not been well-received by the stick-and-rudder guys in the Astronaut Office. By being direct and honest rather than political, I'd shafted myself." All that changed on the last day of February, when Elliot See and Charlie Bassett were killed in an air accident and their Gemini IX backups were pushed into prime position. In mid-March, Lovell and Aldrin were named as the new Gemini IX backups, with a formally unannounced (but anticipated) future assignment as the prime Gemini XII crew.

For Aldrin, whose Nassau Bay backyard bordered that of the Bassetts, it was a devastating way to receive his long-desired flight assignment. Three weeks after the accident, he and Joan visited Jeannie Bassett to tell her the news. "I felt terrible," he wrote, "as if I had somehow robbed Charlie Bassett of an honour he deserved." Jeannie responded with quiet dignity and characteristic grace; her husband, she explained, felt that Aldrin "should have been on that flight all along ... I know he'd be pleased".

In the wake of Gemini XII, on which Aldrin excelled by manually calculating the rendezvous when the computer failed, he found himself assigned with Neil Armstrong and Jim Lovell to the backup crew of Apollo 9 – Frank Borman's original E mission – but circumstances changed rapidly in the summer of 1968. The E mission was cancelled in favour of C-prime and Borman's crew was now rededicated to undertaking the first circumlunar expedition. Moreover, Lovell left the backup crew to replace Mike Collins and Aldrin was 'promoted' from pilot to senior pilot. Rounding out the 'new' backup crew was rookie astronaut Fred Haise. Then, several days after Apollo 8 splashed down, the careers of Armstrong, Aldrin and Collins converged and history was made. Just as Jason and the Argonauts came to be revered by the ancient Greeks as the most famous seagoing crew that ever lived, so the men of Apollo 11 became the most celebrated spacegoing crew ever assembled: 'Moon Team One'.

THE MAN WHO WOULD BE FIRST

For all the fame that came their way in the opening months of 1969, and indeed *ever since*, to this day, some observers have seen the Apollo 11 trio as lacking the camaraderie which other crews had in spadefuls. Whereas the Apollo 12 astronauts – old Navy buddies Charles 'Pete' Conrad, Dick Gordon and Al Bean – were seen speeding around in their matching red, white and blue Corvettes, each with a licence plate bearing his respective crew position (CDR, CMP and LMP), the men of Apollo 11 were more likely to arrive at training separately, to go off to lunch separately, to talk only of technical matters and to act in some ways like passing acquaintances. Mike Collins once referred to himself and his crewmates as "amiable strangers".

Admittedly, their competence as a crew was unquestioned, but during their six months together, a number of issues did arise. One of the most famous and oft-discussed has been whether Armstrong (as the mission's commander) or Aldrin (as lunar module pilot) would be first to set foot on the Moon. Obviously, at the beginning of January 1969, with Jim McDivitt's D flight and Tom Stafford's F mission still in the future, there really was no way to know whether Apollo 11 would be in a position to attempt a lunar landing. However, Deke Slayton had told Armstrong to tailor their training specifically with that goal in mind ... and, from Day One, the media had expected nothing less. When the inevitable question was posed at the press conference to announce the crew, Armstrong diplomatically handed it over to Slayton, who responded that the decision had yet to be finalised. One of the key factors, explained Slayton, was that further simulations in the lunar module were needed to inform this decision. Indeed, Armstrong elaborated, the choice of who would be first depended entirely on how best to execute their timeline on the lunar surface and *not* upon individual whim or desire. He *did* reveal, however, that the First Man would be on the surface for around 45 minutes before the Second ventured outside. In fact, tentative plans were already laid out for a 'Moonwalk' lasting perhaps two and a half hours to erect a flag, collect soil and rock specimens and assemble a set of scientific instruments.

In his biography of Armstrong, James Hansen noted that there was "no doubt" in Aldrin's mind, at least in those early months of 1969, that *he* would be first to walk on the Moon. Aldrin's rationale came from the Gemini precedent, where a commander remained inside whilst his pilot performed extravehicular activity. Some reporters shared this view and several quoted George Mueller as saying that Aldrin *would* be first. However, shortly after Apollo 9 returned to Earth the situation changed markedly and rumour spread that Armstrong would be first. Reasons for this choice, which seems to have been set in stone from April onwards, are both numerous and complex and have been much-debated over the years.

One of the earliest and, for Aldrin, perhaps the most denigrating to his parent service, the Air Force, was that Armstrong was a civilian astronaut and he was assigned the historic role because NASA, a *civilian* organisation, did not want militarism to blight humanity's first footsteps on another world. At the time, of course, America remained embroiled in the Vietnam War, which continued to consume ever more lives and the military was deeply unpopular. Left unsaid in this scenario was that Armstrong had flown combat missions as a naval aviator in Korea. Even Aldrin, despite holding the rank of colonel, had not served in any formal Air Force capacity since being selected as an astronaut six years earlier. Regardless, Aldrin felt that making such a judgement based purely on the civilian/ military question insulted the Air Force by portraying it as "some sort of warmonger". When he brought up the issue with Armstrong, he found his commander to be non-committal. Nevertheless, Armstrong agreed that the decision *was* a historic one and refused to rule *himself* out.

Aldrin next approached fellow astronauts Gene Cernan and Al Bean, both of whom were training for LMP slots themselves, to seek their advice ... and *this* resulted in a consensus throughout the corps that 'Dr Rendezvous' was actively lobbying to be first. In his autobiography, Cernan has proven extremely critical of several astronauts and Aldrin received particular venom: "Buzz had pursued this peculiar effort to sneak his way into history and was met at every turn by angry stares and muttered insults." Cernan wondered how Armstrong managed to put up with such apparent back-biting for so long.

To be fair, each reason that Aldrin presented to his fellow astronauts for why he was best-suited to go outside first was valid and technical and based entirely on procedures and the demands of the checklist. It was part of Aldrin's job to plan the lunar surface activity. As the mission's commander, Armstrong would have his hands full with making the landing – why, pondered Aldrin, should he be saddled with the additional demands of suiting up first and diving headlong into the physiological intensity of humanity's first Moonwalk? On top of this, Aldrin had *performed* a spacewalk on Gemini XII; he knew its difficulties, the workings and idiosyncrasies of the space suit ... and Armstrong was no fitness fanatic. During training for Gemini VIII, he had once joined crewmate Dave Scott in the gym. As Scott pumped iron, Armstrong put the exercise bike into its lowest possible setting and began pedalling slowly, observing that a man only had a finite number of heartbeats and it was best not to waste them!

Other Apollo commanders rolled their eyes and gritted their teeth in the face of Aldrin's lobbying and many of them who were naval officers turned to their seafaring experience for a rationale for Armstrong to get out first. "The Gemini precedent" of the pilot doing the spacewalk, wrote Andrew Chaikin, "didn't apply, because a lunar module sitting on the Moon wouldn't be in flight – it would be *in port*. And as any naval officer knows, the protocol on such matters is clear: When the ship comes to port, the skipper is always first down the gangplank." Deke Slayton made a similar point when he stopped by at Aldrin's office: Armstrong was the senior astronaut and as such he *should* go first, in the same way that Christopher Columbus and other explorers were doubtless first to disembark from *their* vessels.

Over the years, Aldrin has said that his motives were misinterpreted and that the actual outcome did not bother him as much as the need to reach a decision. In Hansen's biography, Aldrin is quoted as admitting that it *would* have been "inappropriate" for him, the junior astronaut, to have gone outside first, uttered the famous first words and collected the priceless first samples, observed by his commander. Still, it is not difficult to speculate that the input of Aldrin's father may have also been a contributory factor in the mix. When he first described the process of 'egress' – going outside – to his father, the older Aldrin reacted with surprising anger, threatened to "do something about it" and tried to pull strings among his high-level friends at NASA and the Pentagon to get the plan reversed. Aldrin Senior, wrote Deke Slayton, "just couldn't seem to leave well enough alone".

At length, the decision came down to pure practicality ... or so it seemed. The interior of the lunar module, as has already been explained, was about the size of a small broom cupboard and the square hatch opened *inwards*, with the hinge on its right edge. This required the astronaut on the right-hand side of the cabin – the LMP – to pull it open and stand back in his corner, whilst the commander got down on hands and knees and reversed himself through the hatch, onto the tiny porch to descend the nine-rung ladder to the surface. For Aldrin to go first would require the two men to swap places in the cabin – a difficult and potentially damaging act given that both would be wearing space suits and bulky backpacks. When faced with the risk of accidentally hitting a switch or circuit breaker, it was more straightforward and safer to go with the design and let Armstrong go first.

Finally, on 14 April, with just three months to go before launch, a press conference at the Manned Spacecraft Center, presided over by George Low, head of the Apollo Spacecraft Program Office, revealed the outcome: Armstrong would be first to step onto the Moon. Next day, an editorial cartoon showed the two men opening the hatch immediately after landing as confused Alphonse and Gaston characters, both politely offering the other the chance to go first – whilst at the same time discreetly muscling their way ahead of each other. Humour aside, stories abounded (including one by a disgruntled public affairs officer) which claimed that Armstrong had 'pulled rank' and demanded that he be first on the Moon. This was a point endorsed by Mike Collins in his autobiography: "Neil ignored [original plans] and exercised his commander's prerogative to crawl out first." Such stories garnered sufficient public interest for George Low to admit that preliminary studies several years earlier, *had* called for the LMP to venture outside first, but that simulations and plans led to a recommendation for the commander to go first.

Buzz Aldrin has argued that he was "fine" with the decision, although other astronauts, engineers and managers have said otherwise. Mike Collins, for one, recounted a distinct element of melancholy and coolness in the air immediately after the announcement; and *this*, it would seem, was the final nail in the coffin for Aldrin being first on the Moon.

Years later, Chris Kraft, who was in 1969 the head of flight operations at the Manned Spacecraft Center, spoke candidly about the fact that Armstrong *was* the best choice to take the historic first step ... not just because of the position of the hatch or because of his seniority or because or his civilian status or even the tasks to be performed on the surface. It was simple: Armstrong had no ego. The first man on the Moon would be remembered forever as the Charles Lindbergh of his generation, "a hero ... beyond any soldier or politician or inventor", and it was Armstrong's very lack of ego, his calmness under duress, his quietness, his understated confidence and his desire *not* to put himself in the spotlight of fame and attention that made him the perfect choice and the best ambassador. (It is ironic that after the mission Armstrong shunned publicity and Aldrin enthusiastically courted it and made vocal contributions to the future of lunar exploration.) The notion that Armstrong was chosen to be first *because* he was Neil Armstrong is supported to an extent by Al Bean, who became the fourth man to walk on the lunar surface. He felt sure that, if NASA really wanted Aldrin to go out first, it would have been relatively easy for men to exchange places in the cabin *prior* to donning their backpacks. Once in place – with Aldrin now on the left and Armstrong on the right – they could don their backpacks and Aldrin could go outside ahead of his commander. There would be no damage to the cabin, no damage to either man's suit ... and *no problem*. Bean's point seemed to be that the choice of Armstrong to be first had *already* been made and that the way in which the hatch was hinged had been used as a convenient excuse by NASA management to end the debate.

Whatever the truth, the *process* of deciding the identity of one of the most famous men the world has ever known is now merely a historical curiosity. Yet the *reality* is that the name of Neil Armstrong – the quiet civilian, the man with no ego, the 'Lindbergh' of his generation – will be remembered for centuries to come.

TRAINING

Of course, the discussion of who would be first to walk on the Moon would be wholly moot if Armstrong and Aldrin could not set their spidery lunar module safely onto the plain of Tranquillitatis. This required them to pilot the craft, 'balanced' on the thrust of its descent engine, in a precise, sweeping arc known as the Powered Descent. It would start in lunar orbit, with the lander's windows facing directly 'down' toward the surface, 15 km below. Armstrong would use the descent engine to brake his speed, causing the lander – which the crew had named 'Eagle' – to start to arc downwards. Four and a half minutes into the descent, under command from its onboard computer, the lunar module would rotate into a 'windows-up' position in order to allow the radar on its underside to begin to determine altitude and rate of descent. Ninety seconds later, the computer would throttle the engine down to 55 percent of its rated thrust to follow the precise profile required to land at the programmed location.

Finally, eight and a half minutes into the descent, and around two and a half kilometres above the Moon, the computer would perform a manoeuvre known as 'pitch-over', rotating the craft into an almost-upright orientation – some 20 degrees off vertical – to enable the descent engine both to continue to brake its forward motion and also to prevent it from falling too fast. At this point, if all went to plan, four minutes would remain until touchdown. Andrew Chaikin has described these adrenaline-fuelled minutes as a period of 'fast time', with no opportunity for the astronauts to read checklists or consult with Mission Control; Armstrong and Aldrin would be on their own and would need to have the flight plan *memorised*. That plan encompassed not only the physical and mental demands of Eagle's descent, but also recognising the landmarks which would guide them to their landing site, including yawning craters, steep-sided mountains and sinuous rilles which were committed to memory from months spent studying Lunar Orbiter photographs.

It was theoretically possible for the computer to fly a perfect trajectory and land the lunar module, but it could not 'see' and hence could not manoeuvre to avoid a vast impact crater or a field of jagged boulders. Armstrong planned to assume a degree of manual control 150 metres above the surface, allowing the computer to operate the descent engine whilst he adjusted the rate of descent and steered Eagle to a satisfactory location on which to set down.

Emergencies arising during this time had the potential to be catastrophic. If anything went wrong – perhaps the descent engine might malfunction or the computer or radar might fail – it would be incumbent upon Armstrong to punch the Abort Stage button on his control panel. This would automatically fire a series of pyrotechnic bolts to separate the ascent stage from the descent stage and ignite the ascent engine to commence the climb back into lunar orbit as the first step in a rendezvous with Collins. However, the time available for Armstrong and Aldrin to do this, and survive, was itself limited. At some point, not much less than a hundred or so metres above the surface, the lander would be too low and moving too quickly for him to call an abort and boost clear of the descent stage. This was called, somewhat darkly, 'the Dead Man's Curve', after an old helicoptering term. Even in the weeks leading up to Apollo 11's launch, engineers debated how close to the wire Armstrong would actually be able to go.

Thirty metres above the lunar terrain, with the computer still controlling the engine, it would be up to Armstrong to execute the final phase of the descent, guided by Aldrin reciting altitude measurements, horizontal speed and rate-of-descent data.

The first contact with the Moon would come when any of three long metal probes extending from the lander's footpads penetrated lunar soil, causing a pair of blue Lunar Contact lights in the cabin to illuminate; one on the control panel in Armstrong's peripheral vision and the other down by the computer display where it would attract Aldrin's attention. Instantly, Armstrong would shut off the descent engine and Eagle would fall the final metre or so to the surface in the weak lunar gravity. To Armstrong, the point, *now*, over who would be First Man would be truly irrelevant: for he and Aldrin would land *at the same time*. In Armstrong's mind, they would *both* be first on the Moon.

Preparing for this audacious mission had consumed all three men since January; in their six months together, they would log more than three and a half *thousand* training hours – split between them, this equated to 42-hour work-weeks focusing on formal training exercises, not counting an additional 20 or so hours per week studying, reading and catching up on paperwork, including ever-changing flight plans and procedures. James Hansen related that the lunar module crew spent around 1,300 hours training *apiece*. Collins, whose activities were mainly centred on operating the command and service modules, clocked a little more than 1,000, of which he spent perhaps 400 in the command module simulator rehearsing his own flight plan, including the 32 hours he would spend alone.

The simulator, Collins related, was a huge, gawdy contraption, accessed through curtains atop a carpeted staircase. During the early days of Apollo, it was rarely in full working order and, in his mind, it would have been easier to get the *spacecraft* to fly than to get the simulator to duplicate it! Nonetheless, it was in this simulator, often alone, that he prepared for the role of the command module pilot; a role which would involve all of the 'normal' responsibilities of transposition and docking and TLI and TEI and managing the spacecraft's systems … and many others, too. Secreted away in what he called his 'solo' or 'rendezvous' book were 117 pages of Collins' own hand-scribbled procedures, computer entries, diagrams and notes that he *would* need and those that he *might* need in the event that he had to rescue Armstrong and Aldrin. No fewer than 18 separate rendezvous scenarios – each dealing with a different situation, from a normal ascent from the lunar surface to a hair-raising rescue from an unstable abort – had to be mastered, as did the overwhelming demands of the guidance and navigation system (loaded with an aptly named program called 'Colossus IIA'), the need to be ready to fix or dismantle a balky docking probe, the need to be prepared to guide the command module manually back to Earth and a multitude of others.

So great were the demands on the three men that serious consideration was given to postponing the launch until August. (Due to the need for favourable lighting and tracking conditions, only a few days were available each month to launch for a particular landing site on the Moon. The 'launch windows' for an Apollo 11 landing at Tranquillitatis were 16–21 July, 14–20 August and 13–18 September.) NASA Headquarters, in particular, was fearful that there was just *too much* still to do to support a July launch. The decision had to be made soon, because the Saturn V had been on Pad 39A since late May and technicians were almost set to begin loading hypergolic propellants aboard the Apollo spacecraft; these were so corrosive that they could not sit in the ship's plumbing for more than a few weeks. "No one likes to delay a launch," wrote Collins, "especially because he has been deemed unprepared; yet neither [Armstrong nor Aldrin] seemed to be speaking up in favour of a July launch. I couldn't tell whether they were being extra cautious or whether they truly would be unprepared to meet certain situations that might arise." Unsurprisingly, the astronauts opposed the idea of postponing until August; they told Deke Slayton they *would* be ready and felt that additional time would make things only marginally better. After a nine-hour flight readiness review on 17 June, Sam Phillips formally announced 16 July as Apollo 11's date with destiny.

With less than a month to go, and Armstrong, Collins and Aldrin declared fit and healthy, the three men were secluded in the crew quarters at Cape Kennedy for the remainder of their time on Earth. They exercised, reviewed flight plans and continued their punishing schedule in the simulator. Their instructors had gradually made the training ever more difficult, throwing a variety of problems at them – stuck-on thrusters, a failed descent engine or a radar gone bad in the case of Armstrong and Aldrin; then failed parachutes, a faulty docking probe or an incorrect re-entry alignment of the command module for Collins. Sometimes they succeeded, but in some cases the problems were so difficult that they were beyond recovery.

One day, late in June, Armstrong and Aldrin were in the simulator, rehearsing their descent to the Moon, when the instructors threw a stuck-on manoeuvring thruster glitch at them. The attitude indicator immediately began to tumble and the men saw the televised picture of the lunar terrain dancing in their windows. At length, Aldrin – then Mission Control, who were by this point also involved in the training runs as part of a so-called 'integrated simulation' to prepare *them* for the mission – advised Armstrong to abort, but it was too late: the image of the Moon froze and everyone knew, if *that* happened for real, Eagle would be scattered across the surface in a million pieces.

Despite 'only' being a simulation, the fact that this unfortunate training event occurred so close to launch left Aldrin ill at ease, particularly since the 'First Man' decision had only recently been concluded. In his autobiography, Mike Collins recalled a decidedly frosty encounter between himself and Aldrin on the night after the simulation. "I could not discern," he wrote, "whether he was concerned about his actual safety in flight, should Neil repeat this error, or whether he was simply embarrassed to have crashed in front of a roomful of experts in Mission Control." The commotion led Armstrong, in his pajamas, to enter the room indignantly. In Armstrong's mind, his actions had been a good test of how far he could push his *own* abilities in the one place (the simulator) where he *could not* kill himself. Moreover, it had been a useful test of the decisiveness of Mission Control, who had *themselves* not responded quickly enough to an emergency. In other words, after a certain point in the descent to the Moon, they would be on their own.

Other simulations, thankfully, eased the tension. Aldrin recalled a frustrating session when he and Armstrong were guiding the lander down to an imitation lunar surface for perhaps the tenth time, looked up at their windows ... and saw an enormous bug on the looming Moonscape. It turned out to be a dead horsefly, stuck on a pin in front of the television camera providing the lunar images. The person responsible for the light-hearted prank was one of their normally diabolical training instructors ...

Early in July, the crew was given a much-needed respite: a weekend at home in Houston with their families. Nevertheless, even part of *that* was given over to a press conference in which the three men – by now well into their mandatory three-week quarantine period – had to wear hospital-style masks and speak from within a three-sided plastic booth with an air conditioner blowing air out over the reporters. Author Norman Mailer likened them to "razorback hogs" and a few journalists even donned masks themselves in order to poke fun at NASA's medical precautions.

When asked if any special precautions had been taken to prevent them from falling ill, Collins retorted, tongue-in-cheek, that his wife and kids had signed a statement promising that none of them had any germs! On a more serious note, NASA flight surgeon Chuck Berry had gone so far as to request that President Richard Nixon should *not* have dinner with the astronauts on the night before launch, on account that he *may* infect them. "Had we been operating behind a germ-free barrier," wrote Collins, "this might have made a modicum of sense, but we were in daily contact with dozens of people." To be fair, admitted Aldrin, there *was* a bad summer flu bug in the air and Apollo 11 definitely did *not* want to risk falling foul of that; although the chance of the president being affected was remote. For his part, Nixon graciously bowed to Berry's request. At the end of the day, he was in a lose-lose situation: if he came to dinner and one of the astronauts fell sick in space, it would be seen as *his* fault, and even if all three sailed through the mission in perfect health, he would be criticised for 'callously' disregarding the flight surgeon's advice.

By Monday 7th, Armstrong, Collins and Aldrin were back at Cape Kennedy, in the simulators or studying procedures and checklists. One personality with whom they had daily contact was their chef, Lew Hartzell, the former tugboat cook, who routinely presented them with 500-calorie desserts, "monolithic" mountains of meat and potatoes, endless salads and rolls. "It did no good," wrote Mike Collins, "to tell Lew that you were on a diet; he took no offence, he simply ignored this irrelevant information!"

Another person who obviously *was* sufficiently germ-free was NASA Administrator Tom Paine, newly appointed to the post since January, and he came to spend a quiet dinner with Armstrong, Collins and Aldrin on the evening of 10 July and made a quite remarkable promise and an offer to them. Don't take *any* unnecessary risks, he told them. If anything should go wrong during the mission, or during the landing attempt, they were to call an abort and head back home; and Paine would see to it that they were assigned the *very next* mission in order that they could try again. The offer was remarkable for being so out of the ordinary ... but, after all, the first Moon landing was no ordinary mission and now, after having been in the public spotlight for six months, the Apollo 11 crew was no ordinary crew.

That this mission was decidedly 'different' was also reflected in its official crew patch. Much discontent had been expressed, not least by NASA's head of public affairs, Julian Scheer, over the flippant callsigns that Jim McDivitt and Tom Stafford had chosen for their spacecraft: Gumdrop and Spider, Charlie Brown and

Snoopy. A modicum of dignity was needed for the historic voyage of Apollo 11. At the press conference in Houston, Armstrong revealed that the command and service modules were to be known as 'Columbia' and the lunar module as 'Eagle'. Neither requires major clarification beyond that they are both nouns of iconic significance to the United States; the former is its female personification – derived, of course, from Christopher Columbus and, incidentally, also an early candidate for the *name* of the country, following the Declaration of Independence – whilst the latter honours its symbolic national bird, the bald eagle.

Mike Collins was responsible for much of the design work that went into Apollo 11's patch; he disliked the emblems of McDivitt and Stafford as being too 'busy' and wanted something more simplistic, more pure and "something which unmistakably said peaceful lunar landing by the United States". At length, it was Jim Lovell, veteran of Apollo 8 and now serving as Armstrong's backup, who suggested the American bald eagle. The image of the *national* bird coming in to land at the United States' most distant and exotic outpost was irresistible for Collins and, after browsing an old copy of *National Geographic*, he found a suitable photograph and traced it onto thin paper. Next, he added an oblique sketch of the Moon, added the Earth in the background and finally the words 'Apollo' and 'Eleven' to cap it off.

Armstrong objected to the spelling of 'eleven', fearing that foreigners would find it difficult to pronounce. When it was noted that so, too, might the Roman numeral 'XI' prove confusing, it was decided to use the Arabic numerals '11'. Additional support came from Tom Wilson, one of the computer instructors on the Apollo 11 training team, who suggested including an olive branch – symbol of peace – in the eagle's beak as it swept down to the Moon. Finally, bordered and lettered in the brightest gold and the deepest royal blue, the design went to NASA Headquarters for official approval. "Washington usually rubber-stamped everything," wrote Collins. "Only this time they didn't, and our design came back disapproved. The reason? The eagle's landing gear, powerful talons, extended stiffly below him, was unacceptable. It was too hostile, too warlike; it made the eagle appear to be swooping down on the Moon in a very menacing fashion." The remedy was straightforward: the olive branch was moved from the eagle's beak to its claws and, hey-presto, all menace vanished. Admittedly, the eagle looked somewhat uncomfortable in his new posture and the astronauts certainly hoped that *their* Eagle would let go of its olive branch before its own mechanical talons came in for landing, but it was finally accepted. It proved so popular and so famous that it was revived a couple of years later on the reverse side of the Eisenhower silver dollar coin, issued during John Connally's term as Secretary of the Treasury. Naturally, the central position of the eagle on the patch made choosing a name for Armstrong and Aldrin's lander an easy one. By extension, 'Columbia' also had connections with the name of Jules Verne's imaginary cannon (the 'Columbiad') and its relationship with the eagle made it the perfect choice.

TO LAND ON THE MOON

As Neil Armstrong savoured his final days on Earth, his mind undoubtedly was drawn to the immense challenge ahead: simply *getting* to the Moon, inserting Apollo 11 into lunar orbit, undocking from Collins and completing a successful Powered Descent was by no means a walk in the park. Perhaps, as he lay in bed at night in the quiet crew quarters, musing over the days to come, he recalled an incident barely a year earlier when, flying an unwieldly, four-legged machine with thrusters spitting and spurting to aid direction and lift ... he had come within a *quarter of a second* of being killed.

Building a device that could properly simulate a lunar landing was a major effort in itself. In their own drive to place a man on the Moon, the Soviets had found that helicopter training offered one of the closest terrestrial analogues; indeed, cosmonaut Alexei Leonov was participating in helicopter and parachute training on the very day that Yuri Gagarin died in March 1968. NASA, on the other hand, soon realised that most helicopters, such as the Bell H-13, simply could not accurately mimic the descent trajectories and sink rates of the lunar module. "Helicopters," wrote James Hansen, "could approximate a variety of final descent trajectories, but to do that often required their flying for substantial periods inside the so-called Dead Man's Curve, the terminal phase ... where it would be impossible to abort safely without crashing into the surface." Nor could the Bell X-14A aircraft, with a vectored-thrust and reaction-control arrangement mirroring the British Harrier, provide a satisfactory alternative.

Instead, the Apollo landing crews – specifically the mission commanders – utilised an ungainly device known initially as the Lunar Landing Research Vehicle (LLRV) and later, in its final form, as the Lunar Landing Training Vehicle (LLTV). Although it offered the only realistic means of simulating a touchdown on the Moon, it gained well-earned notoriety and prompted some NASA managers to call for its cancellation: it was dangerous, unpredictable, sometimes uncontrollable and had come within a whisker of killing several of its test pilots. Yet when Chris Kraft asked Armstrong, point-blank, in the spring of 1969, for his opinion of the machine, the man destined to be first on the Moon spoke glowingly of it. "It's absolutely essential," he told Kraft. "It's just darned good training." Dangerous, yes, but essential. Armstrong's praise for it remains remarkable. A year earlier, on 6 May 1968, he took the LLRV for a spin above Ellington Air Force Base in Texas ... and came within a quarter of a second of death.

The two LLRVs built by Bell Aerosystems each consisted of a tubular frame of aluminium alloy trusses, with a General Electric turbofan engine mounted on a central gimbal so it could direct its thrust downwards. The engine was capable of lifting the entire vehicle to a test altitude of a couple of hundred metres, after which it would be throttled back to support five-sixths of its weight to simulate lunar gravity. Twin hydrogen peroxide rockets managed its rate of descent and horizontal movement, whilst 16 smaller jets, mounted in pairs, supplied roll, pitch and yaw controllability. On no fewer than *three* occasions, between 1968 and 1971, the ejection seat would come in exceptionally handy ...

The notoriously unpredictable Lunar Landing Research Vehicle during an early test flight in 1964. Note the helmeted pilot on the right side of the vehicle. The Apollo commanders spent many hours aboard this machine and, despite the dangers, each would consider it essential to perfecting the techniques for a lunar landing.

Development of the LLRV began in December 1961, just a few months after President Kennedy publicly committed America to landing a man on the Moon. NASA issued Bell with a feasibility study, followed by a $3.6 million contract in February 1963 to construct it. In late October of the following year, NASA research pilot Joe Walker test-flew the four-legged device above Edwards Air Force Base in California for about 60 seconds, reaching a peak altitude of 10 m. 'Balanced' atop the thrust of its downward-directed engine, it flew like a dinner plate on a magician's broomstick. Walker would later compare it to an elevator ride, accompanied by the eerie hissing of the reaction controls, which enveloped the craft in clouds of peroxide steam. Indeed, the nickname 'Flying Bedstead' was quickly superseded by the moniker of 'Belching Spider'. It could be flown in terrestrial or lunar modes; in the latter, the engine was adjusted to reduce its apparent weight and imitate one-sixth of terrestrial gravity. "In the lunar simulation mode," Armstrong told James Hansen, "uncomfortably large attitudes were required for reasonable decelerations." However, the sensitive throttle of the engine made controlling its *altitude* much more benign.

Subsequent flights, piloted by Walker, Don Mallick, Emil 'Jack' Kluever, Joe Algranti and Harold 'Bud' Ream, together with the finalisation of the design of the lunar module itself, filled NASA with sufficient confidence to instruct Bell in mid-1966 to construct three Lunar Landing *Training* Vehicles, each costing around $2.5 million. By the time they were delivered in the winter of 1966–67, the design had shifted somewhat to include a three-axis control stick for the pilot and a more restrictive cockpit view, both of which were characteristic of the Grumman-built lunar module. Ultimately, NASA would employ five vehicles (two LLRVs and three LLTVs) to prepare astronauts for landings on the Moon.

The new 'training' variant was physically similar to the research vehicle, but its systems were specifically engineered to replicate as closely as possible the trajectories and control mechanisms of the 'real' lunar module. In addition to the three-axis stick and enclosed cockpit, it incorporated rate-command/attitude-hold controls, which closely mirrored the handling characteristics of the lunar module, together with a compensation system to sense, 'damp-out' and correct any aerodynamically induced forces and other unwanted motions. The engine was uprated, the tanks for the thrusters were enlarged and the ejection seat was upgraded. Nevertheless, the trainer remained hazardous. One modification was a change from analogue 'fly-by-wire' controls to a digital system like that of the lunar module. "Unfortunately," wrote Hansen, "dead periods existed within the circuitry of the digital system, during which the pilot could not sense the loss of electrical power." In early 1971, NASA research pilot Stuart Present was forced to eject from an LLTV as a result of just such an electrical failure. The machine crashed and exploded; Present, thankfully, survived.

Early in May 1968 Neil Armstrong became one of the first to learn the truly unforgiving nature of the machine. According to eyewitnesses, he took off from Ellington Air Force Base and had the LLRV under solid control for the first five minutes, climbing to an altitude of about 100 metres. As he prepared to perform his vertical descent and landing ... the machine suddenly *dropped*! "Quickly," he told Hansen, "control was non-existent. The vehicle began to turn. We had no secondary control system that we could energise – no emergency system with which we could recover control. So it became obvious as the aircraft reached 30 degrees of banking that I wasn't going to be able to stop it." With limited time available, he temporarily steadied the LLRV and managed to climb a little, but when the craft began to violently flip forward, backward and sideways he had no option but to eject. As he drifted beneath his parachute, he could only watch as the ailing vehicle crashed and exploded. He landed safely not far from the fireball.

Observers reckoned that Armstrong missed death by a fraction of a second. As Buzz Aldrin put it: "If the trainer had tipped completely over and he had fired his ejection seat, the rocket charge would have propelled him head-first into the concrete below. Neil held on for as long as he could, not wanting to abandon an expensive piece of hardware. At the last possible moment, he realised the thruster system had completely malfunctioned and he pulled his ejection handles." According to fellow astronaut Al Bean, the unflappable Armstrong dusted himself down, nursed a lacerated tongue and went back to work in his office ...

The accident investigation board included Joe Algranti and astronauts Bill Anders and Pete Conrad. It found that a poorly designed thruster had led to a propellant leak. The resultant loss of helium pressure in the propellant tanks caused the attitude-control thrusters to shut down and culminated in a loss of control capability. Contributory factors were excessively windy conditions at Ellington that afternoon, which forced Armstrong to use the attitude controls almost continuously. "The way we had designed the system," recalled Gene Matranga of the Flight Research Center at Edwards Air Force Base, quoted by Hansen, "the fuel would go down so far and then there was a standpipe that allowed you to have fuel saved for the lift rockets in case you had to use the lift rockets to recover the vehicle if the jet engine failed. Neil should have shut off the lift rockets and saved his fuel for the attitude rockets and go back to the jet engine. Nobody warned Neil of that. What happened was the helium pressure, which sat on the top of the fuel to pressurise the system, was expended very rapidly and Neil wound up having fuel but no pressurising gas, because the lift rockets were in the open position. In essence, he had no control."

In October 1968, at around the time that Apollo 7 was launched, two reports were published, both urging design improvements, but also recommending that the programme continue. It proved somewhat premature. A few weeks later, on 8 December, Joe Algranti took off from Ellington in an LLTV, but was forced to eject when large lateral-control oscillations developed as he descended from an altitude of a few hundred metres. Algranti, who had had flown the machine more than two dozen times and was an instructor pilot, was unharmed, but the machine itself crashed in flames ...

Not until April 1969, when *another* accident commission, chaired by former astronaut Wally Schirra, released its findings, could LLTV operations resume. Many within the higher echelons of NASA wanted it cancelled outright. Chris Kraft hated it and Bob Gilruth, head of the Manned Spacecraft Center and a virtual father figure to many of the astronauts, felt that a fatality was inevitable. Aside from Stuart Present's brush with disaster, the commander of every lunar landing mission underwent extensive LLTV training before launch. Each flight had sufficient propellant for six and a half minutes in the air, more than half of which was needed to practice landing manoeuvres. In total, besides Armstrong, astronauts Frank Borman, Bill Anders, Pete Conrad, Dave Scott, Jim Lovell, John Young, Al Shepard, Gene Cernan, Dick Gordon and Fred Haise flew the machine. It was retired after Cernan's last flight on 13 November 1972, three weeks before he commanded the final Apollo lunar mission.

As a training aid, admittedly, it was a sluggish affair, requiring each and every manoeuvre to be thought out in advance, and its limited fuel supply and inherent instability meant it gave its pilot just *one* opportunity to make a landing. Yet, in the vacuum of the lunar environment, with round-topped mountains and crater-pocked mare growing steadily larger in the window, the real thing would be no less forgiving of error. The trainer's role in enabling the first manned landing on the Moon was summed up by Bill Anders. "In my view," he said, "the LLTV was a much undersung hero of the Apollo programme."

Undoubtedly, the retirement of the trainer came as a blessed relief for many managers. By January 1971, two LLTVs and one LLRV had been lost. Only one of the second-generation trainers remained. "Having had three accidents and having that one vehicle left," recalled Pete Conrad, "Gilruth asked the guys to figure out how many flights we got on a vehicle before we crumped one. It turned out to be 260 flights or something like that. To finish the training after the third accident [the one involving Stuart Present], they had to fly 240 more flights; and so when Gene flew the last flight in his training ... nobody was *ever* going to fly that thing again as far as Gilruth was concerned!" The last LLRV resides at the Dryden Flight Research Center at Edwards Air Force Base in California, whilst the LLTV hangs from a lobby ceiling at the Johnson Space Center (the old Manned Spacecraft Center) in Houston, Texas.

As June wore into July 1969 and the clock at Cape Kennedy counted down to an Apollo 11 liftoff on the 16th, Armstrong – the man almost killed by the trainer a year earlier – had nineteen LLRV and LLTV flights in his pilot's log. For three straight days in June, he flew the LLTV eight times, performing a variety of tricky descent profiles under different conditions. "No other astronaut," wrote James Hansen, "before or after Armstrong flew the vehicle so much." By launch day, he had spent 417 hours preparing for the Moon landing in either the training vehicles or in the lunar module simulator. As his second launch into space drew closer, every second of that experience, every lurching movement and every thruster spurt from the LLRV or LLTV, every success and every failure, would prove essential in meeting the greatest challenge of his career.

THE JOURNEY BEGINS

If Armstrong harboured anxiety over the days to come, that anxiety must surely have been magnified a hundredfold for his wife, Jan, and sons Rick and Mark. Early in July, Jan had phoned Dave Scott's wife, Lurton – the pair had remained close friends ever since their husbands trained together for Gemini VIII in early 1966 – to ask for help making arrangements to watch the launch, away from the million or so people expected to descend on Cape Kennedy. She had already been invited to view the launch from a motor cruiser owned by North American that was moored in the Banana River and, thanks to Scott, who knew a few senior managers, they were able to fly to Florida together in one of the company's corporate jets.

By early morning on 16 July, the weather at Cape Kennedy was sweltering; indeed, one observer described it as being so hot that the air felt like a silk cloth brushing across his face. Yet the historic nature of what was about to happen was magnetic. "Everybody and his brother wanted to be at the launch," wrote Deke Slayton, "senators, congressmen, ambassadors." Twenty thousand VIPs were on NASA's official guest list, including General William Westmoreland – recently back

Rollout of the Apollo 11 vehicle in May 1969. Note the white room alongside the command module's hatch and the escape tower which would pluck Armstrong, Collins and Aldrin away in the case of an emergency.

from commanding US forces in Vietnam – together with Johnny Carson, Ted Agnew and a direct descendent of Napoleon Bonaparte. Lyndon Johnson, the former president and successor to the man who had committed America to the Moon, was there, as was former NASA Administrator Jim Webb. There were also two thousand journalists in attendance, almost half of them from abroad, representing 56 nations. One Czech writer noted an overwhelming sense of goodwill, even as the ugly cloud of war continued to hover overhead: "*This* is the America we love," he told his readers, "one so totally different from the America that fights in Vietnam." Others took the opposing view, with a handful of pro-communist newspapers operating from Hong Kong expressing criticism of the mission as an attempt to cynically distract the world from the horrors of the conflict and extend American "imperialism" into the heavens.

In his role as head of flight crew operations, it was Slayton's job to keep the media away from the astronauts, but even *they* found themselves taking late-night phone calls from long-lost relatives and old school friends in those final few days.

By launch morning, the headlights of a quarter of a million cars twinkled in the pre-dawn darkness as spectators arose from their backseats, from their tents, from beneath makeshift blankets, from inside their camper vans and even from their boats anchored in the Indian and Banana Rivers. Those fortunate to have actually *been* there would later say that the proceedings *did* exhibit something of a 'carnival' atmosphere – there were snack bars and bermuda-toting and bikini-clad spectators firing up barbecue grills and opening beer-filled coolers – but the sensation was relatively calm. This was particularly true when the three astronauts emerged into the glare of television lights, headed for the launch pad. "You get a feeling," CBS commentator Eric Sevareid told veteran anchorman Walter Cronkite, "that people think of these men as not just *superior* men, but *different* creatures. They are like people who have gone into the other world and have returned and you sense that they bear some secrets that we will never entirely know ... "

The eyes of the world were truly riveted on this event, none more so than in Armstrong's home town of Wapakoneta, where his parents – Steve and Viola – watched the proceedings on a colour set donated by the television network. On the evening before launch, more than two hundred cars circled the area close to their home. The mayor requested that everyone display American flags in their windows, the local dairy sold its own 'Moon Cheeze' and restaurants provided daily supplies of pies, bananas and chips. Local children began to claim *their* father was Armstrong's barber, *their* mother was Armstrong's first girlfriend, and so on.

Notwithstanding this public and media frenzy, some felt that the secrets of the Moon were better left alone, until other, more pressing, more earthly issues had been addressed. In many parts of America, Apollo's $25 billion price tag had been a hard pill to swallow and a hefty proportion of taxpayers felt improving the education system, dealing more effectively with poverty, improving the standard of living and the civil rights of minorities and ending the conflict in Vietnam were far greater national priorities. Reverend Ralph Abernathy, head of the Southern Christian Leadership Conference, was planning a protest with four mules and one hundred and fifty members at the Cape Kennedy gates; his particular focus was upon using

Apollo money to help the poor. Still, even he was awed by the events of 16 July 1969 and the days that followed.

That Wednesday, launch was scheduled for 9:32 am Eastern Standard Time, which dismayed Mike Collins because he had to awaken at the ungodly hour of four in the morning. His last mission, Gemini X, three years before, had blasted off in the late afternoon, allowing Collins and John Young to get up at a more civilised time of the day, "but no such luck *this* time". It was Deke Slayton who came knocking on all three men's bedroom doors and, after showering and dressing, they headed down to the crew quarters' exercise room, where their nurse, Dee O'Hara, waited to perform final medical checks. Next came a final appointment with Lew Hartzell and the traditional 'low-residue' astronauts' breakfast of steak and eggs, toast, fruit juice and coffee – shared with Slayton and Collins' backup, Bill Anders – followed by the laborious process of donning their snow-white space suits.

In his autobiography, Collins related that during the Gemini project they had suited-up in a trailer near the launch pad, but now, on Apollo, NASA had built "an elaborate maintenance, storage and donning facility near the crew quarters". As each man's fishbowl-like helmet was snapped into place, he felt the welcome whoosh of pure oxygen rushing past his face ... and the knowledge that, for the next eight days, he would breathe no more outside air; it would all come from their portable, hand-carried supplies, from the command or lunar module atmosphere or, for Armstrong and Aldrin on the surface of the Moon, from their backpacks. Lumbering outside into the blaze of television lights, all three men *looked* extraterrestrial, sealed as they were in their bulky suits, their protective yellow galoshes adding a slightly comical touch to their appearance.

When the transfer van arrived at Pad 39A, the astronauts ascended to the white room and were greeted by the closeout crew, including Fred Haise, the backup LMP, clad in clean-room garb and hat, who had been there for almost two hours making sure that each of Columbia's switches was set correctly for launch. As Armstrong clambered aboard, he was handed a *bon voyage* gift by pad leader Guenter Wendt – a small crescent-shaped trinket, fashioned from foil-coated styrofoam – and was told that it was the key to the Moon. Unable to take it with him, Armstrong asked Wendt to keep it until he returned home, then gave the self-styled pad 'fuehrer' a mock space-taxi ticket, good between any two planets. It was traditional, Mike Collins wrote, to present Wendt with such light-hearted gifts. "Guenter has spent the past couple of weeks telling me what a great fisherman he is," he explained, "and how he regularly plucks giant trout from the ocean. In return, I have located the smallest trout to be found in these parts, a minnow really, and have had it, uncured, nailed to a plaque and inscribed 'Guenter's Trophy Trout'." Secreted in a suspicious-looking brown paper bag, Collins presented the 'tribute' to Wendt, then took his place in the command module's right-hand seat.

Normally, the CMP occupied the centre position, but Buzz Aldrin's previous stint as backup senior pilot on Apollo 8 made it more practical to continue that way. "Collins had been out for a while," James Hansen explained, following his neck surgery, "so rather than retrain Buzz for ascent, NASA just left him in the centre and trained Mike for the right seat." Finally, Aldrin squeezed himself through Columbia's hatch and dropped into the centre couch.

As the minutes ticked away and the three astronauts steeled themselves for launch, one of Collins' greatest worries was the risk of screwing up on this of all missions, under the spotlight, with maybe a third of the world's population watching or listening. Of key concern was the handle next to Armstrong's left knee, which the commander could twist counterclockwise to fire the Saturn's escape tower and pull the command module to safety in the event of an abort. Looking across the cabin, Collins noticed with horror that the pocket adorning Armstrong's suit leg was uncomfortably close to the handle. (In fact, this was no ordinary pocket: in a few days' time, it would be used to store a so-called 'contingency sample' – the first few grains of lunar soil ever picked up by a human being.) Collins feared that the pocket had the potential to ruin the mission. "It looks as though if he moves his leg slightly, it's going to snag on the abort handle," he wrote. "I quickly point this out to Neil and he grabs the pocket and pulls it as far over toward the inside of his thigh as he can." It was was simply too embarrassing to contemplate the headlines: 'Moonshot Falls Into Ocean', 'Last Transmission From Armstrong Reportedly Was 'Oops'' . . .

Three minutes before launch, the automatic launch sequencer took command of the countdown and began a computerised run-through of each step required to pressurise the Saturn V's internal systems before liftoff. At 50 seconds, the gigantic rocket switched to internal power and four of the nine servicing arms linking it to utilities on Pad 39A were disconnected. Seventeen seconds to go: the final alignment of the launch vehicle's guidance computer was completed and it was transferred to internal power. Throughout each of these methodical steps, the launch commentator continued to report to his global audience what was happening, with growing tension and excitement. "T-15 seconds . . . guidance is internal . . . 12, 11, ten, nine . . . " – then came the start of the ignition sequence, as pressurised RP-1 and liquid oxygen began to enter the combustion chambers of the five F-1 engines – "six, five, four, three . . . " as internal turbines built up the supply of propellants to full flow and brought the first stage up to near-full power. As the final seconds of the count evaporated, all five engines were running in excess of 90 percent of their rated thrust, consuming four and a half thousand kilograms of propellant *every second*. Finally, as the launch pad's deluge system flooded the flame trench with thousands of litres of water to reduce the reflected energy, the Saturn's internal computer carried out its last checks. All was well: " . . . two, one, zero . . . all engines running . . . "

At 9:32 am, the 'launch commit' signal released a series of clamps holding the rocket to the pad and the monster began its climb for the heavens. "Liftoff . . . we have a liftoff, thirty-two minutes past the hour . . . " Twelve long seconds elapsed before the lumbering Saturn cleared the tower and the hundreds of thousands of spectators began to *feel* the vibration and shockwaves pummelling their chests and the soles of their feet. From the commander's seat, Armstrong could be heard announcing the onset of the 'roll program', as the Saturn V's computer began actively guiding it out over the Atlantic Ocean and onto its proper heading for low-Earth orbit and, ultimately, for the Moon.

Lyndon Johnson (centre) and Ted Agnew (to his immediate left) watch the launch of Apollo 11 from Cape Kennedy's VIP area.

To the onlookers, it was nothing short of spectacular, according to Dave Scott, who was watching from the motorboat on the Banana River with Jan Armstrong. Sitting in the blockhouse at the Cape, Deke Slayton, still awaiting his first space flight after ten years as an astronaut, could only watch silently as the rocket thundered into a clear sky. "I think most of us felt like we were lifting it all by ourselves," he wrote later. Tom Stafford, having ridden one of these beasts a few weeks earlier, now found himself in the Cape's VIP area near Lyndon Johnson and Ted Agnew, his chest blasted by the intense staccato crackle. Elsewhere, watching from behind the Iron Curtain, cosmonaut Alexei Leonov – who had long trained to fly such a mission himself – described his emotions as "what we in Russia call 'white envy' – envy mixed with admiration". Nikolai Kamanin's diary is notably lacking comment and for a man who had invested so much of himself into placing the Soviet Union at the forefront of space exploration it must have been an intensely bitter pill to swallow.

For the Apollo 11 crew, encased in their space suits within the very nose of the behemoth, the sensations were unlike their previous experiences. All three men had previously ridden the Titan II rocket, but, rather than the sudden G forces at the instant of liftoff, "there was an unexpected wobbly sway," wrote Buzz Aldrin. "The blue sky outside the hatch window seemed to move slightly as the huge booster began its preprogrammed turn after clearing the tower. The rumbling grew louder, but it was still distant." For his part, Collins felt that the Saturn was "a gentleman" compared to the Titan; despite the shock of staging, the G loads seemed to build no higher than 4.5 and the whole ride proceeded as "smooth as glass, as quiet and serene as any rocket ride can be". In fact, each man spent much of his time monitoring the multitude of gauges on Columbia's instrument panel: keeping track of the attitude indicators, the flight performance data on the computer and regular radioed advice and instructions from capcom Bruce McCandless in Mission Control.

The Saturn behaved with perfection, executing each step of its flight regime precisely: the S-IC stage burned out and separated two minutes and 40 seconds into the climb, followed at three minutes and 17 seconds by the jettisoning of the escape tower, which allowed sunlight to flood the cabin. For almost six more minutes, they rode the S-II stage's five exceptionally smooth J-2 engines, then the single J-2 of the "crisp and rattly" S-IVB. When this stage shut down eleven minutes and 40 seconds after launch, it was travelling at in excess of 28,800 km/h in a circular parking orbit at an altitude of 187 km. By this point, Apollo 11 was barely *four percent* of its weight at liftoff!

"GO FOR THE MOON"

Mike Collins was more than a little relieved when Neil Armstrong finally announced "Shutdown" and the vibrating, buzzing rattle of the S-IVB ceased. Now in an inverted, 'heads-down' orientation, the three men received their first glimpse of Earth in all its grandeur; tantalising, indeed, but brief, for in two hours' time they would reignite the S-IVB to set off for the Moon. "The reason for the heads-down attitude," wrote Collins, "is to allow the sextant ... to point up at the stars, for one of the most important things I must do is take a couple of star sightings to make sure that our guidance and navigation equipment is working properly before we decide to take the plunge and leave our safe Earth orbit."

As the command module pilot, Collins' first minutes of weightlessness were exceptionally busy ones – opening and closing circuit breakers, throwing switches, reading instructions, checking the sextant, the navigation telescope and the cameras – and he moved as carefully as possible to avoid the onset of space sickness. It was apparent now, after eight years of human space flight, that the malaise was notoriously unpredictable in both *who* it would affect and *when* it would hit them; as a result, Collins had spent his last couple of days on Earth flying aerial acrobatics with Deke Slayton in a T-38 jet, trying to condition his inner ear balance mechanism to the unsettling environment of space. Stomach-churning aileron rolls, coupled with rapid side-to-side head movements, normally made Collins sick without any problems and, thankfully, now that he was in orbit, he felt reasonably fine. Nonetheless, aware that *he* would shortly be responsible for executing the delicate transposition and docking manoeuvre to extract the lunar module from the S-IVB, he kept his head motions to a minimum and moved around Columbia's cabin with great care.

In his autobiography, Collins related some low-key irritation with Aldrin, who "seems to have gotten up on the wrong side of bed this morning or at least it seems to me he's more interested in slowing me down than in helping me get through my chores". The LMP questioned a glycol pressure reading on the instrument panel, then sought advice on brackets and lenses for the 16 mm camera. "Lord," wrote Collins, "I didn't have time to discuss camera brackets and lenses now, because it is time to take a couple of star sightings and to realign our inertial platform." To be fair, Collins knew that the minutes were ticking away and, very soon, he would need the camera and other equipment to be set up in order for him to perform the transposition and docking with Eagle. Temporarily ignoring Aldrin, Collins dove down to the navigator's console in the lower equipment bay and unstowed and installed a pair of eyepieces for the sextant and telescope, then jettisoned their protective covers.

Next, he proceeded with the laborious effort to check the navigational stars by which Apollo 11 would chart its course to the Moon – Menkent in the constellation of Centaurus, Nunki in Sagittarius and red-tinged Antares, to name but a few – and by the time the checks were complete, Columbia was drifting serenely above Australia, more than an hour since launch. The purpose of spending a couple of hours making one and a half full orbits around Earth was twofold, as Neil Armstrong related to James Hansen. Firstly, it enabled the crew to ensure that their ship's systems were functioning before committing themselves to the translunar injection (TLI) burn, and secondly it allowed mission controllers to properly analyse and interpret *their* data – because their 'view' by telemetry into the systems was much more detailed than that available to the crew.

It also enabled Collins to find the large 70 mm Hasselblad camera, which had somehow gone missing and sneaked itself away into a nook somewhere in the cabin. The need for the camera arose about 80 minutes into the mission, when the astronauts – busily engaged in their work – finally had chance to watch their first orbital sunrise. Unfortunately, when Armstrong told Collins to get a picture with the Hasselblad, the Swedish-built device was nowhere to be found. Finally, it revealed itself, too late to capture the sunrise, but Collins was glad to have it and secure it before TLI. "Just because something is weightless in space," he cautioned, "doesn't mean it has lost any of its mass. It still contains the same number of molecules and, if thrown, can do just as much damage when it hits something as it would do on Earth."

Finally, two hours and 15 minutes after launch, Mission Control gave them the word that they – and the whole world – were waiting for: a "Go for TLI". (The announcement was expressed relatively trivially, wrote Collins, with "about as much drama as asking for a second lump of sugar!") The burn required them to don helmets and gloves, although none of them imagined how this might prove in any way useful if the S-IVB exploded and damaged Columbia's hull. Further, Armstrong told Hansen, having their helmets on actually impaired their visibility and hearing and reduced their ability to react to problems by restricting their mobility. Still, the three men, with helmets and gloves in place, sat out the almost-six-minute TLI burn as the S-IVB accelerated to the 'escape velocity' needed to climb out of the 'well' of Earth's gravitational influence and head for the Moon.

When the burn got underway at 12:16 pm, Buzz Aldrin noted an intense silence in

The sheer naked power of the Saturn V is amply illustrated in this view of Apollo 11's initial boost into Earth orbit.

the cabin at first, then a gradual shaking as acceleration loads increased from essentially zero gravity to 1.5 G. The Pacific Ocean, he wrote, seemed to 'tilt' noticeably away from them. From his seat, Mike Collins reported seeing peculiar flashes through his window; he assumed them to be flashes of the still-burning J-2 engine, 'behind' them. When he expressed astonishment to Armstrong, the commander took a quick glance through *his* window and replied "Oh, I see a little flashing out there, yes". Armstrong, it seemed, *never* admitted surprise.

Over the years, these strange flashes have been cited by the UFO community as 'evidence' that anything from a single unidentified object to an entire flotilla of alien craft joined Apollo 11 on its mission to the Moon. In fact, each of the men continued to see mysterious flashes at various intervals throughout their flight, causing so much consternation that Aldrin went so far as to mention them in his post-mission debriefings. Surprisingly, when Pete Conrad's Apollo 12 crew flew to the Moon in November 1969, they saw them, too. On the other hand, Tom Stafford's lunargoing crew had seen nothing. It turned out, wrote James Hansen, that they represented "a phenomenon that occurred in the especially dark conditions of outer space *inside the human eyeball*". More recent research has shown that they are most likely the result of cosmic rays entering the spacecraft and that a kind of 'optical-psychological threshold' exists, whereby some astronauts have proven more sensitive to seeing them than others.

Precisely on cue, five minutes and 47 seconds after coming to life, the S-IVB fell silent, leaving Apollo 11 on a near-perfect trajectory to intercept the Moon in less than three days' time. As a result of the manoeuvre, Armstrong, Collins and Aldrin had climbed barely 100 km in altitude, but their increase in *velocity* was what mattered, for they were now moving away from Earth at a blistering 39,000 km/h – *ten kilometres every second*, a speed so enormous that it had only been matched by six other men, the Moon-bound crews of Apollos 8 and 10. "As we proceed outbound," explained Collins, "this number will get smaller and smaller, until the tug of the Moon's gravity exceeds that of the Earth's, and then we will start speeding up again" as they began the long 'fall' into the 'well' of lunar gravity.

Less than half an hour after the completion of TLI, Collins was ready for his major test of the day – transposition and docking with Eagle. He set himself up in the left-hand commander's seat and gingerly uncoupled Columbia from the top of the S-IVB; it was nothing like controlling an Air Force jet or even the nifty Gemini, but six months of training had given Collins the skills he needed to perform the task with a minimum expenditure of propellant. He knew that by moving slowly, he would consume less fuel, but did not want to take *too* long "because I don't want that monster [S-IVB] behind me to be out of sight any longer than is necessary". In the simulator, he had managed to compromise on both counts by separating at a rate of about a kilometre per hour, then swinging Columbia smartly about-face at a rate of a couple of degrees per second. Now, making the manoeuvre for real, he executed a flawless docking with Eagle as Armstrong and Aldrin looked on.

Within minutes of hearing the welcome 'ripple-bang' of the twelve capture latches locking into place, Collins excised himself from his seat and floated down into the lower equipment bay to open the apex hatch to the short tunnel between Columbia and Eagle. His first reaction upon accessing the tunnel was the smell – a strange odour, like charred electrical insulation – but, upon inspection, he could find nothing amiss. All of the exposed wiring was in good condition and none was discoloured. "Perhaps something overheated in the dense lower atmosphere during launch," he speculated later, "or perhaps some of the rocket fumes from the launch escape tower motors were trapped in the tunnel."

Collins moved on with his checklist, verifying that each of the latches was in good shape and plugging in the electrical umbilical which would route a command through the lander to the S-IVB to fire separation pyrotechnics. Then Eagle – now firmly mated nose-to-nose with Columbia – was sprung free from the S-IVB. Shortly thereafter, a three-second pulse of the service module's large SPS engine provided some separation from the now-spent rocket stage, which would shortly vent its remaining propellant and enter perpetual solar orbit. Next, the three astronauts removed their bulky space suits – not easy, wrote Collins, resulting in them "thrashing around like three great white whales inside a small tank" – and donned more comfortable, two-piece white Teflon fabric 'jumpsuits'. Finally, five hours into the mission, they were able to grab a bite to eat.

Apollo 11 was now alone in the virtually unknown region called 'cislunar space'; a region of the most profound emptiness across which it would chart a course over the next three days. "We were kind of bizarre looking," wrote Aldrin of the ship's unusual amalgam of shapes, "with the bullet-like CSM wedged into the cement-mixer LM." To prevent any part from becoming too hot or too cold under the temperature extremes of sunlight and darkness, the entire craft was placed into a 'barbecue roll', positioned broadside to the Sun and rotated slowly, completing a full turn every 20 minutes, like a chicken on a spit. "Most people probably thought Apollo 11 was shooting toward the Moon like a bullet," explained Aldrin, "with its pointed end toward the target. But actually we were moving more like a child's top, spinning on the nozzle of our SPS engine."

As Earth receded 'behind' them, cislunar space came to be regarded by Moon-bound crews as something of a 'twilight zone' in which both *time* and *distance* came to have more meaning than *speed*. "To get a sensation of travelling fast," wrote Collins, "you must see something whizzing by: the telephone poles along the highway, another airplane crossing your path"; but *here*, in contrast, there were no passing landmarks by which to judge their speed. Only distance could be guessed at, based on the sight of Earth, which diminished in apparent size from a vast mass of blue and white 'beneath' them to something akin to a football, then a tennis ball, then a large marble ...

Watching his steadily shrinking home planet, Armstrong found it remarkable that he could discern clear weather patterns: a single cyclonic depression over northern Canada, a low front stretching up from the northernmost extremities of the Great Lakes as far as Newfoundland, a crystal-clear Greenland and a perfect view across the North Atlantic and into western Europe. Casting his eyes south-eastwards over the rust-tinged expanse of Africa, he could also faintly discern the glint of sunlight over Lake Chad. Collins, who had by now become known as the most light-hearted member of the crew, could only remark that "I didn't know what I was looking at, but I sure did like it!"

Sleep at the end of a long day came as a blessed relief; indeed, since a couple of hours after launch, all three men had fought drowsiness, with Armstrong and Collins confessing to each other that they almost dozed off and Aldrin playfully asking them to wake him before TLI. During each sleep period, one member of the crew always remained 'on watch', with a miniature headset taped to his ear, in case Mission Control should call them. Next morning, at 8:48 am EST, when the wake-up call came from Houston, they were alert, refreshed and ready for their second day in space. Among the news stories relayed to them by capcom Bruce McCandless, one in particular was exciting – Vice-President Ted Agnew, recently appointed by President Nixon to head the Space Task Group, had advocated a manned landing on Mars by the end of the century – although Democratic opposition was strong, calling instead for a fundamental refocusing of national priorities onto more pressing issues.

The days in which the menace of Soviet-led communism, the possibility of nuclear invasion and an ingrained fear of 'gaps' in missile technology between the two superpowers were now largely gone. Cordial ties between Russia and the United States were admittedly fragile, but in 1969 it seemed increasingly unlikely that the American populace or its leadership would happily tolerate investing further billions of dollars in the human exploration of space. Earthly needs were simply too strong and too urgent to ignore: the need to end the war in Vietnam, the need to rebuild a nation shaken to its core by civil and racial inequality, the need to tackle and eradicate poverty and the need to improve education and make it universally accessible. From his perch in the Apollo 11 command module, Mike Collins privately hoped that, despite the fact that the budget for human space exploration had been in steady decline since 1967, the excitement of the next few days might yet reinvigorate national opinion and once more enthuse a fickle public and short-sighted administration. Sadly, as the next few years would show, it did not.

Another news story from McCandless that morning had caused more than a little concern for NASA: the Soviets had launched an unmanned probe, Luna 15, toward the Moon a few days earlier and it had reportedly just arrived in orbit. Most sections of the media viewed Luna 15 for what it was – a ploy to beat Apollo 11 to the lunar surface. The intention was for it to land, scoop up some soil and bring it back to Earth before Armstrong's crew could do so. In fact, some people suspected the probe might deliberately attempt to 'interfere' with the Apollo mission and Frank Borman, recently returned from a goodwill tour of the Soviet Union, communicated directly with Mstislav Keldysh of the Soviet Academy of Sciences to find out what was happening and to obtain the precise orbital parameters. There was no need for concern. "The orbit of Luna 15," Keldysh replied, "does not intersect the trajectory of Apollo 11 spacecraft announced by you in flight programme."

Eight days after its launch, according to Tass, Luna 15 "reached the Moon's surface" after completing 52 orbits. 'Reached', it seemed, was a clever choice of word from the Soviets to conceal the ugly and embarrassing truth: for both American and British intelligence sources would quickly verify that Luna 15 had actually tumbled out of control and crashed somewhere in Mare Crisium. By that time, of course, Neil Armstrong and Buzz Aldrin had already walked on the Moon. It did not matter either way, because, as Asif Siddiqi has noted, "Even if Luna 15 had landed, collected a soil sample and safely returned to Earth, its small sample return capsule would have touched down on Soviet territory two hours and four minutes *after* the splashdown of Apollo 11. The race had, in fact, been over before it began."

ARRIVAL AND UNDOCKING

In a strange way, the crew of Apollo 11 attempted to condition themselves to believe that – although the tumultuous Saturn V ride was now behind them and their flight was underway as they charted a course through cislunar space – the main part of the mission had yet to begin. Riding a Saturn into space had been done, transposition and docking had been done and the flight to the Moon was going well. *Their* mission would really start after Lunar Orbit Insertion (LOI) on 19 July and the series of increasingly bold and epochal events thereafter. With this in mind, Mike Collins noted, it was fortuitous that all three men had flown before and were not driven to distraction by the jubilant excitement of a team of rookie spacefarers. They could therefore spend much of the three-day journey to the Moon resting and preparing themselves for the immense challenges ahead.

The resultant quietness of those three days was not helped, in the opinion of mission controllers, by a lack of conversation from the astronauts. "It's all dead air and static," a Houston official complained at one stage of the cislunar coast. The lengthy spells of silence were, however, punctuated by televised shows in which the crew guided a worldwide audience around their ship, revealing the dismantling of the probe and drogue, a shimmy through the tunnel and an 'upside-down' glimpse of Eagle's tiny cabin, with Aldrin, toting dark aviator sunglasses, hard at work. Several of these shows were made by Collins, who enjoyed rotating the camera 180 degrees to turn his Earthbound audience on its head and back again; but, in reality, none of them had much chance to practice with the camera on the ground and its late delivery to Cape Kennedy had not aided matters. "We simply didn't have time to fool around with it," he wrote. "Neil and Buzz didn't even know how to turn it on or focus it, and my knowledge of it was pretty sketchy." With this in mind, they were advised by a helpful instructor that an audience of perhaps a billion or more people would be watching and that screwing up one of their shows was *not* an option ...

The quiet time was interspersed with inevitable chores, mainly performed by Collins: purging fuel cells, charging batteries, dumping waste water and urine, preparing food, dechlorinating the ship's water supply and, notably, performing a midcourse correction burn to refine their path towards the Moon. Twenty-six hours into the mission, and almost 175,000 km from home, Columbia's big SPS engine roared silently into the void for three seconds in what flight controllers lauded as an "absolutely nominal" firing. In his autobiography, Collins related that, for those few seconds, for once, *he* was in active control. Several months earlier, his five-year-old son had asked who had been 'driving' Apollo 8 to the Moon – was it Mr Borman, the ship's commander? No, Collins had replied, it was *Sir Isaac Newton* – or, at least, the influences of the Sun, Earth and Moon, each of which affected the spacecraft's path just as the great scientist's law of universal gravitation had helped predict three centuries before.

The accuracy of the midcourse burn was so good that two subsequent SPS firings were deemed unnecessary and late on 18 July, precisely on schedule, 62 hours into the flight and some 70,000 km from their target, Columbia and Eagle slipped into the Moon's sphere of influence. For the past three days, still under the tug of Earth's gravity, their speed had rapidly decreased from 39,000 km/h immediately after TLI to just over 3,200 km/h; *now*, as the Moon's gravitational pull became dominant, they began to 'fall' towards it, gradually speeding up to 9,000 km/h. Earlier in the evening, Collins had removed the probe and drogue from Columbia's docking mechanism and opened the tunnel to allow Armstrong and Aldrin to enter Eagle and begin checking out its systems.

Both men considered the 'down-up-up-down' trip into the lunar module – as they moved from the 'floor' to the 'ceiling' of Columbia, then found themselves diving headfirst toward Eagle's floor – as one of the most unusual sensations of their mission, although Aldrin described the transition as "perfectly natural", akin to the motions of a swimmer. For two and a half hours, from 5:00 pm to 7:30 pm EST, they verified that the lander was ready to support an undocking and a landing attempt on the afternoon of 20 July and viewers in the United States, western Europe, Japan and most of South America were treated to a televised show of Aldrin performing an equipment inventory inside the tiny cabin.

That evening, back inside the command module, offered some quiet time before the historic events to come. Aldrin recalled asking Armstrong if he had decided what he was going to say when he stepped onto the lunar surface, to which the commander, between sips of fruit juice, replied that no, he was still thinking it over. The crew also noted a second peculiar occurrence as *something* drifted within about 150 km of them; an object, much brighter than one of their navigational stars ... and *moving*. Aldrin pointed it out to his crewmates and, with the aid of the sextant and monocular, they were able to discern a roughly cylindrical main body with "something off to the side", forming a sort of 'L' shape. It was not the spent S-IVB stage – Armstrong checked with Mission Control, whose trajectory specialists confirmed *that* was thousands of kilometres from them – and the astronauts finally convinced themselves that it was most likely one of the four panels which had housed Eagle during its climb to orbit. Those panels had blown away from the top of the S-IVB and, since none were equipped with transponders, their precise trajectories were uncertain. To this day, Armstrong is convinced that the L-shaped object *was* such a panel, but UFO theorists, naturally, have had and continue to have a field day making up their own stories.

Watching strange objects tagging along with them was surely overtaken by the sheer grandeur of the Moon itself – totally different, it seemed, from the pale lamp in the sky that they had watched nightly as they grew from infancy to boyhood to adulthood. "The Moon I have known all my life," Collins wrote, "has gone away somewhere, to be replaced by the most awesome sphere I have ever seen. To begin with, it is *huge*, completely filling our window." It also appeared more menacing than the two-dimensional circle in Earth's skies; now *three-dimensional*, and seeming to bulge towards them, he perceived it to be an intensely unwelcoming and forbidding place, "formidable", "utterly silent" as it hung "ominously" in the void.

Back on Earth, journalists had frequently posed an inevitable question before the flight: was Collins jealous that Armstrong and Aldrin were about to take the first steps on the Moon, while *he* remained in orbit? Collins had responded that in all honesty he was more than happy and content to be flying 99.9 percent of the journey. He would, it is true, be mad to suppose that he had the *best* seat on the mission, but he had already decided that this would be his final space flight; the strain on his wife and children, the constant grind of training and the lengthy spells away from home were too much for them. Shortly before Apollo 11, during a cross-country T-38 flight with Deke Slayton, the man who picked crews had offered Collins the chance to serve as backup commander of Apollo 14 and most likely lead Apollo 17 to the Moon. This would give Collins the chance to walk the lunar surface himself. Collins had declined. Now, as he neared the Moon on this midsummer's evening in 1969 and looked down onto the threatening barrenness of its terrain, then recalled Earth with its waterfalls and valleys and enchanting iridescence of life, he knew he had made the right choice.

Getting into orbit around the Moon on the afternoon of 19 July was a triumph of celestial mechanics and human ability in itself. The Lunar Orbit Insertion manoeuvre actually comprised *two* firings of Columbia's SPS engine – the first, lasting five minutes and 57 seconds, reduced their speed from 9,000 km/h to 5,970 km/h and 'dropped' them into an elliptical orbit of 270 x 97 km with the high point on the nearside; and the second, lasting only 17 seconds, came about four hours later and almost circularised their path at roughly 105 x 87 km. To this day, it remains remarkable that they could be guided so precisely across more than three hundred thousand kilometres of cislunar space and then achieve such a perfect orbit, taking into account the fact that the Moon was itself in motion, circling Earth. "Those big computers in the basement in Houston," wrote Collins, "didn't even whimper, but belched out super-accurate predictions." When one considers that the computing power of one of today's mobile phones would dwarf the entire computing power that guided Columbia and Eagle to the Moon, the act of inserting Apollo 11 into lunar orbit was truly a stupendous achievement.

During the four-hour interval between the two burns, dubbed 'LOI-1' and 'LOI-2', the opportunity arose to closely examine the surface of the strange world upon which Armstrong and Aldrin would shortly take humanity's first steps. Initially, the television camera panned across the terrain and the crew were silent, until Mission Control requested that they describe some of what they were seeing. A group of astronomers from Bochum in West Germany had asked that they take a look at Aristarchus – a prominent, extremely bright impact crater – which had exhibited unusual luminescence over the preceding weeks. "Hey, Houston," radioed Armstrong, after finding the crater, "I'm looking north up toward Aristarchus

A totally different Moon: the rugged and inhospitable lunar farside.

now, and there's an area that is considerably more illuminated than the surrounding area." Aldrin agreed and it was speculated that the luminescence could represent some sort of lunar eruption, fire fountain or even a lava flow.

Other regions and landmarks were enthusiastically identified by the crew by their nicknames – the small hills of Boot Hill and Duke Island, the snake-like rilles of Diamondback and Sidewinder and the twin peaks of 'Mount Marilyn'; the latter unofficially bestowed by Apollo 8 astronaut Jim Lovell in honour of his wife – although, of course, the vast plain of Tranquillitatis was of principal interest. "The Sea of Tranquillity," wrote Collins, "is just past dawn and the Sun's rays are intersecting its surface at a mere one-degree angle. Under these lighting conditions, craters cast extremely long shadows and to me the entire region looks distinctly forbidding." In Collins' mind, it looked far too rugged to set a baby's buggy down, let alone a lunar module.

By the early evening of 19 July, following the LOI-2 burn, which Collins had timed to the split second using a stopwatch, everything was ready for the final checkout of Eagle in advance of undocking and the Powered Descent. Luckily, since Aldrin had successfully lobbied to do much of the checkout a day early, the normally-three-hour task took barely 30 minutes and by 8:30 pm all was in place as the three astronauts bedded down for a night of surprisingly fitful sleep in Columbia.

Next morning, a very groggy Mike Collins responded to Mission Control's wake-up call and, after breakfast and a round-up of the morning news, all three men plunged into their respective checklists. Among the most important tasks were donning their space suits and, in the case of Armstrong and Aldrin, getting into the liquid-cooled underwear which would help to maintain a comfortable body temperature during their time on the lunar surface. Collins shoved the remainder of their gear – "an armload of equipment" – through the tunnel to them, then disconnected umbilicals, reinstalled the docking probe and drogue and sealed the hatch. "I am on the radio constantly now," he wrote, "running through an elaborate series of joint checks with Eagle. In one of them, I use my control system to hold both vehicles steady while they calibrate some of their guidance equipment." Inside the lander, anchored to the floor by bungee-like cords, Armstrong and Aldrin also had their hands full: punching entries into the computer keypad, aligning their S-band antenna with the Earth-based tracking network, checking and cross-checking VHF communications with Collins and deploying Eagle's landing gear.

At length, it was time to bid farewell. "You cats take it easy on the lunar surface," Collins called cheerily. "If I hear you huffing and puffing, I'm going to start bitching at you." A few minutes later, at 1:44 pm EST on 20 July, he flipped a switch to cast Eagle loose. Yet this momentous beginning of Eagle's descent occurred unseen and unheard by Earth, for both craft were *behind* the Moon at the time. Before losing radio contact, the capcom in Mission Control – a young Air Force major named Charlie Duke – had given them the good news that they had a 'go' for undocking. Duke, selected as an astronaut three years earlier, was still awaiting his first space flight, yet he was already recognised as an expert on the lunar module systems; indeed, on Apollo 16 in 1972, he would become the tenth man to walk on the Moon. Impressed by Duke's work in support of Apollo 10, Armstrong had specifically asked him to be the capcom for Apollo 11's descent. His infectious North Carolina drawl and endearing personality would certainly help to lift some of the tension in the hours ahead.

The Mission Operations Control Room (MOCR) at the Manned Spacecraft Center in Houston was packed with virtually everybody involved in the space programme – Wernher von Braun, Tom Paine, George Mueller, Sam Phillips, Chris Kraft, George Low, Deke Slayton and many astronauts, all waiting for more than a decade of hard work to pay off. In addition, an estimated *third* of the world's population was either watching or listening on television or radio.

Still out of direct radio contact with the ground, Collins watched, his nose pressed against one of Columbia's windows, as Eagle drifted serenely into the inky darkness. Armstrong executed a little pirouette, fully rotating the lander to enable Collins to verify that the landing gear was in good condition. This had required Collins to take a trip to the Grumman assembly plant in Bethpage to familiarise himself with the lunar module, its fully extended landing legs and the long sensor prongs affixed to three of its four footpads. (One leg – the one holding the ladder – was originally supposed to have had a sensor, too, but according to James Hansen, Armstrong requested its removal, lest he or Aldrin trip over it during their climb down to the surface.) There were other worries. One side of the descent stage held the Modular Equipment Storage Assembly (MESA), a carrier brimming with a television camera, rock boxes and geology tools, which Armstrong would use on the lunar surface. Was it still firmly secured in place or had it accidentally swung open during the separation process? Collins assured him that all was well.

With such assurances ringing in their ears, all three men could afford a brief moment of light-hearted banter – with Collins telling Eagle's crew that they had a pretty fine-looking machine, despite being upside down, to which Armstrong retorted that, from his perspective, *someone* was upside down ...

At 2:11 pm, Collins fired his thrusters for a nine-second separation burn "to give Eagle some breathing room". At 3:08 pm the first of two firings of the lander's descent engine got underway. Known as Descent Orbit Insertion (DOI), it lasted just under 30 seconds and reduced the lowest point of Eagle's orbit to a height of 15.2 km, at a position convenient for initiating the Powered Descent. The laws of celestial mechanics now became increasingly evident to Collins: in a lower orbit, Eagle was moving *faster* and was actually ahead of Columbia by about one minute.

As they descended, it was necessary for Armstrong and Aldrin to cross-check their instruments, specifically the Primary Guidance, Navigation and Control System (PGNS) and Abort Guidance System (AGS). The former processed data from an inertial platform of gyroscopes and guided the lunar module along a predetermined flight path to its landing site, whilst the latter offered the ability to perform an abort if necessary. "We couldn't *land* on AGS," Armstrong told James Hansen, "unless we got right down close to the surface, because you couldn't navigate the trajectory with it." However, *both* systems had to be operating throughout the descent phase – if an emergency arose, Armstrong might need to switch over instantaneously from PGNS to AGS – and it was imperative that both had the same data. "If tiny errors were allowed to compound," Hansen wrote, "gross errors in computing the LM's course and location could result."

The higher altitude of Columbia meant that Collins was first to regain contact with Houston as the two craft emerged from behind the Moon. The acquisition of signal was what the MOCR had been waiting for. When queried by Charlie Duke over the progress of the DOI burn, Collins responded simply that it had gone "just swimmingly ... beautiful". Ninety seconds later, at 3:49 pm, Aldrin confirmed that the DOI had gone well. Eagle's radar was activated at 15.2 km, and Duke issued a firm 'go' to begin Powered Descent. Brief, but persistent communication dropouts forced Collins to relay this to Armstrong and Aldrin and the lander's descent engine ignited for the second time at 4:05 pm. By the time it shut down, in barely 12 minutes' time, they would be on the Moon.

Although Collins could certainly *speak* to them, he could no longer *see* them;

despite having tied a small black patch over his left eye and squinting through Columbia's sextant, the bug-like lander steadily diminished in size until it looked "like any one of a thousand tiny craters ... except that it is moving". Eventually, it was gone. The best thing Collins could do now was keep quiet and wait.

TENSE TIME

The Mission Operations Control Room was also very quiet. Gene Kranz, flight director of the White Team – one of four shifts dealing with Apollo 11 – had taken over from Glynn Lunney at 8:00 am local time (7:00 am EST) and had already ordered Security to lock the doors in anticipation of the landing; no one would be permitted to disturb the intense focus of himself or his team over the coming hours. By this time, Kranz was one of NASA's most experienced flight directors. He was a crew-cutted ex-fighter pilot and veteran of the Korean War. His no nonsense attitude had prompted his team to nickname him 'General Savage' but, as Andrew Chaikin has observed, he was also "an unabashed patriot" and "could get misty when he heard The Star Spangled Banner". On this emotional day – more so than any other, except the return of Apollo 13 – the stern, steely-eyed Kranz and many of his devoted controllers would quite literally be reduced to tears.

Simply entering the MOCR was like walking into an ever-changing history book; for it *was* a place in which history was being made, day by day, as the central hub of America's manned space effort. When he arrived, Kranz's eyes had difficulty adjusting to the greenish-blue hue of the world map on the wall and he had strained his ears to hear the hushed tones of the flight controllers and the capcom reading the morning news to the astronauts. The air was rich with the scent of coffee and cigarette smoke from dozens of ashtrays and a sort of adrenaline-charged anticipation of what was about to happen permeated the room. A bouquet of roses, sent by some unknown 'admirer' before each and every mission, stood on a low table by the wall. All seemed well that morning as Chris Kraft patted Kranz on the back and wished him good luck.

Since the return of Tom Stafford's crew from the Moon in late May, Kranz's team had worked closely with Armstrong and Aldrin in perfecting their own under-standing of the Powered Descent and the procedures and flight rules that they would need to follow before, during and after the lunar landing. They had been tested by the training staff and instructors just as rigidly as the astronauts; for they would need to know instantaneously how to respond in the event of an emergency, how to judge when Armstrong and Aldrin had entered the Dead Man's Curve and could no longer safely abort ... and they needed to understand that communications took 1.3 seconds each way between Earth and the Moon. Some emergencies, they realised, could spiral out of control before they even knew about them. On the other hand, an abort itself entailed enormous risk and all four Apollo 11 flight directors – Kranz, Lunney, Milt Windler and Cliff Charlesworth – were prepared to stretch the flight rules as far as they deemed safe and practicable. Yet with uncomfortable regularity, in the early days of training, the lander would crash or an overly conservative member of Kranz's team would call an abort at the first hint of difficulty; but by the start of July, with hundreds of simulations behind them, Kranz and his team felt that they were finally ready.

So too did the other teams. "Mission events never fall into neat, equally spaced increments of eight hours," wrote Kranz in his autobiography. "My team must take 32 hours off to synchronise with the lunar trajectory for landing. During this 32-hour period, Charlesworth [the lead flight director and head of Apollo 11's Green Team] will get the spacecraft into lunar orbit, then Milt Windler, the Maroon Team flight director, will have the crew trim the orbit and then perform another interior inspection of the LM. Four flight control teams are being used for the lunar phase of the mission to provide flexibility and, once the LM is on the lunar surface, to support the CSM solo orbital operations." However, planning for each team's spell 'on console' often included so-called 'whifferdills', in which real-time events dictated last-minute changes to the shift pattern. For example, after they had landed, Armstrong and Aldrin asked for their Moonwalk to be moved 'earlier' in the schedule, one such whifferdill was put in place and Charlesworth's team – which had trained for the lunar surface work – was duly called in to supervise it. Some controllers found themselves coming back onto their consoles only eight hours after leaving it, whilst others had a day or more between shifts. "You just have to tell your body to ignore how it feels," Kranz continued, "and get on with the job."

One of the traditions that Kranz always adopted was to wear a white vest, sewn by his wife, Marta; on the morning of Apollo 11's landing he strode into the MOCR wearing a vest made of white brocade with silver thread. Speaking to his two-dozen-strong team – which included Bob Carlton, monitoring the performance of Eagle's descent engine, Jerry Bostick, who kept an eye on the tracking data, and Steve Bales, the lunar module's guidance and computer expert, together with Tom Stafford and Gene Cernan, who knew the plain of Tranquillitatis better than anyone – he told them, on the private Assistant Flight Director loop, that *this* was the final exam: today they were going to land on the Moon. After ensuring that each controller knew his share of the flight rules, he told them that they were the best team he had ever worked with and that he had total confidence in them. He concluded his peptalk: "After we finish the son-of-a-gun, we're gonna go out and have a beer and say 'Dammit, we *really* did something'."

"THE EAGLE HAS LANDED"

For the first 26 seconds after initiating Powered Descent, Neil Armstrong maintained 10 percent of rated thrust, producing a gentle acceleration which allowed the computer to gimbal the engine to ensure that it was directed through the vehicle's center of mass prior to going full throttle. Flying with the engine bell facing the direction of travel and the windows toward the surface, he noticed that they were coming in 'long' – they flew over the crater Maskelyne W a few seconds early, for example – and were likely to overfly their intended landing site. After the mission, it would be judged that very small residual pressures in the tunnel between Eagle and

Columbia during undocking had imparted a slight radial velocity that had perturbed their trajectory prior to the DOI burn. (On future flights, approval for undocking would not be granted by Mission Control until the tunnel's atmosphere had been fully vented.) To the pragmatic mind of Neil Armstrong, it really did not matter on the first landing attempt; as he told James Hansen, "I didn't particularly care where we landed, as long as it was a decent area that wasn't dangerous."

Four minutes after the initiation of Powered Descent, the lander rotated 'face up' so that the radar on its underside was able to acquire the surface and supply altitude and rate-of-descent data. "We needed to get the landing radar into the equation pretty soon," Armstrong told Hansen, "because Earth didn't know how close we were and we didn't want to get too close to the lunar surface before we got that radar." This showed their altitude to be 10.1 km; a kilometre or so *lower* than their PGNS reckoned, because that was tracking their *mean* height above the surface, rather than their *actual* height. Aldrin knew that the radar offered the most reliable calculations and planned to instruct the computer to accept that data, but first he had to wait for Mission Control to verify it. When they did, he keyed a command into the computer to monitor the convergence of the two estimates as Eagle manoeuvred. At this point, a yellow caution light lit on the instrument panel and an alarm tone sounded. "Program alarm," called Armstrong, then glanced down to the computer display, adding, "it's a 1202. Give us a reading on the 1202 program alarm." Neither he nor Aldrin had any idea which of the dozens of different alarms the 1202 represented and certainly had no time to flip through their data books to find out. Fortunately, seated in Mission Control was Steve Bales, the 26-year-old guidance officer (nicknamed 'Guidance') and an expert on the lunar module's computer. He checked with Jack Garman, a colleague in a support room at Mission Control, and assured Kranz that 1202 was an 'executive overflow', meaning the computer was momentarily overloaded, but it would not jeopardise the landing. With typical enthusiasm, Bales yelled into his mouthpiece: "We're Go on that, Flight!"

Bales' call was relayed to Armstrong by Charlie Duke – "We're Go on that alarm" – although it was not to be the end of the 1202: it flashed onto Eagle's display thrice more, but so long as it was only intermittent it did not pose a risk because the computer was able to recover.

Aboard Columbia, Mike Collins was listening in and consulted *his* checklist to find 1202, quickly verifying that, for the command module, too, it represented an executive overflow. "I guess it means the same thing for the LM," he wrote, "as [Massachusetts Institute of Technology] designed both computer programs."

Three minutes before the scheduled touchdown on the Moon, the computer flashed a 1201 alarm. This was another form of executive overflow and was quickly cleared, with Duke telling Armstrong and Aldrin, rapid-fire, "We're Go ... Same type, we're Go".

For Armstrong, the alarms were little more than an irritation and, as long as everything continued to look fine – and it *did*; Eagle was exhibiting no erratic motions – he had every intention of pressing on.

Aldrin, in his autobiography, has stressed that the alarms *were* a potentially serious obstacle in which "hearts shot up into throats" at Mission Control. Even Steve Bales, who effectively 'saved' the landing by quickly diagnosing the alarms and advising Kranz appropriately, had only become familiar with which of the various alarms mandated an abort, and which did not, a few days earlier.

On the afternoon of 5 July, the Apollo 12 backup landing crew (Dave Scott and Jim Irwin) had been in the lunar module simulator in Houston, running practice descents when a 1201 alarm was thrown at Kranz's flight control team. Bales consulted with Garman and could only discern that, although everything *looked* okay with the hardware, there was *something* amiss with the computer. He advised an abort and Kranz duly made the call. Scott punched the Abort Stage button and completed a successful return to lunar orbit, but later that evening Bales and Kranz came under fire from the simulation supervisor who had thrown the problem at them. In particular, Kranz was criticised, firstly, for ordering an abort when it was not needed – if the guidance system was *working*, if the thrusters were *working*, if the descent engine's performance was *good* and if the astronauts' displays were *working*, he should have pressed on – and, secondly, for violating a basic rule of Mission Control: that flight directors had to have *two* independent cues before calling an abort. It was a tough, but valuable lesson. By the time Apollo 11 lifted off, Bales had drawn up a list of those program alarms which would make an abort mandatory and those which would not; neither 1201 nor 1202 were on his list.

When the first alarm flashed up, Charlie Duke – who had been sitting at the capcom's mike during the 5 July simulation – and backroom expert Granville Paules instantly recognised it as "the same one we had in training". Gene Kranz did not want to be stampeded into an abort now that they were flying the mission for real. On the other hand, if the alarms continued, they could bring Eagle's computer grinding to a halt and make an abort unavoidable.

By the time the 1201 alarm appeared, the lander was already descending below 2,200 m, had performed the 'pitch-over' manoeuvre and was now flying tilted backward about 20 degrees off-vertical, such that the astronauts could 'see' the lunar terrain spread out before them. After polling his team, Kranz received a collective "Go for Landing", a message which Duke now passed on to Armstrong and Aldrin. Yet the furore over the program alarms meant that it was another minute or so, not until a few seconds after 4:15 pm, that Armstrong had chance to look at the surface ... and behold a particularly unwelcoming sight: the near slope of a crater, as big as a football field, its environs dotted with boulders the size of small cars. At first, he considered landing 'short' of the crater – later to be dubbed 'West Crater' – then picking a spot somewhere amidst the boulders, although the risk of touching down on a slope or in a tight place quickly changed his mind.

At an altitude of 150 m, a little higher than he had intended, Armstrong selected the semi-automatic mode that would enable him to control attitude and horizontal velocity, whilst the computer – allowing for his commands – operated the throttle. He pitched Eagle almost upright in order to direct nearly all of its thrust downwards and slow the rate of descent, then selected 'attitude hold' and let Eagle fly a shallow trajectory over the obstacles. As soon as he was clear, he began the search for a suitable spot to land.

Trajectory planners had designed the descent path to yield the longest possible shadows to improve the astronauts' ability to discern peaks, valleys, craters and rilles. This required coming in low, shortly after dawn, with the Sun behind the lander. If they had tried to land with the Sun high in the sky, the surface would have appeared flat and essentially featureless. Nevertheless, there were residual concerns that the glare of reflected sunlight might impair the astronauts' depth perception.

As they drew closer, dropping below 60 m and then still lower, Armstrong began to see lunar dust, kicked up by the descent engine, beginning to obscure the surface. The dust, he told Hansen, was not a 'normal' cloud like those encountered in the high desert on Earth, but effectively a 'blanket' – a *sheet* of moving dust which essentially wiped out visibility, apart from several boulders poking through it. Moving almost horizontally, the dust "did not billow up at all; it just moved out and away in an almost radial sheet".

In Mission Control, Gene Kranz's team knew that Armstrong had intervened early, but they did not yet know *why*; they could not have known about the yawning crater and the forbidding field of boulders. "The partnership" between Mission Control and the astronauts, wrote Andrew Chaikin, "had all but dissolved." In this final phase, everyone on Earth had to understand that Armstrong, the man in command and the man 'on the spot', was now running the mission.

At Deke Slayton's recommendation, Charlie Duke called to Kranz: "I think we'd better be quiet!"

"Rog," agreed the flight director. "The only call-outs from now on will be fuel."

Gradually, it seemed, the situation improved and Armstrong was able to begin arresting Eagle's forward and sideways motion with the thrusters; he intended to land in the first clear spot that he could find. He was virtually silent throughout those final minutes, the only voice coming from Aldrin, who called out a steady stream of altitudes and velocity components to guide Armstrong – and a tense, listening world – down towards the new Promised Land.

"Once I got below 50 feet," Armstrong told Hansen, "even though we were running out of fuel, I thought we'd be all right. I felt the lander could stand the impact ... I didn't *want* to drop from *that* height, but once I got below 50 feet I felt pretty confident we would be all right."

Thirty metres above the surface, Aldrin reported "Quantity Light", indicating that only 5 percent of fuel was left in Eagle's descent engine and, in Mission Control, a 94-second countdown started; when this countdown reached zero, the lander would have only 20 seconds left in which to either touch down or abort. Unfortunately, by *that* time, Armstrong would be too far into the Dead Man's Curve to safely call an abort. "I never *dreamed*," Kranz recounted years later, "that we would still be flying this close to empty."

Watching the fuel gauge on his display like a hawk, Bob Carlton reported that only 60 seconds remained – an urgent report passed on to Eagle by Charlie Duke – although the astronauts were too preoccupied to respond. "They were too busy," Kranz said later. "I got the feeling they were going for broke. I had this feeling ever since they took over manual control." In Mission Control, the silence was so pervasive and so enduring that one could have heard a pin drop. Kranz crossed himself and prayed.

Still, the notion that Armstrong may have been going for broke did not mean that he and Aldrin were being reckless; if they had been still *too high* when the Quantity Light came on, there would have been no alternative but to abort. However, at relatively low altitude it seemed safer and more prudent to press on with the landing attempt. After all, during quite a few of his LLTV runs above Ellington Field, Armstrong had landed with less than 15 seconds of fuel remaining in his tanks, so he was not particularly 'panic-stricken' about the low levels.

At 4:17:26 pm, Aldrin called out that they were six metres above the surface and, 13 seconds later, announced "Contact Light" as one of the sensor prongs projecting from Eagle's footpads touched the Moon. Armstrong would later tell James Hansen that he did not react instantaneously when the light glowed blue, thinking it to have been an anomaly and not entirely certain, thanks to the dust, that they had really touched down. As a result, he was a second or two late in shutting down the engine. There was little risk of an explosion, but the potential of damaging the engine on an inconveniently located rock made it essential to shut it down as quickly as possible. Forty seconds had now passed since Charlie Duke's last call, yet post-mission analysis would reveal – due to propellant sloshing around in the descent stage tanks and giving inaccurate readings – that Eagle actually had around 45 seconds of fuel remaining ...

Humanity's first seconds, and *first words*, on a world other than its own were hardly history making:

"Shutdown!" called Armstrong, punching the Engine Stop button. From the right-hand side of the cabin, Aldrin immediately set to work reciting each step of his post-landing checklist and they jointly took the requisite actions to power down now-unneeded systems: "ACA out of detent – Mode controls: both auto – Descent engine command override: off – Engine arm: off." Finally, Aldrin added, "413 is in", which told the AGS to remember the attitude of the vehicle on the surface. Outside, visible through Eagle's triangular windows, dust which had lain undisturbed for maybe a billion years or more began to settle. The altimeter ceased flickering, the surface seemed to shudder, then fell still, and the lunar horizon – which curved into blackness just a few kilometres away – was starker than either man could comprehend.

They had set down on a broad, roughly level plain, whose thousands of craters varied in size, shape and appearance, ranging from several tens of metres down to a few centimetres across. It was later determined that Armstrong landed about six kilometres downrange of their intended spot and their co-ordinates were 0.67409 degrees North by 23.47298 degrees East. Beyond the lander, the colour of the surface seemed to be a mixture of ashen greys, tans and browns and brightened into an intense, chalky white. Some nearby rocks seemed fractured or disturbed by the descent engine; Armstrong thought they looked like basalt. The surreal stillness of the scene and the silence of ages surrounded them and stealthily crept into the cabin. Inside their bulky space suits and bubble helmets, their mouths bone-dry from ingesting pure oxygen for so long, both men were breathing hard; yet they took a few

One of the first views of Tranquillity Base through Eagle's windows: a broad, level plain, pockmarked by craters . . . and starker and more barren than either man had ever seen.

seconds to grin at each other through four-day beards and tired eyes, before Armstrong keyed his mike.

"Houston," he radioed, "Tranquillity Base here. The Eagle has landed!"

Charlie Duke's response was entirely appropriate for his personality, defusing with humour the enormity of what had just happened. "Roger, Tranquillity, we copy you on the ground. You've got a bunch of guys about to turn blue! We're breathing again. Thanks a lot." Prior to launch Armstrong had told both Duke and Aldrin that he intended to change Eagle's radio callsign to 'Tranquillity Base' whilst on the Moon, but it came as something of a surprise to those who did *not* know. Aldrin did not expect him to use it so soon after landing and even Duke seemed tongue-tied when he tried to pronounce it in those euphoric first seconds.

Even 'euphoric' seemed a woefully inadequate adjective to describe the scene in the MOCR. "The whole [room] was pandemonium," wrote Deke Slayton, the man who had chosen and overseen the training of both Armstrong and Aldrin and undoubtedly hoped that one day *he* would be in their position. "It took about 15 seconds to calm down at all." Around the world, the feeling was the same. Walter Cronkite was uncharacteristcally speechless. Seated in the CBS studio next to former astronaut Wally Schirra, he stumbled over his words and stammered to his audience: "Boy ... Man on the Moon!" Firecrackers lit up the sky and drums thundered in celebration. In Russia, Alexei Leonov remembered hearty applause from his fellow cosmonauts – all *military* men, watching the triumph of the rivals they had struggled to beat for so long – and although much of the world remained fractured and torn by conflict, from Vietnam to Northern Ireland, many inhabitants of Earth stood still for a few moments to marvel ... not at an *American* victory, but at a truly epochal, *human* event. To paraphrase the Nobel Prize-winning physicist Robert Hofstadter, the first landing on the Moon may well be the only occurrence from the 20th century to be remembered clearly a thousand years from now.

The youth of 1969 were perhaps the most profoundly affected. It is interesting that, before each new Shuttle crew heads into orbit or another team of astronauts and cosmonauts fly to the International Space Station, they are invariably asked at one of their interviews what inspired them to enter the profession. Many have cited 20 July 1969 and the days which followed, more than any other, as having chiefly influenced their decision. Astronaut Mark Polansky, who led the crew of Endeavour on a station construction flight in July 2009, was 13 years old at the time and clearly remembered being at a baseball game in New York where the public address system announced Armstrong and Aldrin's landing on the Moon. Precisely four decades later, Polansky still remembered the immense feeling of goodwill as the entire crowd in Yankee Stadium rose to its feet and sang 'God Bless America'.

In Mission Control, the lighting of cigars, the waving of flags, the slapping of backs and the free-flowing tears which only Americans could produce would go on long into the night. John Houbolt – the NASA engineer who, a few years before, had advocated lunar-orbital rendezvous for Apollo – recalled Wernher von Braun turning to him, shaking his hand and saying warmly "Thank you, John." For Houbolt, being so honoured by the man who had created the Saturn V, it was one of the greatest compliments of his life.

Another compliment was paid to someone else, later that same evening. For more than five years, John Fitzgerald Kennedy, the president who committed America to landing a man on the Moon before the decade was out, had lain in his grave at Arlington National Cemetery, on the outskirts of Washington, DC. He had barely lived long enough to see the completion of Project Mercury and the early steps toward Gemini when he was cut down by an assassin's bullet in November 1963. Although his intentions when making his bold commitment were chiefly political, it had nevertheless been Kennedy who defined the United States' space ambitions for the Sixties. Today, he continues to hold a nostalgic, even mystical, place in the hearts of space aficionados, as the first major world leader to support a peaceful exploration programme with words, deeds and the clout to secure serious money.

Yet it remains ironic that, in the final months of his life, Kennedy had shown a distinctly different side to his character ... and some observers, indeed, continue to suspect that if his attempts to foster more cordial ties with the Soviet Union had been successful, they could have drastically slowed the drive for the Moon or even cancelled it. In an address to the 18th General Assembly of the United Nations in September 1963, Kennedy had hinted strongly that "there is room for new co-operation" with Russia and pointed specifically to "a joint expedition to the Moon", reinforcing the notion that exploration of the heavens should no longer be an issue of national or international competition. His words were surprising, in view of the reality that the two superpowers had been on the brink of nuclear war a year earlier.

However, one of Kennedy's advisors, Theodore Sorensen, later speculated that the president had become increasingly alarmed by the sheer cost of Apollo and may have been reluctant to continue spending at such an enormous rate. (The $7 billion originally tabled by George Low's committee in early 1961 would have almost quadrupled by the time Apollo 11 touched down on the Sea of Tranquillity.) Maybe, Sorensen theorised, the president simply wanted to cut some of the fat from Apollo by opening it up as a co-operative venture with the Soviets.

The overall picture still seemed to show that Kennedy was broadly supportive of the lunar goal, although taped conversations with Jim Webb, now ensconced in the Kennedy Library, imply otherwise. In November 1962, at a meeting to discuss the space budget, Kennedy categorically told Webb that he was "not that interested in space" and that his stance in support of the programme was based purely on the need to beat the Soviets. In the months following his arrival in office, he had been faced with issues of a perceived 'missile gap' and was then sledgehammered by Yuri Gagarin's orbital flight and the fiasco of the Bay of Pigs. At the time, initiating a programme to land a man on the Moon offered an impressive and necessary – albeit hugely expensive – solution to Kennedy's political problems.

A large percentage of the public were not behind the lunar goal, either, and criticism had steadily escalated in the two years after its conception. In April 1963, Kennedy had asked Vice-President Lyndon Johnson, serving as head of the National Aeronautics and Space Council, to review Apollo's progress. "By asking Johnson to conduct the review, Kennedy was virtually assured of a positive reply," wrote space policy analyst Dwayne Day. "Furthermore, Kennedy's request in effect ruled out cutting Apollo so as not to 'compromise the timetable for the first manned lunar landing'." In his report, the pro-space Johnson advised that, even if cuts *were* to be made to NASA, they ought to be diverted to safeguard Apollo.

A few days after Kennedy's speech to the United Nations in September of that year, Congressman Albert Thomas, chair of the House Appropriations Subcommit-tee on Independent Offices, wrote to the president and asked if he had altered his position on the manned lunar landing. Kennedy replied that the United States could only co-operate in space from a position of strength, but shortly before his assassination he asked the Bureau of the Budget to prepare a report on NASA. A draft of this report, which addressed the question of "backing off from the manned lunar landing goal" was written in early 1964 and still exists. In his analysis, Day posited that this was the very question that Kennedy had asked them to consider.

The bureau's ultimate consensus: the *only* basis for backing away from Apollo, aside from technical or international situations, would be "an overriding fiscal decision".

Nonetheless, the notion that Kennedy wanted to scale back or cancel Apollo has been grasped over the years by dozens of conspiracy theorists, including Hollywood director Oliver Stone, whose 1993 thriller *JFK* posited a changing lunar goal and an angered industrial-military complex may have represented one possible reason for the president's assassination. Yet Day noted that an outright cancellation of Apollo might not have been Kennedy's aim. Since 1962, the president had made approaches to Nikita Khrushchev and other high-level Soviet officials about the possibility of co-operation in space and had received mixed responses: polite noises on some occasions, outright dismissal on others.

At the time, of course, before Gemini had pushed America firmly into the lead, the Soviets remained in pole position in the space race and probably did not believe that Apollo would deliver on its promise. Indeed, it is possible that they perceived Kennedy's UN address as a veiled admission that the lunar goal was doomed to failure – a perception which may have led them to redouble their own efforts in 1964 to ensure that *their* effort to put a man on the Moon would bear fruit. Certainly, Kennedy knew that Apollo could only hope to achieve its goal during a hypothetical second term for his administration (from 1965–1969) and, indeed, NASA had originally aimed to reach the lunar surface sometime in 1967. *That* year, of course, would also be the 50th anniversary of the Bolshevik Revolution in Russia and fear undoubtedly pervaded NASA that the Soviets would not pass up such an important anniversary without attempting a major space spectacular of their own.

With this in mind, it is difficult to decide what Kennedy's next steps might have been had he not been so publicly assassinated in November 1963. It would certainly be ironic to suppose that the man who had so boldly committed his nation to the most audacious engineering challenge in history would, by *being* assassinated, actually prevent the ignominious cancellation of his pet project. Whatever the reality, and whatever the rationale behind his actions in the final months of his life, the result on 20 July 1969 was an inescapable triumph for the human spirit and the ingenuity of our species. It is a pity that Kennedy never lived to see his pledge fulfilled. Yet on that hot midsummer's evening, amidst all the excitement and celebration, an anonymous someone placed a small bouquet of flowers on his Arlington grave with a card that bore a poignant inscription. 'Mr President,' it read, 'the Eagle has landed.'

3

At home, above

FADED DREAMS

In one of history's bitterest ironies, Frank Borman, who commanded Apollo 8, the first manned expedition to the Moon, also became the first American astronaut to be invited to the Soviet Union. When he arrived in early July 1969, he was welcomed with a huge party, held in his honour at Moscow's prestigious Metropole Restaurant. Hundreds of guests flocked to shake his hand. Among them was Alexei Leonov, a cosmonaut who had made history a few years earlier by performing the world's first spacewalk. Borman congratulated Leonov on his achievement and the pair chatted about Apollo 8 and possible future landing sites on the Moon. Most likely, Borman considered Leonov to be showing the interest typical of a fellow space explorer. Little did he realise that Leonov had been training to complete the very same mission for which Borman was now being applauded.

Despite being one of the Soviet Union's most celebrated cosmonauts, Leonov's career at the end of the Sixties seemed laced with misfortune. Had his country pressed on with its own plan to put a man on the Moon, he might very well have commanded either a lunar-circling voyage or perhaps the first landing itself. In Leonov's mind, one of the fundamental reasons why Russia lost the race to the Moon was the death of Sergei Korolev in January 1966. Although the Chief Designer's successor, Vasili Mishin, was an excellent mathematician and a fast-thinking engineer who knew the business, he did not have the clout of Korolev and had poor diplomatic skills. "Lacking [these] political instincts," wrote Asif Siddiqi, "he suffered dearly. Some would argue that so too did the Soviet space programme in the coming years." Notably, Mishin's attempts to hire civilian engineers from the OKB-1 design bureau to fly the new Soyuz spacecraft had infuriated Nikolai Kamanin and alienated many of the veteran military cosmonauts.

In the first few months of Mishin's tenure, Leonov had found himself in charge of a group of men preparing for lunar missions in two quite different spacecraft, known as 'L1' and 'L3'. These spacecraft had arisen some years earlier. In 1964, Korolev had directed one of his deputies, Boris Chertok, to evaluate possible future roles for Soyuz. Of pivotal importance was the need to rendezvous and dock craft together in space – an ability considered vital in the construction of future orbital stations – but this later expanded to cover ship-to-ship extravehicular transfers, long-term missions of weeks or even months ... and a lunar expedition to rival Apollo.

The timing, of course, was particularly convenient. Only a year earlier, during an address to the United Nations in the autumn of 1963, John Kennedy had said that a manned mission to the Moon should be an international affair, rather than one of intense political rivalry. It was a surprising change of tack for a man who had initially pushed for a lunar landing as a means of demonstrating capitalist superiority over communist Russia. Perhaps, as mentioned at the end of the last chapter, the Soviets interpreted this as an admission that Kennedy's pet project was unlikely to succeed and they accelerated their own lunar goal accordingly. Certainly, an enormous rocket had been under development by Korolev since 1959; a rocket which he hoped might someday place a space station into orbit or even send cosmonauts to Mars.

Korolev conceived it as part of a 'family' of new boosters under the designation 'N', for 'nositel', the Russian word for 'carrier'. The largest was the N-1, followed by an N-2 to compete with another launcher (the UR-200), being developed by his rival, Vladimir Chelomei, and finally an N-3 to replace the R-7 workhorse used to carry the Vostok and Voskhod manned craft into space. The N-1 was very much a paper project until several weeks after Kennedy's lunar speech to Congress in 1961, when Korolev was given a small amount of funding to begin its formal development. A Soviet government report, published at around this time, also established a timescale for the N-1's first launch, sometime in 1965. America's public decision to commit itself to a manned lunar landing spurred Korolev and the Soviet leadership in the same direction and the Chief Designer began to push his N-1 as the cornerstone of such a mission, involving multiple launches to assemble a huge 'L3' mother ship and lander in Earth orbit, then boost them towards the Moon. Of course, *multiple* launches, despatched at a rapid-fire pace, served up some recipe for failure, to say nothing of having to execute intricate rendezvous and docking. On the other hand, launching small portions of a lunargoing vehicle imposed fewer stresses and strains on the rockets themselves and lowered their payload mass.

Elsewhere, Vladimir Chelomei's OKB-52 design bureau had proposed an alternate, lower-risk strategy, involving a *circumlunar* mission – without a landing – based around a 'cluster' of his UR-200 'universal rockets' or even a single, much larger UR-500, topped with a piloted spacecraft known as 'LK-1' ('Lunniy Korabl' or 'Lunar Ship'). In August 1964, Korolev's proposal for L3 landing missions was picked as the winner, although Chelomei was instructed to continue his work on the circumlunar plan. (Following the overthrow of Nikita Khrushchev two months later, the incoming Brezhnev regime told Chelomei to fly a circumlunar jaunt in 1967 to mark the 50th anniversary of the Bolshevik Revolution.) As a result, by the close of the Sixties, *two* manned lunar efforts were underway in parallel behind the Iron Curtain: the L1 for circumlunar flights and the L3 for landing missions.

However, also in the summer of 1964, the entire Soviet paradigm for putting a man on the Moon was changed: instead of an Earth-orbital rendezvous approach, featuring several launches to assemble the L3 craft and despatch it to the Moon, it was decided to adopt lunar-orbital rendezvous; a technique almost identical to that being planned for Apollo. Clearly, the realisation was setting in that three or more launches of the N-1, followed by a standard Soyuz craft carrying a crew of cosmonauts, would entail enormous risk, vast expenditure and huge technical difficulty. "Despite the historical significance of the decision," wrote Asif Siddiqi, "the reasons for this abrupt shift still remain obscure." Vasili Mishin has hinted at one possible reason: that Apollo had nudged the Soviet leadership toward "the development of ... vehicles that could support a lunar mission with a single launch". Maybe the industrial-military complex simply wanted a mission profile like that of the Americans – a 'parallel-response' common in weapons programmes – but the decision may also have highlighted a fundamental need for just one rocket to deliver its payload at a far cheaper rate than three or four launches could. Over the years, some have argued that Mishin drove the lunar-orbital rendezvous decision, whilst others attribute it to Korolev himself.

Whatever the truth, the problems quickly began to mount. As 1964 wore on, the lifting capacity of the N-1 into Earth orbit amounted to barely 75,000 kg, woefully inadequate to deliver a translunar injection stage and piloted mother ship and lander to the Moon. One factor was that Korolev insisted that all N-1 stages burn kerosene. By comparison, the Saturn V benefited from a higher-performance cryogenic upper stage burning hydrogen and could lift a payload of 118,000 kg. Moreover, the N-1 was considerably heavier than its American counterpart. These worries pressed Korolev to take steps to reduce the N-1's mass and increase its payload capacity. During the closing months of 1964 and through the following spring, his engineers explored virtually every avenue to shave kilograms – and, in some cases, even *grams* – from the lunar craft. Just as NASA 'rewarded' Grumman for keeping the weight of its lunar module down, so Korolev is said to have doled out bonuses of 50 or 60 roubles to his engineers for each kilogram of mass they saved. One story, cited by Siddiqi, even tells of an engineer who proposed sucking all the air from the rocket, since even *that* had mass ...

Calculations had already shown that the minimum requirement for a mission to the Moon demanded a lifting capability of around 95,000 kg and, aside from minuscule weight savings, it was Sergei Kryukov's team which would alter the N-1 in several significant ways to meet this need. The number of engines on the rocket's first stage, for example, was raised from 24 to an astonishing 30, the Earth-orbital altitude before translunar injection was cut from 300 to 220 km, the launch azimuth was shifted southward to a more favourable 51.6 degrees, the propellant quantity was increased by supercooling prior to loading to improve performance, latticed stabilisers were fitted and the overall thrust was boosted by 2 percent. In addition, the number of cosmonauts was reduced from three to two, one of whom would remain in lunar orbit, whilst the other descended alone to the surface. All this pushed the N-1's payload envelope to 92,000 kg – just barely enough for a stripped-down mission to the Moon – but also increased its mass from 2.2 million kg to 2.75 million kg.

Such profound changes caused much consternation. Among Korolev's staff, mutterings abounded that the lunar effort was "on the brink of fantasy". One senior engineer was reassigned to a different branch of OKB-1 for expressing disapproval and even cosmonaut Konstantin Feoktistov – instrumental in the development of Vostok, the world's first manned spacecraft – felt 92,000 kg was nowhere near enough. Despite what many have interpreted as Korolev's 'autocratic' attitude toward his subordinates, the glaring pitfalls with the lunar plan obliged him to request Mstislav Keldysh, head of the Soviet Academy of Sciences, to form a commission to more closely examine its difficulties. When this reported in February 1965, it was far from kind, decrying Korolev's "nerve" to even contemplate sending a single man to the lunar surface. At this time, of course, only six Vostoks and a single Voskhod mission had been despatched into low-Earth orbit and the physical dangers and psychological unknowns of long-term space exposure, including profound isolation from the home planet, were at the forefront of medical minds in both the East and the West. "Imagine for a minute being *alone* on the Moon," Keldysh had fumed. "That's a straight road to the psychiatric hospital!" More than that, the commissioners feared that the N-1's capabilities were being pushed too far. Korolev's insistence on triple redundancy for critical components meant the rocket had 200 control systems, thousands of kilograms of cabling and in excess of 2,000 black boxes whose documentation was inadequate. In fact, some systems would only be tested two or three times before being committed to the actual rocket. In the commission's collective judgement, these problems and deficiencies pointed to almost certain failure. Still, the commissioners finally bowed to what Siddiqi called Korolev's "headstrong opinions" and also approved the creation of the L3 landing system, under the proviso that OKB-1 would reach consensus on its technical goals and conclude a draft plan before the end of August 1965. It was expected to begin testing the spacecraft and the N-1 late the following year. Delays quickly crept in, as engineers and managers continued to debate technical shortcomings. At the same time, Korolev worked on redesigns to push the payload capability to 93,000 kg and Chertok steadily shaved more and more weight from the lunar lander.

Elsewhere, opposition to the N-1 was led by Valentin Glushko, perhaps the Soviet Union's foremost authority on rocket engine design ... and a fierce rival of Korolev. In 1962, his OKB-456 bureau had begun studies of a new engine, the RD-270. This was fed by unsymmetrical dimethyl hydrazine and nitrogen tetroxide – a toxic mixture disparagingly called "devil's venom" by Korolev, who preferred kerosene and liquid oxygen. Late in 1965, Glushko played his hand: the lunar rocket, he argued, should be completely redesigned to use his RD-270, whose thrust was four times that of the NK-15 engines earmarked for the N-1's first stage, and hence would achieve the same or better performance and neatly avoid the problems of having to synchronise, plumb and manage gas dynamics in 30 separate engines.

Glushko had the support of two of the most powerful figures in the Soviet space industry: Dmitri Ustinov, chair of the Military-Industrial Commission, and Sergei Afanasyev, minister of General Machine Building. However, by this point, the N-1 was three years into its development and manufacturing plants throughout Russia had already received technical specifications and started to build components. In

The gigantic N-1 rocket is rolled out to its launch pad. Some have seen this project as flawed and doomed to failure from the outset, whilst others have argued that – had Sergei Korolev lived just a little longer – it could have beaten the United States to the Moon.

spite of the endorsement of Ustinov and Afanasyev, and notwithstanding Glushko's overt efforts to sabotage Korolev at every turn, the decision was ultimately taken to proceed with the NK-15. Still, this incessant infighting and lack of professionalism or cohesion between the chief designers, coupled with the huge number of technical hurdles still to be overcome, sheds much light on why so little was achieved between 1964 and Korolev's death in early 1966. "Production capacities were inadequate," noted one senior observer. "Plans called for the fabrication of four N-1 rockets in a year's time, but only one and a half were constructed. The chief designers allowed serious deviations from the requirements for the final ground tests."

One particularly severe blow was the termination of the military GR-1 project, a 'global rocket' which was to enter orbit, pass over the South Pole, essentially undetected, deorbit itself and then rain city-flattening nuclear weapons down on the United States, literally obliterating it from coast to coast. The GR-1 never reached fruition and 1967's Outer Space Treaty banned the carriage of nuclear warheads into the heavens. However, the GR-1 would have used engines similar to the NK-15 and many Soviet engineers felt that flight testing it might have overcome some of the problems later faced by the N-1.

It is ironic, therefore, that by cancelling this weapon of mass destruction, one of the loudest death-knells sounded for the Soviet Union's chance to land a man on the Moon. Korolev's efforts to push his smaller N-2 and N-3 boosters through to production, at least to test the engines and other hardware for the eventual lunar rocket, fell on deaf ears in the Soviet leadership. It was clear by 1965 that the days of blank cheques for space spectaculars were gone. In his final years, Korolev considered using high-performance liquid hydrogen in the N-1's third stage, but it was decided to use kerosene for the entire rocket. Gradually, the N-1 moved from drawing board to production and Korolev and Mishin formally signed its first draft plan in November 1965. Soon afterwards, the oft-critical Keldysh Commission granted its support and recommended that the manufacturing of components should begin in earnest.

In its final form, the N-1 would be an impressive 105 m tall, slightly smaller than the Saturn V, but still one of the tallest rockets ever built. Its cluster of 30 first-stage engines produced a million kilograms more thrust than the five F-1 engines of the Saturn V and, for this reason (and despite never having a single successful mission), it retains a place in history for the highest liftoff thrust of any rocket ever flown. It had five stages: three to boost it into Earth orbit and the others to support translunar injection and operations in orbit around the Moon. The first three stages formed a truncated cone, 16.87 m wide at the base; a shape specifically engineered to accommodate its massive kerosene and liquid oxygen tankage.

As noted, the first stage was fed by 30 NK-15 engines, each with a sea-level thrust of almost 140,000 kg. The engines were arranged in two rings: an outer ring of 24 engines to handle pitch and yaw and an inner ring of six which were gimballed for roll manoeuvres. The interior of the ring was open, with air piped into a hole through inlets and then mixed with the exhaust to augment the N-1's first-stage impulse at 4.5 million kg. This exceeded the thrust of the Saturn V at liftoff, but burning kerosene in the upper stages produced a poorer performance overall, and it could insert correspondingly less payload into Earth orbit or onto a translunar trajectory.

The top of the first stage consisted of a lattice-like framework, which served as an 'interstage' and through which the gigantic kerosene tank could clearly be seen. This practice of *suspending* spherical tanks within the rocket made the N-1 quite distinct from other launch vehicles of its time. Its load-bearing configuration and relatively low density resulted in a significant reduction of payload mass, explained Asif Siddiqi, and to work around this obstacle engineers designed the tanks with an unusually low specific mass. Together with the engines, this helped to compensate for the drawbacks of the rocket having a non-monocoque main body. The spherical shape of the tanks also subjected them to lower heating loads and required less insulation, although this may have been a purely practical decision. Soviet metallurgists at the time were unable to produce aluminium sheets more than 13 mm thick. If engineers wanted *integral* tanks, their skins would need to be correspondingly thicker ... and so non-integral tanks were employed instead. In total, there were six tanks inside the N-1's first three stages: two per stage, one for fuel and the other for oxidiser, each of them fabricated from an alloy of magnesium and nickel.

The second stage had eight NK-15V engines, each delivering 162,000 kg of thrust, arranged in a single ring. These were essentially uprated versions of the first-stage engines, but showcased longer, thinner nozzles, optimised to function at higher altitudes. The third stage had four smaller NK-21 engines, clustered in a square, each delivering a thrust of 37,000 kg. All of the engines were built by Nikolai Kuznetsov's OKB-276 bureau (hence the 'NK' label) and employed in-built, impeller-type preliminary pumps and automatic controls with igniters to enhance performance. During the N-1's prolonged genesis, a series of uprated engines was introduced to supersede those from the original design. It is a pity, therefore, that the final design never actually flew.

Kuznetsov was a jet engine designer, with relatively limited experience in rocketry. In fact, this was one of the reasons why Valentin Glushko remained consistently critical of his designs and, it is true, the technical obstacles facing the simultaneous firing of so many engines were enormous. (At a meeting of chief designers in January 1968, Glushko undiplomatically referred to Kuznetsov's engines as being "rotten".) To work around the problem of having to master 30 engines, Kuznetsov introduced a control system known as 'Kord', which would provide for automatic shutdowns if faults were detected in flight. In such situations, the system would shut down both the failed engine and the one opposite in order to maintain the symmetry of the continued thrust. Still, the complexity of building plumbing to feed fuel and oxidiser into the clustered arrangement was both intricate and fragile and contributed in part to all four catastrophic launch failures between 1969 and 1972. The plumbing problems faced by NASA in building the Saturn V's first stage, which, after all, had just *five* engines, must have been exacerbated tenfold for the N-1. By the autumn of 1967, the first unmanned launch was long overdue and some insiders wondered if it would ever fly. Nonetheless, by May 1968, the first flight-ready vehicle was erected at Launch Complex 110 East at Tyuratam, with an expectation that it would carry an unmanned payload to the Moon sometime in September.

This launch complex was, like the N-1 itself, a titanic exercise in engineering. Descending five stories beneath each of the two pads was the flame trench to absorb the punishing impulse of the first-stage engines and expel their exhaust through three massive exit channels. Access to the vehicle was provided by a servicing tower which stood 145 m tall and could be rotated away on rails shortly before launch. The impressively robust fuelling system allowed technicians to load 1.02 million kg of propellant into its tanks.

Within weeks of arriving on the pad, however, hairline cracks were found in the first stage's liquid oxygen tank, forcing a removal of the rocket in June for repairs. Then, in September, a bulldozer accidentally severed the power cables to the pad, pushing the maiden launch back still further to November. By the end of the year, despite a fairly good run of ground tests, the flight had slipped into January, but it mattered little ... for work on the L3 landing system had scarcely begun. On Christmas Day, as Apollo 8 triumphantly orbited the Moon, a small soviet of designers pondered whether to proceed with the N-1 at all. Any hope of beating America to the lunar surface could only hinge on some major disaster or the outright cancellation of Project Apollo, together with a rapid turnaround in the N-1's fortunes.

Unperturbed, a State Commission confirmed 18 February 1969 as the date for the maiden flight. Many wanted nothing to do with it. Tyuratam's base commander, Alexander Kurushin, categorically opposed the launch, citing the many unresolved technical problems. He felt that the booster was unreliable and many of the changes implemented since 1966 lacked proper documentation. Nonetheless, the launch eventually went ahead at 12:18 pm Moscow Time on 21 February. Its payload was an unmanned spacecraft that was to make a week-long circumlunar flight. The mission lasted precisely 70 *seconds*! Shortly after leaving the pad, a rising high-frequency oscillation in the gas generator of one of the first-stage engines caused some components to tear away, leading to a propellant leak at 23 seconds into the flight which ignited a fire at 54 seconds. The Kord monitoring system detected the fire, but incorrectly shut down *all* of the booster's engines at 68.7 seconds. A little more than a second later, the N-1 was remotely blown up and its debris crashed ignominiously into the desolate Kazakh steppe.

In spite of this dismal failure, the effect for the viewers, which included Boris Chertok and Nikolai Kamanin, was spectacular. Chertok recalled that the tongue of flame from the first-stage exhaust was three or four times longer than the rocket itself. Only days later, the United States scored an impressive success with the mission of Apollo 9, a manned shakedown of their complete lunar craft in Earth orbit. As engineers, technicians and managers at Tyuratam watched the better part of a decade's work go up in a ball of smoke, their mood was as melancholy and wintry as the time of year.

Remarkably, the resilient Soviets pressed on. An accident report in March set in motion changes to the Kord control system, rerouting its wiring and better insulating it with asbestos to ensure that it performed satisfactorily in future. A second launch was scheduled for midsummer. On the other hand, more pressing worries – most notably the need to build a stand at Tyuratam to properly static-test the first stage – were ignored due to lack of money. "When they were all fired up at the same time," Alexei Leonov wrote of the first stage engines, "a damaging and destabilising vacuum was created between the two [rings]. This had not been discovered prior to launch because we had no facilities to test all 30 engines together." The one saving grace of the N-1's maiden flight was that its escape tower worked as advertised, plucking the lunar craft away from the inferno and parachuting it safely to the ground. *That* was also the only positive result on the night of 3 July 1969, when the second N-1 suffered an even more abysmal failure: the oxidiser pump on one of its engines ingested a fragment of slag and exploded 0.25 seconds after liftoff. A fire quickly broke out and raged as the rocket thundered clear of the tower ... and after all but one of the engines shut down at 12 seconds, the vehicle *came back down*, bellyflopping directly onto the pad and completely destroying the launch complex ...

One of the duty officers that fateful night was Lieutenant Valeri Menshikov of the Strategic Missile Forces and his memories were noted by Asif Siddiqi. Almost everybody involved with the N-1 project was present. Menshikov recalled that within seconds of the rocket lifting off the pad at 11:18 pm Moscow Time, "the steppe was trembling like a vibration test rig, thundering, rumbling, whistling, gnashing ... all mixed together in some terrible, seemingly unending cacophony. The blast wave from the explosion passed over us, sweeping away and levelling everything. Behind it came hot metal, raining down from above. Pieces of the rocket were thrown 10 km away and large windows were shattered in structures 40 km away ... [and] a 400 kg spherical tank landed on the roof of the installation and testing wing, 7 km from the launch pad ... "

Menshikov's harrowing account is testimony to the nightmare of the Soviet Union's final chance to ensnare a pyrrhic victory to dampen the triumph of Apollo 11. It is remarkable that none of the hundreds of assembled soldiers, technicians, managers, engineers and cosmonauts were killed. The effect of two and a half million kilograms of exploding propellants illuminated the steppe for dozens of kilometres, blowing out windows, sending spectators scurrying for cover and literally raining kerosene from the sky. Thirty-five kilometres to the south, in the city of Leninsk, eyewitnesses reported seeing a bright 'burst' in the heavens as the Soviet answer to the Saturn V vanished in a fireball.

At first light the next morning, as teams of technicians and military personnel ventured to the scene of the carnage – for it could no longer be termed a 'launch pad' and no longer possessed the structure of one – they were astounded at the devastation. The largest and most powerful explosion ever recorded in the history of spacegoing rocketry had left an utter ruin. The 145 m tower had been displaced from its rails and all ground support equipment and even a lightning arrester had been totally destroyed. "Windows and doors were smashed out, the iron entrance gate was askew," one observer wrote. "Equipment was scattered about with the light of dawn and was turned to stone ... the steppe was literally strewn with dead animals and birds. Where so many of them came from and how they appeared in such quantities at the station, I still do not understand."

Not surprisingly, recriminations followed hard on the heels of the disaster. The designer of the launch complex fumed that he would never again permit the launch of a rocket which could shut down its engines immediately after liftoff and wreak such havoc. Nikolai Kuznetsov, for the second time in five months, was directly in the line of fire: his Kord system had commanded several engine shutdowns, but these failed to occur. Ultimate blame settled on a faulty liquid oxygen pump and a defective rotor. The embarrassment and humiliation was intensified by the presence of American astronaut Frank Borman in the Soviet Union at the time. Two days after the disaster, he visited Star City and narrated a slide show to a glum and demoralised audience of Nikolai Kamanin and the assembled cosmonauts ...

Three weeks later, as Russia smarted and the world marvelled at Armstrong and Aldrin's landing on the Moon, the blame – and the *pain* – were both still acute. This 'culture' of blame-shifting and avoidance of responsibility on the part of chief designers and top-level political operatives had, to some extent, doomed the Soviet lunar effort to failure. In his diary, which was absent of comment in the six weeks following the 3 July catastrophe, Kamanin argued that a lack of centralised organisation, the paucity of *real* government support, improper and inadequate funding and constant infighting and manoeuvring between chief designers, Keldysh's Academy of Sciences and the Ministry of Defence had left a once-proud space effort in shreds. The deaths of Korolev, Vladimir Komarov and Yuri Gagarin, he added, had undermined morale and, in Kamanin's mind, the incompetence of Vasili Mishin was astonishing. Decades later, one can still hear the seething anger in the words of this aging, 61-year-old general, who had worked constantly and tirelessly, day and night, for a decade to beat the Americans to the Moon.

Yet as technicians, engineers and soldiers picked over the charred fragments of the N-1 and explored the post-apocalyptic scene that had once been Launch Complex 110 East, some things did not change. The short-sighted pettiness and old propaganda tricks of the Soviet leadership returned drearily to the fore. In August 1969, when the investigation of the second N-1 disaster was published, recommending delays to future missions until corrective actions had been taken and new, improved Kuznetsov engines built, a quite remarkable directive came from the Central Committee of the Communist Party: a manned lunar landing *must* occur in time for the centenary of Vladimir Lenin's birth in April 1970! On that date, it was mandated, a Soviet cosmonaut would plant a flag bearing the Hammer and Sickle into the lunar soil and unveil a monument celebrating communism. Once again, this underlined the wide gulf between the thought processes, rationale and sensitivities of the men who built the spacecraft and rockets and the political motivations of the men who made the decisions and allotted the money. Blood, of course, cannot be extracted from stones and, not surprisingly, the notion of a cosmonaut on the Moon to commemorate Lenin's birthday never happened. Not until June 1971, one of the darkest months in the Soviet manned space programme, would another N-1 fly ... with similarly catastrophic results. Improvements to the Kord control system, new filters in pumps to prevent foreign objects being ingested into engines, better sensors, upgraded instrumentation and hardier cabling would do little to bring the ultimate goal of a Soviet citizen standing on the surface of the Moon any closer.

One man who may have had an equal or greater claim than others to be first on the Moon was Alexei Leonov, who led the lunar detachment of cosmonauts. When formed in 1966, his team comprised veterans Valeri Bykovsky, Andrian Nikolayev and Pavel Popovich, unflown pilots Anatoli Voronov, Yevgeni Khrunov, Viktor Gorbatko and Yuri Artyukhin and civilians Valeri Kubasov, Oleg Makarov and Nikolai Rukavishnikov. More than a year earlier, Nikolai Kamanin had established a plan to train five or six crews for the landing missions over a 30-month period and, significantly, the Moonwalker would make a spacewalk to transfer from the mother ship to the landing craft and back again. With this in mind, it is not surprising that Leonov – the only man with actual spacewalking experience – should have been placed in charge of the group.

The choice of cosmonauts for these audacious missions was by no means arbitrary; like the selection of men to fly Grumman's lunar module, candidates were picked based on expertise and, crucially, also on weight and height. They could be no heavier than 70 kg and no taller than 1.75 m, in order to fit inside the cramped confines of the lander's cabin. "Vasili Mishin's cautious plan," wrote Leonov, "called for three circumlunar missions to be carried out with three different two-man crews, one of which would then be chosen to make the first lunar landing." Over the years, he has hinted strongly that he and Makarov would probably have flown the first circumlunar mission, originally timetabled for June or July 1967, and *also* the first landing, scheduled to take place a year later in September 1968. Other crews under consideration included Bykovsky and Rukavishnikov for the second circumlunar mission and Popovich and civilian Vitali Sevastyanov for the third.

Government documentation from February 1967 supports this timetable, even though it was ludicrously tight. Unmanned L1 circumlunar missions would have been undertaken in May and June, followed by piloted flights in July, August and September and the first test launches of the N-1 with the full L3 lunar payload possibly at the end of the year or, more likely, in the spring of 1968. At least two manned missions would fly into lunar orbit in April and June of that year carrying unmanned landers, as preparation for the first landing. A final manned lunar-orbital mission, in which the lander would undock and touch down on the Moon in an unmanned capacity, was scheduled for August. Barely a *month* later, the first manned landing was scheduled to take place. These plans were taken with a healthy pinch of salt by many insiders, notably Kamanin, who felt that circumlunar missions – at a push – might be possible in 1967, but that actual landings would probably not be achievable before 1969. When Vladimir Komarov was killed aboard Soyuz 1 in April 1967, all such plans moved still further over to the right.

The craft aboard which Leonov and Makarov might have become the world's first lunar landing team was a truly enormous vehicle, measuring 43.2 m long and consisting of two stages. Its first segment, Blok G, was equipped with a single NK-19 engine and would have burned for 480 seconds to accelerate the L3 to around 40,300 km/h, in order to propel them out of low-Earth orbit and onto a free-return trajectory to the Moon. The L3 itself comprised three distinct parts: a Lunniy Korabl (LK) or 'lunar ship', which would perform the landing, a Lunniy Orbitalny Korabl (LOK) or 'lunar orbital ship', which would remain in orbit around the Moon while the lander was away, and a final propulsive stage called Blok D. The latter served as the fifth and final stage of the N-1 and was to execute two or three midcourse corrections on the way to the Moon, then insert the spacecraft into lunar orbit and later on perform the early stages of a Powered Descent to the surface. Designed by Mikhail Metnikov, it was 5.7 m long and 3.7 m in diameter.

Atop Blok D sat the LK lander and, above that, the LOK mother ship. Early designs for the craft which would perform the Soviet Union's first manned landing on another world had been *tabled* in 1964, but its specifications were not *finalised* for another four years – after numerous changes to braking angles, considerations of delta-V, a guidance system capable of supporting landings away from the lunar equator and a problematic radar. Like the Grumman lander, its design was dictated by the need to shave away as much mass as possible; in fact, its pilot would have had barely enough room to stand inside his cramped cabin. According to computations, the entire lander could weigh no more than 5,500 kg, making it a third as heavy as the American craft. "Given the generally heavier microelectronics components and relatively poor capabilities of Soviet computers," added Asif Siddiqi, "this was indeed a tall order for Korolev's engineers."

Having a single cosmonaut aboard the LK for descent, landing, surface operations and ascent back into orbit was one result of the need to keep the weight of the craft down, but it also prompted worries about safety; worries which Mstislav Keldysh's commission had already hotly criticised in 1964. Additional concerns arose from the fact that, unlike Grumman's two-stage lander, the LK had just one engine for both descent and ascent. Several designs were proposed, rejected and changed during the machine's genesis, but in its final form – which the Western world would not see for more than three decades – it comprised three parts: a landing aggregate and takeoff apparatus, fulfilling the roles of both descent and ascent stages, and a propulsive unit known as 'Blok E'.

The landing aggregate was a 2.27 m frame, shaped like a pair of truncated cones with their bases fused together, and included a suspended instrument compartment fitted with the 'Planeta' ('Planet') landing radar, a battery of surface science gear, including a drill and 'operational manipulator', a pair of parabolic communications antennas, storage batteries and a fold-down ladder by which the cosmonaut would descend to the lunar soil. Supporting the weight of the craft was a four-legged Lunar Landing Unit, designed to allow the LK to drop safely from a height of a metre at a lateral velocity of perhaps a metre per second, whilst also preventing it from capsizing. A solid-propellant 'hold-down' motor at the upper end of each of the legs would fire to cushion the touchdown and ensure that the LK did not topple over. The lander was designed to be able to touch down safely on an incline as steep as 20 degrees from the horizontal. It is interesting that Soviet and American ideas for developing this landing apparatus were remarkably similar. The decision to employ a more stable combination of four landing struts came after several other proposals had been discussed and discarded: some engineers advocated a 'tripod' arrangement, whilst others even suggested that a supportive 'ring', akin to a giant inner tube, might serve the purpose more effectively.

Mounted atop the landing unit was the Lunar Takeoff Apparatus, a spheroidal pressurised cabin with a dome-enclosed instrument compartment tacked onto one side. Like Grumman's machine, it paid little heed to aesthetics, possessing a crazy amalgam of attitude-control thrusters and a quartet of omni-directional and rendezvous antennas. The cabin measured 2.3 m by 3 m, barely enough for one man, fully suited and harnessed to the floor, to stand upright. Slightly to his right was an instrument panel, which would allow him to guide the lander to a touchdown on alien soil. His visibility during descent, ascent and rendezvous with the mother ship was aided by a pair of tiny viewports. Apparently, there was also a joke that the cabin would also have contained a bust of Lenin, on a shelf behind the cosmonaut's head …

Original plans called for a pure oxygen atmosphere inside the cabin, but the weight demands that this would impose led designers to opt instead for an 'ordinary' terrestrial composition of oxygen and a slightly reduced amount of nitrogen. Mass constraints made an airlock out of the question and so, like NASA's lander, the entire cabin would have to be depressurised before the cosmonaut could venture outside. All in all, the LK's life-support system and batteries could sustain an independent mission no longer than 72 hours. This meant that, theoretically, stays of up to two days on the lunar surface could be attempted on later landings.

Had the Soviets beaten the Americans to the Moon, their cosmonaut would have endured an uncomfortable ride: clad from start to finish in the 'Krechet' ('Gyrfalcon') space suit, a cumbersome, semi-rigid structure weighing 90 kg and capable of supporting him for up to six hours. Developed by Gai Severin, it would have had its work cut out on an L3 mission. Not only would it have needed to sustain the cosmonaut on the Moon's surface, but it would also have been worn during two spacewalks as he transferred from the LOK to the LK and back again in lunar orbit. Furthermore, on the last of these spacewalks, he would have been laden with rock boxes, core tubes and other priceless samples …

Unlike previous suits, the Krechet was entered through a 'door' in its rear and provided a wide range of utilities to keep its occupant functional: its backpack maintained thermal control, pressure, air purification and dehumidification and a beacon to enable ground controllers to determine his exact location relative to the lander. The bulk of the Krechet also dictated the shape of the LK's hatch: an *oval* aperture – the first time such a shape had been adopted in a Soviet manned spacecraft. It is a minor detail, but offers an intriguing footnote that the designs of the American and Russian landers encountered many of the same engineering obstacles and found remarkably similar solutions.

The Krechet's shoulders and wrists included ball-bearing joints, permitting almost 360 degrees of rotation, and its 'rear-entry' mode of ingress and egress – a design attributed to engineer Anatoli Stoklitsky – was considered easier and much more reliable than zippers. A control and instrument panel was mounted on the cosmonaut's chest and there was a metal 'ring' on his back, shaped like an oversized hula-hoop, to allow him to roll onto his side and use his arms and legs to pick himself up, should he fall on the lunar surface. Like the Apollo suits, the Krechet had a pair of snap-down visors, one of them gold-tinted in order to reflect harmful solar ultraviolet radiation.

Had Leonov and Makarov's September 1968 mission actually occurred, they would have become the world's first lunar landing crew, almost a year ahead of Neil Armstrong and Buzz Aldrin. They would have launched from Tyuratam aboard their LOK spacecraft, at the very tip of the mighty N-1, and been quickly inserted into a 220 km orbit around Earth. However, unlike Apollo, they would have spent 24 hours checking their ship's systems prior to igniting the Blok G for translunar injection. This stage would then be jettisoned and Leonov and Makarov would settle down for a 101-hour cruise to the Moon, punctuated by a couple of firings of the Blok D to fine-tune their trajectory.

The 10 m long LOK craft which the two cosmonauts would have called home during their journey to and from the Moon was, appearance-wise, very similar to the Soyuz-7K-OK in that it, too, consisted of a spheroidal orbital module, a beehive-shaped descent module and a cylindrical instrument module; but it was mated with the Blok I propulsion unit. Measuring 1.5 m long, this carried 300 kg of unsymmetrical dimethyl hydrazine and nitrogen tetroxide in six tanks, supporting four sets of engines and, uniquely, the 'Volna' ('Wave') fuel cell. Its purpose was to ensure an electrical supply throughout the mission and boost the cosmonauts out of lunar orbit and onto a transearth trajectory. The use of a fuel cell also meant that, unlike Soyuz-7K-OK, the LOK did not need solar panels and was not so constrained in the orientations it could adopt. At the opposite end of the spacecraft, atop the orbital module, was a docking apparatus known as 'Kontakt' ('Contact'), which would provide a mechanical – although not an electrical – docking with the LK lunar lander.

Consequently, four days after launch, the cosmonauts would have braked themselves into lunar orbit at an altitude of around 150 km, then adjusted their path to some 100 × 20 km. Next would have come the tricky part. Unlike NASA's mission profile, in which the astronauts transferred to their lander through an internal tunnel, Leonov was required to spacewalk from the LOK over to the lander, which was still encased inside its protective launch fairing. He would open an outer hatch, then an inner one, and finally enter the cabin of the lander itself. After Leonov was satisfied with his systems, Makarov would undock the LOK and the LK would start the descent to the surface. The Blok D would be jettisoned at an altitude of 1.5 km and the LK's main engine – the Blok E, a single-chambered, *throttleable* engine which would allow Leonov to hover and manoeuvre to select a safe landing point – would take over for the final minute or so. The Blok E simply *had* to work and early documentation suggested a reliability of 99.976 percent, a virtually unheard-of (and extremely doubtful) level in Soviet rocketry. Fed by unsymmetrical dimethyl hydrazine and nitrogen tetroxide, its weight dictated that it be installed as low as possible within the Lunar Landing Unit to ensure maximum stability. In addition to the primary engine, a two-chambered backup was provided. At the instant of touchdown, the four hold-down engines on the legs would fire to aid stability ... and the first man to walk in space would be just minutes away from also becoming the first man to walk on a world other than his own.

Although the overall mission is evident, little has emerged in the space flight literature over the years, even in the post-Soviet era of greater openness and transparency, to suggest the extent of scientific exploration of the Moon during however many LK landings were planned. Subsequent missions were expected to spend as much as 48 hours on the surface and travel perhaps 3 km from the lander, but scientific objectives were probably never formalised. Asif Siddiqi has hinted that the first landing, like that of Armstrong and Aldrin, would have been a 'flags and footprints' affair: dominated by the planting of the symbolic Hammer and Sickle, the deployment of a small set of scientific instruments and the taking of a few photographs. A geological extravaganza as detailed as, say, Apollo 15, is unlikely ever to have been planned and weight constraints aboard the LK would have made it impossible to carry a heavy load of scientific gear. *That* would have to await another generation of explorers. The Soviet *need* to land on the Moon was purely dictated by a desire to beat America at all costs. Indeed, even for the Americans, the scientific exploration of the Moon was totally subordinate to Apollo's 'real deal': the need to score a technological and ideological triumph over communist Russia, in full sight of the world.

The piloting challenges involved in bringing a spidery LK onto the surface of the Moon were daunting. Yet, as far as we are aware, the Soviets possessed no Lunar Landing Training Vehicle of their own. Instead, the commanders of *their* landers practiced using helicopters, notably a modified Mi-4 Hound. This had originally been designed to match the American H-19 Chickasaw during the Korean War and Leonov, for one, considered it the closest available analogue for the real thing. In fact, he spent much of 1966 and 1967 getting himself qualified as a helicopter pilot and was actually undergoing training on the day of Yuri Gagarin's death in March 1968. "The flight plan of a lunar landing mission," he wrote, "called for the landing module to separate from the main spacecraft at a very precise point in lunar orbit and then descend towards the surface of the Moon until it reached a height of 110 m from the surface, where it would hover until a safe landing area could be identified. The cosmonaut would then assume manual control of its descent."

The relatively close terrestrial analogue of using helicopters and other vertical takeoff and landing (VTOL) machines had also led the LK engineers to consult specialists at several aviation design bureaux across the Soviet Union during their craft's genesis. Even in the wake of the first N-1 failure in February 1969, Leonov and other military cosmonauts preparing for lunar landings continued to master the art of landing helicopters with limited fuel reserves in the shortest possible time. Like the Apollo commanders, their available 'hover time' over the Moon's surface would be counted in *seconds*, not minutes. It had long been recognised that, in the event of a cabin depressurisation or some other major malfunction, the cosmonaut, encased inside his bulky Krechet suit, would need to operate the lander's controls with gloved hands and differing strategies for completing this safely and practicably were discussed and perfected.

In spite of the traditional overwhelming focus by Soviet spacecraft designers on automated systems, the military cosmonauts saw landing on the Moon as an intensely *personal* flying challenge and they expected to have manual control. "As commander of a spacecraft," wrote Leonov, "what I needed once a flight was in progress was as little communication as possible from the ground – since it served mainly to distract me from what I already knew was necessary – and only manual, not automatic control." One crucial tool in the ability to exercise effective manual control was being able to properly orient the spacecraft, a task that demanded an advanced knowledge of star positions and an understanding of the ship's sextant and stellar sensors. Throughout the late Sixties, as the Soviet and American space programmes recovered from their tragedies with Komarov and Apollo 1, Leonov regularly led teams in Moscow's planetarium to examine stars in the northern hemisphere, then flew to the Mogadishu region of Somalia to study those in the southern hemisphere.

Leonov's time on the Moon would have been limited by consumables aboard the lander and it is amusing that, after re-entering the cabin from his Moonwalk and repressurising the atmosphere, he was supposed to remove his helmet and "begin a rest period". The prospects of resting after having just completed the first, adrenaline-driven walk on another world would probably have been as remote as asking a child to rest after having just opened presents on Christmas morning ...

At length, the Blok E would have been reignited to begin the climb into orbit to dock with Makarov in the LOK mother ship. Unlike the command module pilot aboard Apollo – upon whom fell the responsibility of actually *flying* the craft alone in lunar orbit – the role of the flight engineer during his solo time was more of a 'supervisory' one. In fact, the proximity operations and docking with the ascending lander would be entirely automatic, being performed by the Kontakt rendezvous device, and Makarov would probably have not needed to even touch the controls, unless an emergency demanded his attention. Vasili Mishin had long disregarded the need for 'real' piloting during space missions, believing that automated systems were perfectly adequate.

Yet another excursion in his Krechet suit would await Leonov, as he crawled hand-over-hand from the lander to the safety of the LOK, towing his priceless lunar samples. Weight constraints, imposed whilst Sergei Korolev was still alive, had long since determined that an *extravehicular*, rather than internal and 'shirt-sleeved', transfer was necessary. Indeed, the extravehicular transfer of cosmonauts Khrunov and Yeliseyev between spacecraft, originally planned for Soyuz 2/1 in April 1967, formed a logical first step in preparing for the lunar-orbit transfer. By the time they finally performed their transfer on Soyuz 5/4, of course, such techniques were effectively obsolete and the chance of beating the Americans to the surface had diminished. Unlike the Soyuz system, which employed a 'pin-cone' configuration and could support multiple dockings and undockings, the Kontakt arrangement had a pin, a honeycomb plate with 'claws' and a set of simple shock absorbers and it permitted only a single docking. This would have provided Leonov and Makarov with a *mechanical* mating mechanism and no electrical, fluid or power transfer.

Finally, after 38 orbits and three days around the Moon, the mother ship's attached Blok I engine would have made the transearth injection burn to set the triumphant cosmonauts on course for home. An 80-hour return journey would have ended with a parachute-assisted touchdown on land, a week after launch.

Had it happened, the Leonov/Makarov mission would have marked the longest Soviet manned space flight yet undertaken. Indeed, until Soyuz 9, the most experienced cosmonaut was Valeri Bykovsky, who had chalked up almost five days at the end of his Vostok 5 mission in June 1963. The Americans, it is true, had long since doubled and nearly tripled that achievement, but Phillip Clark summed up the truly daunting nature of both a circumlunar and a landing mission for the Soviets. They "would have been committing a ... crew to the longest manned mission of a Soyuz-derived craft to that date," he wrote, "and furthermore to the longest Soviet manned mission of *any* type undertaken to that date." With this in mind, it is remarkable to think that but for a few twists and turns of fate and cruel luck, the week that Richard Nixon would later refer to as "the greatest since the Creation" might very well have occurred in September 1968 ... and belonged to the Soviets.

STRAINED RELATIONS

As an increased spirit of entente with the United States developed in the early Seventies, the Soviet Union's relationship with one of its closest physical and ideological neighbours deteriorated, literally, to the extent of armed conflict. Although both Russia and China espoused communist doctrine, their principles differed markedly. The roots of disagreement stretched back to the Fifties, for China, with no great urban working class of its own, had difficulty in applying traditional Marxist/Leninist thought to creating its own egalitarian paradise and instead adopted a model of 'peasant revolution' and a series of Five-Year Plans to achieve economic prosperity. In spite of close ties between Joseph Stalin and Mao Tse-Tung and an uncomfortable stance of togetherness during the Korean conflict, the Fifties gradually pulled their respective ideals further and further apart.

One reason for this divergence was Stalin's successor, Nikita Khrushchev, who openly denounced his predecessor's policies and de-emphasised the core communist themes of the need for worldwide conflict between capitalism and socialism. Attempts to engage Dwight Eisenhower in dialogue in 1959 and a decidedly pro-Indian position in the 1962 Sino-Indian War irritated Mao, who felt Khrushchev was retreating from 'true' communism and becoming far too conciliatory with the West. It certainly appeared, in Mao's eyes, that a friend was degenerating into a foe. At first, their criticisms centred on each other's attitude toward other communist neighbours – for example, Soviet support for Yugoslavia's Josip Broz Tito and the Chinese endorsement of Albania's Enver Hoxha – but by the end of 1960 Khrushchev and Mao had quarrelled openly at numerous major conferences. They compromised to avoid an all-out ideological split, but when Russia broke diplomatic ties with Albania their differences were expanded from political parties to nation states. When Khrushchev withdrew, badly humiliated, from the Cuban Missile Crisis in the autumn of 1962, Mao denounced his "capitulationism" and within two years the two countries had effectively severed diplomatic links.

Chinese hopes of a thaw in relations with the arrival of Leonid Brezhnev were disappointed and in January 1967 Red Guards even besieged the Soviet embassy in Beijing. This was rapidly followed by a mass mobilisation of Soviet troops along the 4,300 km border with China and in March 1969 there were several armed raids along the Ussuri River. Shortly thereafter, the situation was calmed when Alexei Kosygin visited Beijing to speak to his counterpart Zhou Enlai and, by October of that year, the two countries had entered into border-demarcation talks. Although these proved unsuccessful, they at least restored some formal diplomatic dialogue. Nevertheless, an uneasy relationship would ensue: by 1973, Soviet forces on the border were almost double the size that they had been four years earlier and the risk of a Chinese invasion was feared above that of an attack by the United States.

By 1975, Soviet relations with America would have reached their zenith in the form of a joint manned space mission, Apollo-Soyuz. It is ironic, therefore, that the 'new enemy' of the Soviet Union was not so much the capitalist West, but a close neighbour with which it purportedly shared similar communist principles.

ZOND

When Neil Armstrong and Buzz Aldrin touched down on the Sea of Tranquillity on that hot midsummer's evening in 1969, Alexei Leonov and doubtless millions of others had reason for despair; not so much because they had themselves worked tirelessly for many years to achieve the same goal ... but because their own leadership and the Soviet Union's stifled, government-run media prevented such enlightening and historic news from being properly disseminated. No public comment was made by Leonid Brezhnev's hard-line regime, although an official letter of congratulation was sent to the Nixon administration and the Soviet official mouthpiece, the newspaper *Pravda*, restricted its coverage of the actual landing ... to an *inside page*. Russian audiences saw only brief, televised clips from the mission, which both angered and saddened Leonov.

"Not showing live coverage of the ... landing was a most stupid and short-sighted political decision, stemming from both pride and envy," he wrote. "Our country robbed its own citizens by allowing political considerations to prevail over genuine human happiness at such events." The Soviets were not alone; China, too, refused to broadcast Armstrong and Aldrin's Moonwalk live.

Still, even in the wake of their second N-1 failure, dreams of planting a Hammer and Sickle into the lunar soil continued for some time. It is hardly surprising that Leonov, whose own promised land had been snatched from him by the Americans, was vocal about continuing with the plans to accomplish a circumlunar mission and, eventually, once the N-1's woes had been rectified, a landing of their own. He took his words as high as the Central Committee of the Communist Party and, at first, received warm approval ... but this approval diminished after Apollo 11's triumph. This was a *race* – and an unashamedly political one, at that – and Brezhnev's regime had no interest in the scientific exploration of the Moon. Like an inquisitive dog urinating against a fence in order to mark its territory, the simple need to be *first* and to stake a claim to the lunar surface was all that mattered ... and, for the Soviets, *that* had been lost. As early as May 1969, two months before Apollo 11's landing, one of the earliest declarations in what would become a standard and dreary diatribe from the Soviets over the next two decades was made: that they *never* had *any* *intention* of landing a cosmonaut on the Moon.

Most observers in the West refused to believe such claims. In the United States, the CIA had known about the N-1 since at least the spring of 1967 and NASA Administrator Jim Webb had been so insistent about its existence that the mysterious super-booster was nicknamed 'Webb's Giant' by the press. George Mueller, Associate Administrator for the Office of Space Flight, added fuel to the fire by speculating in the summer of 1967 that the Soviet rocket was larger than the Saturn V "by a factor of two". Little could be proved, of course, except through military intelligence dossiers and spy satellite imagery, some of which revealed new blackened scars on the desolate Tyuratam steppe after each disastrous launch. The fact that the Soviets' lunar dream could *not* be proven offered a perfect shroud for one of the most elaborate cover-ups in the history of space exploration. The media in the West were undecided: *Aviation Week and Space Technology* told its readers in the autumn of 1966 that Russia was "showing increasing signs of having conceded the manned lunar landing race to the US", whereas loose-lipped cosmonauts Leonov in Hungary and Vladimir Komarov in Japan had explicitly stated to journalists that piloted Moon missions *were* on the cards as part of the next 'Five-Year Plan'. Still other journalists were even sceptical of what little intelligence there was. "Like the Loch Ness Monster or Soviet submarines seen off the East Coast when the American Navy's budget is under review," wrote one, "[it] tends to be motivated by witnesses who are considered unreliable or prejudiced."

Notwithstanding the faults and failings of the N-1, the Soviet dream of reaching the Moon with men also had a second strand: the L1 circumlunar effort, utilising a modified Soyuz spacecraft later known as 'Zond', boosted aloft by an uprated version of Vladimir Chelomei's UR-500 Proton rocket. Unlike the N-1, Chelomei's launcher had shown greater promise ... though, unfortunately, the Zond had not proven reliable enough to be entrusted with a human crew.

Certainly, Nikolai Kamanin, in his position as commander of the cosmonaut team, had favoured the Proton since at least 1966. In December of that year, he and a group of other high-level military officers had visited Chelomei's OKB-52 facility and been impressed by its competence, its organisation, the quality of its project planning and the sophistication of its rocketry and spacecraft designs. He was shown hardware for the UR-100, UR-200 and UR-500 Proton boosters, together with blueprints for a gigantic UR-700, capable of sending a cosmonaut directly to the lunar surface and with a payload capacity which outstripped even the Saturn V. In his diary, Kamanin grumbled that Korolev's problem-plagued N-1 was now five years out of date and that the UR-700, by contrast, utilised modern technology and could be significantly upgraded. In 1964, Chelomei had been given the task of developing the LK-1 circumlunar craft, but after the fall of Nikita Khrushchev responsibility for such a mission had passed to Korolev and Chelomei was ordered in November 1965 to cancel work on his project. Nevertheless, Chelomei continued developing the craft quietly at his own risk.

Korolev's Soyuz-derived circumlunar vehicle, the L1, was to be boosted by Chelomei's UR-500 Proton rocket and use the Blok D stage from his own N-1 for translunar injection. Asif Siddiqi has pointed out that by flying the Blok D 'earlier' as part of this system, any problems could be ironed out. However, when told to collaborate with Chelomei and utilise *his* booster to launch the L1, Korolev was "dismayed" and their working relationship in the final months of the Chief Designer's life was hardly harmonious. A fundamental problem was that the propellants for the UR-500 were highly toxic. For the safety of the crew, and perhaps also illustrating his distrust of the reliability of Chelomei's booster, Korolev proposed using the Proton to launch an unmanned L1 spacecraft with its Blok D stage into Earth orbit, then a standard Soyuz carrying a crew of two cosmonauts. The spacecraft would dock, the crew would transfer to the L1 and boost themselves towards the Moon. However, when Vasili Mishin was appointed as Korolev's successor in May 1966, he concluded that the L1 could be lightened sufficiently to launch safely with a UR-500/Blok D combination and deleted the need for an additional ferry flight.

Like the LOK, the L1 was a modified version of Korolev's standard Soyuz, with the major differences being the deletion of the spheroidal orbital module and the addition of a pair of solar panels to the service module. The two-man crew would spend the entirety of their circumlunar mission, anywhere between seven and ten days, enclosed in a habitable volume of only 2.5 m^3. Also removed from L1 was the reserve parachute, to accommodate a hatch in the side of the descent module. This major safety deviation, together with the absence of space suits and no emitted signal to confirm parachute deployment, worried Nikolai Kamanin and many of the military cosmonauts when shown a mockup of the craft in April 1966. Government documentation from February of the following year proclaimed that, following a series of automated flights, a manned circumlunar mission, which Alexei Leonov presumed would be crewed by himself and Oleg Makarov, would occur in the summer of 1967. It seemed an impossible schedule to achieve, but it started reasonably well. In mid-March, under the cover name of Cosmos 146, one such spacecraft was boosted into a highly elliptical orbit by the first of Chelomei's four-stage UR-500 Protons. A full lunar mission was not intended and there were no plans to recover the craft, but the rocket and the Blok D performed as advertised ... and infused everyone with a misguided sense of complacency. Three weeks later, in April, another L1 craft, dubbed Cosmos 154, reached orbit, but its Blok D failed to fire and it later ignominiously burned up in the atmosphere.

The situation deteriorated rapidly. Two weeks after the Cosmos 154 failure, Vladimir Komarov was killed returning from orbit aboard Soyuz 1, a disaster that essentially hamstrung any chance that a pair of Soviet cosmonauts would circle the Moon before the end of 1967. Nonetheless, with immense pressure from above to conduct a circumlunar mission and ensure that it coincided with the 50th anniversary of the Bolshevik Revolution in November, the efforts continued. A 'full-up', circumlunar L1 was launched from Tyuratam on 27 September, but its Proton failed, crashing 60 km from the pad. Film from the launch site indicated that one of its six first-stage engines was to blame. Still, the L1 itself was plucked clear by its escape tower and landed intact. "When rescuers arrived," wrote Asif Siddiqi, "they were greeted by a strange scene: from one end of the horizon to the other, there was an eerie, yellowish-brown cloud of nitrogen tetroxide and unsymmetrical dimethyl hydrazine all over the steppes."

Despite demands for a lengthy delay to better understand the cause of the failure, another L1 was launched on 22 November . . . but one of the engines of the Proton's second stage failed to start. The other three engines ignited, but were shut down by ground command when tracking showed a deviation from the planned trajectory and the vehicle crashed. Again, the escape tower saved the spacecraft, although the landing parachute dragged it along the steppe for half a kilometre. This did not fill Leonov, who later travelled to the landing site, with confidence for the survival of a cosmonaut crew. Indeed, in January 1968, follow-on tests went badly wrong when the parachute failed to inflate properly.

The next mission was launched on 2 March and its Blok D successfully achieved a translunar coast, at which time it was named Zond 4. Problems with the spacecraft's orientation system prevented an initial mid-course correction burn, although it completed its mission and returned to Earth. Unfortunately, its guidance system failed during re-entry; it hit the atmosphere at precisely the calculated time, but a planned 'double-skip' manoeuvre to bring its descent module down into the Soviet Union was not possible and the spacecraft was remotely destroyed. The double-skip was a necessary component in executing a successful re-entry at lunar-return velocities. "The speed at which a spacecraft would travel . . . upon returning from the Moon," explained Leonov, "would be 11.2 km per second, which would have to be reduced to 8 km per second in order for it to re-enter . . . for a safe landing. To lower the speed, the spacecraft would have to enter the Earth's atmosphere for a short period, then bounce off, leave it and re-enter again. The key to this difficult manoeuvre was the angle of re-entry." Nikolai Kamanin later criticised the decision to remotely destroy the craft, arguing that Zond 4 could still have been recovered intact by Soviet naval vessels. A decision was therefore made, wherever possible, to attempt recovery of future Zonds.

With depressing regularity, it seemed, yet another L1 failed in April 1968. This time, a short-circuit in the malfunction detection system incorrectly indicated a problem with the Proton. Again, the descent module was plucked to safety by its escape tower. It was the *next* Zond which caused the American and Soviet lunar efforts to converge, for *this* flight led to NASA adapting Apollo 8 to undertake its mission to the Moon. Storm clouds of foreboding seemed to be gathering over Tyuratam when three technicians were killed by the explosion of the Blok D's oxidiser tank in July. Fortunately, both the Proton and Zond were relatively unscathed. When Zond 5 was launched on 14 September, it began the first fully successful unmanned circumlunar flight. Four days into its mission, it swung around the Moon at a distance of 1,950 km with a cargo of live turtles, wine flies, meal worms, plants, seeds and bacteria and snapped high-quality photographs. A glitch with its gyroscopic platform just prior to re-entry did not compromise a satisfactory splashdown in the Indian Ocean. The near-perfect flight had lasted seven days.

A second success, Zond 6, began on 10 November and duly looped around the Moon. However, during its return journey, a gasket failed, which depressurised the descent module, killed all the biological specimens and certainly would have doomed a human crew, had one been aboard. Ironically, its double-skip re-entry worked well and it landed barely 16 km from its launch pad . . . although 'landed' is perhaps an inappropriate term. As a result of the depressurised cabin, the craft's altimeter issued a false command to release the parachute, whose container also depressurised at an altitude of 5.3 km. "Without a parachute," wrote Siddiqi, "the ship simply plummeted down to the ground and smashed into pieces."

A successful lunar loop, and even a successful double-skip re-entry, could not compensate for a depressurised cabin and – a year after Komarov's death – *another* problem with parachutes. It is hardly surprising, therefore, that no Soviet mission was authorised by Vasili Mishin to coincide with their *next* lunar launch window in early December. To observers in the United States, including Deke Slayton, this omission implied that something had gone awry. Years later, Leonov would tell Tom Stafford that he and Makarov were prepared, regardless of Zond 6's problems, to take the risk and ride their own Zond for seven days to the Moon and back. So was their backup crew, Valeri Bykovsky and Nikolai Rukavishnikov. So too was Vitali Sevastyanov, the flight engineer of the Popovich crew, who is supposed to have written a letter to the Politburo to get permission to try a circumlunar shot in December. All requests were met with silence. In his diary, Kamanin wrote on 26 November that he felt "haunted" by American plans to fly Apollo 8 to the Moon. Publicly, in the days after the mission of Borman, Lovell and Anders, Academician Leonid Sedov noted that "there does not exist at present a similar [circumlunar] project in our programme. In the near future, we will not send a man around the Moon". These propagandist lies were espoused by others, too. Cosmonaut Gherman Titov announced that it was "not important" *when* the first manned Moon shot took place. The steamroller of disinformation was moving.

For Leonov, the key obstacle to beating the Americans had always been Vasili Mishin; a man who, despite engineering excellence, was fatefully lacking in the art of senior politics. "A lot depended on the way he presented circumstances to the Politburo," Leonov wrote. "If he had argued firmly that everything was ready, perhaps their decision would have been different. But because the chief constructor did not seem adamant, it was easier for them to cancel the programme."

The situation did not improve in the following spring. Barely a month before the first failed launch of the N-1, another unmanned L1 mission was parachuted back to Earth when one of the second-stage engines of its Proton failed. By this time, any hope that the Soviet Union might once have nurtured to become the first to send men around the Moon was gone. Leonov and Makarov, who until the end of the previous year had been working feverishly towards a possible March 1969 circumlunar mission, were now stood down from training. So too were the second and third circumlunar teams who had been targeting May and July launches. "It was decided after the American success," Mark Wade noted on his website, www.astronautix.com, "to cancel any 'second-place' Soviet manned circumlunar flights." In Leonov's mind, the following months were bittersweet: losing his chance to travel to the Moon was an understandably crushing blow, but seeing *human beings* travelling to our closest celestial neighbour, in a way, transcended any political considerations that he might have harboured. When Neil Armstrong and Buzz Aldrin actually *walked* on the lunar surface a few months later, Leonov wrote, their achievement "filled me with pride for all humanity".

Another unmanned Zond in early August 1969 marked the only completely successful circumlunar mission of the programme which *could* have returned a crew uninjured to Earth. By this time, of course, three weeks after American astronauts had *landed* on the Moon, a successful *circumlunar* flight, manned or not, was too little and far too late. Of course, publicly, it did not matter. Since May, the Soviet propaganda machine had been churning out its disinformation dribble, focusing on the maxim that a manned lunar effort had *never* been on their 'to-do' list. Nikolai Kamanin, whose diary suspiciously notes that he was on a *six-week* vacation – at precisely the time of Apollo 11's mission – would write in September that poor management, over-reliance on automated systems and Mishin's incompetence had directly caused this catastrophe for the Soviet Union.

Nor had Leonid Brezhnev's regime helped matters, by offering limited support when it was acutely needed and then making absurd demands whenever a politically motivated launch date neared. One incident in the autumn of 1967, mentioned by Asif Siddiqi, came in the run-up to the 50th anniversary of the Bolshevik Revolution, when Mstislav Keldysh insisted that *two* cosmonauts should ride an LK lander on the *very first* flight of the N-1. If *that* was impossible, Keldysh added, then *one* cosmonaut should be flown ... but still on the *very first flight*! Brezhnev, whose lack of understanding of scientific and technological matters once famously prompted him to suggest beating the Americans by landing on the *Sun*, apparently supported such idiocy. These "ludicrous demands," wrote Siddiqi, "underline ... the incredible gap between the people building the spacecraft and those who controlled the purse strings".

The 'troika flight' of Soyuz 6, 7 and 8 in October 1969, despite the minimal propaganda impact that it afforded, was seen as a jaundiced and unsubstantial Soviet reply to the American lunar triumph. A final unmanned Zond in October 1970 would also succeed ... but, 15 months *after* Apollo 11, it was virtually ignored and the reliability of the Proton was considered insufficient to push the project further. Besides, by this point, the Soviet programme had long since begun to move in a quite different direction. The Moon would be abandoned to a generation of robotic explorers, including a pair of automated rovers known as the 'Lunokhods'. The Soviets would now concentrate their efforts on near-Earth studies, including the establishment of the world's first 'true' space station. Named as a 'salyut' (or 'salute') to Yuri Gagarin, the station would offer a much-deserved return to space for Alexei Leonov, who began training to command one of its missions after the demise of his lunar flight. It is ironic, therefore, that after so much bad luck, a slight twist in the crew selection process should have contributed directly to saving his life.

"THE MAIN ROAD FOR MAN INTO SPACE"

Academician Leonid Sedov's announcement in August 1969 that the Soviet Union would not be sending men to the Moon – *robots*, he explained, could carry out sample-return missions; there was no need to risk human life – was a feeble attempt to pour cold water on an American victory. Today, his words can be seen for what they were: just one component of a cynical disinformation drive. However, back *then*, in the weeks after Armstrong and Aldrin's landing, the Western world remained undecided and very much in the dark over exactly what kind of advanced projects were being pursued behind the Iron Curtain. For at least three years previously, journalists had shared their own ideas; some saying the Soviets *had* a manned lunar programme, others countering with equal vigour that they *did not* have one. Spy satellite imagery and speculation in aeronautical publications convinced many that the Soviets' lunar rocket *was* real and that it *was* intended as an answer to the Saturn V. On the other hand, the flight of the unmanned Luna 15 probe to the Moon in July 1969 lent credence to at least some of Sedov's words.

Behind closed doors, three options for the future had been tabled: a manned Mars mission, 'advanced' lunar landings and a large, Earth-orbital space station. The reality that the Soviets could not successfully bring to operational status a rocket with only 70 percent of the Saturn V's payload capacity rendered the first two options largely moot and these were never openly discussed with the world. Indeed, wrote Asif Siddiqi, the fact that plans for a Mars mission existed at all "is testament to the often unrealistic ambitions of both space industry officials and the chief designers". The concept of a space station, on the other hand, and the steadily growing maturity of Soyuz and its ability to rendezvous and dock with other vehicles in orbit seemed a more practical response to Apollo. In the face of an overwhelming and humiliating public relations disaster in the summer of 1969, the Soviets cleverly turned their own weakness to their advantage: now *America* could be presented as the imperialist superpower in search of the glory and the spectaculars. The beautiful socialist state, on the other hand, could now depict itself as pursuing peaceful, scientific goals closer to Earth; an ironic depiction, considering the shameful Soviet invasion of Czechoslovakia the previous year and the reality that three of their early space stations would be almost exclusively military in nature ...

By the return of the Soyuz 6, 7 and 8 crews to Earth in mid-October 1969, plans for a large orbital space station were already advanced. For at least six years, Vladimir Chelomei had formulated plans for a military station, later renamed 'Almaz' ('Diamond'), which he planned to launch atop his UR-500 Proton booster. Following the overthrow of Nikita Khrushchev and the arrival in power of the largely anti-space Brezhnev and Kosygin, Almaz might have breathed its last, but for one thing: in August 1965, Lyndon Johnson gave formal approval to the United States Air Force to develop its Manned Orbital Laboratory. This two-man 'space station' and modified Gemini capsule would have conducted month-long reconnaissance and intelligence-gathering missions over the Soviet bloc, perhaps as often as four times per year, with the first flight due in late 1968. It is with little surprise, therefore, that just eight weeks after Johnson's announcement, the official order was signed for Chelomei's Almaz project to go ahead and its first paperwork was drawn up the following year.

According to early blueprints, Almaz would have comprised four parts – a manned re-entry vehicle, a work compartment, a reconnaissance facility with the means to take long-focus photographs and a propulsion module – and, like the

MOL/Gemini it would have been launched into space with its crew riding a capsule atop it. "This eliminated the task of developing a rendezvous and docking system," wrote Grujica Ivanovich. However, there remained an acute need to reduce the station's mass in order to accommodate more scientific and military equipment. As a result, in 1967, the State Commission authorised a two-launch scenario – one for the station itself and a second for the manned craft – to better enable the complex to 'grow' and more effectively exploit the 20,000 kg payload capacity of Chelomei's Proton. This prompted the decision to mate the manned re-entry craft with a Functional Cargo Block (FGB) and carry the crew and their scientific and military gear into orbit together. It led to the development of a large crew-and-cargo-carrying vehicle known as the Transport and Supply Ship (TKS). "Crews would be exchanged," continued Ivanovich, "at intervals of two or three months and the station would have an operational life of up to two years, being unoccupied only during the short intervals between one crew departing and next one arriving."

Military and civilian experiments would have been juxtaposed aboard Almaz, with observations of politically sensitive targets using high-powered imaging equipment combined with scientific and ecological monitoring of bushfires and the state of river pollution. The station's three-man crew, interestingly, were to have worked around-the-clock on staggered, eight-hour shifts. When Chelomei was denied his bid to build the Soviet Union's manned circumlunar spacecraft, Almaz became his focus. In its final form, the station measured 11.6 m long and 4.15 m across at its widest point and weighed 18,900 kg. It had an internal living and working volume of around 90 m^3 and comprised, from the 'front', a crew compartment, a large work area and an unpressurised segment containing the propulsion system. The crew area, measuring 3.8 m long and 2.9 m wide, would have bristled on its exterior with the Igla ('Needle') rendezvous antenna, solar orientation sensors, television cameras, laser devices and infrared sensors. Inside, the cosmonauts would have enjoyed a large space in which to rest, eat, exercise on a treadmill and perform medical experiments. The much larger work compartment would have been dominated, particularly at the rear, by the Agat-1 ('Agate') unit, a large optical telescope in a hermetically sealed conical section whose aperture was in the Almaz 'floor' and extended to a series of imagers, almost at ceiling height. Important data from the telescope could be analysed onboard the station or returned to Earth in a special re-entry capsule attached to the rear of Almaz. Finally, the propulsion system consisted of a pair of engines and a series of smaller thrusters to execute orientation and stabilisation control. A pair of solar panels with a total area of 52 m^2 would have provided electrical power. For its time, the station was extremely complex and two of the most fundamental problems facing its designers were developing equipment capable of operating reliably over a two-year lifespan and building the TKS ferry. As a result, the project fell badly behind schedule and plans to make operational flights in 1969 quickly became untenable.

The TKS difficulties soon led to the decision to employ Soyuz as an interim crew ferry for Almaz. In fact, Sergei Korolev had earlier envisaged two modified variants of his spacecraft, known as 'Soyuz-P' and 'Soyuz-R', to be capable of military missions, including rendezvous, inspection and possibly even the destruction of American spy satellites. The Soyuz-R was officially endorsed by the Soviet government in June 1964 and would have required two separately launched unmanned craft; these would have docked and established a kind of miniature 'space station' with a length of 15 m, a habitable volume of perhaps 31 m^3 and a mass of around 13,000 kg. A manned Soyuz, carrying two cosmonauts, would have been launched next to occupy the facility and perform a series of military observations and experiments.

When the American Gemini began flying in 1965, the Soviet government saw it as a threat, because in addition to its role in the Manned Orbital Laboratory it could be used to intercept their own satellites. Almaz and the Soyuz-R complex would not be ready for three years and another plan was hatched for a manned reconnaissance spacecraft called 'Soyuz-VI' or 'Zvezda' ('Star'). This would have measured 8 m long and 2.8 m in diameter, with an internal volume of 12 m^3 and a mass of 6,600 kg. Interestingly, had Soyuz-VI flown, it would have been powered by radioisotope thermoelectric generators, flown month-long missions – twice as long as Gemini could achieve – and, unlike Korolev's plans for the standard Soyuz, would have had two cosmonauts clad in full pressure suits. It also incorporated the descent module *above* the orbital module, which necessitated cutting a hatch into the base of the former. This opened up a whole plethora of worries about the thermal integrity of its heat shield during re-entry. "Although dynamic tests ... showed that the hatch was safe," wrote Ivanovich, "there were lingering doubts." Finally, to provide protection from a perceived American 'satellite killer', the Soyuz-VI was provided with a rapid-rate-of-fire Nudelman cannon; a device capable of firing 0.2 kg projectiles at 690 m/ sec and hitting and destroying a target within five seconds. The Soyuz-VI design was conducted under Dmitri Kozlov, chief designer of OKB-1's Branch No. 3 in Kuibishev (today's Samara).

Cosmonaut training for Soyuz-VI had begun in September 1966, with Pavel Popovich as team leader, but their preparation time was relatively short: within 16 months the project was cancelled, despite the support of Nikolai Kamanin, due to a combination of political manoeuvring and intrigue involving Kozlov and Vasili Mishin. A substitute was offered, based on the old Soyuz-R design and known as the Orbital Research Station (OIS), although its priority declined as work continued on modify the standard Soyuz after the death of Vladimir Komarov and plans to send cosmonauts around the Moon aboard Zond shifted into high gear. In fact, in the spring and summer of 1968 the priority for the OIS was so low that its cosmonaut trainees did not even have their own spacecraft simulator and 'preparation' was restricted to theoretical, physical and survival training. The short-lived OIS dream officially ended in February 1970, the assigned cosmonauts were transferred to Almaz and Popovich would go on to command the first military mission in 1974.

It has already been noted that Sergei Korolev himself had envisaged launching a space station as one of the goals of his enormous N-1 booster and these ideas crystallised into something called a Multi-Role Space Base Station (MKBS), which continued under Vasili Mishin as late as 1969. It would have comprised a 20 m long core segment, connected by a series of 60 m 'pylons' to a nuclear power source and plasma electric engine. The mammoth project would have weighed 250,000 kg and carried up to eight 'modules' and a substantial quantity of scientific instrumentation. Interestingly, its altitude – Ivanovich mentions 400–450 km – and an inclination as high as 91 degrees (a Sun-synchronous orbit) would have enabled its cosmonauts to survey the *entire Earth* on a daily basis. It would have had a crew of six and an operational lifetime of a decade or more. In a sense, the MKBS was an ahead-of-its-time forerunner of many of the concepts later employed by the Soviets on their Mir station; additionally, it employed some features of Ronald Reagan's Star Wars programme ... *and*, of course, the six-man crew complement was in keeping with current trends aboard today's International Space Station.

In August 1969, a few weeks after Apollo 11, a group of designers, led by Boris Raushenbakh, suggested to Boris Chertok that the propellant tank of a Soyuz rocket should be converted into a makeshift space station. At this stage, of course, *something* was needed to respond to the American success and Almaz, clearly, was some way from becoming a reality. At first, Chertok was sceptical; Soyuz systems were limited by issues of mass and more powerful engines would be needed to maintain the craft's orbit for several months in space. Others, including Konstantin Feoktistov, suggested using already extant Almaz prototypes standing idle at Chelomei's factory in Moscow. However, Chertok and Feoktistov both knew that Chelomei would oppose any attempt to 'requisition' his spacecraft and Vasili Mishin would dislike a proposal to use a competitor's hardware. "Their strategy," explained Grujica Ivanovich, "was to avoid anyone who might raise an objection and to go straight to Dmitri Ustinov, who was on the Central Committee of the Kremlin and was in overall control of the Soviet space programme." This was easier said than done, because protocol dictated that an approach to Ustinov should be made through Mishin, as head of OKB-1. He was away on leave at the time and Chertok and Feoktistov asked his deputy, Konstantin Bushuyev, but the latter did not want to approach Ustinov without the permission of his boss.

Consequently, it was the "disobedient" Feoktistov who arranged the meeting. He knew that Ustinov wanted the Almaz station to be operational and that it would not be ready until 1972 at the earliest. This was around the same time that the Americans were planning to launch their own space station, Skylab. Then there was the key difference between the two projects: Skylab was a civilian scientific outpost, whereas Almaz was a military one and, by necessity, was shrouded in secrecy. "If the first Soviet space station could be portrayed as a civilian space project," wrote Ivanovich, "and it was given lavish coverage in the newspapers, then it would serve to mask the true role of the subsequent Almaz stations – about which much less information would be released." Launching a scientific station *first*, he concluded, was a perfect *maskirovka* ('deception') behind which the *real* deal could be hidden.

The plan moved rapidly. One day in mid-October 1969, in his Kremlin office, Ustinov listened as Feoktistov proposed equipping the core of one Chelomei's stations with Soyuz solar panels, guidance and control systems. Within a year, Feoktistov felt, a station could be launched, well ahead of the American Skylab. Bushuyev and Chertok, also at the meeting, verified that Soyuz' systems were of

A model of a Soyuz spacecraft (left) is shown about to dock with the DOS-1 station.

sufficient reliability, having been tested on more than a dozen unmanned and piloted missions, and efforts to devise a docking system with an internal means of crew transfer was underway. There were concerns, though. Mstislav Keldysh worried about the impact on the still-undecided future of the N-1 and Bushuyev wanted the station plan to be formalised quickly, since to be discovered working on it would prove embarrassing.

Thankfully, Ustinov was warmly supportive of the proposal, primarily because it offered a perfect means for the Soviet Union to pull ahead of the Americans in one arena of manned space flight and be in orbit by 1971. "An 18-month station," added Mark Wade on his website, "[to] be in space in time for the 24th Party Congress, seemed alluring." It also offered Ustinov an opportunity to gain the upper hand over Chelomei, a man he despised, by effectively 'stealing' his Almaz design. The 'conspirators' were doubtless surprised and happy to have successfully convinced Ustinov, who ordered them to prepare a project timescale and officially endorsed the plan in January 1970, but in truth such ideas had been floating around in Soviet senior political circles for some time. When the crews of the troika mission returned to Earth in October 1969, even the nominally anti-space Leonid Brezhnev had announced that his nation's ultimate aim was to establish "space cosmodromes" in the heavens, enabling men to set sail for other planets.

This representation of the Soviets as following a more logical and gradual 'grand plan' for mastering near-Earth space was reinforced by Brezhnev's speech at the Kremlin Palace of Congresses on 22 October. "Our country has an extensive space programme," he began, "drawn up for many years. We are going our own way: we are moving consistently and purposefully. Soviet cosmonautics is solving problems of increasing complexity. Our way to the conquest of space is the way of solving vital, fundamental tasks – basic problems of science and technology. Our science has approached the creation of long-term orbital stations and laboratories as the decisive means to an extensive conquest of space. Soviet science regards the creation of orbital stations with changeable crews as the main road for man into space." It was clear from this address, wrote Asif Siddiqi, that Brezhnev was implying the Americans were chasing "an empty, politically motivated enterprise, [whereas] Soviet cosmonauts were doing their all for the advancement of science and ultimately for the benefit of humankind". Sending *men* to the Moon was expensive in comparison to the automated orbiters and landers being regularly dispatched by the Soviets.

It almost seemed that a sort of Five-Year Plan was being formalised and, far from being a politically inspired enterprise, it would be seen as being ordered, organised, properly financed and executed with precision; the perfect cover, if ever it were needed, for the hastily-cobbled-together reality. In a major sense, this dramatic turnaround from the humiliation of two failed N-1 launches and the crushing disappointment of Armstrong and Aldrin's landing on the Moon was a masterstroke.

Other high-profile personalities in the Soviet space effort now came to the fore to reinforce Brezhnev's pronouncement; Mstislav Keldysh, for one, told Swedish journalists on 24 October that "we no longer have any scheduled plans for manned lunar flights". Many observers in the West fell for it. However, in the Soviet Union, feelings were quite different. One Moscow journalist noted with sarcasm in 1990 that Brezhnev's speech was little more than a desperate bid to "come up with an alternative space project to save face, as well as the badly tarnished myth of Soviet superiority in space ... Designers, cosmonauts and thousands of other people probably laughed up their sleeves, knowing full well that the General Secretary was lying."

Additionally, the CIA suspected that the 'sudden' advancement of the space station concept into Soviet strategic thinking was a direct consequence of their loss of the Moon race to Apollo. "The implication," read an intelligence report issued in January 1970 and declassified in 1998, "in light of public statements like those of Brezhnev and Keldysh ... is that the Soviets have downgraded their manned lunar landing program[me] and have placed new emphasis on space stations ... One reason for the public stress on space stations and de-emphasis of manned lunar landings may be the success of the US Apollo program[me] ... " In conclusion, the report suggested that the failings of the N-1, the effect of whose disastrous July 1969 launch had been clearly photographed by a Corona spy satellite, profoundly shifted the Soviet direction in space towards an orbital station, which offered "potential short-run economic benefits ... in contrast to the longer-run potential gains from exploration of the Moon ... "

In spite of the support from the Kremlin for this 'short-run' crash project, utilising Soyuz systems and a slightly modified version of Chelomei's Almaz to create a spacecraft now known as the Long-Duration Orbital Station (DOS), it is hardly surprising that Vasili Mishin was livid at what he saw as a 'conspiracy' among his subordinates. He is even said to have raged that if anyone else in his bureau dared to involve themselves in the DOS, he would "send him to hell". Mishin knew that the lunar programme, already on shaky and uncertain ground, had been dealt another blow and both privately and publicly he attacked the DOS effort. Other senior managers, who had invested almost a decade of their professional lives in the Moon project, did the same.

Perhaps in response to Mishin's hostility, Yuri Semenov was appointed in late December 1969 as the lead designer for DOS. His father-in-law was Andrei Kirilenko, the Politburo's ideology chief and one of the most powerful men in Russia at the time; such close ties with the Kremlin undoubtedly helped Semenov ward off any attempt by Mishin or Chelomei to undermine the station. Finally, on 9 February 1970, Decree No. 105-41 was issued by the Central Committee and the Council of Ministers. Among its directives: all pertinent documentation and all existing hardware, including manufactured Almaz station cores, should be transferred to the new DOS programme. Mishin's OIS proposal would be scrapped and, although the Almaz military effort would continue, it would be subordinate to the civilian station. This proved easier said than done in March 1970, for in one of Semenov's first meetings with Chelomei, the latter overtly fumed that his design was being stolen from him. "The proud Chelomei evidently gave Semenov an earful," wrote Asif Siddiqi. Only after speaking to Sergei Afanasyev was Semenov able to persuade Chelomei to release four Almaz cores to the DOS effort. (Ultimately, a total of eight cores were transferred.) Chelomei's bitterness is understandable: in late 1965, his manned circumlunar project had been reassigned to Korolev and his perception was that, *now*, after five years of planning, his Almaz station was being shifted aside and essentially 'stolen' by DOS. Other opposition came from Nikolai Kamanin, who felt that political manoeuvring was again derailing a logical space programme and by backing DOS, the military Soyuz-VI and Almaz projects would be effectively stopped in their tracks.

Many of the original conspirators who had so angered Mishin were heavily involved in the definition of what would become the world's first true space station, DOS-1. After reviewing the documentation, it was Konstantin Feoktistov who drew up its specifications, fully exploiting the payload capacity of Chelomei's Proton booster. The station would be 14 m long and 4.15 m wide, having a mass of around 19,000 kg and an internal volume of 100 m^3. With ten times as much living space as Soyuz, DOS-1 had the ability to support a crew of up to three cosmonauts for an extended period of time. Yet it was also complex: it boasted around a thousand instruments and some 350 km of wiring and one of the requirements was that its scientific apparatus should be accessible to the crew for repairs and maintenance. Other problems surrounded its altitude: the higher it operated, the less propellant it needed annually to maintain its orbit, but, at the same time, this also required the Soyuz ferry to expend more fuel to reach it and exposed the cosmonauts to increased radiation hazards.

In physical appearance, DOS-1 comprised three main segments: a 3 m long transfer compartment, a working area and an aggregate section. The first was located at the forward end of the station and was fitted with a passive docking 'node' for receiving Soyuz. It contained life-support and thermoregulation equipment, together with the Orion ultraviolet telescope, cameras and biological instruments. It also included a small side-hatch, with an aperture of some 80 cm, through which cosmonauts could perform spacewalks (although no extravehicular suits were assigned to the first station). A pair of Soyuz solar panels were mounted onto the transfer compartment's exterior with a total span of 11 m. Next came the work area, which consisted of two cylinders, measuring 6.5 m long in total, and featured a control panel and facilities for working and eating, heating food and recreation. In the rear cylinder, the work area was dominated by the conical structure of the 'scientific apparatus compartment', featuring the OST solar physics telescope with an aperture 2 m in diameter. Other instruments for analysing celestial gamma rays and charged particles were also aboard. The overwhelming emphasis on astrophysical and Earth-resources studies was balanced by a battery of medical and exercise equipment, including a stationary running track capable of 10 km/h. The psychologists had been hard at work on the interior design of the station, too, although the colour scheme was rather dismal. In order to prevent the cosmonauts from becoming disorientated, the 'front' and 'rear' walls were light grey in colour, one side was apple-grey and the other yellow and finally the 'floor' was dark grey. Finally, at the rear of DOS-1 was the aggregate section, which was not accessible to the crew and measured just 1.4 m long. It contained a main engine that was derived from the KTDU-35 aboard the Soyuz and produced a thrust of 417 kg. Another pair of solar panels, identical to those on the transfer compartment, were mounted to the hull of the aggregate and according to Siddiqi the overall impression was almost bird-like.

To say the very least, it is surprising that such an obviously political project should actually have benefited from 'real' scientific input in the final months of its design and definition. In July 1970, Vasili Mishin, by now less irked by the space

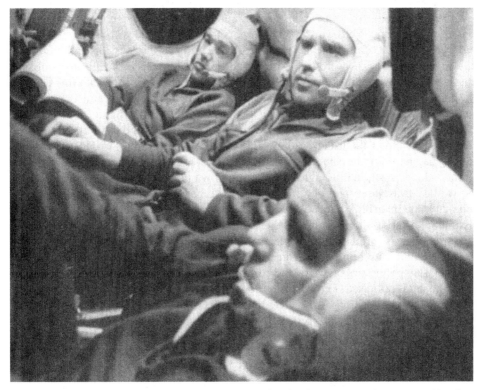

The Soyuz 10 crew during pre-flight training in the simulator. From the rear are Alexei Yeliseyev, Vladimir Shatalov and Nikolai Rukavishnikov.

station, wrote to numerous scientific and probably military organisations to enquire as to the kind of 'goals' or questions that they believed the DOS-1 programme should be able to answer. Early the following year, he met with one of Keldysh's deputies at the Soviet Academy of Sciences and the requirements were established: an astronomy capability, resources for creating optical and radio telescopes in Earth orbit, a space-based parabolic antenna with a diameter of up to 50 m and a large-mirrored telescope for up to eight years of operation. "While all of these were not intrinsically related to the DOS," wrote Siddiqi, "the space station programme seems to have served as a catalyst for this new co-operation between the scientific and space communities."

Shortly after the release of Decree No. 105-41, the first mutterings of exactly who might fly the early DOS-1 missions were made. One of the first names was veteran engineer Alexei Yeliseyev, who had flown twice the previous year and had actual experience of rendezvous, docking and spacewalking. Another was a sour-faced rookie engineer named Nikolai Rukavishnikov, who was one of the men training for circumlunar L1 flights. The situation now moved quickly. In late April, Nikolai Kamanin met with Deputy Chief Designer Yakov Tregub and Sergei Anokhin, head of the cosmonaut group at OKB-1 (now renamed TsKBEM), and was informed that plans called for the construction of not one, but *two* identical stations, both to be occupied by a pair of three-man cosmonaut crews. The first crew consisted of rendezvous veterans Yeliseyev and Vladimir Shatalov, teamed with Rukavishnikov, and were aiming to undertake a world-record-breaking 30-day mission in the DOS-1 station. The Americans were not expected to be able to achieve such a feat until at least 1972. The second crew to occupy DOS-1 would comprise veterans Georgi Shonin and Valeri Kubasov, together with rookie Pyotr Kolodin, and they would push the envelope further by attempting 45 days in orbit. The third and fourth crews – consisting of Boris Volynov, Konstantin Feoktistov and Viktor Patsayev, then Yevgeni Khrunov, Vladislav Volkov and Vitali Sevastyanov – would serve in a backup capacity for the Shatalov and Shonin crews, then fly to the DOS-2 station. Of these dozen cosmonauts, eight had already flown into space (and, in June 1970, so too would Sevastyanov) and had been exposed directly to rendezvous and docking, manoeuvring in orbit, spacewalking and had performed various experiments not dissimilar to those planned aboard the DOS stations.

Kamanin was immediately displeased with some crucial elements of this crew set-up. He had long opposed having Feoktistov, who he regarded as an "invalid", as part of a spacegoing crew. In the run-up to the Voskhod 1 mission in October 1964, he had criticised Feoktistov's poor training experience and his physical condition, to say nothing of the fact that he was *not* a member of the Communist Party and that he was in the process of getting divorced from his wife. "In fact," wrote Grujica Ivanovich, "given that Feoktistov had gone behind Mishin's back to get the DOS programme started, it was ... surprising that the TsKBEM's Chief Designer had allowed his name to go ahead at all!" Boris Volynov's Jewish heritage has already been discussed in the run up to his Soyuz 5 mission and it once again became a problem. In addition, Kamanin insisted that two members of each crew should be military cosmonauts. There were other issues, too. For example, Yevgeni

Khrunov had recently been involved in a car accident and had left the scene without helping an injured bystander; as a punitive measure, he was temporarily removed from flight status. Finally, Kamanin disliked the idea of the Soviet Union's two most rendezvous-savvy cosmonauts (Shatalov and Yeliseyev) being on the *same* crew.

Bearing these concerns in mind, 'revised' crews were formally announced on 6 May 1970. The first crew to visit DOS-1 would consist of Georgi Shonin, Alexei Yeliseyev and Nikolai Rukavishnikov and the second would be Alexei Leonov, Valeri Kubasov and Pyotr Kolodin. Backups to these two crews, and candidates for the DOS-2 station, would then be Vladimir Shatalov, Vladislav Volkov and Viktor Patsayev and a new and relatively inexperienced line-up of Georgi Dobrovolski, Vitali Sevastyanov and Anatoli Voronov. These changes reflected the growth of marginally better relations between Kamanin and Mishin, although Shatalov proved no shrinking violet in his expression of bitterness at losing the command of the world's first space station mission. However, these crews underlined the reality that at least some level of compromise had been reached between Kamanin and Mishin in their long-standing military-versus-civilian cosmonaut debate: for Crew One would now comprise a military pilot (Shonin) and two civilian engineers (Yeliseyev and Rukavishnikov), whereas Crew Two would have *two* military pilots (Leonov and Kolodin) and *one* civilian (Kubasov). Crews Three and Four would follow a similar pattern.

Unlike previous station plans, DOS-1 proceeded with relative smoothness throughout the summer of 1970, notwithstanding the poor physical conditions of Nikolayev and Sevastyanov after their 18-day Soyuz 9 mission. According to Mishin's deputy, Sergei Okhapkin, by July DOS-1 was targeted for launch in the spring of 1971 and its DOS-2 successor would follow sometime in 1972 for a variety of terrestrial, solar and astronomical experiments.

Not everything was in perfect order, though: a delay of perhaps two months to the DOS-1 schedule already seemed likely, in view of the need to modify the Igla rendezvous device, which had experienced difficulties during the troika mission of October 1969, and problems with constructing a 0.8 m diameter hatch in the Soyuz craft for a tunnel to access the station. It was expected that Shonin's Crew One would launch aboard Soyuz 10 about eight to ten days after DOS-1 had reached orbit, arrive at the station 24 hours later and spend a month aboard. Twenty-five days after the return of Shonin's crew, the Leonov team would blast off aboard Soyuz 11 and attempt a stay of 45 days. However, the adverse reactions of the Soyuz 9 crew prompted Kamanin to recommend flights of no more than three weeks until a clearer perspective could be gained on how weightlessness affected the human body. Nevertheless, and further highlighting the Politburo's real motivation for endorsing the station project, it was directed that DOS-1 *had* to be launched in time for the 24th Congress of the Soviet Communist Party in March 1971. Privately, Kamanin doubted that any such mission would be ready to launch before April or May.

The speed with which DOS-1 moved from project conception to reality astounded Alexei Yeliseyev. "It took only a few days to build a wooden mockup," he said. "With Rukavishnikov, I went to see the mockup of the first station. Compared to the Soyuz, it looked like a giant – it was more than 10 m from one end to the other. There was room for several people to work, without hindrances ... Engineers were working continuously, checking every detail of the documentation. Every revision was tested on the mockup, with a detailed inspection. If the change was acceptable, then it was made to the station. We were involved in testing the positioning of the controls, instrument panels and the apparatus for visual monitoring. We were also consulted on how the crew should work and rest."

By the autumn of 1970, all of the cosmonauts with the exception of Dobrovolski's Crew Four were training intensely for their missions; the latter trio began formal classes in January of the following year, after wrapping up their duties on the now-essentially-defunct L3 lunar landing effort. Vitali Sevastyanov's extended time for recuperation after Soyuz 9 also meant that he began his Crew Four training slightly later than anticipated. "As a result," wrote Ivanovich, "the fourth crew did not begin serious training for DOS until January 1971 and expected to have at least 18 months before making their flight." Little could they possibly have realised that one of their number, Dobrovolski, would end up commanding the very first mission aboard a space station ... and would die only minutes before returning to Earth.

Much of what happened in the spring of 1971 to change the crew set-up for the first DOS missions surrounded Georgi Shonin. As already discussed, his increased dependency on alcohol in the months after he commanded Soyuz 6, the first of the troika flights, ultimately led Nikolai Kamanin to remove him from consideration. Indeed, in early February 1971, Shonin's drunkenness had resulted in his missing an important airborne training session on the 'Svinets' ('Lead') experiment, which it was expected his crew would use to monitor intercontinental ballistic missile launches. Vasili Mishin was furious and told Kamanin that Shonin would "never fly again in my spaceships!" Shonin was apologetic and implored Kamanin to retain command of Soyuz 10, but to no avail. Mishin then attempted to manipulate the situation to his advantage by insisting that the first DOS-1 crew should now be reshuffled to consist of *three* civilians – Yeliseyev in command, teamed with Kubasov and Rukavishnikov – but Kamanin vetoed it. So it was that Vladimir Shatalov, the most experienced military cosmonaut in terms of rendezvous and docking and the man pointed at the first DOS-1 mission right from the start, now re-entered the game in pole position. At Kamanin's insistence, he was installed to command Soyuz 10 in Shonin's place. Shatalov would lead Yeliseyev and Rukavishnikov on what promised to be a mission which would score two impressive triumphs for the Soviet Union: firstly, they would become the *first* crew to occupy a 'true' space station and, secondly, by more than *doubling* the American endurance record, they would undertake the *longest* manned space flight to date. In the reshuffle, Georgi Dobrovolski moved up to command Crew Three and another military cosmonaut, Alexei Gubarev, was appointed to lead Crew Four.

Training, particularly in those final weeks in the spring of 1971, was feverish. In his autobiography, Alexei Leonov, whose own Crew Two was untouched by the Shonin episode, related the enormous pressure of leaving his previous duties in the circumlunar and landing project and learning the intricacies of the enormous DOS-1 effort. "Besides the intensive physical training," he wrote, "I had to take technical drawings and detailed plans for the programme home to study at night. The strain was so great that my hands sometimes used to shake." Preparing for a long-duration mission aboard a space station was quite different to Leonov's 24-hour Voskhod flight and the seven days or so that he and Oleg Makarov might have endured on their loop around the Moon and back. Physical conditioning was paramount and its importance was further highlighted when Nikolayev and Sevastyanov returned from Soyuz 9 in a pitiful state, physically and mentally exhausted. Early medical opinion in the Soviet Union had been that weightlessness might actually prove *beneficial*, since it imposed less stress on the heart and other bodily organs, but this was drastically reassessed in the run up to the launch of DOS-1. It was now clear that the strange environment actually posed tremendous risks and had the potential to cause weakening of muscles, bones and even the immune system over a prolonged period of time.

With this in mind, continued Leonov, steps were taken to ensure that exercise equipment aboard DOS-1 was utilised frequently to maintain the strength of the cosmonauts' muscles and bones. Increased emphasis was placed on biological and medical training and all four crews received intensive instruction in how to perform anything from first aid to the extraction of teeth and how to administer and interpret electrocardiograms and encephalograms to taking blood samples from fingers and veins. This physically and mentally taxing work was on top of preparing for launch and re-entry, rendezvous and docking and a range of other scientific experiments, including solar and astronomical observations. "We co-operated at astrophysical observatories with leading academics in the field," wrote Leonov of their gruelling schedule. "After completing each course, we had to pass a series of exams."

As the training entered its final stages, the launch of DOS-1 seemed to move further and further to the right. The original plan to launch on 5 February 1971 and despatch the Soyuz 10 crew on the 15th was quickly seen to be hopelessly optimistic; in fact, Kamanin had doubted this schedule for several months. The station's environmental control system would not be ready in time for February, the spacecraft simulators were inadequate for the cosmonauts' needs and indecision even existed over which to launch *first*: DOS-1 or Soyuz 10! This proved an organisational nightmare, Kamanin wrote, when training the crews.

Delays in the production of the station hardware and the validation of its systems were simply unacceptable to the Soviet leadership. As long ago as October 1970, Sergei Afanasyev had personally visited the production plant and demanded that engineers complete the flightworthy DOS-1 within six weeks. Only through extensive overtime was the station finished by the end of November. However, 'finished' did not mean that all of the problems were resolved: vibration tests were repeatedly delayed, there were glitches with the communication system and four Igla rendezvous devices for the Soyuz ferries were performing badly. By this time, the station's launch was not anticipated before March.

Against this backdrop, DOS-1 arrived at Tyuratam on 1 February 1971 and completed its checkout a month later, with launch expected no earlier than 15 April. Vladimir Shatalov, Alexei Yeliseyev and Nikolai Rukavishnikov would blast off in Soyuz 10 four days later and dock with the station. Many physicians and managers felt that spending a month aboard posed a grave health risk. In his diary, Kamanin noted that Soyuz 9 cosmonauts Nikolayev and Sevastyanov had, thankfully, landed virtually in the laps of the doctors, but if they had been called upon to make an emergency landing in the ocean or in the wild Siberian taiga they would have been in real trouble. For Kamanin, every extension past 20–22 days was a severe risk not only to the *health* of the crew, but also to their very *lives*.

The DOS cosmonauts pressed on with their training, regardless. On 9 March, Shatalov, Yeliseyev and Rukavishnikov completed a simulated 15-hour 'flight' aboard the DOS-1 trainer and practiced responding to emergencies aboard the station. In Shatalov's mind, after consulting with Yeliseyev and Sevastyanov, a 30-day mission *was* feasible and clearly the weightless environment affected different individuals in different ways. Yet the mission would be daunting. Not only would they be operating a very different and much larger spacecraft than ever before, but even the Soyuz ferry which would deliver them was different to the one that Shatalov had flown in 1969. The main difference was that Soyuz-DOS crews would have to transfer *internally*, essentially in 'shirt-sleeves', without extravehicular activity. By the spring of 1970, a new design was underway: an 'active' docking unit incorporating a projecting rod that would be compatible with a cone on DOS-1's passive node. Also, since the Soyuz 'ferry' would not need to remain in 'autonomous' flight for very long, it would have a simplified life-support system. The Igla rendezvous hardware was transferred to the orbital module and one radio link was removed. The reduced propulsion requirements allowed the elimination of the toroidal propellant tank around the engine unit of the service module. Nevertheless, at 6,700 kg the new variant of Soyuz was a little heavier than its predecessors. It was capable of six days' worth of flight time, three of which were for autonomous operations, and to expedite the effort it was decided to carry a crew on the craft's *very first* mission. The new Soyuz could accommodate three cosmonauts and, following tradition, *none* of them would wear pressure suits ... and *this* decision, perhaps more so than any other, would prove tragically regrettable in the months ahead.

Vasili Mishin's intention to rename DOS-1 as 'Zarya' ('Dawn') was foiled by the fact that the Chinese had already claimed this name for one of their 'secret' manned space programmes. "It is interesting," wrote Asif Siddiqi, "that ... the Soviet space community was aware of this in 1970 – the existence of the nascent Chinese manned space project of that name was not revealed publicly in the West until 2002!" Aware of the fact that the station's launch would almost coincide with the tenth anniversary of Vostok 1, it was decided instead to call it 'Salyut 1' as a 'salute' to Yuri Gagarin, the first man in space. The name of Zarya did not die; for it had already been emblazoned on the station itself and on the Proton rocket's nose fairing ... and it was too late to change it. None of the pictures of the booster bearing this nomenclature would be released to the West for almost three decades.

On 15 April, atop Chelomei's three-stage UR-500 Proton, the world's first space station was duly rolled to its Tyuratam launch pad. Four days later, at 4:40 am Moscow Time, Salyut 1 headed into an orbit of 222 × 220 km, inclined 51.6 degrees to the equator. Not everything went to plan, however. By the end of its first orbit,

The Proton rocket, with the Salyut 1 space station aboard its nose fairing, is raised on the pad at Tyuratam. Within days of launch, the station would be visited by three cosmonauts. *They* would fail to dock, but Salyut would be triumphantly occupied just six weeks later.

ground controllers discovered that the large cover for the OST telescope had not jettisoned properly, significantly jeopardising the scientific mission. Moreover, six of Salyut 1's eight environmental control fans failed. One story, told by Alexei Leonov, actually led *him* to be blamed by Mishin for the problems with the fans ... and not lightly, either. "All my possessions – my underwear, pyjamas, sketch pad and coloured pencils – had already been stored aboard Salyut 1," he wrote in his autobiography. "For some time people thought my belongings had caused the [environmental control] system to malfunction; that the crayons and the thread that held them together had become entangled with some operational part of the spacecraft." At one point, Mishin phoned Leonov and, in all seriousness, it seems, blamed him outright for damaging the environmental control system with his crayons ...

Nevertheless, preparations continued for the launch of Shatalov, Yeliseyev and Rukavishnikov in the early hours of 22 April. The cosmonauts were inserted into the Soyuz 10 descent module early that morning and all seemed to be proceeding normally until a minute before the scheduled blastoff, when it was discovered that one of the masts on the service tower refused to retract as planned. The mast in question supplied electrical power to the rocket's third stage and had failed to detach because rainwater had accumulated in the connector and frozen in place. It was feared that the launch escape system might be spuriously activated, perhaps causing an explosion, and with reluctance Vasili Mishin agreed to postpone until the following day.

Temperatures at Tyuratam dropped precipitously to -25°C during the night and when the cosmonauts arrived at the pad shortly after midnight on the 23rd they wore thick black leather coats over their flight suits to keep out the intense cold. After again being strapped into Soyuz 10's cabin, they were astonished by a repeat of exactly the same problem. On this occasion, however, Mishin opted to push ahead regardless and at 2:54 am Moscow Time, Soyuz 10 lifted off and entered orbit a few minutes later, 'leading' its quarry by 3,456 km. Despite the successful launch, the prognosis for a good mission remained low: the cover on Salyut 1's scientific compartment still refused to budge, threatening the loss of at least 90 percent of the crew's scientific objectives, and the failed ventilation fans raised the prospect of a station atmosphere filled with carbon dioxide and other "harmful materials".

The Western media had already put two and two together and judged that the large Salyut 1 was almost certainly a Soviet orbital station and that Soyuz 10 was carrying its first long-duration crew. "Observers saw the two ships," *Time* told its readers on 3 May, "shining as brightly as first-magnitude stars, crossing the night skies of northern Europe." Yet the position of Soyuz 10 'ahead' of its target was presumed in the West to be an error and Shatalov's space-to-ground comments – "Looks like you threw us up a bit too high. Well, it doesn't matter, we'll fix it" – only served to reinforce such notions.

In fact, Swedish radio analyst Sven Grahn, who was at the time involved in military exercises in Uppsala as a member of the Signal Corps, also noticed the unusual positioning of Soyuz 10 *ahead* of Salyut 1 and he expressed surprise that Shatalov, Yeliseyev and Rukavishnikov had not been inserted into a *lower* orbit to 'chase' the station. However, wrote Grahn, "the reason for this *modus operandi* . . . is quite obvious if one looks at the orbital geometry of Salyut 1. Its ground track was almost stabilized around 348.5 degrees West northbound equator crossing longitude for the orbit passing closest to the launch site on the day of launch of Soyuz 10. The reason for this was that the initial [orbital] period of the station around 88.6 minutes, dropped down through the value for a 16-revolution repeating ground track (occurs at 88.39 minutes). Since the initial northbound equator crossing longitude of a Soyuz is about 347.2 degrees West it was necessary to let the Soyuz be $((348.42–347.2) \div 15) \times 88.4 = 7.2$ minutes (corresponding to $7.2 \times 60 \times 8 = 3,456$ km) ahead of Salyut in order to have co-planar orbits." Based upon this data, Grahn noted that the two craft were "approaching rapidly" and Soyuz 10's orbit "needed badly to be lowered to reduce the closing rate for a rendezvous the following morning". During Soyuz 10's fifth and sixth orbits, difficulties were encountered in modifying its orbit to match that of the station. "The reason the first time was that there was some kind of error in the programming logic for the command," wrote Grahn. "The reason for the second delay in the correction was that the cosmonauts were given the data for the correction manoeuvre *very late* and did not have the time to enter it into the keyboard on the instrument panel." The situation was compounded by other errors, including contaminated optics which prevented the use of Soyuz 10's ionic orientation system. At length, Shatalov took charge and manually made the 17-second manoeuvre at 1:35 pm Moscow Time. The crew then opened up their orbital module, advising Rukavishnikov – 'the rookie' – to remain in his couch in the descent module and prevent sudden head movements, hopefully avoiding space sickness. By the early hours of the 24th, they were trailing the station by around 16 km.

Back on Earth, at the Yevpatoria control centre in the Crimea, the situation seemed quite comical. Armen Mnatsakanyan, anxiety written across his face, waited to see if his Igla radar would actually work, whilst Vasili Mishin argued constantly with General Kerim Kerimov, head of the State Commission, and distracted flight controllers with endless questions and demands. At one stage, General Pavel Agadzhanov, head of the mission management team, was so frustrated that he shouted into the microphone: "We understood you – the distance is 10 km. Do not interfere!" Unfortunately for Agadzhanov, the first part of his message was intended for Shatalov's crew and the second part to quieten Kerimov and Mishin, but he left the microphone keyed as he conveyed the whole message! The cosmonauts, unaware of such an unprofessional furore in the Crimea, retorted with surprise and one controller muttered, only half in jest, that it would be a miracle if he survived the morning without suffering a heart attack ...

Much to Mnatsakanyan's relief, the oft-troublesome Igla successfully brought them to within 180 m of their quarry and Shatalov again took manual control. "All the dynamic operations of the ship were conducted without any problems," he noted later. "The only issue appeared at the time that the Igla took control of the approach: the ship would oscillate from side to side periodically, requiring the firing of the correction engines. At a distance of 150 m, I took manual control. It was simpler than on the Soyuz 4 mission. The station grew bigger and bigger – in space it appeared to be much larger than it had on the ground! When we were very close, Alexei and Nikolai carefully inspected its docking mechanism, antennas and solar panels." Docking came at 4:47 am Moscow Time. The first men ever to visit a space station were about to open it up for business, or so they thought. What should have been a moment of euphoria for a space programme that had lain prostrate in the shadow of Apollo for more than two years quickly succumbed to the ugly reality that the docking was not, in fact, a secure one.

The three cosmonauts heard and felt vehicle motions and a slight scraping as Soyuz 10's probe slid into the cone-like receptacle of Salyut 1 and then began to retract in an action designed to draw the two vehicles together in a metallic embrace. However, all was not well. Shortly before five in the morning, and nine minutes after the initial contact, Shatalov radioed Yevpatoria that the docking indicator on his instrument panel was unlit, suggesting a problem with the coupling mechanism. Telemetry indicated there was a 9 cm 'gap' between the vehicles. Nothing Shatalov tried had any effect – not even firing Soyuz 10's engines in a brute-force attempt to bring them firmly together. The spacecraft was connected to the station only by small latches at the head of its probe. "As the probe penetrated the drogue," explained Grujica Ivanovich, "the spacecraft had been deflected and the control system had tried to eliminate the angular deflections. However, the ship was no longer free to manoeuvre and instead of rotating about its centre of mass, as the control system expected, it swung on the end of the probe and this broke part of the mechanism." The cause was twofold: firstly, Soyuz 10's control system was configured to remain 'active' after the initial capture and secondly, the docking sequence was automated. "Yeliseyev, who had participated in the development of the control system, had realised that the control system was jeopardising the docking process," continued Ivanovich, "but had no way to intervene. He was a frustrated spectator." Any attempt to retry docking was now futile and the only option was to separate from Salyut 1. The lack of space suits aboard Soyuz 10 also meant that none of the crew could transfer to the station by extravehicular means.

As his crewmates glumly monitored their instruments, Rukavishnikov moved into the orbital module to verify the electrical contacts of the docking mechanism and ensure that the retraction had not been halted by something as simple as an erroneous signal. Unfortunately, all of the connectors were as they should be.

After four frustrating orbits in this 'soft docked' state, the cosmonauts were instructed to separate from Salyut – a process which proved incredibly difficult because the designers of the docking apparatus had assumed that a spacecraft wishing to *undock* would be successfully *docked* with the station in the first place. Since Soyuz 10's docking mechanism had not fully engaged with Salyut 1's drogue, it was quite possible that the spacecraft might fail to separate. Indeed, on Shatalov's first attempt, he fired the thrusters ... and the spacecraft simply swung around on its damaged probe. Back in the circus of the Yevpatoria control room, General Andrei Karas shouted sarcastically: "Well, congratulations! You've developed a docking system in which Mum doesn't release Dad!"

There were two emergency options to release Soyuz 10 from the station and neither of them held much promise for the future of Salyut 1. The first was to cut loose the docking mechanism at the front of the spacecraft and the second was to shut the descent module hatch and separate from the *orbital module*, leaving that hanging uselessly on the front of the station. In either case, access to the station – which, after all, had only *one* docking port – would be blocked. Fortunately, Vsevolod Zhivoglotov, a member of the docking mechanism team, suggested an alternative and instructions were radioed to Rukavishnikov. The cosmonaut entered the orbital module and reconnected a number of cables to 'deceive' the mechanism into assuming that a release command had been issued by the station itself. That command was issued at 10:17 am, less than six hours after the initial docking ... and it *worked*. Soyuz 10 was finally free and pulled slowly away from Salyut 1.

Shatalov maintained close formation with the station while ground controllers – not yet appreciating that the probe mechanism had been damaged – debated whether to attempt a second docking. An analysis of the state of Soyuz 10's gyroscopes, propellant levels and oxygen supplies made this impossible and Shatalov was ordered to prepare for an emergency return to Earth early on 25 April. That night, Shatalov and Yeliseyev snoozed fitfully, but Rukavishnikov floated, wide awake, his eyes glued to the window, snapping photographs. It was not just that he was awestruck by the beauty of his home planet; he was also cold. "At a temperature of 20 degrees," he grumbled later, "it is impossible to sleep in the flight suits. During the first night, we slept only two or three hours. Instead of sleeping, we sat and shivered. It is necessary to carry sleeping bags." The nocturnal re-entry, which began at 1:59 am Moscow Time, was quite spectacular; the descent module was enveloped in glowing plasma and the three men likened their ride home to being inside a neon tube, with colours constantly changing.

Touchdown at 2:40 am was 120 km north-west of Karaganda, close to a lake, after a flight lasting a few minutes shy of two full days. Naturally, the problem for the Soviet leadership was how to report the progress of the mission to the outside world. The solution: that Soyuz 10 was simply *testing* the rendezvous and docking hardware and that the cosmonauts had *absolutely no intention* of entering or occupying Salyut 1. Speaking on *Radio Moscow* a few days later, Shatalov – who was widely tipped to succeed Nikolai Kamanin as commander of the cosmonaut team and so had to be careful with his words – said that the mission was "not extensive in duration, but tense and magnificent in its tasks". Some observers in the West were not entirely fooled. Tom Stafford, who was at the time NASA's chief astronaut, related in his autobiography that "we knew *that* was bull: you wouldn't send a crew to make *that* kind of test, then bring them home after 48 hours". Others in the Western media were reaching similar conclusions. On 10 May, *Time* told its readers that the "delay in the launch of Soyuz 10 ... stirred more suspicions" about difficulties with Salyut 1 and that "officials were apparently deciding if it was worthwhile trying to rendezvous and dock with a craft that would not long remain in orbit".

The reality would not be known for many years. Even in the early Nineties, it was still widely suspected that some kind of difficulty in equalising pressure between the Soyuz and Salyut was to blame. This possibility was aired by Phillip Clark in his study of the early Soviet space programme. There was also speculation that Rukavishnikov suffered debilitating space sickness and that *this* contributed to the premature return to Earth. Former cosmonaut Boris Yegorov, the first physician to fly in space, was quoted as saying Rukavishnikov experienced "unusual and rather unpleasant feelings" as a result of the increased blood flow to his head – a normal consequence of entering the weightless environment – and even that he had suffered severe vertigo and been unable to move into the large interior of Salyut. In truth, Rukavishnikov's biomedical data confirmed that he coped well and he actually felt better than either Shatalov or Yeliseyev, but in the absence of any other explanations, such stories persisted for more than two decades.

SO NEAR ... YET SO FAR

When Shatalov and Yeliseyev attended the 24th Congress of the Soviet Communist Party on 30 March 1971, they did so as 'flown-and-known' cosmonauts. Little did many of the assembled journalists at the event suspect that both men were only days away from launching into space again – nor did they expect that they would be joined by a third cosmonaut: a sour-faced and frequently unsmiling civilian named Nikolai Nikolayevich Rukavishnikov, a man for whom technical matters were a way of life and who loved to repair old apparatus and build new machines. "His ambition," Shatalov joked in the days leading up to the Soyuz 10 launch, "is to convert a refrigerator into a vacuum cleaner!"

Rukavishnikov had been born in the western Siberian city of Tomsk on 18 September 1932 and grew to admire geography, mathematics and physics from a young age. After graduating from school, he enrolled at the Moscow Institute of Engineering and Physics in 1951 and received a diploma six years later, specialising in dielectrics and semiconductors. His early work came at the Central Scientific Research Institute in Podlipkah, near Moscow, where his focus was the development of the Ural computer and testing automatic control and protection systems for nuclear reactors. By the end of 1959, he was working for Sergei Korolev's OKB-1 bureau and received his first introduction to space matters by helping to design controls for unmanned interplanetary probes.

Like several other civilian employees of the bureau, Rukavishnikov passed initial cosmonaut screening in the early summer of 1964 as part of the effort to select an engineer for the Voskhod 1 mission. Two years later, a formal group of civilians was picked, but when four of their number failed the Soviet Air Force's advanced medical examinations, it was decided to nominate two more candidates and Rukavishnikov and Vitali Sevastyanov were chosen. The first few months of cosmonaut preparations were difficult. "I had to catch up on all the training that other cosmonauts had already passed," he recalled later. "This included thousands of hours of intensive training, centrifuge, altitude chamber, simulated weightlessness flights and parachute training." Rukavishnikov's short, skinny stature surprised many of his peers, including Alexei Yeliseyev, who expected him to be dismissed on health grounds. He could not possibly have known that this serious technician would become his crewmate in a few years' time. Moreover, he could also not have foreseen that in 1979 Rukavishnikov would also become the first civilian cosmonaut to actually *command* a Soviet space mission.

As a member of Russia's elite spacefaring fraternity, Rukavishnikov quickly established a reputation for himself as a hard-working and totally committed individual, routinely staying at OKB-1 both day and night until his tasks were completed. In addition to his technical knowledge and passions, he enjoyed motorbikes and travelling and during summer vacations he would venture off alone into the hills to explore. Very soon after completing initial training, he was assigned to the L1 circumlunar project and, had this not been cancelled, would probably have served as Valeri Bykovsky's flight engineer. Then, in the spring of 1970, he received a short-lived assignment to a Soyuz flight that was to evaluate the Kontakt rendezvous hardware for subsequent lunar landings. By midsummer, his career had taken a quite different turn and he was detailed as 'research engineer' aboard the first flight to a space station. It is ironic, therefore, that despite flying three times into the heavens (and actually being assigned to a fourth mission in 1984), Rukavishnikov would *never* get to see the inside of a space station ...

THE CREW THAT NEVER FLEW

Two weeks after the return of Shatalov's crew, an investigation into the cause of the failure, headed by Boris Chertok, settled on the Soyuz docking apparatus. It seemed that the shock-absorbing coupling mechanism was subjected to 160–200 kg of force, when in fact it was designed to withstand no more than 130 kg. This had broken one of four levers surrounding the base of the probe. Engineers had not expected the docking to impose more than 80 kg of force on the mechanism. The increased force had been partly caused by the failure to halt the motion of Soyuz 10 after its 'soft' docking. The haste with which the investigation was concluded (its documentation was prepared within 24 hours and physical corrections to the docking mechanism were made within a week) points to the need for a major Soviet success aboard Salyut 1, *before* the end of the station's operational lifetime.

Chertok's panel recommended that the speed of the Soyuz at the point of contact with the station should not exceed 0.2 m/sec, that the retraction of the docking probe should not begin until the craft was stable and, most importantly, that the cosmonauts should have the ability to exercise manual control. Special levers were installed around the pin of the docking probe to evenly distribute the potential loads caused by oscillations of the Soyuz and the levers which had suffered breakages on Shatalov's mission were strengthened and reinforced.

Meanwhile, Vasili Mishin also pushed strongly to still fly *two* long-duration missions to Salyut 1: Alexei Leonov's Soyuz 11 on 4 June and another flight, Soyuz 12, commanded by Georgi Dobrovolski, on 18 July. He also advocated reducing the first crew to just two cosmonauts (probably Leonov and Kubasov), to accommodate space suits inside the Soyuz and facilitate an extravehicular inspection of Salyut 1's docking node *and* perhaps the removal of the station's still-jammed scientific cover. Nikolai Kamanin objected, arguing that none of the cosmonauts had been trained for spacewalks on these missions and that space suits could not be certified in time. Furthermore, Kamanin doubted that Salyut 1's consumables could last long enough to support the Dobrovolski crew throughout the entirety of their mission; as he understood it, the station could house them until the beginning of August at the very latest. As a consequence, it was decided that the duration of Dobrovolski's tour would be dependent upon the satisfactory outcome of Leonov's mission and the consumables remaining aboard the station.

All of these plans were thrown into disarray on 3 June, when physicians from the Institute for Biomedical Problems in Moscow found a swelling in Valeri Kubasov's right lung – a dark spot about the size of a chicken egg – and *the entire crew* was replaced with their backups. It was feared the swelling might represent the onset of

Had Valeri Kubasov (top) not been misdiagnosed only days before launch, these three men might have been the first occupants of a true space station. For Alexei Leonov (centre), it was a bitter blow, since he had also lost the chance to fly to the Moon several years earlier. However, it must have hit Pyotr Kolodin (right) the hardest, by eliminating his chance to fly into space.

tuberculosis ... and a regrettable misdiagnosis was made. "It turned out later," wrote Leonov, "that [Kubasov] was allergic to a chemical insecticide used to spray trees". The flight rules dictated that if a cosmonaut fell ill *before* departing for the launch site, he should be simply substituted for his backup; in other words, the *rest* of the crew would stay together. However, the fact that Leonov, Kubasov and Kolodin had *already* been at Tyuratam for more than a week invoked the second part of the rule, which stated that "carrying out the replacement of the individual *at the cosmodrome* is not possible. In case of such a necessity, it is only possible to carry out the replacement of the crew ... "

Not surprisingly, Leonov and Kolodin were furious. Surely, wrote Leonov, if they too had been infected by Kubasov, then they would have already fallen ill. More importantly, they had been training for a year and knew the station and its systems thoroughly. It made no difference; after much discussion, the State Commission decreed that all three would yield to their backups, Dobrovolski, Volkov and Patsayev. Appeals to Mishin and Kamanin were useless; Mishin pointed out that Leonov may have drunk from the same cup as Kubasov and could be similarly infected, whilst Kamanin told Leonov that, as the mission commander, his arguing was not "reacting in the correct manner" to a decision which had already been made. Pyotr Kolodin, after eight years as a cosmonaut and seemingly only hours away from what would have been his first mission, returned to the hotel at Tyuratam that night and got himself heavily drunk on vodka. He is said to have cried bitterly with rage and frustration at losing his chance to fly into space.

Had Kubasov *alone* been replaced, then Leonov might have been obliged to accept Vladislav Volkov as his new flight engineer. As it turned out, he was not given the choice, but Grujica Ivanovich has described a complex relationship between the two cosmonauts and structured a convincing case that, perhaps, neither man *wanted* to fly with the other; Leonov due to Volkov's outspoken criticism of his military colleagues and Volkov for Leonov's belittling attitude towards the civilian's status as a member of only the *third* Salyut crew. Either way, the removal of Kubasov meant Volkov was assured of a place on the Soyuz 11 crew. Ivanovich argued that perhaps Volkov *wanted* the entire crew swapped because if he flew on the Dobrovolski team he could enjoy the status of being the only veteran amongst rookies. Such backbiting over crew assignments had characterised the American manned space programme and there is no reason to suppose that it did not also haunt the ranks of the Soviet corps. There may also have been a simpler explanation that was more rational for all parties: even though Dobrovolski, Volkov and Patsayev had only been together for a few months, they were at least a *crew*. They knew each other's skills and shortcomings and, with two days to go, they were more cohesive as a team.

Yet even Dobrovolski's crew were displeased and aghast at the decision – for *they* had been training for a Salyut mission for barely five months. Originally, they had expected to fly to the DOS-2 station sometime in 1972, but their mission had been advanced by a year following the Soyuz 10 failure and Mishin's desire to still have two crews visit Salyut 1. Now their tour had been brought forward even further ... with a launch in just *two days' time*! More significantly, they would become the very first crew to attempt a Salyut docking with the new hardware and, unlike Soyuz 10, their only veteran crewmember (Volkov) had no docking experience. The commander, Dobrovolski, who held ultimate responsibility for ensuring that *this* docking ran as advertised, would be making his first flight. In Nikolai Kamanin's mind, the crew was fully trained and had achieved good scores in their exams. It did little to soothe the tensions or ease the sense of foreboding. "If you look at photographs taken of the replacement crew just before the launch," wrote Leonov, "they even look a little scared." Late on the evening of 3 June, he did a pencil sketch of Patsayev and would name the portrait 'Patsayev's Eyes' because the civilian engineer, about to embark on his first space voyage after only a fraction of the required training, looked distinctly troubled. At some point that evening, the considerate Patsayev walked up to Leonov and offered his heartfelt apologies for what had happened.

By the time the crew change was announced, the Soyuz 11 launch had been rescheduled to the early morning of 6 June. On the evening of the 4th, the six cosmonauts whose futures had been so abruptly changed – Dobrovolski, Volkov, Patsayev, Leonov, Kubasov and Kolodin – sat, stone-faced, before the State Commission as Nikolai Kamanin formally introduced them. Dobrovolski stood up and verified that his crew was ready, whilst Leonov, head held low, could only express regret over what had happened. Journalist Mikhail Rebrov noticed an intense silence in the room.

It was ironic in the wake of the decision that Kubasov had been misdiagnosed and was simply suffering from an allergic reaction to insecticide ... whereas the pre-flight medical screening had failed to detect that Patsayev had a chronic kidney inflammation. The result: "a healthy cosmonaut being grounded," wrote Ivanovich, "and one with a chronic medical problem being launched into space!"

In a roundabout sort of way, Leonov and Kubasov achieved some measure of revenge on the day preceding launch. When instructed by Kamanin to take the traditional walk with the prime crew in homage to their predecessors, both men refused, not out of bitterness towards Dobrovolski's team ... but out of bitterness against what had been an ill-judged decision and against those responsible for making that decision. "If I am healthy, then I must fly," Kubasov told Kamanin. "If I am sick, I should not be there!"

A CONVOLUTED PATH TO 'PRIME CREW'

"He flies calmly" was one observer's description of Lieutenant-Colonel Georgi Timofeyevich Dobrovolski of the Soviet Air Force. Indeed, the man himself, born in the Black Sea coastal town of Odessa on 1 June 1928, grew up among peers whose own plans for the future were dominated by maritime and naval matters, rather than aviation. Yet for Dobrovolski, whose name, fittingly, meant 'a man of goodwill', the happiest years of his life were spent alone in the cockpit of an aircraft. "I cannot imagine myself without flying," he once said. "I am still in love with the sky. I'm not saying 'I love it', rather 'I'm *in love* with the sky." The ultimate kind of flying, in his mind, was travelling in space and seeing the Blue Planet from on high.

Before he had any chance of achieving that goal, he had to pick his way through a hard, formative life ... a life punctuated in its teenage years by the bitter conflict with Nazi Germany. In fact, at the age of 16, as bombs dropped on Odessa and its environs, Dobrovolski became a member of the underground resistance, carrying ammunition, providing information on enemy movements and, one day, setting out with a Baretta revolver in his pocket, intent on killing an SS officer. He was caught, beaten with a rubber truncheon, had his fingers broken and was sentenced to 25 years of hard labour. In fact, his name and the details of his punishment appeared on a Nazi news sheet issued on 23 February 1944. "By the standards of the time," wrote Grujica Ivanovich, "he had been treated leniently: all adult saboteurs arrested with a weapon would have been summarily executed." When one reads accounts like this, it is interesting to ponder how many other would-be cosmonauts were not so lucky during a war which robbed the world of so many millions of people ...

Thankfully for Dobrovolski, a member of his family managed to bribe a guard and within weeks of his arrest, the Red Army liberated Odessa. His mother pushed him to continue his studies and, after unsuccessfully applying to join the Odessa Nautical School, he turned to aviation. Completion of initial training in 1946 brought enrolment in the Chuguyev Military School for Air Force Pilots, in which he shone: achieving maximum scores for his flying skills, but falling short on theory work. At length, in 1950, he graduated as a naval fighter pilot and subsequently flew MiG, Yak and Lavochkin aircraft in all weather conditions and performed no fewer than 111 parachute jumps. (In fact, when chosen as a cosmonaut in 1963, one of Dobrovolski's earliest duties was as a parachute instructor.)

Marriage to Ludmilla in 1957 and the birth of their daughter was followed by the completion of a correspondence course at the Military Aeronautical Academy for Command and Navigation Staff. He later served as a deputy squadron commander and, in January 1962, was called into his boss' office and asked, point-blank, if he was interested in preparing for space missions! Dobrovolski had heard of the early cosmonauts, but was six or seven years older than most of them and doubted that he would be a realistic candidate. However, in the spring of the following year, he was selected. The Soviet cosmonaut corps had changed markedly since the selection of Yuri Gagarin's group and advanced flying experience, higher technical qualifications and greater seniority of rank were now preferable. As part of this second group of cosmonauts, Dobrovolski rubbed shoulders with men alongside whom he would someday train: Shatalov, Filipchenko, Yuri Artyukhin, Lev Dyomin, Alexei Gubarev and Vitali Zholobov. Indeed, Vladimir Shatalov would describe him as standing tall amongst his peers with his "extraordinary intensity", his "responsibility", his "modesty" and his extreme self-discipline.

Experience and qualifications aside, the new recruits were still required to pass a battery of tests, including the notorious 'chamber of silence' for psychological assessment. "He entered the chamber with his body covered with biosensors," wrote Ivanovich of Dobrovolski's first time in the chamber in February 1964, "and spent the next ten days alone in total silence, reading books on astronomy, mathematics and the German language." He even took a piece of wood along and carved a doll for his daughter. A slightly negative impression of him came through the pages of

Nikolai Kamanin's diary: early in January 1965, the general noted that Dobrovolski's performance in some of his exams was unacceptable. He had, apparently, indicated that the thickness of Vostok's heat shield was 440 mm (the correct answer was 140 mm) and had misidentified the Krug homing beacon as belonging to a search aircraft, when it fact it was part of the spacecraft itself. How significantly such misdemeanours and errors would have tarnished Dobrovolski's career is unknown, although one would assume that he would not have ended up in an important command post if his abilities were anything short of top-notch.

Initial training led to assignment to the L1 circumlunar project and, for a time, Dobrovolski and civilian engineer Georgi Grechko were assigned to the fourth mission. Unfortunately, by the autumn of 1968, it was decreed that only three circumlunar Zonds, crewed by Leonov/Makarov, Bykovsky/Rukavishnikov and Popovich/Sevastyanov, would be necessary. It is interesting, though, that the Dobrovolski/Grechko pairing would have been the only all-rookie Zond crew; a testament, perhaps, to the respect in which these two men were held within the cosmonaut corps. Dobrovolski next trained on the reserve crew for Soyuz 4, then for Almaz and finally for a mission in Earth orbit which would have tested the Kontakt docking system. This gave him substantial rendezvous training and was perhaps a contributory factor in Kamanin's decision in the early summer of 1970 to give him command of the fourth DOS crew. By the following spring, after Georgi Shonin's removal from Crew One and the change of assignments, Dobrovolski was named to command Crew Three, which was expected to fly a mission to the *second* DOS station in 1972.

With the removal of Kubasov from *his* mission and the decision to replace the *entire crew*, Dobrovolski found himself, with barely two days to go, in charge of a mission which would not only prove historic ... but which would truly make or break the future Soviet manned space programme in the eyes of the world.

THE QUIET EXPLORER

Seated to Dobrovolski's right in the cramped Soyuz descent module was the mission's only veteran, Vladislav Volkov, who had made his first journey into space nearly two years earlier as part of the troika mission. His relationship with Alexei Leonov has already been mentioned and, indeed, whilst training and flying Soyuz 7, his commander on *that* crew, Anatoli Filipchenko, is said to have had to "restrain" the "energetic" flight engineer on several occasions. Observers have speculated over the years that disagreements flared between Volkov and Dobrovolski during the course of Soyuz 11 – disagreements, it seems, between the rookie military commander and the veteran civilian who was his subordinate. Ivanovich, for one, points to suggestions that Volkov considered himself *de facto* commander of Soyuz 7, owing to his perceived technical superiority over Filipchenko and Gorbatko ... and there is evidence to suppose that a similar situation may have arisen with Dobrovolski less than two years later.

The third member of the Soyuz 11 crew, positioned in the left-hand seat of the

The ill-fated Soyuz 11 crew in relaxed mood before the flight. From the left are Viktor Patsayev, Soyuz 9's Andrian Nikolayev, Vasili Mishin, Georgi Dobrovolski and Vladislav Volkov.

cabin, was Viktor Ivanovich Patsayev, a quiet, serene, almost nervous-looking 'research engineer'. Like Volkov, he was an accomplished technician with a sharp mind, but by complete contrast to the burly, gregarious and sporty Volkov, Patsayev was a slightly built, reserved and humble man. He surprised many in the cosmonaut corps with his complete lack of interest in sports. Born in Artyubinsk on the banks of the Ilek River in northern Kazakhstan on 19 June 1933, Patsayev was self-educated and an explorer at heart: Ivanovich recounted how he taught himself to read, enjoyed the wildness of the steppe and actually started school a year early. His life changed when his father went off to war and was killed defending Moscow in 1941; Patsayev, it is said, became more serious and reserved in nature. Nonetheless, he completed high school several years later, excelling in mathematics and physics, and for a time considered a career in geology. Unfortunately, his score was not high enough for admission into a geological institute and he moved to the Penza Industrial Institute. It was here that he was introduced to the young world of calculators and computers (at that time called 'analytical machines') and it was *these* which finally changed his mindset from geology to technology.

Despite establishing himself on this path, Patsayev seems to have been one of the 'Renaissance Men' in the cosmonaut corps, as his interests were wide: he enjoyed literature, music, the arts and history, as well as the more scientific topics that were his main passion. He graduated as a mechanical engineer in 1955 and worked for a time at the Central Aerological Observatory of the National Hydrometeorological Service in the town of Dolgoprudny, just north of Moscow. His activities encompassed the design of atmospheric research instruments for high-altitude balloons and sounding rockets and, very soon, his commitment led to the post of senior engineer. Two major people influenced his life at this time: firstly, his wife,

Vera, whom he married in 1956, and secondly, Chief Designer Sergei Korolev, whom he met at a lecture in September of the following year, just before the launch of Sputnik. A face-to-face meeting shortly afterwards led to a firm appointment to the OKB-1 bureau in November 1958.

In the years ahead, as civilian candidates were realistically considered for cosmonaut training, Patsayev put his name forward in 1966, but without being accepted. The following August, however, he was chosen, alongside two others. Simulated weightlessness in the Tu-104 aircraft, parachute jumps, flight training in the Yak-18, practice sessions in the bulky Krechet lunar-surface suit and test work on the LK lander followed during the course of 1968 and his intensity impressed many of his fellows. "We saw immediately how well this quiet and unpretentious man could work," Yevgeni Khrunov once said. "His modesty was incredible. I remember, during the first months, I sat next to him in the cafeteria and, apart from 'hello' and 'see you later', he did not say anything during a period of two weeks. Later, I understood that he remained silent simply because he did not wish to make much noise!"

Work on the Kontakt docking system in 1969 led Patsayev to meet Dobrovolski, the man with whom he would later fly in space, for the first time. Early the following year, both men were assigned, along with Volkov, to the third DOS crew. One might suppose that his previous nature as something of a loner would have rendered him incompatible with the teamwork necessary for a spacegoing crew, but this proved not to be the case. "Your position in the crew," he once said, "flight engineer, researcher, physician or commander, isn't important. In order to work well together, we have to believe in and respect one another and we must celebrate the achievements of our crewmates. *That* is the foundation of a crew." Patsayev's words bring into stark relief what may have been one of the motivations behind *not* simply substituting Volkov for Kubasov and retaining Leonov and Kolodin in those unhappy first few days of June 1971. The *cohesion* of Dobrovolski's crew, even though their training had been brief, was already in place and to disrupt it could have imposed greater stress on what already promised to be a tough mission. Yet on the eve of Soyuz 11's launch, no one could possibly have known precisely *how* stressful and how tough the next few weeks – and months – would be.

LONG HAUL

Early on 6 June 1971, Dobrovolski, Volkov and Patsayev were awakened for the final preparations for their journey into space. By five that morning, the bus bearing the three cosmonauts, each clad in a grey cotton flight suit and pilot's cap bearing the Soviet coat of arms, arrived at the base of the launch pad to a small crowd of well-wishers. Dobrovolski made a short speech to General Kerimov, assuring him of the crew's preparedness to fly the mission, and proceeded to lead his men to the foot of the elevator which would take them to Soyuz 11. "This was another contrast to the NASA way," wrote Grujica Ivanovich, "whose astronauts don their suits in a building 8 km from the pad and, upon emerging, simply wave to friends and reporters on their way to the van which drives them to the pad ... [at Tyuratam], the departing crew walks through the crowd, speaking to individuals, even joking."

At the top of the steps, the cosmonauts paused before entering the elevator. The disbelief and troubled sense of foreboding which had plagued all three men just a few days earlier seemed to have vanished; even the normally anxious-looking Patsayev was smiling. All three removed their caps and waved them at the crowd below, then disappeared into the elevator. Volkov entered Soyuz 11 first, dropping down through the orbital module into his seat on the right side of the cabin and switching on the lights and ventilators. Next came Patsayev, seated to the left, and finally Dobrovolski in the centre couch. An hour still remained before launch and the final preparations proceeded like clockwork: first the retraction of the service structure 'arms', then the final topping-off of the rocket's propellants, then the pressurisation of the kerosene tanks. All three men listened as commands crackled over their headsets from the control bunker.

Precisely on time, at 7:55 am Moscow Time, the four-chambered engine of the rocket's central core and those of the quartet of strap-on boosters roared to life. Within nine minutes, orbital insertion had been achieved at an altitude of 185 × 217 km, inclined 51.6 degrees to the equator. At Kettering Grammar School in England, science master Geoff Perry picked up the signal at around the time of orbital insertion and was able to announce the launch a full hour ahead of *Radio Moscow*. "In fact," wrote Ivanovich, "because the Kettering team had noticed that after a period of silence lasting about five weeks, Salyut had recently made several manoeuvres and started to transmit signals, they had been *awaiting* the launch."

For the two rookies in Soyuz 11's cabin, the launch was as 'normal' as a 'normal' journey into space could be. There were, Dobrovolski later wrote in his diary, very few vibrations during the boost to orbit and the separation of the rocket's third stage and the onset of weightlessness was accompanied by an intense silence, a brightening of the cabin and the presence of "a great deal of dust", some of which was collected by the craft's ventilator and some by wet tissues. He also recounted an initial unpleasant sensation of weightlessness – "just as if someone was trying to pull off our heads; we felt our neck muscles strain" – but returned to normal shortly thereafter. The sensation was that 'everything' in their bodies had moved: a natural reaction, since the internal organs normally held in place by gravity were now free to migrate upwards inside their chests and fluids began pooling in their upper bodies. This helped to explain their feelings of a swollen head and neck.

The cramped descent module, with its internal volume of just 2.5 m^3 and two small portholes, was no match for the spacious orbital module, opened up by Volkov. Within this 'space aquarium', Ivanovich noted, the four men had 4 m^3 and four large portholes, arranged at 90-degree intervals, to really experience the novelty of weightlessness for the first time. Moreover, they were granted their first proper glimpse of the grandeur of Earth: the final stage of the rocket had shut down north of China, moving eastwards, so their vista encompassed the enormous, deep-blue expanse of the Pacific Ocean. Sun-glint on the water lent an electrifying, three-dimensional impression to the scene and far off in the distance, toward the sleeping Americas, the horizon curved into darkness.

Six hours and four orbits into the mission, Dobrovolski executed the first manoeuvring burn to begin their day-long chase of Salyut 1. Back on Earth, the families of the two rookies were finding out, *for the first time*, that they were related to a cosmonaut in space. Not surprisingly, Dobrovolski and Patsayev's young children had little idea that their fathers belonged to such a profession, only being aware that they had "some important and very serious work" to do. The men's parents, too, could only respond with dumbfounded astonishment when they listened to the broadcast on *Radio Moscow* and were heralded by neighbours as relatives of the newest group of space heroes.

A prolonged, ten-hour sleep period, which began mid-afternoon, had been specifically designed to cater for the crew's early-morning awakening and the intensity of the docking with Salyut 1, which was scheduled for early on 7 June. During that first 'night' in space, Dobrovolski and Volkov slept in the orbital module, whilst Patsayev stretched across the three couches in the descent craft. At six the following morning, the control room in Yevpatoria was packed as the final approach got underway. Leonid Brezhnev's pledge that his nation was following "the main road for man into space" was about to be fulfilled. More importantly, some of the fire was about to be stolen from Apollo.

As the distance between themselves and Salyut 1 decreased, the cosmonauts switched on the Igla rendezvous device to commence the final approach and docking. This duly occurred at a distance of 7 km and Volkov confirmed its satisfactory operation by radioing to colleague Alexei Yeliseyev in Yevpatoria that the milestone known as 'radio-capture' had been passed. The Igla guided Soyuz 11 perfectly to within 100 m of Salyut 1; a point at which Dobrovolski took manual control of the ship. At a separation of 50 m, Yeliseyev asked them to visually inspect the station's docking cone for any evidence of damage as a result of the ill-fated Soyuz 10 attempt and was assured that it was "clean" and "clearly visible". At this stage, agonisingly the spacecraft passed over the Pacific Ocean and out of direct radio communication. The ground would have to wait over an hour, until 8:56 am, before they heard any more.

Dobrovolski was, by this time, firmly in control of his ship. After switching to manual control, he noticed that Salyut 1 began to drift slightly to the right in the field of view of his periscope. "I had the feeling that the left hand controller was insufficient," he recounted later, "so I switched to the right one and slightly raised the ship. Then, with the left controller, I succeeded in reducing the lateral speed. At 60 m, I reduced the speed to 0.3 m/sec. Mechanical contact at 7:49:15 [am Moscow Time]. We were stable. The [hard] docking occurred at 7:55:30. There were no vibrations or shaking. We almost did not feel the final contact."

The first confirmation received by Yevpatoria that docking had been accomplished came not from the lips of the cosmonauts themselves – in fact, Yeliseyev's entreaties for them to respond on their status carried a tone of frustration – but from the operator responsible for handling television signals from Soyuz 11. It was *he* who announced to the packed control room: "There is *television*! Docking *achieved*! The picture is *outstanding*!" The first words from the spacecraft, by complete contrast, almost seemed anti-climatic and rather matter-of-fact, as the cosmonauts reported that there had been "no oscillations" during the final minutes of approach, that they were preparing to "check the hermetic seal and equalise the pressure according to the programme" and that all three men were presently in the orbital module. Then came the excitement from Soyuz 11: "The programme is *complete!*"

Stifled applause sounded briefly in Yevpatoria, but stopped just as quickly, for there was still much to be done before the cosmonauts could actually enter Salyut 1. The hermetic seal between the Soyuz and the station had to be verified, the docking tunnel had to be pressurised and the hatches opened. Then, of course, there was the question of whether the atmosphere inside Salyut – now on its 795th orbit and some six weeks after its launch – was habitable. Had the problem with its ventilation fans turned the air toxic? After final checks confirmed that nothing was amiss, Volkov requested permission to open the hatch. An initial command to do this was issued at 10:32 am, with the crew keeping a crowbar handy in case some extra force was needed. The opening indicator did not light, but the hatch opened perfectly ... and within minutes, at 10:45 am, Viktor Patsayev, callsigned 'Yantar-3', became the first human being ever to inhabit an Earth-orbiting space station.

Shortly thereafter, everyone in Yevpatoria, and the cosmonauts themselves, were surprised to hear that Leonid Brezhnev himself – "the First", as he was called – was on a telephone line from the Kremlin and wanted to speak to the crew. This had to be deferred by one orbit when Patsayev remarked on a "strong smell" inside Salyut and retreated back to Soyuz 11 to don a face mask. Without any further data, and with the now-linked spacecraft again heading out of radio contact, the physicians could offer no advice or recommendation on how to proceed. They need not have worried, though, for seconds before the resumption of voice contact the black-and-white television screen in Yevpatoria came to life with a view of Patsayev and Volkov *inside* the station! *Now*, at last, the control room exploded with applause.

Amidst the congratulations, the cosmonauts were advised to take a meal and get some rest, in anticipation of starting the research programme first thing on the morning of 8 June. Patsayev activated a system to cleanse Salyut 1's air, which, it turned out, had gone 'stale' over six weeks due to burned insulation on two of the failed ventilation fans. Alexei Leonov was relieved that, despite Vasili Mishin's accusation, his paints and brushes were not to blame when Dobrovolski confirmed that they were still sealed up and taped in their box. At length, Patsayev and Volkov restored all eight fans to operational status and retired to Soyuz 11 to sleep whilst the station's atmosphere was thoroughly cleansed.

Now, at last, firm plans could be made for the future. That evening, the State Commission convened and decided that a landing on 30 June would produce the maximum allowable duration – nearly 24 days – to ensure a return to Earth in daylight. Nikolai Kamanin was doubtless satisfied that, after the pitiful state of Andrian Nikolayev and Vitali Sevastyanov a year earlier, the endurance record was being pushed at a more reasonable rate. Judging from the greater volume of Salyut 1 and the medical equipment, pushing the record by 25 percent should not excessively overtax the crew. If all went well, it was expected that Leonov and Kolodin, possibly teamed with Nikolai Rukavishnikov in Kubasov's place, would launch toward the station sometime in July to attempt a second long-duration mission. *How* long their mission would be was, of course, entirely dependent upon the post-flight medical assessment of Dobrovolski's crew and how long Salyut 1's resources were likely to last: many engineers and managers did not expect it to sustain a crew beyond the first few days of August. It is ironic, therefore, that the tragedy which would overtake Soyuz 11 in just a few weeks' time would sound the final knell on Pyotr Kolodin's chances of ever reaching space. Had *this* crew flown, it might also have put an immediate stop to persistent rumours, already circulating in the West, that Rukavishnikov had reacted badly to weightlessness on Soyuz 10.

The chances of Leonov's crew occupying Salyut 1 brightened in the middle of June, as the Soyuz 11 cosmonauts began an ambitious programme of scientific and medical research. The station in which they lived and worked was, by their own admission, "huge ... there seems to be no end to it!" The crew enjoyed four meals per day and 'Ration No. 1' gives an idea of the kind of delicacies served aboard Salyut: a breakfast of sausages, black bread, chocolate and coffee with milk; a 'morning tea' of Russian cheese, Rizhski rye bread and cookies; then a main meal of a mixed-vegetable soup called 'shchi', chicken, bread, plum jam with nuts and blackcurrant juice and a dinner of Caspian roach, puree, bread and honey cake. Two litres of water per day were mandated for each man, although they actually consumed no more than about 1.2 litres.

The main working area was towards the rear of the station and was dominated by the so-called 'Main Scientific Equipment' (ONA) – the large white cone which extended from the 'floor' almost to the 'ceiling' – and which comprised a variety of solar and astrophysical instruments, including the OST-1 solar telescope, the RT-2 X-ray telescope, the ITS-K infrared telescope and spectrometer, the OD-4 optical viewer and a range of photographic and processing apparatus. Sadly, the failure of the protective cover to jettison from the exterior of Salyut meant that this hardware was effectively inoperable. In the upper corner to one side of the ONA were strung sleeping bags for the cosmonauts, although they could rest inside the Soyuz if they preferred. A set of chest expanders, a treadmill and an exercise bike were provided and special 'Pingvin' ('Penguin') suits, designed to stimulate the muscles and prevent wasting, were scheduled to be worn periodically during the course of the mission. Behind the ONA cone was the euphemistically named 'sanitary and hygienic unit', including a makeshift toilet which drew urine into a collector and separated it into fluid and gaseous components. Solid wastes were stored in hermetic tanks. Wet and dry towels provided a means of personal hygiene and a series of handholds on the walls aided movement and orientation. Finally, the unpressurised rear compartment, which was inaccessible to the crew, housed the main propulsion system; known as 'KTDU-66', this was based upon the Soyuz engine, but with larger propellant tanks containing 1,490 kg of unsymmetrical dimethyl hydrazine and nitric acid.

Early on 8 June, the cosmonauts finished breakfast and made a start on checking the station's life-support system; when all was confirmed to be well, they began shutting down Soyuz 11. Its interior would be ventilated from Salyut and its 'autonomous' time of only three days meant that its own power reserves would be needed later in the mission for undocking, retrofire, re-entry and landing. Very quickly, Dobrovolski, Volkov and Patsayev settled into their routine – a routine dominated by six major daily objectives, focusing on the flight programme, personal hygiene, physical exercise, four meals, individual rest time and an eight-hour sleep period. To break these objectives down, the flight programme encompassed 140 specific experiments, radio and television broadcasts and photographic sessions, whilst exercise demanded at least two hours daily and a minimum of 30 minutes 'light-walking' on the treadmill. "Many lessons had been learned from the 18-day flight of Soyuz 9," noted Grujica Ivanovich, "and the complex for physical training was more substantial than the one available on that mission." Each man had two hours per day of 'leisure time', to be spent reading, resting, observing Earth or preparing for future tasks. Every seventh day was set aside for a 'weekend' and sleep patterns were adjusted so there would always be one cosmonaut on duty and at least one resting at any time.

Unfortunately, by 9 June, it became apparent that exercising on the treadmill produced unwanted vibrations which transmitted through the station's structure and caused the solar panels and antennas to 'flap' with an amplitude of about 5 cm at their tips. Consequently, the cosmonauts were asked to use the treadmill sparingly. On the other hand, the Penguin suits quickly became their favourite working attire. "[They] are referred to as penguin suits," *Flight International* told its readers on 24 June, "because of the penguin-like movement by an Earthbound wearer." The garments were designed to impart loads on certain muscles to simulate the forces experienced in everyday life on Earth and, although the men initially had some difficulty in moving their arms and legs, they soon found them so comfortable that they wore them all day and even slept in them. Having said this, Nikolai Kamanin doubted their effectiveness as the mission wore on: their elastic restraining straps tore and they did not adequately duplicate gravitational effects.

The cosmonauts also trialled an innovative piece of equipment known as 'Veter' ('Wind'), which consisted of a 'waist' fastened to Salyut's wall and a pair of rubberised leggings. When one of them had hermetically sealed himself into the apparatus, some of the air was extracted from the leggings by means of a pump, drawing blood into the lower body, just as if he were standing upright on Earth. "In weightlessness," explained Ivanovich, "the feet do not require so much blood and therefore the cardiovascular system rapidly adapts by transferring 1.5 litres of blood to the upper body – in particular to the chest and head, which is why on their first days in space the cosmonauts felt swollen-headed." In time, much of this excess was removed by increased urination, but it had the potential to cause great stress on the cardiovascular system when the cosmonauts returned to Earth. "The reduced amount of blood that is circulating in the upper part of the body," Ivanovich continued, "drains to the feet, imposing a considerable pressure on the vessels … When a cosmonaut stands up after returning to Earth, his weakened cardiovascular system is unable to supply blood to his head, the brain is temporarily starved and there is a risk of fainting." As a possible countermeasure for this so-called 'orthostatic intolerance', the Penguin and Wind garments helped to 'train' the men's cardiovascular systems to adapt to a state approximating that of terrestrial gravity.

Every five days, the cosmonauts undertook detailed medical checkups, which included blood samples, electrocardiograms and analysis of bone tissue. The procedures were considerably more sophisticated than earlier missions. "For instance," wrote Ivanovich, "whereas only the rate of breathing had previously been measured, now this was augmented by measurements of the volume and speed of inhalation and exhalation and the overall lung capacity." After Soyuz 11's return to Earth, these measurements helped physicians to determine how levels of sugar, urine and cholesterol varied in their blood. This, Ivanovich continued, "was normal in the blood samples taken during the first and third weeks, but increased in the fourth week just before the cosmonauts left the station. There was an increase in the level of urine in the blood of all three men owing to the manner in which their kidneys adapted to weightlessness. There was no detectable change in the level of cholesterol". Generally speaking, the data from each of these investigations showed heightened heart rates and increases in arterial pressure and the blood's exchange rate in both Dobrovolski and Patsayev, but was more stable in Volkov, possibly because he had flown in space before.

Major scientific investigations, with a main focus on solar physics and astrophysics, began in earnest on 11 June. One particularly important tool was the Anna-3 gamma-ray telescope, whose operation required Volkov to orient the entire station to point the instrument at its targets. Elsewhere, in Salyut 1's transfer compartment, was the Orion telescope, which, with a focal length of 1.4 m and a pair of mirrors, was capable of acquiring spectrograms of stars in the range of 2,000–3,000 ångströms. Typically, Dobrovolski would control the station whilst Patsayev operated the telescope and he had to do so with relative haste because there was only a 35-minute 'window' on each orbit during which observations could be made passing through the Earth's shadow. A total of six spectrograms were acquired for the star Agena (beta Centauri) in the southern sky and nine of Vega (alpha Lyra) in the north. Observations of cosmic-ray particles were also carried out with the FEK-7 photo-emulsion camera.

However, despite the seemingly harmonious tone of conversations with the ground and the completion of tasks, an unmistakable sense of tension began to creep into the air during the cosmonauts' second week aloft. In part, it was due to flaws in mission planning and poor use of brief periods of communication with Yevpatoria, but also irritability from the unnatural 'circadian rhythms' and sleep patterns – typically, Volkov would start his working day at 10:00 pm, then Dobrovolski would join him at 2:00 am and Patsayev would replace Volkov at 6:00 am. Veteran cosmonaut Alexei Yeliseyev, at the Crimean control centre, was the technical flight director, and later described the difficulties. "The programme was planned," he recounted, "in such a way that all important crew activities would be carried out while the station was in range of the tracking stations . . . However, due to the timing of the orbit it was impossible to retain the normal terrestrial duration of 24 hours for the crew and their cycle was 25 minutes shorter. We thought that they would soon accommodate themselves to the planned circadian rhythm." A meeting with an expert in biorhythmology warned Yeliseyev that daily deviations of the rhythm of life from the norm were imposing pressure on the cosmonauts and "would cause nervous disruption, if not worse". At first, Yeliseyev ignored the expert's advice, but over the coming days the psychological stress on the cosmonauts became increasingly clear. In his notebook, Dobrovolski recorded that the mission was turning into an endurance marathon: making reference to "a general absence of everything, no interesting things, no happiness, the monotonous sound of the ventilators, strong smells, numerous experiments … "

Circumstances were not aided, it seemed, by relations within the crew. Volkov, the veteran, tended to indirectly threaten the authority of his commander, a man who had spent much of his life in a firm and uncompromising military chain of leadership. Volkov had been open in his scorn for military authority and he had already caused Anatoli Filipchenko some 'problems' on his first flight. The fact that the long-duration Soyuz 9 crew had trained for a year and Dobrovolski's team had only had five months to prepare did not help matters, either. Back on Earth, neither Vasili Mishin or Nikolai Kamanin gave much thought to these psychological difficulties and it seems surprising with the benefit of hindsight that little consideration was given to the potential downside of sending Dobrovolski's relatively new (and inexperienced) crew on such a long flight.

Meanwhile, the experiments continued. Biological investigations focused on how weightlessness affected plant growth, using a small hydroponics chamber which contained Chinese cabbage and bulb onions. Genetic tests accumulated mutations in tiny fruit flies, tadpole embryos, yeast cells, chlorella and the seeds of linen, cabbage and onion. Additionally, the development of fertilised frogs' eggs was closely monitored by the crew. Observations of atmospheric features, snow and ice cover and clouds over Russia's mighty Volga River continued, as did spectroscopic analyses of sections of the Caspian Sea, in conjunction with two specially equipped aircraft from Leningrad State University and the Soviet Academy of Sciences. Large-scale weather phenomena, including a cyclone brewing near Hawaii and moving westwards almost as far as the east coast of Australia, were closely tracked by Patsayev and Dobrovolski.

As the mission wore on, it became clear that consumables aboard Salyut 1 should be sufficient to support a second 30-day expedition, beginning sometime between 15–20 July. Yet again the unfortunate Valeri Kubasov was out of luck. By now physicians had determined that his malady was merely an allergy, but when the crew for Soyuz 12 was formally announced, it comprised Alexei Leonov, Pyotr Kolodin … and Nikolai Rukavishnikov. Nor did Kubasov's name appear on the list for the follow-on DOS-2 missions in 1972: Alexei Gubarev, Vitali Scvastyanov and Anatoli Voronov were tipped to fly Soyuz 13, followed by Pyotr Klimuk, Yuri Artyukhin and an as-yet-unnamed civilian TsKBEM engineer on Soyuz 14. It was expected that the Leonov crew, with Gubarev's team backing them up, would complete their training on 30 June … the very day that Dobrovolski, Volkov and Patsayev were due to land. The two new crews would then fly to Tyuratam early in July in readiness for a launch three weeks later.

FIRE!

Meanwhile, aboard Salyut 1, the regular televised transmissions flooding nightly into Russian homes gave absolutely no hint of a growing personality conflict between Dobrovolski and Volkov. The occasional irritation of an unwanted facet of weightlessness – a pencil floating away and vanishing into the dark recesses of the station, for instance – seemed almost comical and the commander even wrote in his notebook early on 16 June that he was "writing with Viktor's pencil … I lost mine a long time ago; almost *all* of our pencils have gone!" However, later that very same day, the situation aboard Salyut took a distinctly uglier turn.

Shortly before the onset of another communications session with Yevpatoria, Volkov noticed the aroma of smoke coming from somewhere at the aft end of the station. The Soviets' insistence on ridiculous secrecy was still very much alive and his reaction was tempered by the need to remember what code word to use over the radio. "Aboard the station," he reported carefully, "is the curtain." 'Curtain', in this context, represented the code word for 'fire' or 'smoke'. Any fire in space is serious, but the controllers had forgotten what 'curtain' meant and innocently asked Volkov for an explanation … at which he blurted over the radio: "There is a *fire* on board! We are now entering into the [Soyuz]!" In their haste, they forgot to take with them the emergency undocking procedures and now demanded that Yevpatoria read them over the radio.

It subsequently became clear that the smoke had come from behind some panelling on the aft wall of Salyut. Despite the good sense of evacuating the station, Alexei Yeliseyev and fellow cosmonaut Andrian Nikolayev decided to keep the "agitated" men calm with the suggestion that they switch off all of the onboard scientific instruments. Perhaps one of those had overheated. Before any recommendations could be relayed up to the crew, however, the station drifted out of radio contact. In the Crimea, ground controllers were in a quandary: had the crew closed the hatches and, indeed, had they *already* undocked? Plans were hurriedly put in place to automatically check Salyut 1's oxygen, its atmosphere and its power supply.

When contact was re-established, it was *Volkov* – not Dobrovolski, the man who should have reported such emergency situations – who responded. He told them that the crew was back inside the station, that "smoke isn't being produced anymore, but there is still smoke in the station … We have headaches." Volkov's voice was strained, but he was able to tell Yeliseyev that the smoke had come from the control panel of the scientific apparatus on the wall at the rear end of the main work compartment. It seemed likely that one of the instruments had failed or overheated. Thankfully, the problem was *not* one of Salyut's critical systems. Yeliseyev calmly read the crew through their procedures checklist for an emergency undocking, but cautioned that they should *only* attempt this after receiving a command from the ground.

The reaction of Volkov to the emergency was clearly an instinctive response to the situation and, on the face of it, would not imply a considered attempt to undermine Dobrovolski's authority … but it did not go unnoticed. Years later, Vasili Mishin remarked in an interview that "I had a complex conversation with Volkov. He declared himself to be in command. When the cable burned, they lost their heads and wanted to depart the station." Mishin told Volkov that Dobrovolski was in charge and that he must follow instructions, but the excitable flight engineer retorted that it was the crew's collective responsibility to decide things together. Nikolai Kamanin, the no-nonsense general for whom a military chain of command was law, expressed more overt displeasure over Dobrovolski's performance. Volkov, despite being the only veteran, was, after all, only the *flight engineer* and Kamanin considered it utterly objectionable that he was dominating the conversations with Yevpatoria and, more importantly, that Dobrovolski was doing little to stop him. In his notebook for the hours in the wake of the fire, Dobrovolski wrote, with more than a hint of sarcasm: "If *this* is harmony, then *what* is divergence?"

Work aboard the station, however, continued. Early on 17 June, the faulty scientific apparatus which had caused the fire was explored in greater depth and isolated from its power supply. It turned out that a seizure in a fan had caused it to overheat and begin producing dense black smoke. "Although there had been no flame as such," wrote Grujica Ivanovich, "this was the first case of a 'fire' on a manned space mission." Periodic checks of the station's air confirmed that it was now stable and the oxygen supply was normal. General Agadzhanov told the State Commission that no the danger was over. The planned flight duration of 24 days was still possible. Nevertheless, all scientific equipment was shut down until 18 June to give Boris Chertok's team time to perform a full status check on Salyut 1's electrical system.

Little was known of this 'fire' on the ground. Even the cosmonauts' families and, to a greater extent, their young children, had no idea that it was happening at all. Yet some Western radio listeners had picked up Volkov's encrypted transmission and the mention of fire quickly aroused their suspicions. However, even a decade or more after the event, the details remained imprecise. "The Kettering group monitored the activation of Soyuz 11 during the day," wrote Phillip Clark in 1988, "as if it were being prepared for a return to Earth. The Soviets merely said that some 'minor correction work' had been undertaken that day. There have been rumours that a fire of some sort broke out on Salyut ... "

As the mission proceeded past its midpoint, experiments resumed in earnest, with Patsayev carrying out ultraviolet observations with the Orion telescope ... and celebrating his birthday in orbit on the 18th. To alleviate the tension, a conversation was arranged between Patsayev and his wife and children. Aboard Salyut, the cosmonauts prepared a birthday meal of cold veal, cookies and tubed blackberry juice. Volkov had smuggled along fresh onions and lemons in anticipation of the birthday and, said Patsayev, "we sliced it into three parts and shared it". The celebration marked another Soviet first: the first time a spacefarer had celebrated a birthday *off the planet*. Little did Patsayev realise that his 38th birthday would actually be his last.

TRIUMPH TURNS TO TRAGEDY

Two weeks into their flight, the crew were drawing near to the 18-day record set by Nikolayev and Sevastyanov a year earlier. Their acceptance of the weightless environment seemed to be increasing, with Patsayev noting both positives and negatives in his diary: they could "swim like fish in an aquarium" through Salyut's vast compartment, their legs flailing behind them, but "loose objects will float out of reach if you don't attach them to something". It was clear to him that there would be a need for artificial gravity on future space missions. He firmly anticipated that his return to terrestrial gravity after more than three weeks would be difficult.

A couple of days later, on 20 June, during a scheduled period of rest for the crew, Volkov told mission controllers that he enjoyed watching the cloud-covered Earth and "the real splendour of space". Both he and Dobrolvolski had begun to sport beards – the latter only after requesting permission from Nikolai Kamanin – and Volkov described their appearance as Tatarian or Mongolian. Patsayev opted to shave each morning. By now, the personalities of all three men were clear to the flight controllers on the ground, with Alexei Yeliseyev telling journalists that Volkov was the most vocal, Dobrovolski somewhere in the middle and Patsayev quiet.

As the mission entered its final days, the first real concerns were raised over exactly how the three men might readapt to normal terrestrial gravity. None of them had exercised as much as intended and their evident exhaustion and lack of concentration was blamed variously on a poorly organised flight programme, the unfamiliar circadian rhythms and an excessively heavy workload. Nikolai Kamanin felt that Volkov would encounter the greatest difficulty, since he had not drunk enough water, had refused to eat meat and was showing a tendency to make mistakes. Others, however, felt that his love of sports and experience as a veteran cosmonaut might make his readaptation easier than that of Dobrovolski or Patsayev.

On the evening of 24 June, the crew officially broke the Soyuz 9 record and it was during those final days that they began exercising feverishly, using every spare minute of their daily schedules to do so. "We actually *want* to overload ourselves," Dobrovolski told the Yevpatoria controllers. After one particularly intensive exercise session, Volkov announced that he had exhausted himself, but that he had enjoyed the experience. On the ground, Nikolai Gurovski, one of the Soviet Union's leading aerospace physicians, was convinced that the Soyuz 11 crew would return to Earth in much better condition than Nikolayev and Sevastyanov.

Two days later, the laborious process of shutting down Salyut 1's experiments and unneeded equipment began, as did the transfer of experiment samples and logbooks to the descent module. Doctors also readjusted the cosmonauts' sleep patterns, for the first time allowing all three men to sleep at the same time in order to ensure that they would be sufficiently rested for landing. The cosmonauts did not warm to the change, with Dobrovolski cautioning that it would effectively leave Salyut without anyone on duty, but they pressed on with their programme. The station was cleaned and 'mothballed' and rubbish was stowed inside Soyuz 11's orbital module, which, of course, would be jettisoned just before re-entry.

Meanwhile, concerns were being raised on the ground about wind speeds in the recovery zone. "If the descent module were to land on its side, as often happened," wrote Grujica Ivanovich, "and there was a strong wind, then it might roll after landing and even on a flat surface this would be unpleasant for the men inside, especially if they were feeling weak." The Soyuz 11 Landing Commission prepared two possible sets of instructions for the cosmonauts, the first for the primary landing opportunity and the other for a secondary landing opportunity.

Late on the evening of 29 June, Volkov, then Patsayev and finally Dobrovolski departed Salyut 1 for the last time and took their places in the Soyuz 11 descent module. Yeliseyev pointed out that Volkov had forgotten to switch on the station's noxious gas filter; the flight engineer initially blamed the ground for this oversight, but a check of the previous afternoon's communication logs verified that he was, in fact, in error. After rectifying the situation, and with all three men securely inside the descent module, Dobrovolski closed the hatch and sealed it by means of a rotating grip. However, he was surprised, then alarmed, to notice that the 'hatch open' indicator on his panel remained lit. This told him that the hatch was not hermetically sealed, posing the risk of a pressure leak during re-entry and – since none of them had been provided with space suits – the death of the crew.

It was Alexei Yeliseyev's calm voice from Yevpatoria which solved the problem. He talked Dobrovolski and Volkov through the procedure: reopen the hatch and turn the grip fully to the left, swipe the hatchway with a tissue to ensure that no foreign debris had blocked it, then close the hatch and turn the grip several times to the right. The two men tried this several times, but to no avail; the indicator remained illuminated. Further troubleshooting eventually isolated the cause of the fault. "As the hatch closes," Yeliseyev recalled later, "it pushes the sensors and they produce signals. All the sensors were in working order, but the guys found that the hatch hardly touched one of the buttons, with the result that it did not push down sufficiently to send the signal." After visually checking that the hatch was, in fact, closing tightly and correctly, Dobrovolski was told to stick a bit of insulation tape over the sensor to "generate the signal artificially". When the crew tried again, the indicator blinked off, much to the relief of all concerned.

Minutes later, at 9:25 pm Moscow Time, the separation commands were initiated and Soyuz 11 successfully undocked from the station. When one considers the nightmare that was to follow in the next few hours, it is remarkable that no further sense of tension was noticeable in the cosmonauts' words. Like the astronauts of Columbia, three decades later, laughing and joking as they began their perilous dive back through the atmosphere, as far as Dobrovolski's crew was aware, a perfectly nominal return to Earth lay ahead of them. A long, challenging and successful mission was finally coming to an end and they would soon be able to see and touch and speak to their loved ones again. The strain was gone from Volkov's voice, as he joyfully announced the undocking and the stunning view of the world's first space station drifting away into the inky blackness. They would be the last humans ever to see Salyut 1 up close.

Shortly after one in the morning on 30 June, three hours since undocking and flying serenely above the Pacific Ocean, close to Chile, Dobrovolski and Volkov oriented Soyuz 11 such that its main engine was facing into the direction of travel. The last words to or from the spacecraft have been hotly debated over the years – officially, Nikolai Kamanin spoke to them at around 12:16 am and a controller signed off with wishes of good luck for a soft landing, though space historian Peter Smolders suggested later in 1971 that Dobrovolski did make one final call to confirm that he was "beginning the descent procedure". Adding more confusion to the mix was Alexei Yeliseyev himself, who recorded that Volkov jokingly asked for flight controllers to "prepare cognac" – a traditional welcome-home gift – and signed off with "See you tomorrow!"

In order to provide maximum visibility at the landing site, Soyuz 11's retrofire occurred on its third orbit after undocking from Salyut. The engine fired for 187 seconds, as planned, beginning at 1:35 am, after which the orbital and instrument modules were jettisoned and the descent module manoeuvred into the correct orientation for its fiery plunge back to Earth. The disaster that would engulf the three men must have unfolded rapidly and with scarcely any warning, for in his diary entry for that day, 30 June, Kamanin noted that no progress reports on either the retrofire or the separation of the orbital and instrument modules were heard from the crew. "There was an oppressive silence in the [control] room," he wrote. "There was no communication ... or any data about Soyuz 11. Everyone understood that *something* had occurred aboard the spacecraft, but no one knew what ... " Nor did anyone aboard the tracking ships *Bezhitsa* or *Kegostrov*, respectively positioned off the coast of Africa and in the South Atlantic, which would normally have been monitoring the retrofire. It would appear that a breakdown of communications and a change of plans to perform the retrofire on the *third*, not the *second*, orbit after undocking effectively meant that Soyuz 11's position at the point of retrofire was out of range of both tracking vessels.

Nine minutes after the completion of the retrofire burn, at 1:47 am, the circumstances that would precipitate disaster began to unfold. As the spacecraft passed high above central France, a dozen explosive charges jettisoned the orbital module and six others set the instrument module adrift, as planned. The bell-shaped descent module was now on its own, continuing on its trajectory towards a touchdown on Soviet soil, a couple of hundred kilometres east of the city of Dzhezkazgan. In the control centre at Yevpatoria, everyone was still very much in the dark ... and concerns were rising. No one knew if the retrofire had been completed and, even if it had *not*, there should have been some form of communication from the crew. Both Kamanin and Vladimir Shatalov tried repeatedly to call the cosmonauts over VHF radio, but to no avail. An awful, yawning silence pervaded the airwaves.

Soyuz 11 continued to plummet Earthward, passing over eastern Germany and Poland and finally entering Soviet territorial airspace at 1:54 am. Radar installations had actually detected the incoming descent module a few minutes earlier as it passed to the north of the Black Sea, but, since it was sheathed with super-heated plasma, it had been temporarily out of radio contact. The Yevpatoria controllers, upon hearing this news, were somewhat encouraged and speculated that perhaps a radio failure was to blame for the cosmonauts failing to respond to their calls.

Perplexed as to why the cosmonauts have not yet responded to radio calls, would-be rescuers peer inside the Soyuz 11 descent module after it had made a perfect touchdown. The horrific realisation that all three men had died during re-entry came just a few seconds after this photograph was taken.

All seemed to be going well. The small drogue parachute automatically deployed on time and at 2:02 am, just a few kilometres above the ground, so too did the main canopy. Three minutes later, recovery crews aboard an Il-14 aircraft and four helicopters reported that they could see the descent module, swinging gently beneath its red-and-white parachute and had detected signals from it, although they were unable to speak directly to any of the cosmonauts. One of the helicopter commanders radioed a commentary to Yevpatoria and an overjoyed Alexei Yeliseyev took up the story: "Finally, we heard a report ... that they could see the parachute. It was wonderful ... Then, the report from [the helicopter commander]: 'It has landed. Our helicopters are landing nearby.' Well, it seemed that was all. Next, they would report the general state of the crew and with that we would finish our work. Only a few minutes more ... "

The touchdown itself was entirely automatic – from the parachute deployment to the firing of the soft-landing rockets in the descent module's base – and the Soyuz 11 mission came to an end at 2:16:52 am, following a record-breaking flight of 23 days, 18 hours, 21 minutes and 43 seconds. Dobrovolski, Volkov and Patsayev had soundly smashed not only the Soyuz 9 duration, but, more importantly, had come close to *doubling* the American endurance record *and* had demonstrated that a true space station was feasible in low-Earth orbit.

Perhaps the worst kind of tragedy is one that cannot be anticipated; one for which there has been absolutely no warning or expectation or explanation, one which comes at the end of an event that has, in all other respects, been enormously successful. Soyuz 11 was no ordinary mission. The dullness with which previous flights had been documented in both the West and the East did not apply in this case. For the last three weeks, Russia had been abuzz with the names of Georgi Dobrovolski, Vladislav Volkov and Viktor Patsayev; the men had their own slot on a Soviet television programme called *Cosmovision*, young girls had transformed Volkov from an anonymous engineer into a teen idol and pin-up star and their triumphant return home had been accompanied by nothing less than a 'carnival' atmosphere. *Everyone* in the Soviet bloc seemed to know their names and feel a personal affinity with the bearded commander, the enthusiastic flight engineer and the quiet and serene research engineer. None could possibly have been more shocked than Nikolai Kamanin and Alexei Yeliseyev, who, after waiting more than an hour with only silence on the radio from the landing site, were finally made privy to the devastating news. It came in the form of just three numbers – '1-1-1' – and what those numbers represented was both inescapable and tragic: for they were the code to denote *the death of the entire crew*.

A system of five numbers, Yeliseyev later explained, ordinarily ran from 5 down to 1 to explain a cosmonaut's health: these respectively denoted that he was in excellent condition (5), that he was in good condition (4), that he had suffered injuries (3), that the injuries were of a severe nature (2) and that they were fatal (1). There was one number for each member of the crew. The triple '1' meant all three were dead. Immediately, Yeliseyev, Kamanin and Shatalov were flown directly to the landing site to see for themselves what nobody could quite believe to be true.

For this author, one of the most enduring images of the tragedy which befell the Soyuz 11 mission is a photograph from the desolate steppe, showing the descent module lying on its side. A group of would-be rescuers are pictured around the hatch and two men, clad in short-sleeved shirts, squat down next to it and peer inside. It is one of those photographs which, like images snapped at the very second at which aircraft hit the World Trade Center or the very second of Challenger's destruction, etches into one's memory *that instant* when tragedy occurred or was first realised. It is difficult to discern expressions on faces in the photograph – one of the kneeling men looks toward the camera with a slightly perplexed and troubled look in his eyes. Did they know at that instant what had happened? Did they wonder why none of the cosmonauts was responding to their cheerful calls of welcome? Did they realise, indeed, that these three heroes were dead?

Vasili Mishin certainly could not believe it. Only days earlier, his world had been thrust into a new nadir of depression and gloom as the *third* N-1 lunar rocket vanished in a fireball, setting back to square one any attempt to land a Soviet citizen on the Moon. Nor could Sergei Afanasyev, the 'space minister', believe it, and Kerim Kerimov, chair of the State Commission, was pale and visibly shaken when he heard the news. "Two minutes after the landing," he told the stunned control room, "members of the recovery team ran from the helicopters to the descent module, which was laying on its side. Outwardly, there was no damage whatsoever. They knocked on the side, but there was no response from within. On opening the hatch, they found all three men in their couches, motionless, with dark-blue patches on their faces and trails of blood from their noses and ears. They removed them from the descent module. Dobrovolski was still warm. The doctors gave artificial respiration. Based on their reports, the cause of death was suffocation ... "

Since the craft had landed on its side, the extraction of the bodies was difficult and attempts at CPR continued for some time, doubtless explaining the lengthy absence of communication between the landing site and Yevpatoria. Yet the warmth of Dobrovolski's corpse surely made them try even harder to save the men's lives. It would be determined in the subsequent investigation that an air vent had been jerked open during the separation of the orbital and descent modules and that all three men had been dead for over half an hour. Moreover, for at least 11 minutes of this time, they had been exposed to vacuum. "Humans and experimental animals had sometimes suffered rapid decompression in terrestrial laboratories or on scientific balloons at high altitude," wrote Grujica Ivanovich, "but the Soyuz 11 crew were the first humans to suffer the vacuum of space at an altitude in excess of 100 km. Cardiopulmonary resuscitation is only likely to be effective if given within six minutes of the cessation of the heart, since after this the brain is permanently damaged. The rescuers stood no chance of reviving the cosmonauts."

By the time Shatalov and Yeliseyev arrived, the bodies had already been removed from the landing site. As he stood out there on the barren steppe, Yeliseyev – the man who had spoken to them so often over the past three weeks – was struck and driven to tears by the absurdity of it all: a picture-perfect landing, a descent module in good shape, an outstanding mission, excellent weather, a flat field ... "and the guys dead". Like many others, he blamed himself. Had the hermetic hatch sealed properly after undocking? He could not help asking himself: had *he* doomed the crew by giving them incorrect information?

Also wondering if he could have done more was Alexei Leonov, the man who might otherwise have been in Dobrovolski's position. In his autobiography, Leonov noted that he had advised the Soyuz 11 crew to close a series of air vents between the descent and orbital modules and to reopen them when the parachutes deployed. "Although this deviated from the flight regulations," he wrote, "I had trained for a long time for the mission they were flying and in my opinion this was the safest procedure. According to the flight programme the vents were supposed to close and then open automatically once the parachute had deployed after re-entry. But I believed there was a danger, if this automatic procedure was followed, that the vents might open prematurely at too high an altitude and the spacecraft depressurise."

The crew did not take Leonov's advice.

"THEY WOULD HAVE SURVIVED"

In the United States, the response was equally shocking, but for different reasons. Since the late summer of 1969, Tom Stafford had been chief of NASA's astronaut corps and among his duties he had helped to supervise the direction that the manned space flight effort would take after the Apollo lunar landings. America was expecting to launch a large orbital station of its own, called Skylab, in the spring of 1973. Crews of three men would spend between one and two full months aboard this outpost, performing a variety of scientific and biomedical studies in weightlessness. At the time of the Soyuz 11 disaster, Stafford was visiting Europe with his wife and daughters and was due to make an appearance at the International Aeronautical Federation Conference in Belgrade.

"Before I reached Belgrade," he wrote, "I heard the news that the ... cosmonauts had died on their return to Earth. My first worry was that the stress of a long-duration flight had killed them and wondered what that would mean to our Skylab crews. Clearly we needed to know more than what was in the news." Back in Houston, Deke Slayton was of the same opinion: a year before, Nikolayev and Sevastyanov could hardly *stand* after their 18-day mission and now Dobrovolski, Volkov and Patsayev had returned *dead* from orbit after 24 days. "Was there something about being weightless that long that could kill you?" Slayton wondered. *Flight International*, too, speculated on 8 July that "it is possible ... that the degrading effect of weightlessness increased exponentially". The news from the Soviet Union, not surprisingly, offered a sketchy and none-too-helpful insight, but could not sidestep the inescapable truth that this was a tragedy of immense proportions. Yet even this truth was restricted to the periphery of an initial report which emphasised the flight's strengths and tried to downplay the calamity which had befallen it. The report began tersely with "Tass reports the deaths of the crew of the spaceship Soyuz 11 ... ", followed by a lengthy discussion of the extraordinary success of the mission and its re-entry, before ending abruptly with "upon opening the hatch ... [the rescuers] found the crew of the spaceship in their couches without any signs of life. The causes of the crew's deaths are being investigated ... "

Throughout Russia, the disaster brought about an unprecedented wave of mourning. People wept openly in the streets for three men who for over three weeks had appeared nightly on their television screens – cosmonauts who were being presented as human beings and not cold, distant, faceless supermen – and who had offered a clear response to Apollo that the Soviet Union was back in the manned space business and firmly in the lead. Now, instead of three heroes, bearing broad smiles and bedecked in medals and garlands of flowers, all the Soviet people had was ... three *funerals*.

Those funerals were to be a day of mourning and despair as waves of lament swept over an entire nation in memory of a broken dream. All three men, the autopsies at Moscow's Burdenko Military Hospital found, had died of haemorrhages in the brain, subcutaneous bleeding, damaged eardrums and bleeding of the middle ear. "Nitrogen," wrote Grujica Ivanovich, "was absent from the blood; it, together with oxygen and carbon dioxide, had boiled and reached the heart and brain in the form of bubbles. The formation of gas in the blood was a symptom of rapid decompression. The blood of all three men contained enormous amounts of lactic acid, fully ten times the norm, which was an indication of terrible emotional stress and anoxia." A day or so after the disaster, their bodies were laid in state in the Central House of the Soviet Army, each clad in a civilian suit and each resplendent with the gold star of a Hero of the Soviet Union. Only Patsayev's body showed visible evidence of the trauma he had endured: a dark-blue mark, similar to a large bruise, covered most of his right cheek. Tens of thousands of grief-stricken Muscovites, together with the cosmonauts' families and an emotional Leonid Brezhnev, who covered his face with his hand at one stage, filed past the open coffins to pay their final respects.

At around this time, Tom Stafford received word from Malcolm Toon, the American ambassador to the Soviet Union, that *he* was to represent President Nixon at the funerals. Four years earlier, the Johnson administration had offered to send an astronaut to Vladimir Komarov's funeral, but this overture had been rejected on the grounds that it was a 'private' affair. Now, relations between the two spacefaring nations seemed somewhat more cordial and, upon his arrival in Moscow, Stafford travelled in the limousine of Georgi Beregovoi directly to the funeral ... and even acted as one of the pallbearers for the massive urns which carried the cosmonauts' ashes for interment in the Kremlin Wall. Alongside the three-man crew of the Osoaviakhim-1 balloon, who had plunged to their deaths in 1934 after setting a new altitude record, Dobrovolski, Volkov and Patsayev – now record-setters themselves – represent one of only two 'group burials' ever to have taken place in the Kremlin necropolis.

By the time of the funeral, it was becoming increasingly clear that the deaths of the cosmonauts were due to a mechanical problem within their spacecraft and had nothing to do with their individual physiological states or their prolonged exposure to weightlessness. Initially, NASA physician Chuck Berry was so sure that nothing physiological could be to blame that he pointed to the release of a toxic substance inside the descent module as one possible cause. Decompression and its effects seemed the most reasonable explanation and this was verified by the post-mortem examinations of the cosmonauts. One of the bitterest ironies is that if Dobrovolski, Volkov and Patsayev had been provided with space suits, they would have survived the decompression. This became clear when the 12-member State Commission, chaired by Mstislav Keldysh, wrote its report a few weeks later. Ten subcommittees were established to investigate every aspect of the Soyuz which could have contributed to the disaster, although physicians had judged almost immediately after finding the bodies that the most likely cause of death was decompression. Specialists from Moscow who arrived at the landing site on 1 July had verified that there were no cracks or holes in the hull of the descent module.

Based on data from the onboard memory device, the orbital and instrument modules separated at an altitude of around 150 km and lasted just 0.06 seconds. "The pressure in the descent module," wrote Ivanovich, "began to fall rapidly at that moment. At 1:47:26.5 am [Moscow Time], two seconds prior to jettisoning the orbital module, the pressure in the descent module was 915 mm of mercury, which was normal. But some 115 seconds later the pressure had dropped to 50 mm and was still falling. In effect, there was no longer any air in the cabin!"

Decompression could have been caused by either the premature opening of one of two valves at the top of the descent module or a leakage from the hatch. When Vasili Mishin testified before Keldysh's panel on 7 July, he presented diagrams which seemed to endorse the first of these possibilities. It seemed unlikely that an incorrect command had caused the valve to prematurely open, because *both* valves operated from the *same* circuit. When one considers the 2 cm thickness of the valve tube, the volume of the descent module itself and the fact that the air would have escaped *at the speed of sound*, it is easy to see how the cabin pressure could have diminished nearly to zero and killed the crew in under a minute. The positions of the bodies in the descent module suggested that Dobrovolski and Patsayev had tried to unstrap in order to close the valve, but had been unable to act quickly enough. At the instant of separation of the orbital and instrument modules, the cosmonauts' pulse rates varied broadly: from 78–85 in Dobrovolski's case to 92–106 for Patsayev and 120 for Volkov. A few seconds later, when they first became aware of the leak, their pulse rates shot up dramatically – Dobrovolski's to 114, Volkov's to 180 – and thereafter the end had been swift. Fifty seconds after the separation of the two modules, Patsayev's pulse had dropped to 42, indicative of someone suffering oxygen starvation, and by 110 seconds all three men's hearts had stopped.

It would be a blessing to suppose that their deaths, though mercifully rapid, were also painless ... but high-altitude decompression and exposure to the vacuum of space does not produce painless results. The official autopsies from the Burdenko Military Hospital remain classified to this day, as do several other documents pertinent to the disaster, but a number of conclusions *have* been made. Dobrovolski, Volkov and Patsayev would first have experienced strong pains in their heads, chests and abdomens, after which their eardrums would have burst and blood would have begun streaming from their noses and mouths. "Due to outgassing of oxygen from the venous blood supply to the lungs," wrote Ivanovich, "the men would have remained conscious for 50–60 seconds. However, they could have moved about and tried to remedy their plight only during the first 13 seconds; this being the 'time of useful consciousness', corresponding to the time that it took for the oxygen-deprived blood to pass from the lungs to the brain." Dobrovolski and Patsayev were best positioned to reach up and try to close the valve, but could not be certain as to the source of the leak ... and only had a matter of seconds to find it. Remembering their earlier problems with the hermetic seal on the forward hatch, *this* – rather than the valves – would probably have been their first port of call, likely wasting what precious few seconds they had left. Maybe, at length, they heard the whistling of air and decided that it was, in fact, one of the valves, but time would have run out for them before they had a chance to close either of them. (After the disaster, Alexei Leonov tried manually closing just *one* of the valves ... and it took 52 seconds!)

Of course, the 'blame game' got underway as rapidly and with just as much vigour as air had been sucked from Soyuz 11's cabin. Of central significance to Keldysh's commissioners were the two primary causes of the tragedy – the valves themselves and the lack of space suits; for if Dobrovolski, Volkov and Patsayev had been attired in pressurised garments they would have survived. The decision to do away with suits stemmed from Sergei Korolev's decision to fly the Voskhod 1 crew in October 1964 as a 'stunt' for Nikita Khrushchev to squeeze three men into a one-man Vostok capsule. When Soyuz was developed, the volume available in the descent module was not much bigger and Korolev, feeling that wearing a pressure suit in a spacecraft would be just as impractical and uncomfortable as wearing a wetsuit in a submarine, opted to do away with them. Nikolai Kamanin objected strongly to the idea and several cosmonauts had written to Khrushchev, Brezhnev and Dmitri Ustinov over the years to plead their cases. All were ignored. This 'normalisation of deviance', about which NASA would be criticised in the wake of Challenger and Columbia, affected the Soviet manned space programme in a hauntingly similar way. Konstantin Feoktistov, one of the leading designers of Soyuz, felt that his hands were as bloodied as Korolev's for the deaths of the three men. "The feeling of guilt persists," he said resignedly two decades later.

To be fair, since 1966, Soyuz had been subjected to no fewer than a thousand tests and had encountered absolutely no problems with regard to decompression. When the Soyuz 11 descent module arrived in Moscow for inspection, it was subjected to powerful shocks and vibrations in an attempt to 'prove' that the valves would jerk open, but they remained firmly shut. Next it was decided to simulate their performance under conditions of vacuum in an altitude chamber and, again, they refused to open. It was only when conditions were adjusted to *simultaneously* impose a number of different dynamic loads on the module that they were finally shaken open. On 10 July 1971, Keldysh's commission issued a 200-word statement, asserting that the loss of the crew had been caused by "a loss of the ship's hermetic seal", even though there had been "no failures in its structure".

A pressure leak without mechanical failure implied to many Westerners that the cosmonauts themselves may have been at fault and Victor Louis, a Moscow-based journalist for the *Evening News* in London, claimed that an improperly closed hatch on the descent module had caused the disaster. Although this had some basis in reality, for the cosmonauts *had* encountered difficulties when closing the hatch prior to undocking from Salyut 1, it was firmly discounted by Keldysh's commissioners as a reason for their deaths. Not until October 1973 – more than two years later – would the Western media, in the form of the *Washington Post*, finally be made aware of the fact that the valve was to blame. In the meantime, by early August 1971, when Keldysh's report was completed, a number of recommendations were made for future missions. Firstly, the valve needed to be more stable with regard to shock loads. Secondly, there needed to be quick-action chokes to shut the valves manually in a matter of seconds. Finally, and of pivotal importance, space suits were to be worn for *all phases* of a mission in which depressurisation was a possibility.

In response to the last requirement, the 'Sokol-K' ('Space Falcon') suit was developed under the auspices of Gai Severin as a lightweight garment which could be individually tailored to each cosmonaut and was compatible with the seat liners aboard the Soyuz. A prototype was completed within weeks of the disaster and by the spring of 1972 had been fully tested and signed off as ready to fly. Since the Soyuz 12 mission, which finally flew in September 1973, the suit and its descendents have been worn by every cosmonaut during launch, docking, undocking, re-entry and landing. "In the event of decompression on the Soyuz," wrote Rex Hall and Dave Shayler, "the [Sokol] is automatically isolated from the cabin environment and supplied directly with either pure oxygen or an oxygen-rich mixture from a supply in the cabin or from self-contained systems." It included a soft helmet which could be pushed back over the head when not in use, a removable, white-topped 'skull-cap' for communications headgear and pressure-sealed gloves. The Sokol could also be used in the emergency transfer of cosmonauts from one spacecraft to another, with the aid of small hoses connected to the spacecraft's life-support system or through a portable backpack, although this has never been done. Testimony to the success of the Sokol is that, since 1971, no other cosmonaut has lost his or her life through the decompression of their spacecraft; indeed, the hardware has proven so reliable that there have been no other instances of depressurisation aboard a Soyuz.

However, in order for the suits to be properly accommodated in the confines of the spacecraft, the third crew seat – that of the research or test engineer – would be eliminated and its place taken by a system which could automatically pump air into the cabin in the event of decompression. The presence of this equipment pushed the weight of the Soyuz slightly above the rocket's payload capacity, in turn necessitating the removal of the solar arrays and their replacement with lightweight chemical batteries which could support an independent flight of around two days and were rechargeable from power generated by Salyut's electrical system.

The decision to reduce Soyuz crews from three to two meant that many of the military engineers originally scheduled to fly were quietly pushed aside. Not until November 1980 and the arrival of an upgraded version of the spacecraft would another three-man cosmonaut crew venture into orbit. Therein resides perhaps the most visible *other* 'casualty' of the Soyuz 11 mission: poor Pyotr Kolodin, the man who might have filled Patsayev's shoes had it not been for the misdiagnosis of Valeri Kubasov's lung ailment. Within a week of the disaster, all future crews, including the Leonov/Rukavishnikov/Kolodin team aiming for a second Salyut occupancy, were stood down pending repairs to the spacecraft. Kolodin *was* assigned to a subsequent mission in 1978, but was again dropped shortly before launch. The irony was not lost on him: by *losing* the chance to ever fly in space, his own life had actually been *spared*. Twenty years after the Soyuz 11 tragedy, and doubtless to this day, he ponders the question of 'what might have been'.

To a great extent, the Soyuz 11 tragedy also played on NASA's conscience. Within hours of learning of the disaster, astronauts and managers alike were wondering if exposure to the space environment for three weeks had caused the deaths of Dobrovolski, Volkov and Patsayev. When decompression and a lack of proper space suits were blamed, a change was made to the Apollo 15 lunar mission, due to launch just a few weeks later in July 1971. It was decided that astronauts Dave Scott and Jim Irwin would wear their space suits during ascent from the lunar surface. "The decision," read a NASA press release, dated 19 July, "was based on a re-evaluation of the requirements for crew members to wear pressure suits during different phases of the mission. The evaluation was conducted following the Soyuz 11 accident ... " Nor was this simply a knee-jerk reaction: the 're-evaluation' encompassed reviewing the design and testing of windows, hatches, valves, fittings and wiring in both the lunar and command modules. "In addition," the release continued, "studies were performed on re-entry effects on crew and cabin with a completely failed window, structural loading during lunar module jettison, cabin pressure decay caused by various-sized holes, suit-donning times and post-landing emergencies." Although the results established a high level of confidence in the Apollo hardware, this re-evaluation is notable in that it shook to the core not only the Soviet manned space programme, but also that of the United States.

THREE FAILURES

With the deaths of the Soyuz 11 crew and the suspension of further missions for the foreseeable future, the final curtain was drawn on the first Salyut space station. Its consumables could probably have sustained another team of men through part of August – and, indeed, Alexei Leonov said he, Rukavishnikov and Kolodin were prepared to take the risk – but realistically any such notion was out of the question. During the late summer of 1971, it was manoeuvred into a higher orbit and, finally, on 11 October, after it became clear that all chances of another expedition were gone, its engines were fired one final time to ensure a safe, destructive re-entry over the southern Pacific Ocean. It had remained in space for 175 days.

In the wake of the disaster and the need to extensively modify Soyuz before future flights could be undertaken, Vasili Mishin continued his bureau's focus on lunar landing missions and the establishment of a large-scale space station. The former had suffered yet another brutal blow on 26 June, just a few days before the deaths of Dobrovolski, Volkov and Patsayev, when the third N-1 rocket exploded shortly after liftoff. Admittedly, some progress had been made toward the lunar goal in the past two years; firstly, an unmanned shakedown of the LK lander in Earth orbit in November 1970 and then a test of the Blok D translunar upper stage a few weeks later. The drama of Apollo 13 offered some impetus for a resumption of work on the N-1. If Apollo were to be cancelled after such a close brush with disaster, the road to the Moon would be left clear for the Soviets. However, it was *not* cancelled and the spectacular Apollo 14 mission in January 1971 essentially nixed any further political support for the lunar landing project.

Nikolai Kamanin agreed that a radical re-examination of what he and many others perceived to be an inherently flawed booster was necessary: he wanted to cancel the N-1, replace it with Vladimir Chelomei's UR-700 and then attempt a series of lunar landings in the mid-Seventies. Yet with the Moon race lost, the N-1 no longer had a clear *purpose*, although efforts continued to perfect it. Perhaps, Mishin hoped, after 18 months of improvements, new filters to prevent foreign-object ingestion, new systems to cool its engines and a completely redesigned Kord unit, the third flight would be charmed. It wasn't.

Liftoff occurred at 2:11 am Moscow Time on 26 June, with all 30 engines firing normally ... for about *five seconds*. Then, to the horror of the onlookers at Tyuratam, the rocket began a slow axial rotation. As it cleared the tower, this reached a rate of eight degrees per second. The N-1's gyroscopic platform commanded an engine shutdown, which the control system opted to block until 50 seconds into the ascent, whereupon it plummeted back to Earth and crashed some 30 km downrange of the pad.

A year later, the N-1 still seemed riddled with problems and nowhere near a successful launch. Four roll-control engines were introduced to prevent a repeat of the difficulties experienced in June 1971, a digital computer was added to provide improved guidance, a freon anti-fire unit was aboard, a huge telemetry network with no fewer than 13,000 sensors, better shielding against excessive temperatures and an uprated Kord engine-management system. None of this dispelled the overriding sense of gloom that the rocket itself was jinxed and further launches were cursed to fail. Even Vasili Mishin, when asked about its chances, responded that another launch simply *had* to go ahead ... in order to properly test its guidance system! It seems quite remarkable that the *fourth* N-1 launch was approved when even the Chief Designer himself was uncertain as to whether it would work. It did not work. At 9:11 am on 23 November 1972, the final N-1 lumbered off the launch pad and, surprisingly, survived for nearly 107 seconds, coming very close to the scheduled burnout of its first stage. However, the programmed shutdown of several engines in order to prevent overstressing the vehicle ruptured a propellant line and the last N-1 ever to take flight was gone in a savage ball of fire.

Nevertheless, this final launch had actually come closest to *working*; the first stage had only about seven more seconds of thrusting ahead of it and had by far exceeded the performance of its predecessors. Mishin felt vindicated and assured senior leaders that the *next* N-1 would work. Two weeks later, Dmitri Ustinov asked if it was worthwhile continuing with the project and, certainly, several chief designers wanted it scrapped. After all, they asked, what was the point of landing a Soviet man on the Moon, now that the Americans had been there? Others countered that as satellites and manned spacecraft – including orbital stations – grew larger, heavier and more complex, a launcher with the lifting capacity of the N-1 *would* be necessary before the end of the Seventies.

With this mind, in March 1974, Mishin and Nikolai Kuznetsov prepared a memo for Leonid Brezhnev himself, appealing for support. Brezhnev promptly handed it over to Ustinov, who in turn passed it to the defence ministries. At length, Ustinov convened a number of the most prominent chief designers for their reactions. Among them was Valentin Glushko, a long-standing critic of the design of the Kuznetsov engines, which he felt had doomed the project from the start. Mishin did not help himself by accepting partial blame for the N-1's woes and it was *this* admission, in combination with failures in the DOS station effort during 1972–73, that hammered the final nail into his professional coffin. He was fired as Chief Designer of TsKBEM in May 1974 and replaced by Glushko, whose new and enormous empire – he was now *de facto* master of all manned and unmanned civilian Soviet space projects – would henceforth become known as 'NPO Energia'.

Years later, Mishin would admit that the decision to fire him came "as a complete surprise", but Glushko was far from sympathetic and even spitefully revoked his predecessor's permit to set foot on TsKBEM premises. Amongst

Glushko's first moves were terminating the last remains of the lunar project. By the late summer of 1974, the N-1, having swallowed more than 3.6 billion roubles, was cancelled and in February 1976 a new resolution ordered work on another superbooster named 'Energia'. (This vehicle would subsequently launch the Soviet shuttle Buran, but precious little else.) In Boris Chertok's mind, had Sergei Korolev lived, the N-1's woes *would* have been resolved; he would have weathered a couple of failed launches, then turned to Glushko to develop new engines. Asif Siddiqi has noted that the demise of the N-1 may have been a cynical response to the fear that maybe – just maybe – it *would* have worked on its next mission. In May 1976, a brief attempt was made by some N-1 workers to implore the 25th Party Congress to reinstate funding, but their letter went unanswered. The Soviet Union's last attempt to steal some of Apollo's thunder had vanished, metaphorically and literally, in a fireball.

Notwithstanding the end of the lunar dream, in his final months as head of TsKBEM, Mishin did have other ambitions for the N-1 booster, one of which was the establishment of a large, Earth-orbital station for astrophysical and fundamental scientific research. The Multi-Module Orbital Complex (MOK) and its central element, the nuclear-powered Multi-Role Space Base Station (MKBS), were expected to be launched using the N-1 in the mid-Seventies – after a next-generation DOS series of stations and Vladimir Chelomei's Almaz had acted as a sort of proving ground in the nearer term. Still, in the wake of the Soyuz 11 disaster, these 'nearer term' prospects seemed bleak. The slowdown of the N-1 effort had already prompted Mishin to cancel the Soyuz flights that were to have tested the Kontakt docking system in Earth orbit and at a meeting with Dmitri Ustinov in August 1971 it had been made abundantly clear that the *next* station must be a civilian DOS and not one of Chelomei's military Almaz. Time was of the essence, for America's Skylab was tentatively scheduled for launch in the spring of 1973 and *that* station would have an interior volume – and probably scientific capabilities, too – that were several times greater and more sophisticated than the Soviets could muster.

Consequently, plans were initiated for no fewer than *three* DOS stations, the first of which (DOS-2) would be placed into orbit sometime in the spring of 1972 and occupied by as many as three Soyuz crews. It would be followed by DOS-3 at the end of the year and DOS-4 in late 1973. The launch of DOS-2 was supposed to coincide with – and steal glory from – the Apollo 16 lunar mission and in October 1971 Vladimir Shatalov, now in charge of the cosmonaut team following the retirement of Nikolai Kamanin, began picking crews for this station. Unsurprisingly, his prime candidates for the first flight were Alexei Leonov and Valeri Kubasov. However, the DOS-2 launch was repeatedly postponed into the summer and eventually the autumn of 1972. "The delay," explained Asif Siddiqi, "may have had less to do with the station ... which was almost identical in design to the first Salyut ... than problems with requalifying the Soyuz for flight."

Vasili Mishin had already flown an automated Soyuz, under the cover name of Cosmos 496, to evaluate the ability of the improved life-support system to support the new Sokol-K pressure suit. It was launched on 26 June 1972 and completed a highly successful six-day mission. Afterwards, the cosmonauts earmarked for the first manned flight arrived at Tyuratam. By this time, the rumour mill had been running for several months, with some sources hinting that another station would be occupied for perhaps 30 days. In April, Shatalov himself told Czech journalists that new manned missions would fly "probably this year". The State Commission of General Kerim Kerimov formally set the launch of DOS-2 for late July. Leonov and Kubasov would follow four weeks later. Then, in mid-October, Vasili Lazarev and Oleg Makarov would begin a month-long stay.

All these plans ground to a halt immediately after DOS-2's launch at 6:20 am Moscow Time on 29 July. Less than three minutes into the ascent, the control systems on the second stage of the Proton booster failed. "US over-the-horizon sensors evidently monitored telemetry from the launch attempt," wrote Siddiqi, "prompting subsequent news reports that one of the four second-stage engines had stopped firing." Despite the profusion of such reports, since the failure had occurred so early and the station had not even reached a preliminary orbit, its launch was not officially declared and remained a state secret for many years. Picking through the debris at the crash site, one member of the military team sent to 'secure' the wreckage found a pair of sleeping shorts, bearing the initials of Leonov, the man who would have been DOS-2's first commander. The shorts were returned to him with a light-hearted quip that at least his *sleepwear* had launched successfully! The despirited Leonov was in no mood for banter. For the second time in two years, his mission had been taken from him. There was more bad luck to come.

By the late summer of 1972, there were two flight-ready Soyuz vehicles with no station to visit and the State Commission decided in August that one of them should be despatched on a 'solo' mission in Earth orbit with a crew onboard to test the Sokol-K suits and redesigned systems. Two teams of cosmonauts – a prime crew of Alexei Gubarev and Georgi Grechko and their backups, Pyotr Klimuk and Vitali Sevastyanov – were hastily assembled, trained for a month and successfully passed their final exams. However, at this point, Kerimov's commissioners began to suffer cold feet, particularly since a brief shakedown flight in Earth orbit would seem trivial in comparison to America's final lunar landing, Apollo 17, timetabled for December 1972. Consequently, the mission was cancelled.

Salvation came in the form of Vladimir Chelomei, who, despite the order two years earlier that he must hand over his prototype Almaz cores to Mishin for the DOS project, had continued to work on his military effort. In spite of the overwhelming resources devoted to the civilian stations, Chelomei had managed to test an updated Almaz control unit and an improved power and propulsion system and had run a series of hulls through stress, vibration and heat evaluations. Even as the DOS-1 project got underway in the summer of 1970, Chelomei secured approval from the minister of defence to proceed not only with Almaz, but also the large Transport and Supply Ship (TKS). When docked in space, these would weigh 40,000 kg. However, delays made it unrealistic to expect a first TKS launch before the middle of the decade and in 1971 Chelomei signed an agreement with Mishin to employ variants of Soyuz to transport cosmonauts to and from Almaz. One point that he did *not* like was the decision to rename 'his' station 'Salyut 2', but it was explained to him in no uncertain terms that this would conceal its military nature

A civilian Salyut space station in the assembly building at Tyuratam. The insert shows the station inside its payload shroud, atop the Proton booster.

behind the façade that it was simply an improved version of DOS-1. However, Chelomei ordered that the name be written on the ring of the Proton's third stage so that after this ring was jettisoned in space only 'CCCP' (the Cyrillic acronym for 'Union of Soviet Socialist Republics') in big red letters would be visible.

By the late summer of 1972, four crews of military cosmonauts were working on the first Almaz missions: Pavel Popovich and Yuri Artyukhin would fly first, then Boris Volynov and Vitali Zholobov, Gennadi Sarafanov and Lev Dyomin and finally Vyacheslav Zudov and Valeri Rozhdestvenski. All were officers in the Soviet Air Force, with Rozhdestvenski a veteran of naval operations. The Almaz station duly arrived at Tyuratam on 15 December and was launched into orbit on 3 April 1973. Initial checkouts were pleasing: the electricity generating solar arrays and antennas deployed as planned and the internal atmosphere was stable. Its orbit was steadily increased to 260 × 296 km. However, on 15 April it was evident that its telemetry system was not working. When the backup system was activated, ground controllers realised that the internal pressure had dropped precipitously and that venting oxygen was disturbing its orientation. One by one, all of the station's systems began to fail.

The plan had called for Popovich – who had flown the Soviet Union's fourth manned Vostok back in August 1962 – and rookie Artyukhin to launch aboard Soyuz 12 a day or so after Salyut 2 reached orbit, then spend perhaps 15 days aboard the station. However, "a technical problem" with the craft's parachutes led to their launch being rescheduled for 8 May. (In fact, many insiders were of the opinion that Mishin was deliberating delaying the Soyuz tests to ensure that Chelomei missed his launch window for Salyut 2. Chelomei was so enraged that he had written to the Soviet leadership in February 1973 to complain about Mishin's unpreparedness for flight.) Now, by mid-April, all plans for a manned mission to Salyut 2 had ground to a halt. Early theories supposed that the oxygen leak was caused by a problem with the station's supply system, which was housed inside the propulsion unit, and this judgement was initially accepted by the accident commissioners. In the meantime, the media were coming to their own conclusions. One report, citing 'classified sources', suggested that Salyut 2's hull had been breached so violently that the solar arrays, rendezvous antennas and radio transponders had been ripped away. Certainly, *Aviation Week and Space Technology* reported that Salyut 2 had disintegrated as early as 14 April and that many of its fragments had burned up in the atmosphere. Tracking by the United States showed as many as 17 'objects' re-entering the atmosphere, though some of them came *before* 14 April. Since Salyut 2 was functional until at least that time, the fragments could not have come from the station itself. Nor could they have come from the separation ring on the third stage of the Proton booster, since an automatic television camera aboard Salyut 2 had recorded that as having separated cleanly and drifted away. The ultimate conclusion was that the third stage had separated with 290 kg of residual propellants aboard and a malfunction had caused it to explode, peppering Salyut 2 with shards of debris travelling at a velocity of perhaps 300 m/sec.

In fact, this was one of the theories explored by *Flight International* on 19 April, which had already written that the arrival of the crew was "expected" within four days of launch. Pointing to the higher and more elliptical orbit of Salyut 2 than its predecessor, the magazine argued that "this orbit may have resulted from poor performance by the Proton launch vehicle ... Although this largest-known Russian rocket has had a reasonable success rate, it has had several failures and has, in fact, never been used for a manned mission". In conclusion, the magazine pointed to the multitude of fragments as indicative that "the Proton upper stage had exploded after separation, although it is not known whether the station was affected". Clearly, a correct judgement that the Proton was to blame for the loss of Salyut 2 was very quickly reached by the Western media within days of the loss.

"Based on a model of the explosion," explained Grujica Ivanovich, "a ballistic analysis verified that 21 of the objects that were tracked by the Americans could have been pieces of the third stage. It was also found that the orbits of *five* of these pieces intersected that of the station." With such data in hand, the State Commission concluded that Salyut 2's hull had been punctured by debris and, ironically, that until this calamitous impact it had operated perfectly. Nor did the name 'Salyut 2' offer any secrecy or cover for the station's military nature, for Western analysts quickly confirmed that it operated at a frequency of 19.944 MHz ... a frequency normally used by Soviet spy satellites. "Because the name Almaz was a secret," concluded Ivanovich, "the stations became known in the West as 'military Salyuts' ... which is precisely what the Kremlin had hoped to avoid!"

Secrecy, however, was still paramount and on 18 April, only days after the loss, unofficial sources in Moscow reported that no manned visits had *ever* been planned for Salyut 2; then, the news agency Tass added on 28 April that the station, "having checked the design of improved onboard systems and carried out experiments in space, had completed its flight programme". The omission of the word 'successfully' from the Tass release certainly raised a few eyebrows in the West, many of whose space observing community were more sceptical than they might have been in years gone by. In particular, in a March 1975 article for *Flight International*, David Baker quoted a lame Soviet excuse that the 'loss' of Salyut 2 had actually been part of "a planned pressure test which had resulted in the destruction of the vehicle". Nevertheless, on 28 May 1973 – the very day that America's first Skylab crew blasted off for a 28-day stay aboard that space station – Salyut 2 quietly decayed from orbit and burned up in the atmosphere.

So it was back to the third DOS station, a *civilian* affair, which seemed next in line to reach orbit ahead of Skylab. In fact, both DOS-3 and Almaz had shared preparation facilities at Tyuraram since the spring of 1973. One fundamental difference between DOS-3 and its civilian predecessors was that its electricity generating solar arrays were *steerable*; that is, the previous stations had fixed-alignment panels at the front and the rear and this had required them to manoeuvre to keep their arrays illuminated by sunlight ... and *this* manoeuvring consumed a substantial amount of propellant. DOS-3, by contrast, had *three* much larger arrays, arranged in a 'T' shape and attached to the narrower part of the main compartment and they were each capable of rotating to track the Sun. With a total collection area of 60 m^2, the arrays could produce four kilowatts of power – more than double that of the two previous stations. To compensate for the bulk of three large arrays, DOS-3's propellant was reduced, which in turn meant that its operating altitude had to be raised to 350 km, since it would be able to fire its engine more sparingly to prevent orbital decay. A new navigation and orientation system provided finer control of the station's movement in space and, thanks to a condenser in the air-conditioning system, the water supply for the crew was partially recyclable. The scientific payload, too, was greater than previous stations and included a Roentgen spectrometer and the ITS-K infrared telescope. Overall, DOS-3 had the capacity to support crews for as long as two months at a time and it was planned to send three teams, starting with Leonov and Kubasov.

An initial launch attempt on 8 May was postponed when a vent in one of the six oxidiser tanks on the Proton's first stage developed a leak just 20 minutes before the scheduled liftoff. Vasili Mishin is said to have reacted "emotionally" and, after an altercation with Vladimir Chelomei, refused to allow 'his' DOS-3 to fly aboard the booster; only the intervention of senior TsKBEM leadership changed his mind. The station duly flew at 3:20 am Moscow Time on 11 May 1973 and entered an initial orbit of 218 × 266 km. Then, with horrifying suddenness all control was lost. "On the very first orbit," General Kerimov recalled years later, "the attitude-control rockets began working irregularly. As a result, all the fuel reserves were burning up." Analysis would confirm that human error had caused the rockets to fire at their *maximum* thrust, rather than at their *minimum* thrust. In Yevpatoria, controllers could do nothing as DOS-3 literally emptied its propellant tanks and any chance of sending a crew aloft on 14 May was gone. For the second time in less than a year, Leonov and Kubasov were stood down ... and the Soviets could only watch as the American Skylab speared into orbit and, despite problems of its own, was nursed to a full recovery and successfully occupied on three occasions throughout the summer and autumn of 1973 and into the spring of the following year.

It is interesting – humorous, even – that, since the failure occurred so early, the Soviets were able to disguise the mission's identity by renaming it Cosmos 557. The investigation which followed, chaired by Vyacheslav Kovtunenko, was closely scrutinised by the KGB, who may have suspected sabotage in this second space station disaster within a month. Needless to say, the question of blame was quickly tackled and it was decreed that the flight controllers could have averted the failure if they had reacted more quickly. Yakov Tregub, the DOS-3 flight director, was sacked and his position taken by Alexei Yeliseyev, who set about the reorganisation of mission operations. By 1975, a new control centre had been established in Kaliningrad and Yevpatoria would henceforth only be used in support of military missions. The loss of three space stations within one year, together with his admission of blame for the N-1, spelled the end of Vasili Mishin's tenure as Chief Designer of TsKBEM; in May 1974, he was replaced by Glushko.

In the meantime, the useless DOS-3 station fired its main engine one last time on 22 May in an attempt to raise its orbit, but due to an improper orientation it re-entered the atmosphere and its debris plunged into the Indian Ocean. With the return of Skylab's first crew a few weeks later, the Soviets once again found themselves lagging behind America and in an even worse position than they had been in 1969: for they were now second-place runners-up in terms of both lunar landings *and* space stations ...

THE SOLO FLIGHTS

The only opportunity left to close what was now approaching a two-year hiatus in manned missions was to perform a 'solo' flight, Soyuz 12, in Earth orbit. As 1973 wore on, Vasili Mishin also inserted a second solo mission, Soyuz 13, for a series of astrophysical and other experiments which had been undermined by the loss of the DOS stations. Before either of them could fly, a final unmanned craft, under the cover name of Cosmos 573, was launched on 15 June and spent two days in space, demonstrating the new Soyuz design equipped with chemical batteries instead of solar panels. Its completion reinvigorated the Soviet effort and cosmonauts entered full-time training in July for what would effectively be 'test' missions in readiness for a return to space station operations sometime in 1974.

In command of Soyuz 12 was Vasili Grigoryevich Lazarev. Born in Poroshino, in the Altai region of Russia, on 23 February 1928, he was one of only a few cosmonauts to combine an impressive resume of a pilot with equally impressive academic credentials; for Lazarev was both a lieutenant-colonel in the Soviet Air Force *and* a medical doctor. After graduation from Higher Air Force School, he had received his doctorate in 1951 and served as an aviation physician. In the early Sixties, he had participated in the Volga high-altitude ballooning project, which involved the testing of prototype pressure suits and parachute jumps from as high as 32 km. In May 1964, Lazarev was considered as one of four finalists for the doctor's seat on the Voskhod 1 mission. There were heated discussions between Nikolai Kamanin and his superiors throughout the summer before the seat ultimately went to Boris Yegorov. On 27 August, for example, Kamanin suggested to his boss Sergei Rudenko, the deputy commander-in-chief of the Soviet Air Force, that the most qualified and fittest crew for Voskhod 1 was Lazarev, Boris Volynov and Vladimir Komarov, but his recommendation was overruled. This, indeed, was one of Kamanin's main gripes: that the Soviet leadership was 'hand-picking' cosmonauts, instead of following a fair and rational crew selection process.

As late as mid-September 1964, only weeks before the launch, Kamanin was still pushing for Lazarev to be aboard Voskhod 1, considering him to be "a qualified and fit flight surgeon, a qualified pilot as well as a physician with 15 years of research experience in aviation medicine". Sergei Korolev, however, was opposed to Lazarev; his reasons are unclear, but certainly he wanted two *civilian* passengers aboard Voskhod 1 and the presence of a military surgeon may have been the deciding factor for him. Korolev's reluctance to fly Lazarev extended into the following spring, when plans were laid for further Voskhod missions involving physicians. Despite the resubmission of Lazarev's name, the Chief Designer still remained "opposed" to flying him.

In September 1966, some months after Korolev's death, Lazarev was assigned to the initial Almaz military station project and later supported both the Soyuz 9 and 11 missions as the Soviet equivalent of a 'capcom' in the Yevpatoria control centre. When he was teamed up with civilian TsKBEM engineer Oleg Makarov in mid-June 1971, Lazarev could have anticipated an assignment to one of the Earth-orbital Soyuz missions to evaluate the Kontakt lunar-orbital rendezvous hardware. After these missions were cancelled by Vasili Mishin, he and Makarov remained together for not one, but *two* flights. The first of these, Soyuz 12, was a relatively short and quiet test flight; but the second mission, which took place in April 1975, was even shorter ... yet would turn into one of the most hair-raising and difficult missions ever attempted and one from which both men were lucky to return alive.

Oleg Grigoryevich Makarov, the frail-looking flight engineer aboard Soyuz 12, would go on to make history as the first Soviet cosmonaut to chalk up four space flights ... though not four *orbital* flights, for his harrowing second mission with

The Soyuz 12 crew: Vasili Lazarev (left) and Oleg Makarov. The pair also flew the near-disastrous Soyuz 18A mission a little more than a year later.

Lazarev in the spring of 1975 would fall decidedly short of reaching orbit. Makarov was born on 6 January 1933 in Udomlya, a few hundred kilometres north-west of Moscow. He proved an outstanding student of engineering at the Bauman Higher Technical School, graduating in 1957 and initially worked on the development of the Vostok spacecraft at Sergei Korolev's OKB-1 bureau. "His first job," wrote *Guardian* columnist Pearce Wright in his obituary to Makarov in June 2003, "was to develop the cosmonaut's control panel and instruments for the first manned flights." He was one of only a handful of civilian engineers to pass the first round of preliminary screening for the Voskhod selection in May 1964.

A little more than two years later, with the bureau now under Vasili Mishin's control, Makarov was picked as a cosmonaut and almost immediately detailed to the L1 effort. By September 1968, had been teamed with Alexei Leonov for the first circumlunar mission. In spite of the American success with Apollo 8 and their impending landing on the Moon, candidates were still being selected for L3 missions as late as the following summer and Makarov was one of them. By the middle of 1971 the situation had changed markedly and he found himself teamed with Lazarev, initially for an Earth-orbital flight of the Kontakt rendezvous radar and, following its cancellation that September, as the prime crew for a long-duration DOS mission and finally Soyuz 12.

Lazarev and Makarov were duly launched into space at 3:18 pm Moscow Time on 27 September 1973, climbing perfectly into an initial orbit of 193 × 248 km, inclined 51.6 degrees to the equator. Seven hours later, they fired Soyuz 12's main engine to alter their orbital parameters to 326 × 345 km, apparently to simulate the early portion of a rendezvous. Obviously, on *this* occasion, there was no station to visit, but in view of Western rumours of the failed docking on Soyuz 10, the Soviet press was careful to announce during the first day of the mission that Lazarev and Makarov would *only* be spending two days in orbit and that their flight was *only* a test of the new configuration of the craft.

Very little was carried out in terms of scientific experimentation, although Makarov took a series of terrestrial photographs using the LKSA multispectral camera, designed and built at Moscow State University, while Lazarev obtained photographs of the same targets using a standard camera. Back on Earth, other researchers took similar photographs from aircraft, apparently to identify the spectroscopic signature of the atmosphere. A handful of small biological payloads were also aboard the orbital module, but no details were released by the Soviets as to their precise nature.

In the wake of the Soyuz 11 disaster, one of the main tasks for Lazarev and Makarov was to thoroughly evaluate the new Sokol-K suit and the modified life-support system. It was during their attempt to do this that the cosmonauts encountered difficulties, although their severity has never been revealed. "At some point during the mission," wrote Asif Siddiqi, the cosmonauts "depressurised part of their ship to test these suits. On the second day, however, there were 'serious defects' in the life-support system, followed by a failure in the ship's attitude-control system." Nor were the men quiet and passive in their complaints, for it turned out that they had "candidly and bluntly" written about their difficulties in their onboard journals, forcing officials to "muffle" these issues after landing. The fact that their careers did not suffer suggests, perhaps, that whatever problems they raised were quickly swept under the carpet.

Nevertheless, Soyuz 12 appears to have functioned normally and the descent module touched down some 400 km south-west of Karaganda at 2:34 pm Moscow Time on 29 September. The mission had lasted slightly less than two full days. It was, concluded Siddiqi, "the first Soviet piloted mission in more than three years that had *fully* achieved its objectives". As a follow-up, another unmanned test was launched as Cosmos 613 on 30 November and completed a 60-day mission in which it climbed to a 'working orbit' akin to those planned for future stations and then powered down to simulate the kind of conditions under which a Soyuz would operate whilst docked to a DOS for a prolonged period of time.

It came as something of a surprise when Soyuz 13 was launched at 2:55 pm Moscow Time on 18 December 1973, for what turned out to be an eight-day solo mission with a suite of scientific gear. Confusion was rife in the Western media for some time about whether this flight was related to Cosmos 613, with some analysts speculating that the latter may have been another failed Salyut and that cosmonauts Pyotr Klimuk and Valentin Lebedev were a 'rescue crew'. However, it quickly emerged that Soyuz 13 *was*, indeed, an independent flight. Its main cargo seemed to be the Orion-2 ultraviolet telescope, which had been designed by Grigor Gurzadyan of the Armenian Academy of Sciences and was originally intended to fly aboard one of the civilian Salyuts. This was installed in place of the docking apparatus, at the very nose of Soyuz 13, and the orbital module itself was completely transformed into

Pyotr Klimuk (left) and Valentin Lebedev in the simulator. Both are wearing training versions of the mandatory pressure suit implemented in the wake of the Soyuz 11 disaster.

a kind of miniature laboratory for scientific research. To accomplish a mission of longer than two days, the spacecraft was fitted with solar panels.

Actually, the crew, Klimuk and Lebedev – who, incidentally, became one of the youngest cosmonaut pairings ever to fly; both aged just 31 at the time – were originally the *backup* team and should not have flown this mission at all. Since July 1973, two other cosmonauts, Lieutenant-Colonel Lev Vorobyov in command and civilian TsKBEM engineer Valeri Yazdovski, had been preparing intensely for the flight. Unfortunately, some stories tell that their relationship was at a distinctly low ebb as the year wore on and rumours even abounded that they refused to sit at the same table in the cafeteria during lunch breaks. By the beginning of December, Vladimir Shatalov decided to replace them with their backups.

In fact, Vorobyov had gotten himself into hot water in the past. "Having joined the cosmonaut corps [in January 1963]," Asif Siddiqi explained, "he and another trainee, Eduard Kungo, publicly criticised the Communist Party. When asked to make a speech in front of a local Party meeting, Kungo had evidently told a senior official: 'I will not speak to a party of swindlers and sycophants!' He was promptly expelled from the cosmonaut team." The fact that Vorobyov was a fully registered member of the Communist Party saved him from the same fate. He eventually went on to train for both Kontakt and Almaz. His would-be crewmate, Yazdovski, was a long-time employee of Korolev's OKB-1 (later TsKBEM) bureau, having worked since 1957 on the development of the Vostok and Voskhod spacecraft; interestingly, he had played a pivotal role in the design and construction of the Orion-2 astrophysical apparatus that would be aboard Soyuz 13. It is bitterly ironic, therefore, that his poor relationship with Vorobyov should have squelched his only chance of accompanying it into orbit.

According to Rex Hall and Dave Shayler, however, it was the outspoken nature of the two would-be cosmonauts which led to their fall from grace. Citing unspecified sources, these authors obliquely pointed out in 2003 that both Vorobyov and Yazdovski were "very principled cosmonauts who did not suffer fools gladly and spoke their mind". No further detail was provided, but this would seem to be in keeping with Vorobyov's implication in the 1964 Eduard Kungo incident. "They consequently made enemies," concluded Hall and Shayler, "who decided that they should not fly under any circumstances." Indeed, images of Vorobyov and Yazdovski training for Soyuz 13 do not outwardly suggest any hostility between them. Maybe a few senior managers – Vasili Mishin, perhaps, or possibly Shatalov – had long and unforgiving memories.

Their replacements were not only rookies, but they also experienced the shortest-ever period between selection and making their first flights. Born on 10 July 1942 to a peasant-stock family in the Komarovka district of Brest, close to the Polish border of the then-Belorussian Soviet Socialist Republic, Major Pyotr Ilyich Klimuk officially became the first Belorussian spacefarer and entered cosmonaut training in October 1965, aged just 23. Siddiqi has described him as "something of a child prodigy" and, indeed, Klimuk would fly no fewer than three missions before his 36th birthday, would become assistant head of the cosmonaut corps shortly thereafter and would ultimately secure the headship for himself before he turned 50.

Klimuk's father died in the Second World War and, after graduation from middle school in 1959, he entered Primary Aviation School and later the Leninski Komsomol Chernigov High Aviation School. Completion of initial flight instruction in 1964 was followed by enrolment in the Soviet Air Force, from whose ranks he would eventually retire as a general. Shortly after joining the cosmonaut team, he began preparing for the L1 project and in September 1968 was named as a backup crew member for the second circumlunar mission, supporting cosmonauts Valeri Bykovsky and Nikolai Rukavishnikov. Three years later, he was formally paired with military engineer Yuri Artyukhin and a civilian TsKBEM engineer for a DOS station visit, before receiving the backup command of Soyuz 13 in July 1973.

Valentin Vitalyevich Lebedev's rise from a wet-behind-the-ears cosmonaut selectee to hardened veteran spacefarer was even shorter; in fact, he had only been chosen from the ranks of the TsKBEM engineers in March 1972 ... and flew into orbit just 21 months later! Born on 14 April 1942 in Moscow, he completed high school in Naro-Fominsk and studied for a year at the Higher Air Force Navigators School, near Orenburg. However, due to the reduction in the number of personnel in the armed forces, he was discharged and continued his studies at the Moscow Aviation Institute, graduating in 1966. Lebedev then worked as an aircraft designer, specialising in structures and materials, before joining the cosmonaut team a month shy of his 30th birthday. However, his involvement in space-related matters had actually begun a few years earlier, when he served aboard an expedition of the Eighth Naval Squadron to the Indian Ocean and was later based in Bombay (today's Mumbai) to support rescue operations for two unmanned Zond missions.

Both Klimuk and Lebedev had trained extensively with the Orion-2 hardware at its *alma mater*, the Byurakan Observatory in Armenia. Unlike Orion-1, which had been housed *inside* Salyut 1, its successor sprouted, almost menacingly, like a cluster of miniature cannon, from the nose of the spacecraft's orbital module. During its eight days in space, using NASA-supplied film, it would acquire no fewer than 10,000 ultraviolet spectrograms of more than 3,000 stars – some as faint as 13th magnitude – in the constellations of Taurus, Orion, Gemini, Auriga and Perseus. One particular observation of planetary nebula IC2149 revealed spectral lines of aluminium and titanium, which had not been seen previously.

The Orion-2 hardware was mounted on a three-axis-stabilised platform with a pointing accuracy of two to three seconds of arc and it was operated by first orienting the spacecraft in the general direction desired and then refining this by using a set of electric motors in the instrument. It also featured an X-ray detector for solar physics research. Although Lebedev was primarily responsible for operating the telescope, both he and Klimuk also participated in a range of biological, Earth-resources and navigational tasks. In particular, they studied the movement of blood to their brains, investigated protein mass in space with the Oasis cultivator in an early attempt to understand the requirements for a future 'closed-loop' life-support system and observed Earth's horizon with the RSS-2 spectrograph.

They also became the first Soviet cosmonauts to be in orbit at the same time as a team of American astronauts, namely the final Skylab expedition of Gerry Carr, Ed Gibson and Bill Pogue. Neither crew had the opportunity to talk to the other, although Carr was quoted in *Time* a week or so later as having wished the cosmonauts a smooth mission. Klimuk and Lebedev landed safely – albeit in a snowstorm – some 200 km south-west of Karaganda at 11:50 am Moscow Time on 26 December after an eight-day flight which, in Siddiqi's words, was "an unqualified success" and offered convincing proof that the Soviet programme had "bounced out of the dismal dregs of the past few years". However, there was still a long way to go, for the United States was halfway through a record-setting 84-day space station mission. If the Soviets were to make their mark in the Seventies, then requalifying Soyuz was only part of the battle. After three DOS failures, their *next* space station in 1974 simply *had* to work.

A MILITARY SALYUT ...

In view of the classified nature of Almaz, it is hardly surprising that very little information about it emerged from behind the Iron Curtain until quite recently. As late as 1988, in fact, Phillip Clark admitted that when attempting to describe its features and functions, the space analyst was "at a great disadvantage", the Soviets having revealed "not a single sketch which hints at the correct design". Next to nothing was known about its orbital operations and these military Salyuts had only been discussed in a very general – almost 'poetic' – way, by remarking that their solar-cell arrays closely resembled "a bird in flight".

One point which Clark *was* able to make, however, was that the stations carried a re-entry capsule, which he presumed was for the return of classified experimental results and photographic canisters and which had not been aboard the 'civilian' Salyuts. It was not for several years that the design of this capsule was revealed in more depth: it apparently consisted of a two-part device, measuring 1.35 m long and 85 cm wide and included a heat shield and a small payload container capable of holding samples or films weighing up to 120 kg. Small capsules would have been inserted into a cylindrical airlock and it was reported that a 'manipulator' on the exterior of the station would have moved samples to the re-entry craft. When fully loaded, the capsule was pneumatically jettisoned at a 60-degree angle towards Earth and against the station's direction of motion. A timer 'spun-up' the craft and a solid-fuelled retrorocket was fired to parachute it to a soft landing on Soviet soil.

As for the Almaz itself, judging from the dimensions of the ill-fated Salyut 2 and the successful Salyut 3 which followed, it measured approximately 14.5 m long and 4.15 m wide and weighed in excess of 18,900 kg. Internally, it had a volume of around 92 m^3. It was fitted with two large solar panels, capable of rotating 180 degrees to track the Sun, and its main habitable compartment was divided into control, work and living sections, with a 'corridor' along the port side. 'Floors' were painted in darker colours and lighter tones were used for 'ceilings'. Other features included four windows in the living quarters, a special 'sofa' for medical experiments, fixed and swinging 'beds', hot and cold water provisions, a table for eating, a shower and toilet, a small library and a tape recorder for music.

One of the station's most notable features was a defensive *weapon* to protect itself against interception by American military satellites. It was based on the Nudelman single-barrelled aircraft cannon, had a range of up to 3 km against co-orbital targets and could fire a 23 mm projectile at 690 m/sec. Less dramatic, but no less important, were Almaz' intelligence-gathering and photographic reconnaissance equipment. Of this, the Agate-1 device had a focal length of 6.3 m and possessed the highest resolution of all the instruments. Elsewhere was the OD-5 telescope, officially for "detailed reconnaissance of areas of Earth's surface in interest of national economy [and] science", of which its primary foci would be meteorological phenomena, the observation of breeding grounds, forest and steppe conditions, the extent of changes caused by natural disasters . . . and, presumably, a significant chunk of its time would have been spent snooping on the United States, possibly China and their assets . . .

Other tools included a panoramic survey unit for studying large portions of Earth's surface, a circular observing periscope for tracking targets, an infrared device for monitoring explosions and other "high-temperature events" – in other words, missile launches and nuclear detonations – and an impressive suite of cameras and imaging equipment. Cosmonaut Pavel Popovich, who commanded the only successful mission to Salyut 3, recalled in an interview that there were a total of 14 cameras aboard the station. Naturally, the Soviets presented its capabilities as being in support of hydrological, meteorological, geographical, oceanographical and Earth-resources investigations, but the fact that it transmitted on military frequencies left many in the West in little doubt that its equipment would have almost entirely been handed over to more covert reconnaissance-gathering objectives. It has been suggested over the years that Almaz was even intended to carry side-scanning radar, although this apparently was not completed in time and was never flown.

The first successful Almaz station, whose name was released to the Western media as 'Salyut 3' in an attempt to conceal its military nature, was launched atop a three-stage Proton booster at 1:38 am Moscow Time on 25 June 1974. It achieved a satisfactory orbit of 219 × 270 km and was followed, at 9:51 pm on 3 July, by the Soyuz 14 spacecraft and cosmonauts Pavel Popovich and Yuri Artyukhin. On his website, www.svengrahn.pp.se, the Swedish radio analyst Sven Grahn noted that within a few hours of the Soyuz launch, he was able to beat Radio Moscow's official announcement in identifying Popovich as the commander. "I had never really picked up clear voice from a Soviet spacecraft before," Grahn wrote, "so now was my chance … At [12:54:25 am Moscow Time on 4 July], voice signals came through from Soyuz 14 and after a while the crew could be heard calling 'Zarya ya Berkut, pryom' ('Dawn, this is Golden Eagle, over'). I thought that I remembered who had this callsign and it did not take me long to look up that 'Berkut' had been used by Pavel Popovich!"

Pavel Romanovich Popovich was already one of the Soviet Union's most celebrated cosmonauts, having flown the fourth Vostok mission in August 1962. In fact, chosen as one of the original 20 pilots, alongside Yuri Gagarin, he had been one of six finalists and might have been the first man in space. Now a full colonel in the Soviet Air Force, Popovich was making his second voyage into the heavens. Born on 5 October 1930 in Uzyn, within the Kiev Oblast in the north of the then-Ukrainian Soviet Socialist Republic, he is today recognised and revered as the first ethnic Ukrainian spacefarer.

During his early teens, Popovich apparently so loathed the Nazi occupation of Russia that he refused to learn German at school, instead stuffing cotton in his ears and getting himself expelled as a result. He was, it is said, even dressed in old frocks by his mother and passed off as a girl to avoid being sent to a Nazi labour camp. After the Second World War, Popovich worked as a herdsman, before obtaining a diploma from a technical school in the Urals and entering the Soviet Air Force. Whilst assigned as a fighter pilot in Siberia, he met his future wife, Marina, a woodcutter's daughter who became a high-ranking officer and engineer. She was also an accomplished stunt pilot and outspoken researcher into UFOs, an interest which her husband, too, later embraced. In fact, in 1984, after retiring from the cosmonaut corps, Popovich headed the UFO Commission of the Soviet Academy of Sciences. He was a voracious reader, an admirer of Ernest Hemingway and Marie-Henri Beyle, better known by his *nom de plume* of Stendhal, and often quoted the works of the Soviet poets Sergei Yesenin and Vladimir Mayakovski. In the isolation chamber, Popovich was very much the opposite of steely Andrian Nikolayev: he was more light-hearted and jocular, often relieving the tedium by dancing and singing operatic arias with such gusto that scientists and engineers gathered to listen.

Yet this extroverted and ebullient individual had raised more than a few eyebrows in the aftermath of his three-day Vostok mission. During a state visit to Havana shortly after Christmas 1962, for example, Popovich had publicly announced that the Soviet Union would support Cuba on Earth and in space … an announcement

Pavel Popovich (left) and Yuri Artyukhin with a mockup Soyuz descent module during pre-flight training.

which quickly got him into hot water since his remarks were completely at loggerheads with state policy. The timing of his words – just weeks after the Cuban Missile Crisis had been defused and a humiliating Soviet withdrawal effected – must have left some senior Kremlin officials red-faced. Then, in April 1966, at a public event, he had struck his wife and, in turn, had been punched by her brother and given a black eye. (Several of the other cosmonauts' wives even wrote to his commander, insisting that Popovich should be thrown out of the corps.) Nonetheless, he remained on active duty and from the autumn of that year his spaceflying career took a definite change in direction towards military missions: firstly, he was assigned to the Soyuz-VI project, along with Artyukhin, and then to Almaz. He was also, as we have seen, detailed to the L1 project in 1968 and probably would have commanded the third circumlunar mission sometime the following year.

Lending genuine credence to the likelihood that Salyut 3 was an exclusively military affair was the presence of a military flight engineer aboard Soyuz 14. The stern-looking Lieutenant-Colonel Yuri Petrovich Artyukhin was born in Pershitino in Moscow on 2 June 1930. He obtained a doctorate in engineering from the Soviet Air Force Institute, specialising in military communications systems, before becoming a cosmonaut in early 1963. Artyukhin was one of the second group of pilots and engineers who, as well as being generally older than those of the previous team, were far more experienced in terms of flying skills and academic credentials. Initially assigned to Soyuz training duties, he was considered briefly in the spring of 1965 for a place on a long-duration Voskhod mission, paired with Vladimir Shatalov. When these plans failed to crystallise, Artyukhin moved into the world of military flights and supported Soyuz-VI. At the same time, interestingly, he also appears to have been a candidate for the L3 lunar landing project. By 1971, he was teamed with Pyotr Klimuk for a lengthy DOS mission; and finally, in the wake of the Soyuz 11 disaster, found himself paired with Popovich for an Almaz flight.

Less than 26 hours after launch, with Popovich having removed his space suit gloves so that he could grip the controls more carefully, the crew performed a perfect docking with Salyut 3 and Artyukhin entered the station and turned on the lights. Years later, Popovich recalled that the launch vehicle had delivered Soyuz 14 to within 600 m of Salyut 3 and that from a distance of 100 m he and Artyukhin were able to take manual control of the final approach. On his website, Sven Grahn pointed out that Soyuz 14 had been launched when the station had established itself in an orbit which almost exactly repeated after every four days or so.

"In this way," wrote Grahn, "launch opportunities to rendezvous and dock with the station occurred every four days. An additional factor important for launching a ferry to Almaz was the equator-crossing longitude for the space station revolution passing nearest [to] the launch site. If a direct ascent (within a few revolutions) is made to the space station, the longitude of the station equator crossing should be very close to that of a Soyuz launched from [Tyuratam]. Soyuz ferries to early space stations used about one day to reach the station. The altitude of the ferry would be gradually raised to coincide with that of the station. The higher up the station is, the larger the average difference in orbital altitude during the rendezvous phase between the ferry and the station. Thus, to reach the station after 24 hours, the initial phase difference along the orbit between the ferry and the station needs to be larger the higher up the station is ... but to let the orbital planes coincide at launch, the equator-crossing longitude of the station at the revolution passing closest to the launch site on the day of the launch of the ferry needs to be further east the higher the station is. In that way the proper orbital phasing is achieved when the station's orbital plane drifts westward until the time the ferry is launched."

Popovich and Artyukhin spent around 15 days aboard Salyut 3, performing experiments using "remote-sensing equipment" – in other words, the military surveillance hardware – and making a series of observations and photography of "Earth's surface". As to the precise locations of these terrestrial targets, little was revealed, with the exception of an area of Central Asia. This would seem to tie in with reports in *Aviation Week and Space Technology*, which hinted that a set of 'targets' were laid out near Tyuratam and photographed to evaluate the imaging capabilities of Salyut 3's surveillance gear. Other experiments focused on the cosmonauts themselves and included studies of blood circulation to the brain and blood velocity in the arteries before and after physical exercise. A 'universal trainer' for speedwalking, running and jumping was used by both men each morning and evening. The cosmonauts also examined the horizon with the RSS-2 spectrograph and made celestial navigation tests. Television 'shows' from Salyut 3 were given, although the interior was darkened and the audience found it difficult to discern any detail. One touch of humour, later in the mission, came when the air-conditioning system comically blew Artyukhin from one wall to the other ...

The cosmonauts undocked from the station at midday, Moscow Time, on 19 July and touched down, some 140 km south-east of Dzhezkazgan in at 3:21 pm. The mission had lasted almost 16 days, making it the third-longest Soviet space flight to date. Popovich and Artyukhin's spirits remained high throughout and *Flight International* speculated that their mission had "proved that a more Earthly working schedule (eight hours experimental work, eight hours recreation and housekeeping and eight hours sleep) ... can result in consistently high crew performance". This seemed to be at loggerheads with a more difficult situation aboard the American Skylab late in 1973, whose third crew had gone on strike in protest at flight controllers imposing an excessively heavy workload.

It is likely that the next mission to Salyut 3, which was launched at 12:58 am Moscow Time on 27 August, would have attempted to eclipse the duration set by Popovich and Artyukhin; indeed, some sources have argued that Soyuz 15's Gennadi Sarafanov and Lev Dyomin were to spend perhaps 25 days aloft. The spacecraft's initial orbital parameters were satisfactory, at 230×180 km, but the mission quickly ran into problems the following day when it proved impossible to dock with the station. A fault with the automated reaction-control system and an apparent failure of the Igla rendezvous device caused the spacecraft to perform excessively long manoeuvring burns. On their first approach, Sarafanov and Dyomin passed within 7 m of Salyut and on a subsequent try they missed their target by 30–50 m. Since the 'ferry' variant of the Soyuz was not equipped with solar panels, its chemical batteries limited its independent lifetime to just two days. With nothing in reserve, the hapless cosmonauts had little option but to execute an immediate return to Earth. (Vladimir Chelomei had complained several years earlier that the lack of reserves or systems to facilitate repeated *manual* docking attempts exposed the Soyuz to just this sort of weakness.)

It was Soyuz 15's landing which helped Soviet officials 'explain away', albeit in a pathetic manner that aroused scepticism from the outset, what had actually happened. Sarafanov and Dyomin fired their retrorockets as planned and landed in particularly adverse weather conditions, 48 km south-west of Tselinograd, at 11:10 pm Moscow Time on 28 August. The recovery team extracted them from the descent module within minutes, but the flight allowed Vladimir Shatalov to announce that they had actually been 'testing' an automated docking system for future station tankers. *Of course*, Shatalov continued, Sarafanov and Dyomin *could* have docked ... but there was *no* need for this, since Soyuz 15 was only a test mission! As for the

Polar opposites: the unusual crew of Soyuz 15. Space grandfather Lev Dyomin is on the left, with his much younger commander, Gennadi Sarafanov, on the right.

landing, it had simply demonstrated the ability to touch down in 'different' conditions and "overcome the restrictions imposed on space flight by sunlight". This time, the poor Soviet explanations stretched credibility a little too far. *Time* told its readers a week later that "the landing, made at night and in bad weather, seemed to underline the urgency of the return. What had gone wrong?" Was it a failure with the hardware? Had Sarafanov and Dyomin, both rookie cosmonauts, simply screwed up? Whatever the reason, lying about it was particularly unwise, for the Apollo-Soyuz Test Project (ASTP), a joint American-Soviet docking mission between a two-man Soyuz and a three-man Apollo, was scheduled for July 1975 ... and the Americans were watching each new mission with greater than normal interest. Lame excuses about 'testing' docking exercises and risking the life of a crew would not do; the Western journalists' bullshit detectors were considerably sharper than those of the state-controlled Soviet media. If something was amiss with the Russian docking hardware, the rendezvous radar or even the Soyuz itself, the Americans wanted to know about it.

First amongst them was Tom Stafford, who had been training since January 1973 to command Apollo 18, the American participant in the mission. "It was ridiculous," he wrote in his autobiography, "to believe that the Soviets had sent a crew to fly around and inspect a station they had previously occupied." A few weeks after Soyuz 15, Stafford met Vladimir Shatalov in Washington, DC, and told him, point-blank, that the Soviets must come clean about whatever problem Sarafanov and Dyomin had experienced. Shatalov continued to claim there had not been a problem. "Washington isn't Moscow," Stafford explained bluntly. "Everything leaks to the press. What's 'secret' today winds up on the front page of the *Washington Post* tomorrow; 'top secret' will be in the *New York Times* in a week. If *you* say you didn't have a problem, and somebody from an intelligence agency knows differently and Congress leaks that you really *did* have a problem, ASTP is dead!"

Shatalov understood. Both men knew that ASTP was no ordinary space mission. It was an important political symbol to foster better relations between the Soviet Union and the United States. Digesting Stafford's words, Shatalov went pale for a moment, then went off to make some calls. Three days later, an official announcement came from Moscow: there *had* been a malfunction on Soyuz 15 which had prevented a link-up with Salyut 3. Then, in mid-September, Shatalov and Stafford sat side-by-side at a Houston press conference and laid it all bare for the world's media.

It was a small detail, perhaps, since the rendezvous and docking systems for ASTP were of a slightly different design, but it underlined the need for at least *some* element of transparency and trust on this first occasion of co-operation between the United States and the Soviet Union in manned space flight. Some senior members of Congress, including Senator William Proxmire, a noted critic of NASA, were already concerned that the joint project was taking chances with the lives "of our astronauts for the sake of some untangible diplomatic benefit". To be fair, though, Shatalov's admission showed how difficult the Soviets found it to make even minor admissions of fault or failure in a space programme that they regarded as one of the few remaining pillars of national pride.

Not for two decades was the full story revealed by Boris Chertok, a member of the Soyuz 15 investigation panel. Apparently, Sarafanov and Dyomin had reached a distance of 300 m from Salyut 3 and the Igla had failed to switch to its 'final-approach' rendezvous mode. Instead, it implemented a sequence which should normally have been executed at a much larger distance of perhaps 3 km from the station. As a result, Soyuz 15 was accelerated more rapidly towards its quarry, at one stage reaching 72 km/h. It would seem that the cosmonauts failed to notice the problem and allowed the Igla to reacquire radar contact and perform two further attempts ... both of which failed and, more worryingly, both of which narrowly missed colliding with Salyut. By the time ground controllers finally deactivated the Igla, no more propellant remained for a third attempt. However, comments from Sarafanov apparently indicated he was determined to complete a manual docking, but it remains unclear as to whether or not he tried to do this. Modifying the Igla would take several months and plans for another occupancy of the station in October – perhaps by Boris Volynov and Vitali Zholobov – were scrapped.

In the meantime, NPO Energia – the new name for TsKBEM – refused to accept that Igla was at fault and the two cosmonauts took a verbal roasting. An investigative board in September 1974 reprimanded them for "cutting off the flight programme" and neither man went into space again. Certainly, Hall and Shayler have suggested that they had suffered from a lack of sleep on their first night in orbit and, consequently, the day of rendezvous was long and arduous. They were an unusual pairing, partly because of the difference in their ages – Sarafanov was barely 32, Dyomin a sprightly 48 – and also because the latter happened to be the world's first 'space grandfather'. Speculation surfaced over the years that two men of such vastly different ages had been deliberately put together as a psychological exercise in crew dynamics, although this has never been verified.

Turning firstly to the commander, Lieutenant-Colonel Gennadi Vasilyevich Sarafanov had been born in the Saratov region of Russia on New Year's Day 1942. He joined the Soviet Army, but at the age of 18 transferred to the Air Force and graduated from Balashov Military Pilots' School in 1964. His final assignment before becoming a member of the cosmonaut corps the following year was as a fighter pilot in the Guards Regiment. Colonel Lev Stepanovich Dyomin came from Moscow, where he was born on 11 January 1926. Like Yuri Artyukhin, he held a doctorate in engineering when he was chosen as a cosmonaut in 1963. His performance seems to have been exemplary: as early as January 1965 his 'scores' in cosmonaut training were among the highest and by March of that year he had entered training as a candidate for a later Voskhod mission. He then started working on the Almaz effort in the autumn of 1966.

The fate of Salyut 3 is intriguing in itself. Years later, it would be revealed that the defensive gun was test-fired on 24 January 1975 against the station's velocity vector to shorten the "orbital life" of its shells. When it became clear that Salyut 3 re-entered the atmosphere to destruction the *very next day*, some observers in the West speculated that the cannon had so altered the orbit of the station that it was effectively a failure. "It wouldn't do," said one commentator, "to fire at an attacker, only to discover you have deorbited yourself!" In reality, it seemed unlikely that the firing could have *caused* or even *contributed* to the end of Salyut 3's life, but the incident is an interesting footnote concluding an interesting mission. Also completed in the wake of Sarafanov and Dyomin's flight was the ejection of the sample-return capsule from the station on 23 September 1974; although it suffered heat damage during re-entry, it landed safely beneath its parachutes on Soviet soil.

Even after more than three decades, so little is known – and perhaps no more will *ever* be known – about precisely what Popovich and Artyukhin did during their two weeks aboard Salyut 3 and what additional research Sarafanov and Dyomin might have completed. Without doubt, there are documents, hidden in the deepest recesses of the Kremlin archives, which will almost certainly never see the light of day. Some spectators have argued convincingly over the years that the only reason the Soviets imposed such secrecy on their entire space programme was as a kind of 'reverse psychology' that went right back to the perceived, though erroneous, 'missile gap' with the United States back in the early Sixties.

In essence, the Soviets *knew* that they held no huge technological advantage, they *knew* that America's miniaturised computing technology was ahead of their own and they *knew* that constant infighting between their design bureaux and their chief designers and political manoeuvrings from above made it a constant uphill struggle to sustain an effective and rational space programme. When one considers the psychological impact of this cloak-and-dagger stance on America, the *absolute lack* of reliable information and transparency from behind the Iron Curtain allowed unfounded stories to grow, unanswered questions to fester and mysteries and rumours to linger. In this way, by revealing nothing, the Soviets cleverly played the situation to their advantage; for all the mystery, suspense and fear which surrounded the N-1, it was, in fact, a shadow of its American counterpart. The much-lauded Salyut stations, too, had barely a fraction of the capacity of Skylab ... and doubtless their living conditions and the standard of their experimental hardware were similarly inferior.

Nevertheless, the lure of the unknown and the insatiable desire *to know* has always appealed to the human spirit. The reality that so few details and hard facts have ever been released surely adds to the fascination of these secrecy-enshrouded missions. The Soviets, as masters of psychological and ideological warfare, clearly played on such xenophobic fears and suspicions. Perhaps the truth about Almaz and the military Salyut programme, if it is ever revealed, will come as something of a disappointment. Without even realising it, perhaps the Western rumour mill had done the Soviets' job for them – and done it spectacularly well – by inflating the importance and technological superiority of what is more likely to have been a sickly, jaundiced reply to what American spy satellites had already achieved.

... AND A CIVILIAN SALYUT

By New Year's Day 1975, not one, but *two*, Salyuts were in orbit. The first was the military station, visited successfully by cosmonauts Popovich and Artyukhin and missed by a matter of metres by their fellows, Sarafanov and Dyomin. The second was Salyut 4, a civilian DOS station, launched at 7:15 am Moscow Time on 26 December. Physically, it was similar to DOS-3, lost as Cosmos 557: it measured 15.8 m long and 4.15 m in diameter and weighed some 18,900 kg. It was something of a quantum leap over the experience of Dobrovolski, Volkov and Patsayev, possessing a larger suite of scientific and biomedical instrumentation and three steerable solar panels with a total surface area of some 60 m^2.

Its focus, quite clearly, was on science. Amongst its 2,000 kg of research equipment, Salyut 4 carried the OST-1 solar telescope, built by the Crimean Astrophysical Observatory, together with a pair of X-ray instruments, swivel chairs and lower-body negative pressure gear for vestibular and cardiovascular studies, a bicycle ergometer and integrated trainer and a series of cosmic-ray detectors, multispectral cameras for Earth-resources studies, a teletypewriter and an autonomous navigational system. The interior of the station was divided into propulsion, transfer and work compartments; the first housing fuel tanks and orientation engines, the second containing navigation windows and the 'Raketa' ('Rocket') vacuum cleaner and the third providing living accommodation and work space. Home comforts came in the form of a small cupboard for plates, knives and forks, whilst two small heaters allowed the cosmonauts to warm their daily soup or coffee.

At 12:43 am Moscow Time on 11 January 1975, two weeks after Salyut 4's launch, the first crew headed to the new station. In command was Lieutenant-Colonel Alexei Alexandrovich Gubarev, one of the second generation of older, more experienced and better-qualified cosmonauts picked in January 1963. He was born on 29 March 1931 to a peasant-stock family on the eastern bank of the Volga River in the Samara region. After the death of his father, the family moved to a collective farm near Moscow and in 1950 Gubarev graduated from middle school and entered the Naval Aviation School for Aircraft Mechanics. He later served with the Soviet Air Force and after completing advanced studies was detailed in 1962 to join an aviation unit in the Black Sea; shortly thereafter, he drew the attention of recruiters for the second group of cosmonauts. Alongside Vladimir Shatalov and Lev Dyomin, Gubarev was one of the bright stars from this class and seems to have worked on the Soyuz-VI military project, the L1 circumlunar effort and the DOS-2 civilian station between 1966 and 1971. Together with Vitali Sevastyanov and Anatoli Voronov, he was

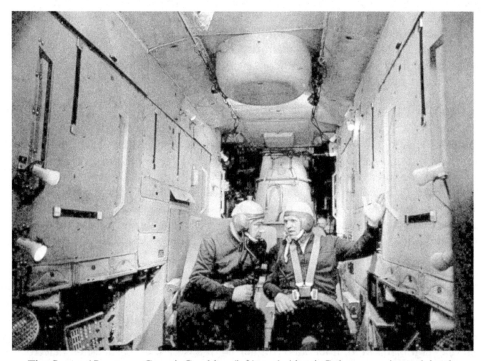

The Soyuz 17 crew – Georgi Grechko (left) and Alexei Gubarev – pictured in the working compartment of the Salyut station trainer. The large white cone in the background is the OST-1 solar telescope.

named in mid-June 1971 to back up Alexei Leonov's expedition to Salyut 1, but his training was abruptly suspended by the deaths of Dobrovolski, Volkov and Patsayev during their return to Earth.

The flight engineer on Soyuz 17, the bubbly and jovial Georgi Mikhailovich Grechko, was almost exactly the same age as his commander. Born on 25 May 1931 in Leningrad, he was one of the handful of OKB-1 civilian engineers who passed preliminary medical screening for a Voskhod crew. In September 1966, after a slight relaxation of the rules on health requirements, he was selected as a cosmonaut trainee. Grechko seems to have been used as something of a pawn between Vasili Mishin and Nikolai Kamanin later in the decade: in the days preceding the Soyuz 7 mission, for example, Mishin pushed for Grechko to fly instead of Vladislav Volkov, but Kamanin nixed the idea. Six months later, it was *Kamanin* who wanted Grechko to fly Soyuz 9 instead of Vitali Sevastyanov and, this time, *Mishin* objected. In time, he would become one of the Soviet Union's most experienced and respected cosmonauts, taking the absolute space-endurance record in the spring of 1978 and later becoming the then-oldest cosmonaut to enter space when he flew to the Salyut 7 station in the autumn of 1985.

The fact that Salyut 4 had been established in a relatively high orbit of around 350 km and the need for Gubarev and Grechko to perform two manoeuvres of their Soyuz in order to reach it implied to many Westerners that the mission was predominantly astrophysical in nature. (Having said this, Konstantin Feoktistov told journalists on 14 January that the higher altitude would also cut by as much as half of the propellant required by Salyut 4 to maintain its orbit.) The primary scientific instrument was the conical OST-1, located in the aft portion of the station, and during the course of two long-duration missions – the first crewed by Gubarev and Grechko, the second by Pyotr Klimuk and Vitali Sevastyanov – it acquired detailed spectra of the Sun. Observing astronomical targets at X-ray wavelengths were the Fillin-2 spectrometer and the RT-4 telescope, whilst the ITS-K instrument analysed infrared radiation emitted by the Earth and celestial objects.

Surprisingly, after the successful docking of Soyuz 17 with the station at 4:25 am Moscow Time on 12 January, the normally secretive Soviets issued almost daily reports on the cosmonauts' progress ... including significant details about the work they were undertaking. Gubarev and Grechko undertook observations of the Crab Nebula, together with supernova remnants in the constellations of Vela and Puppis, white dwarfs, neutron stars, suspected black holes and the background radiation of the entire galaxy along the Milky Way. The OST-1 telescope – a version of which could not be operated on Salyut 1 because of a jammed cover – produced some pleasing results, although the Sun itself was relatively quiescent at the time. However, not all went well. Apparently, the telescope had been operated autonomously in the two weeks before the crew arrived and a problematic pointing system caused the Sun to 'blind' its main mirror. Experts at the Crimean Astrophysical Observatory successfully revived it by asking the cosmonauts to manoeuvre Salyut 4 such that the optical axis of the telescope was pointing directly towards the Sun. This proved no easy task, because it required the crew to measure the time it took for the rotating mirror to move from one support to another in its normal mode of operation and then calculate where it had to be stopped to assist the main mirror. The only way to do this was by listening to the mirror's movements, which the crew did with a stethoscope from their medical kit. This not only restored the device to service, but it once again proved the usefulness of having humans in space. Although the main mirror was recessed in a conical niche to protect it from micrometeorites, on 3 February Grechko resurfaced it by spraying a new reflective layer. The Russians were delighted that the process worked, for it was a deciding factor in their plans for future stations. If the surface could not be recoated, there would be little point in sending up other telescopes for long-duration studies.

Other experiments focused on Earth resources – which, unlike the exclusively military targets of Popovich and Artyukhin, had more scientific goals. Primary areas of interest were the Caspian depression, the wide swathes of Central Asia, the southern part of European Russia, the Far East and Kazakhstan. The ITS-K infrared detector, although used for astrophysical research, was primarily for studying Earth's atmosphere. In addition, atomic oxygen levels in the upper atmosphere were examined using the Emissia system, part of a project to provide valuable meteorological data and assist with the definition of orbital dynamics of future satellites. The cosmonauts' productivity was substantially enhanced by the presence of the 'Stroka' teletypewriter, which operated very much like a newspaper teletype, with messages issued on strips of paper. This allowed flight controllers to send instructions to the crew without distracting them from other tasks.

The physiological and psychological performance of the cosmonauts was carefully monitored and maintained with a range of exercise equipment and the Penguin and lower-body negative pressure garments. Typically, Gubarev and Grechko would work for six days and rest on the seventh. They consumed four small meals per day, shaved with safety razors or electric ones and 'washed' with moist gauze napkins. At length, after loading their logbooks and scientific data results aboard the descent module, Soyuz 17 undocked from the station around mid-morning on 9 February and landed at 2:03 pm Moscow Time in a fierce snowstorm. In total, Gubarev and Grechko had spent just over 29 days in space, soundly surpassing the previous Soviet record set more than three years earlier by Georgi Dobrovolski's ill-fated crew. Unfortunately, by this time, the Skylab effort had scored a number of rapid-fire successes and three teams of American astronauts had spent 28, 59 and 84 days in orbit during the last half of 1973 and into the spring of 1974. The Soviets were steadily catching up, but it would take another three years before they would take the definitive lead.

Certainly, the next step after Soyuz 17 was to extend the Soviet endurance record yet further, perhaps to two full months. *That* would appear to have been the aim of Salyut 4's next crew, old comrades Vasili Lazarev and Oleg Makarov, who were duly launched at 2:04 pm Moscow Time on 5 April 1975. The ascent proceeded normally for almost five minutes; then, at an altitude of 145 km, the situation went badly wrong. Ordinarily, two sets of pyrotechnics should have fired to cut the central core of the booster from its upper stage and six latches should have blown. It did not happen. "An excessive degree of vibration caused the relay in half of the upper sequencer to close down and to signal three of the six latches to fire prematurely," explained Rex Hall and Dave Shayler, "with the lower core and upper stages still firmly attached. This was activated only seconds prior to the planned separation, but with the latches armed. The connection *that* triggered was in the same location as the electrical link between the upper and lower segments of the structure. Therefore, when the electrical contacts were severed due to the premature explosions, so also were *all* links to the *lower* latches, causing an uneven linkage between the core and upper stage as the vehicle continued to climb." The upper stage ignited on time ... but the central core *was still attached!* Although the thrust of the upper stage engine broke the remaining locks and separated the two stages, within seconds the unanticipated strain threw the booster off its intended trajectory. The onboard gyroscopes detected a deviation beyond the mandated ten-degree safety limit and automatically commanded an abort. Two hundred and ninety-five seconds into Lazarev and Makarov's mission, the escape tower fired to pull the orbital and descent modules away from the booster.

"Inside Soyuz," wrote Hall and Shayler, "Lazarev felt the vehicle pitch and roll and reported a heavier pitch than on Soyuz 12. At that moment, the Sun suddenly disappeared from sight and a loud siren sounded. On the instrument panel, the red 'Booster Failure' light flickered on. For a few moments, the crew wondered what was happening as the sound of the booster stopped and for a second or two they became weightless as the forward velocity faltered." Now on the very fringes of space, the cosmonauts could only grit their teeth as the abort programme automatically released the bell-shaped descent module to plunge back to Earth. Normally, they could have expected perhaps 15 G in such an emergency, but since Soyuz 18 was already pointing *directly downward*, its rate increased and the two men were subjected to no less than 21.3 G of deceleration. "We began to experience a creeping and unpleasant pull of gravity," Lazarev recounted later. "It increased rapidly and its rate was much greater than I had expected ... Some invisible force pressed me into my seat and filled my eyelids with lead ... Breathing was becoming increasingly more difficult ... " The two men could barely communicate with one another, scarcely uttering a few guttural grunts, wheezes and puffs under the immense loads.

The descent module's parachutes deployed as intended and it came down in the snow-covered Altai Mountains, about 830 km north of the Chinese border, landed on a slope and began rolling towards a sheer precipice. Thankfully, the parachute snagged on some vegetation. In temperatures of -7°C, the cosmonauts donned their cold-weather clothing and clambered outside. Lazarev feared they had landed in China – he had already radioed a request for tracking information during the descent, but was met with silence on the airwaves – and one story tells that he burned a pile of classified papers for a military experiment which he would have performed aboard Salyut 4. (It was apparently so secret that even *he* did not know its purpose!) In fact, their flight path passed to the north-west of Xinjiang, which had earlier entered the headlines when two Soviet helicopters landed there in error and their pilots were captured by a Chinese patrol. At length, Lazarev and Makarov encountered friendly locals who spoke Russian and a helicopter rescue team quickly made radio contact.

"Oleg calculated the landing place almost precisely," Lazarev said later of the flight engineer's navigational measurements during the descent. "We landed a bit to the side of the place he had indicated. It was painfully disappointing and somewhat unpleasant. We had prepared ourselves for the mission for so long, only to wind up like this ... The very fact of failure was quite discouraging." They had actually touched down close to the town of Aleysk, about 200 km south of Novosibirsk, in southern Siberia – well within Soviet territory – but it was the *next day* before the rescue team, battling the difficult terrain, chest-high snow drifts and altitude, reached them. Plans to drop physicians onto the mountain were called off by Lazarev, a veteran parachutist himself, on the grounds that it was too dangerous. Next morning, it was hoped to extend a ladder from a helicopter, but the instability of this option made it impractical; another effort led to a group of would-be rescuers getting themselves stuck in an avalanche. Finally, a civilian helicopter dropped a forest guide next to the two cosmonauts to render assistance, after which a military helicopter arrived and extracted all three of them.

Tass revealed the shocking truth the following day, when it announced that the booster had "deviated from the pre-set values" and "an automatic device produced the command to discontinue the flight". Vladimir Shatalov had, it seems, learned little from his 'tell-the-truth' conversation with Tom Stafford a few months earlier, for he quickly made a statement asserting that both men were in good shape and ready to fly another mission. Not surprisingly, this was far from reality. "Makarov says that under the G forces which they experienced, they could easily have first lost vision and then consciousness," wrote Hall and Shayler of the incident. "Although this did not happen, they experienced black-and-white vision and then tunnel vision." Makarov, it is true, *would* make a full recovery and undertake two further missions, but Lazarev suffered undisclosed internal injuries and never flew again. Rumour abounded in the days after the flight that both men had died in the accident; these rumours only subsided when Lazarev and Makarov were ordered to play *football* with some Americans to convince them that they were still alive ...

The cosmonauts' flight was effectively suborbital, lasting barely 21 minutes. With the ASTP joint mission looming in July 1975, some American congressmen were more vocal than ever before about the Soviets' safety record, prompting Konstantin Bushuyev to tell his NASA counterpart Glynn Lunney that Lazarev and Makarov had been launched atop an *older* version of the Soyuz booster. Would it happen again? "We were willing to bet that it wouldn't," observed Deke Slayton, "but not because we believed *that* excuse!"

Although Lazarev and Makarov's unlucky flight was never given a numerical designation by the Soviets, who referred to it simply as 'the 5 April Anomaly', it became known as 'Soyuz 18A' in the West because at 2:58 pm Moscow Time on 24 May the Soviets launched the backup crew of Pyotr Klimuk and Vitali Sevastyanov in what the Soviets designated Soyuz 18 and Westerners decided to call Soyuz 18B.

"The docking began on the 19th orbit," *Flight International* told its readers a week later, "after a 12 m/sec approach to the space station and took place with the two vehicles in the Earth's shadow. The final approach speed was 0.3 m/sec, but several minutes before docking the craft went out of radio contact with ground control and details of the operation's climactic moments were not available until the orbit had been completed." Fortunately, Klimuk handled the docking perfectly and he and Sevastyanov boarded Salyut 4 at around 12:30 am Moscow Time on the 26th. It might be expected that their goal was to complete a 60-day mission – which, indeed, they did – but at first the new mission was only scheduled for a month. When it became clear that Salyut 4 had not yet outlived its usefulness and also because no more Soyuz vehicles were available to transport additional crews, it was decided to extend their flight by an additional month.

In contrast to its predecessor the mission ran like clockwork, although it was reported that the cosmonauts needed fully ten days to acclimatise to the space environment. In view of this, managers decided that they should initiate their preparations for return to Earth ten days earlier than originally planned. Klimuk and Sevastyanov conducted one of the most comprehensive programmes of scientific and technical experiments ever attempted. The station's complement of ultraviolet, X-ray and infrared instruments yielded observations of objects in the constellations Scorpio, Virgo, Cygnus and Lyra. Photography of the entire European and Asian portions of the Soviet Union – including the eastern section of the Baikal-Amur railway – was undertaken and a series of navigational tasks were completed. Also, on the night of their arrival at the station, a laser beam was targeted at Salyut 4 as part of a high-accuracy satellite-tracking test. Unlike the activities of Popovich and Artyukhin a year earlier, this crew's daily work was readily described and easily available to Western observers. Phillip Clark noted that more than 14 days were devoted to work with the solar physics and X-ray equipment, 11 days to Earth-resources observations and photography, nine days to biomedical and physiological activity, six days to 'technical experiments' and two days spent on atmospheric studies.

As Klimuk and Sevastyanov's mission drew to a close, the joint ASTP flight was launched, marking only the second occasion – after 1969's troika mission – that three crews had been in space simultaneously. This must have come as a surprise to some in the West, for *Flight International* reported on 5 June that "it is now felt ... that the Russians will not allow Soyuz 18[B] to steal ASTP's thunder and that the flight will be terminated after about six weeks, just a week before the beginning of the joint mission ... " Quite the opposite occurred and the Soviets won a public relations bonanza of which even Tom Stafford and Deke Slayton were impressed: the ability to operate a complex long-duration station visit *and* an equally complex international docking mission *at the same time*. The resumption of missions in the wake of the Soyuz 11 tragedy had marked the introduction of the new flight control facility near Moscow. It was possible to operate Salyut 4 in parallel with ASTP by running it from the old facility at Yevpatoria.

When the Soyuz 18B crew finally landed at 5:18 pm Moscow Time on 26 July, they had spent 63 days in orbit. It was a new record for the Soviets ... though still three weeks short of the longest Skylab mission. Yet according to suggestions made by Clark in 1988, it is possible that Salyut 4 was now being pushed a little too far past its expected lifetime. "The Soviets have hinted," he wrote, "that there were problems with ... the environmental system which was being asked to support cosmonauts for two months *longer* than scheduled." Obviously, if they had made it to the station, Lazarev and Makarov would have completed their own 60-day mission early in June. By effectively extending Salyut 4's lifetime into late July, reports arose that Klimuk and Sevastyanov had to struggle to see through the portholes and even that the interior of the station was covered with mould. However, Clark cautioned that "Soviet statements ... do not confirm this situation".

Whatever the reality, a long and hard trail had been trodden since the end of the Sixties and the Soviets had dragged their space programme from the depths of despair and onto a road of recovery. In fact, Leonid Brezhnev's words, uttered shortly after the troika mission in October 1969, now seemed to have a ring of truth about them ... for the Soviet Union *had* now "approached the creation of long-term orbital stations" and it *had* now shown that their creation "with changeable crews" was "the main road for man into space". The United States had dreams of its own, it is true, and those dreams were ambitious ones: a reusable Space Shuttle capable of carrying up to seven people at a time *and* flying once every week. However, in the early summer of 1975, *that* dream was at least three years away and by the time the Shuttle finally flew, its reusability, its enormous payload capacity, its versatility and the frequency with which it could operate were balanced by a steady realisation that it was expensive, complex, the result of too many design compromises ... and intrinsically and fatally flawed. Some observers would wonder whether the United States had made the right decision: was flying a dozen or more times each year *really* a better alternative to the Soviet model of having cosmonauts *living* in orbit on a long-term space station for many months at a time?

When we look back at American achievements in the late Sixties and early Seventies, it could be argued that, despite the unquestionably vital role that the Shuttle has played in the construction of today's International Space Station, in some ways it was a sideways step which sent more people into orbit, and more frequently, but effectively limited itself to low-Earth orbit and delayed human exploration of the Moon by half a century. Its effort to 'internationalise' space by carrying scientists from different nations had already been done to a lesser extent by the Soviets as part of their Intercosmos project. Indeed, between 1978 and 1981, over half a dozen international cosmonauts – from Mongolia to Cuba and Vietnam to Romania – would have flown into orbit with Soviet crewmates.

Russia's intent to effectively colonise low-Earth orbit on a longer-term basis was prospering and would attain new heights as the disco-fever Seventies wore into the techno-pop Eighties. Cosmonauts would frequently spend four, five, six, then *seven* and even *eight* months in orbit, nearly tripling the endurance times achieved aboard Skylab. By the end of the Eighties, the number of international 'research cosmonauts' who had flown aboard Soviet stations had risen to more than a dozen: many from communist countries, to be fair, but a handful from capitalist and Western-aligned nations, too, including France and India. When the first elements of the International Space Station were placed into orbit in November 1998, the name for this American-led project was something of a misnomer; for the Soviets had operated not one, but *several*, 'international' space stations of their own ... *two decades earlier*.

Despite the promise and the abilities that the Shuttle afforded, the American dream in space quickly became one of faded nostalgia and efforts to reassert the bold dreams of the Sixties always seemed just beyond reach. There would be no more expeditions to the Moon, no space station and flights to Mars were out of the question. The Americans had closed down the production line for the Saturn V – a rocket which, despite its enormous cost, could have achieved so much more – and moved instead to what was promoted as a cheaper means of getting people into space. The Shuttle would gain respect and denigration and win itself both friends and foes during its 30-year monopoly of America's space programme.

In a sense, therefore, the summer of 1975 marked the end of an era for the United States and the beginning of one for the Soviet Union. The two Soyuz crews orbiting Earth that summer – one on a high-profile international mission, the other engaged in long-duration research aboard an aging space station – would open the door on half a decade dominated by Russian spectaculars. On the reverse side of the coin, the Apollo piloted by Tom Stafford, Vance Brand and Deke Slayton would be the last of its kind and would mark the onset of a long hiatus in American human space flight. Even when that hiatus was over, and the Shuttle finally flew, the chances of returning to the Moon or building a space station were slim ... and years away. In 1984, when Ronald Reagan set Space Station Freedom in motion, its completion was not expected for at least ten years; similarly, George W. Bush's 2004 pledge to return people to the Moon has been ingloriously abandoned by his successor, Barack Obama.

Astronaut John Young once called politicians "a strange bunch of critters" and he was right; for many successive administrations have shown little real support for space. Richard Nixon's short-sighted words to the Apollo 17 crew as they headed home from the Moon in December 1972 come to mind. It was highly unlikely, the leader of the United States announced, that astronauts would return to the Moon before the end of the century; effectively, he closed the door on any space dreams which might have been nurtured by a generation of youngsters. (One such youngster was an 11-year-old African-American boy named Barack Hussein Obama.) Yet, sadly, Nixon was right about the Moon, and so it is that in 2010, as the Shuttle enters the twilight of its life and completes the audacious challenge of building the International Space Station, it has proven an incredible workhorse ... but has left us no closer to planting bootprints back on lunar soil.

Barely a dozen men have trodden the flat plains and pristine foothills of Earth's closest celestial neighbour. The names of the *next* team of lunar explorers are unknown to us; they may be astronauts already or they may still be at school. Hopefully, they are, at least, *alive*. However, the fact remains that as humanity prepares to enter the second decade of the 21st century, we are *still* unable to leave low-Earth orbit. As I approach my mid-thirties, two things are abundantly clear to me: I am getting older and, year by year, I am losing more hair. One thing that is *not* clear is when I will see a human being standing on the surface of the Moon. I can only hope that we *will* return there, and perhaps visit Mars, too, in my lifetime.

4

Luna cognita

ASHORE ON A SILENT SEA

On the hot midsummer's evening of 20 July 1969, more than three and a half billion people occupied the blues, browns, greens and snow-covered whites of Planet Earth. Some lived at the bottom of oceans, others on remote islands, a few atop mountain peaks, some deep within valleys, ravines or canyons and millions in the endless hustle and bustle of cities; though all shared something in common, for all were the descendents of thousands of generations of human beings for whom Earth was – and continues to be – the only home they had ever known. Everything that makes us human is embodied in what the Apollo explorers could only compare to a beautiful, glittering marble in the unfathomable blackness and incomprehensible depth of space; a blob so incredibly fragile and delicate that it could be covered at will by an astronaut's thumb. Friend and foe, family and stranger, warmth and bitterness, kindness and hatred, war and peace, religion and reason: *all* resided there as they had done, unchangingly, for thousands of years.

But on *this* night, something *had* changed and something *was* different: something profound, unprecedented and quite remarkable had drawn human minds away from their daily labours, their trials and their tribulations, their petty squabbles, their newspapers and their coffee ... to three distant voyagers, who had travelled further by far than da Gama or Columbus or Cook or Lindbergh or even the ancient Polynesian navigators of the Pacific could have ever dreamed. For on this evening of 20 July 1969, the human race comprised three and a half billion souls on Earth ... *plus three*. One of those three, the gregarious Mike Collins, a man who described himself as lazy and ineffectual, a lover of good wine and a pruner of roses, waited anxiously in his command ship, Columbia, in orbit around the Moon.

The other two – the quiet, unflappable Neil Armstrong and the ambitious, outspoken Buzz Aldrin – were somewhere even more exotic. Breathing hard, their mouths dry from ingesting pure oxygen, they stood side by side in a cabin the size of a broom cupboard; a cabin whose monotonous grey hues possessed all the aesthetic beauty of a ship's boiler room. For now, though, their eyes were riveted on just one thing: the view. Each man peered inquisitively through a small triangular window directly in front of his face ... and beheld a stark scene that was stranger, more barren, more alien and more hostile than anything ever seen by human eyes. Like many explorers before them, Armstrong and Aldrin had become the first to make landfall on the shores of a distant sea, but with one exception: *this* sea – Mare Tranquillitatis, the Sea of Tranquillity – lacked water, air and the slightest trace of life. These two men had just achieved what humanity had spent thousands of years dreaming about: they had become the first of their kind to land on the Moon.

One of the strangest sensations was the feeling of weight. For the first time in four days, since leaving Earth, Armstrong and Aldrin could *feel* something of their body weight ... for the Moon's gravity, though weak, was nevertheless present. It was barely a sixth as strong as it would be on Earth, but for now it enabled Aldrin to perform a rite which was important to him: a celebration of Holy Communion. Opening a personal stowage pouch, given to him by his Presbyterian minister, Reverend Dean Woodruff, he pulled out a small flask of wine, a tiny chalice and some wafers and placed all three onto the computer keypad before him. Perhaps a little nervousness filled him, for the Apollo 8 crew had been criticised for broadcasting the opening lines of Genesis to a listening world in December 1968. Deke Slayton had warned Aldrin not to make any religious observance over the radio waves and he knew he must be careful with his choice of words.

"This is the LM pilot speaking," he announced at 6:57 pm Eastern Standard Time (EST), some two hours after landing. "I'd like to request a few moments of silence. I'd like to take this opportunity to ask every person listening in, whoever and wherever they may be, to pause for a moment and contemplate the events of the past few hours and to give thanks in his or her own way." Without uttering another sound, keeping his religious faith to himself, Aldrin upturned the flask and watched as the wine, in one-sixth gravity, curled its way with the sluggishness of slow-flowing syrup into the chalice. Next, again in silence, he read a passage from the Book of John (15:5): "I am the vine and you are the branches / Whoever remains in me and I in him will bear much fruit / For you can do nothing without me."

The words, of course, form part of the traditional Presbyterian ceremony of Communion, but Aldrin knew in the days before launch that any event with a bias towards religion was bound to provoke some controversy. After Apollo 8, the celebrated atheist Madalyn Murray O'Hair had sued the federal government over the Bible readings and, in the weeks preceding Apollo 11, she had accused NASA of purposefully withholding facts about Neil Armstrong's atheism. For the astronauts themselves, it was a private act of personal religious devotion and O'Hair's accusations were seen by many as the actions of someone with little better to do with her time. For his part, Armstrong told Aldrin before the mission that he had no problem with the Communion. "I had plenty of things to keep busy with," he related to James Hansen. "I just let him do his own thing."

Since landing, there had indeed been much to do. Of paramount importance was making sure that Eagle, their home on the Moon for the next day or so, was undamaged. "If we had problems that indicated that it was not safe to continue staying on the surface," Armstrong told Hansen, "we would have had to make an immediate takeoff." For this reason, NASA had devised a system of three early times that Eagle could liftoff and achieve a satisfactory trajectory to rendezvous with Mike Collins. The first, known as 'T1', came two minutes after touchdown, T2 followed eight minutes later and T3 would occur after another two hours, when Columbia completed another circuit of the Moon. Each 'T' would be declared after a consensus had been reached on the performance of the LM's systems. Of course, emergencies could arise *outside* these times, but it would be up to Armstrong, Collins and Aldrin to find some means of bringing their respective craft together for docking.

Precisely on time, two minutes after landing, at 4:19 pm, Flight Director Gene Kranz's team declared 'Stay for T1' and Charlie Duke, the capcom, passed the welcome news up to the astronauts. Next came a second confirmatory call: Stay for T2. So far, so good, it seemed; Eagle's performance was satisfactory, her systems seemed undamaged and she had not touched down on an unacceptably steep slope or partially in a crater. In fact, Armstrong's main worry was that too much heat and pressure may have been building up in her fuel lines – it was, after all, early morning, local lunar time – and hydraulics experts had advised him prior to the mission of this possibility. "If we closed all valves and trapped fluid in certain lines," Armstrong told Hansen, "then we were sitting on a two-hundred-degree surface of sunlight with lots of reflected heat coming up towards the bottom of the LM and it's heating up the pipes. The fluid pressure might really build in that time and then we'd have a problem ... [but] it wasn't an uncontrollable situation. We had a couple of options ... so we were not too concerned about it."

Sure enough, shortly after touchdown, a sharp increase in pressure became apparent. Armstrong and Aldrin, as planned, vented Eagle's fuel and oxidiser tanks but the evaporation of residual propellant caused it to continue rising. Mission controllers considered it a dangerous situation, fearing that a line or tank could burst, spraying the descent engine with fuel and perhaps – however unlikely in a vacuum – starting a fire. At length, the continued venting calmed the situation down. Armstrong was unruffled. He felt the *worst* that could possibly have happened was a ruptured fuel line ... and if *that* happened, it would not seriously affect the rest of their mission, because the descent stage's job was now done. In order for them to return to orbit, only the ascent stage *had* to work.

The astronauts' next step was to run through a complete dress rehearsal for their liftoff from the Moon, scheduled to take place the following day. "This required aligning the LM platform," Armstrong recalled, "which was a first because no one had ever done a *surface* platform alignment before. We used gravity to establish the local vertical and a star 'shot' to establish an azimuth; in that way, we got the platform aligned and ready for takeoff." Despite this being a simulation, the criticality of the following day's events meant that Armstrong and Aldrin and the flight controllers in Houston treated it exactly as they would the real thing. By the time they were finished, Columbia and Mike Collins had passed overhead and the final early stay time, T3, was declared. If there was any possibility that the astronauts could relax and unwind on this utterly alien and totally inhospitable world, it came after the announcement of T3, because they could now plan ahead to the most anticipated part of their mission ... the world's first Moonwalk. Yet no one knew precisely where Eagle had landed. The man best-placed to find out exactly *where* Armstrong and Aldrin would be walking on the lunar surface, Mike Collins was unable to see anything but craters, even when he was using Columbia's 28-power sextant to inspect the position identified by flight controllers on the basis of radio tracking of the LM during its descent.

Collins had lost sight of the lander as a "minuscule dot" about 70 km from the landing site, when it seemed to fade into nothingness against the backdrop of craters. Although the sextant was a powerful tool, with a high magnification, it also had a very narrow field of view, akin to looking down a gun barrel, which meant that finding Eagle in the limited time available during each orbital pass was virtually impossible. He never did find the lunar module during *any* of his orbits and, whilst the exact landing co-ordinates mattered little to the geologists, for *they* were happy that a landing had been effected *anywhere* on the Moon, it did matter to trajectory planners thinking ahead to the rendezvous. As many as nine more lunar missions were on the cards, each of which would require setting the LM down at a precise point. For Apollo 12, Pete Conrad would have to land on the Ocean of Storms within walking distance of Surveyor 3, in order that he and Al Bean could retrieve parts from it. Some planners were already talking of subsequent landings in the lunar mountains, in hard-to-access craters and even on the unseen far side of the Moon.

After the confirmation of T3 and eating a meal, the conservative flight plan called for Armstrong and Aldrin to take an abbreviated, four-hour nap before beginning preparations to venture out onto the lunar surface; in fact, virtually everything in their planned 21 hours on the Moon had been timed and practiced on Earth. There would be two hours and four minutes for post-landing checks, the flight plan decreed, followed by a 35-minute lunch break, then a four-hour sleep period, an hour-long breakfast, two hours of EVA preparations, then the two-and-a-half-hour Moonwalk itself, and after ingress there would be a further 90 minutes of housekeeping work inside Eagle, then another five hours of sleep and finally a little over two hours to eat and prepare their craft for ascent. To ensure that neither man forgot what they were to do whilst outside, they even had timed checklists sewn on the gauntlets of their gloves.

Having said this, it had long been recognised that expecting Armstrong and Aldrin to rest so soon after landing on the Moon was about as likely as a child returning to bed on Christmas morning. In the weeks before launch, the idea of skipping this brief sleep period and proceeding directly into what was formally termed 'EVA Prep' in NASA-speak had been discussed and when Armstrong formally requested it shortly after 6:11 pm, Charlie Duke came back almost immediately with Mission Control's approval.

"Houston, Tranquillity?"

"Go, Tranquillity. Over."

"Our recommendation," Armstrong began, "at this point is planning an EVA, with your concurrence, starting about 8 o'clock this evening, Houston time [an hour earlier on the east coast]. That is about three hours from now."

"Stand by," replied Duke, turning to Flight Director Gene Kranz.

"Well, we'll give you some time to think about that," added Armstrong.

Notwithstanding the 2.6-second lag as radio signals cracked back and forth across the translunar gulf, Duke's approval reached the ears of the astronauts just *nine seconds* after Armstrong made the request:

"Tranquillity Base, Houston. We thought about it. We will support it. We're Go at that time. Over."

It was not simply a case of excitement which prompted Armstrong to ask for the advancement of the EVA; it also made sense from a public relations standpoint. To the mission commander, the descent and landing were the most challenging phases and an early sleep period to ensure that both men were rested for the excursion outside was certainly wise, but the early sleep period was inserted into the flight plan for another purpose. "If we scheduled the surface activity as soon as we could after Columbia's first revolution," Armstrong told Hansen, "and then didn't make it on time, the public and the press would crucify us. That was just the reality of the world, so we tried to finesse things by saying that we were going to sleep and then we would do the EVA." In the real world, behind closed doors, the crew had already told Deke Slayton and Chris Kraft that they would ask for an early EVA if all was well. "We knew it would create a change that people weren't expecting," recalled Armstrong, "but we thought that was the better of the two evils."

As a result, Apollo 11's two-part press kit, published by NASA on 6 July 1969, mentions only that the Moonwalk was scheduled to begin ten hours after touchdown, *after* a meal and sleep period. Almost immediately after the meal had been gulped down and Aldrin had celebrated Communion, they set about the laborious process of donning the parts of their space suits which would give them sure footing and keep them alive and comfortable on another world. This took longer, and proved much more difficult, than it had during their weeks of training in the lunar module simulator; for on Earth they were able to suit-up in a clean and organised cockpit, rather than one filled with checklists, data, food packages, a monocular, a stopwatch and other things. It took the two men an hour and a half just to prepare their equipment, then three hours to don the gear: a pair of massive rubber-soled lunar overshoes, the Portable Life Support System (PLSS) backpack, double-locked oxygen hoses, umbilicals to circulate cooling water through tubes in their underwear, the large outer helmet with its polycarbonate shell, gold-plated Sun visor and pull-down smoked visor, a chest-mounted control unit, and so on. At length, they snapped on their gloves and switched on the fans and pumps in the PLSS; instantly, they could hear the hum of the high-tech machinery and the whoosh of oxygen past their faces that would keep them alive in the most hazardous environment ever visited by human beings. The suit was essentially a miniature spacecraft in its own right and on the missions that would follow it would keep astronauts alive as they ventured several kilometres away from their landers. Rusty Schweickart had tested it on Apollo 9 and, although it was a clumsy ensemble, Neil Armstrong, standing in Eagle's cabin in lunar gravity, found it surprisingly comfortable. By the time they were ready to depressurise the cabin, they felt like a pair of fullbacks trying to change positions inside a Cub Scout tent. This reinforced in Armstrong's mind, at least, the rationale for letting him go outside first. "It was pretty close in there with the suits inflated," he admitted to James Hansen, and both men found it difficult to avoid bumping into each other.

The depressurisation of Eagle's cabin through a small vent in the main hatch progressed more slowly than expected. When the reading reached almost zero, Aldrin eagerly tried the hatch handle ... but it stayed firmly shut. "Neither man wanted to tug on the thin metal door for fear of damaging it," wrote Andrew Chaikin. "Finally, Aldrin peeled back one corner to break the seal; that did it." Instantly, the last vestiges of air rushed out in a flurry of ice crystals. Now the men were exposed to vacuum and dependent upon their suits and backpacks to keep them alive. All they could hear from within their high-tech cocoons was the harshness of their own laboured, excitement-tinged breathing.

As Aldrin pulled the hatch fully open, Armstrong clumsily dropped to his knees, his head facing the rear of the cabin, his feet inside the yawning square gap that now literally marked the threshold to what had been only a dream for thousands of years. Following verbal cues provided by Aldrin, he slid steadily backwards – "the backpack extended quite a long way above your head" and "you needed to get quite low, but then you also had things on the front of you that you didn't want to damage" – onto Eagle's tiny porch.

Armstrong found the procedure fairly straightforward, just like training, but in doing it for real he forgot to pull the lanyard to deploy the Modular Equipment Storage Assembly (MESA), which would expose a black-and-white camera to transmit images of his descent to the surface. The capcom on duty for the Moonwalk, rookie astronaut Bruce McCandless, reminded him and Armstrong moved back and pulled the lanyard. From his perch on the right-hand side of Eagle's cabin, Aldrin closed a circuit breaker and a fuzzy image appeared on television screens in Mission Control.

The image is poor and almost ghostly in its quality, yet it is the only record of our species' first footsteps into the Universe around us. Armstrong is difficult to see at first, within the deep shadow of the lunar module, although the bright, sunlit plain of Tranquillitatis and the unfathomably black sky beyond is clearly visible. After a few seconds, the observer can see his arms, both hands tightly gripping the ladder, and the silhouette of a left foot seeking the next rung. The view, indeed, may not have been spectacular or stunning, or even in *colour*, but what the image *meant* – what it *represented* – was utterly profound.

Descending the ladder was by no means dizzying or precarious for Armstrong; rather, he felt so light and fell so slowly and gracefully in one-sixth gravity that there seemed no danger. Eagle had touched down extremely gently on the Moon and its landing legs had not compressed as much as they should have done; as a result, the last rung was a good metre or so above the bowl-shaped footpad and Armstrong appeared to *drop*, seemingly in slow motion, down into it. As McCandless and the world listened, he briefly sprang back up to the last rung – just to make sure he could do it – and then returned to the footpad. "It takes a pretty good little jump," he informed Aldrin. Now came the opportunity to describe for a watching audience of millions the stark landscape around him: "The surface," he began at

Obviously, no one was on the surface to capture the fullness of Neil Armstrong's descent to the alien soil of the Sea of Tranquillity. However, this image of Buzz Aldrin squeezing his way through Eagle's hatch and onto the porch – a little like being born, one astronaut would recount – offers the reader an impression of what those electrifying moments must have been like.

10:55:38 pm EST, "appears to be very, very fine-grained as you get close to it. It's almost like a powder ... " Forty seconds later, the first man set foot on the Moon.

SMALL STEPS

Back on Earth, an audience of six hundred million or more watched and listened in awestruck astonishment. CBS News anchorman Walter Cronkite was almost lost for words. "Armstrong is *on* the Moon," he intoned. "Neil Armstrong ... a 38-year-old American ... standing on the surface of the Moon. On this July 20th, nineteen hundred and sixty-nine ... " What amazed him equally as much as the technical achievement was the ability to get the television pictures and sound back to Earth, in more-or-less real time, enabling terrestrial audiences to *see* what was happening. After all, when Christopher Columbus landed in the New World, he had barely a handful of locals for his audience and Queen Isabella of Spain did not learn of his achievement for six months. Years later, Cronkite considered the pictures to have been just as much a miracle as the landing and the Moonwalk themselves. When Armstrong viewed them after the mission, their appearance "gave it sort of a superimposed unreal image" and even offered some ammunition to those cynics who would later argue that the whole thing was a *Capricorn One*-style set-up.

To be fair, there was little that either Armstrong or Aldrin could have done to improve the grainy television pictures. They could possibly have moved from a small antenna to the larger S-band erectable dish, stored aboard Eagle, which might have enhanced the quality a little, but the fact remains that *these images* are our species' record of a truly remarkable moment in history. No previous event of *this* magnitude had ever been captured for a worldwide television audience. That alone makes them both iconic and utterly priceless.

What happened next would be timed, according to NASA's official flight transcript, as having occurred four days, 13 hours, 24 minutes and 20 seconds into the Apollo 11 mission – for at 10:56:15 pm EST, Armstrong lifted his left boot over the rim of Eagle's footpad and planted it firmly onto the lunar soil. Seconds later came the immortal first words on a world other than our own, taught to virtually every schoolchild in virtually every science lesson in virtually every country across the world: "That's one small step for man ... one giant leap for mankind".

In years to come, observers and historians would wonder about Armstrong's words. The obvious question was: *where* had they come from? Even during the journey to the Moon, when Collins and Aldrin asked him pointedly, Armstrong had told them that he was still mulling over what he would say. In his mind, it mattered relatively little, though he appreciated that it mattered greatly to others. "My own view," he told James Hansen, "was that it was a very simplistic statement: what *can* you say when you step off something? Well, something about a step. It just sort of evolved during the period that I was doing the procedures of the practice takeoff and the EVA prep. I have never thought that I picked a particularly enlightening statement ... it was a very simple statement."

Simple, indeed, but significant, historic and highly appropriate. Coming from the lips of a man who, despite his intelligence, was not known as a wordsmith, for whom silence was a response to a question and for whose wife the word 'no' represented an argument, it is interesting to explore Armstrong's motivation for saying what he did. It *was* a small step ... but an important one, too, and some have speculated over the decades that its root may have been an excerpt from J.R.R. Tolkien's *The Hobbit*, in which Bilbo Baggins leaps over the villainous Gollum; indeed, Armstrong would move to a farm in Lebanon, Ohio, in 1971 and call it 'Rivendell' after the elvish valley of Lord Elrond. Unfortunately, this explanation for the words does not appear to have been accurate, for Armstrong did not even read any of Tolkien's work until *after* his return from the Moon. Other suggestions came from the piles of mail he received in the months before launch – some of whose senders offered him biblical passages to read and lines from Shakespeare – or, perhaps, from within NASA itself. Certainly, Willis Shapley, one of the agency's senior administrators, had written in an April 1969 memo that a series of symbolic activities should be carried out on the Moon to highlight a *historic step forward for all mankind*. Some have speculated that this memo worked its way through George Mueller and Deke Slayton to Armstrong, but the man himself had no memory of ever seeing the memo and the mystery about where he got his inspiration endures to this day.

Perhaps Armstrong actually enjoyed some of the mystery and the ambiguity, too, for scholars have questioned over the years whether or not his statement was meant to read 'a man' or simply 'man'. In his analysis of the words, Andrew Chaikin believed that the indefinite article 'a' was forgotten and *not*, as some have suggested, lost in the crackly radio transmission to Earth. When one listens to the actual recording, a distinct heightening of pitch is detectable in Armstrong's voice; although this may reflect the missing 'a', it seems more likely that it was simply forgotten. (Armstrong, for many years, has insisted that the 'a' should be inserted in parentheses.) Four decades later, in 2006, journalist Peter Shann Ford claimed to have located a missing 'a' in the waveform of Armstrong's voice; he theorised that the word *had* been said, but had been lost due to the limitations of contemporary transmission technology. Nevertheless, grammarians have countered that the word 'man', whilst normally referring to the species, *can* also refer to an individual, even *without* the indefinite article. The opinion of this author is that the 'a' *must* have been intended: when one bears in mind that the second half of Armstrong's statement included the word 'mankind' anyway, the use of 'man' as a synonym for 'mankind' in the first half would have been wholly unnecessary. To me, it seems probable that Armstrong first referred to his *own* step (intending to say "for *a* man") and secondly to the giant leap made by all of humanity ("for *mankind*").

Today, of course, such trivia is purely academic, but the fact that such debates have continued for so long illustrates the value that we humans attribute to our symbolic gestures, our words and our deeds. Aside from the appropriateness of his statement, Neil Armstrong was an engineer and a test pilot and probably did not realise the extent to which the public *expected* him to utter a profound and thought-provoking comment. After all, even into the modern era, the Moon is regarded by many cultures as an object of wonder, a protective influence from enemies, a virgin land, a final resting place of ancestors ... and, for some, an intensely holy place. By expecting Armstrong to treat his first steps upon it with dignity, reverence and respect, our modern society was actually hailing age-old principles; that, beneath all the technology and the techno-babble, the *human* journey – our species' first-ever pilgrimage to the Moon – was of equal, if not greater significance.

In this way, Armstrong, the quiet man with absolutely no ego, *was* the right person to be First. His first few minutes on the surface were spent testing his weight and describing as best he could the consistency of the lunar soil, which he could pick up loosely with his toe and which adhered to the soles and sides of his boots like layered charcoal. His footprints only indented the surface a matter of millimetres, but left clear impressions, and moving around in one-sixth of terrestrial gravity felt perfectly natural and comfortable. On Earth, he and Aldrin had rehearsed every minute of their planned two-and-a-half-hour Moonwalk in a box of pulverised lava at the Manned Spacecraft Center in Houston, but the real thing was far easier: Armstrong's mother, watching on television, would later describe his motion as "buoyant" and he seemed to be "almost floating as he walked".

One astrophysicist who was left somewhat red-faced by the ease with which the First Man was able to move around was Thomas Gold, an Austrian-born academic from Cornell University. Several years earlier, as a member of the President's Science Advisory Committee, he had argued that the Moon was covered by a layer of rocky powder, perhaps several metres thick, into which the lunar module and its astronauts might sink. With the flights of unmanned Surveyor landers to the Moon, Gold's prediction was proven wrong and he adjusted his theory to determine that the astronauts' boots would sink just 3 cm into the surface. When the Apollo 11 crew returned with lunar soil which *did* prove to be powdery in nature, Gold felt vindicated to an extent, although he continued to be ridiculed by some areas of the scientific community.

Moving around Eagle, and thankfully *not* sinking up to his knees, Armstrong was able to make a few general observations about its appearance and state. The descent engine, he told Mission Control, had not left an appreciable crater, but it was possible to discern a few erosive rays on the surface, caused by the impulse just prior to touchdown. Next came the opportunity to begin photographing the landing site and Aldrin, in Eagle's hatch, hooked the large Hasselblad 70 mm camera onto a device known as the Lunar Equipment Conveyor (LEC). The latter had been nicknamed 'the Brooklyn Clothesline' by the astronauts and provided a means of transferring equipment down to the surface at the start of the excursion and subsequently lifting rock boxes up to the ascent stage. In his conversations with James Hansen, Armstrong recalled that, during training, he and Aldrin had evaluated moving equipment, cameras, geology tools and rock boxes up and down the ladder *by hand* and found it "very cumbersome" since it was "difficult to manhandle all that stuff and get it in the proper position so that the other person ... could pick it up".

Unhooking the Hasselblad from the LEC, Armstrong affixed it to a bracket on his chest-mounted control unit. He had designed the bracket himself, realising that the sheer size of the camera and the fact that it was *not* automatic – there was a motor to wind the film, but the shutter speeds, f-stops and focus were *all* manual – made it impractical to operate with one hand. He certainly did not want to set it down in the abrasive lunar soil whenever he needed to work on something else, so the bracket came in exceptionally useful. Unfortunately, it would also explain why so few images were taken of Armstrong *himself* on the lunar surface.

After activating the Hasselblad, he quickly became so engrossed in shooting a panoramic sequence of images of the local landscape that he almost neglected a key item on his checklist ... the so-called 'contingency sample'. It had long been recognised that he should obtain at least a tiny sample of lunar soil, just in case an emergency develop in the first few minutes of the Moonwalk which would oblige him to retreat into Eagle. Bruce McCandless reminded him a couple of times, as did Aldrin. The device used to obtain the sample looked similar to a dog's pooper-scooper – a collapsible handle with a removable bag on the end – and Armstrong found that the loose consistency of the soil made it surprisingly easy to pick up. However, when he attempted to dig a little deeper, the scooper met resistance.

One of the most iconic images from the Apollo 11 mission: Buzz Aldrin's pristine bootprint in the ancient lunar dust of Tranquillity Base.

"Looks like it's a little difficult to dig through the initial crust," observed Aldrin.

"This is very interesting," Armstrong responded. "It's a very soft surface, but here and there where I plug with the contingency sample collector, I run into a very hard surface ... but it appears to be a very cohesive material of the same sort. I'll try to get a rock here ... just a couple ..."

By now, Armstrong had been outside for around ten minutes and paused again to simply marvel at the sheer beauty of his surroundings. It was untouched. Not a whip of wind disturbed the soil which had lain undisturbed, quite possibly for millions of years; not a tree or blade of grass or *anything*, in fact, of colour or life could be seen; and everywhere the dominant colours were greys, browns, tans and, as Armstrong looked towards the Sun, an intense, chalky white. "It has a stark beauty of its own," he remarked, to no one in particular. "It's much like the high desert of the United States," maybe thinking back to his experiences as a test pilot flying over Edwards Air Force Base in the late Fifties. "It's different, but it's *very* pretty out here."

Armstrong detached the sample bag, stuffed it into the pocket on the left thigh of his space suit and, almost without a second thought, threw the handle away in a long, lazy sidearm movement ... and it glided with the utmost grace on a perfect, impossibly long trajectory on an airless world, before coming to rest.

In these few moments of reflection before Aldrin joined him on the surface, Armstrong distinguished himself as one of the few pilot-astronauts to develop an active interest in geology. "Be advised," he reported, "that a lot of the rock samples out here – the *hard* rock samples – have what appear to be vesicles in the surface. Also, I am looking at one now that appears to have some sort of phenocrysts." *This* proved a significant piece of evidence to enthuse the hot-Mooners – those scientists who argued that ancient volcanism *was* responsible for the lunar maria – for vesicles are the spherical imprints of gas bubbles which form during the steady cooling and hardening of lava and phenocrysts are small crystals embedded in a fine matrix. Later in the Moonwalk, Armstrong would change his mind about the vesicles and decide, correctly, as it turned out, that what he was seeing were tiny craters made in the rocks by small impacts. Still later, he recognised and collected some genuinely vesicular basalts. Indeed, the fact that basalts – *volcanic* rocks – were brought back to Earth by Armstrong and Aldrin from the Sea of Tranquillity proved extremely encouraging for the hot-Mooners.

Sixteen minutes into the EVA, it was finally Aldrin's turn to venture outside and this gave Armstrong and the chest-mounted Hasselblad the opportunity to acquire some of the most iconic images from the mission. These show Aldrin squeezing his way through the square hatch – "making sure not to lock it on my way out!" – and kneeling on all fours on the porch, then gingerly descending each rung down to the footpad. (Locking themselves out, realistically, was an unlikely prospect, since the hatch could be opened from the outside as well as the inside, and Aldrin's gallows humour was not lost on his commander. Having said this, it was known that materials inside the cabin would continue to de-gas and cause pressure to build; furthermore, leaving the hatch partly open helped to minimise thermal stress from sunlight.) When he finally set foot on the Moon, even Dr Rendezvous was spellbound by the view. It was a view that he managed to verbalise in just two words: Magnificent desolation.

"Nothing prepared me for the starkness of the terrain," he recalled later. "It was barren and rolling and the horizon was much closer than I was used to. Earth's diameter is such that its inhabitants have no personal awareness of the curvature; it's easy to understand why, for centuries, it was believed to be flat ... but on the smaller Moon, my impression was that we were on a ball, or on the knoll of a hill that extended more than 2 km and was neatly rounded off. I even felt a bit disorientated because of the nearness of the horizon ... "

Moving around on the surface to assess his mobility and stability was one of Aldrin's early tasks and it was one for which both men had trained exhaustively before launch, on the sandbox in Houston and aboard a KC-135 aircraft which flew parabolas to simulate one-sixth gravity. As they walked on the regolith – as geologists call the rocky fragments that form the lunar soil – Armstrong found that the most comfortable and 'natural' gait was a sort of loping motion, in which he alternated feet, pushed off with each step and floated ahead, before planting the next foot. Others included a kind of 'skipping stride' and a rather playful 'kangaroo hop'. Although the *weight* of their backpacks was reduced by five-sixths on the Moon, its effect on their balance meant that they were always slightly pitched forward as they walked; and when Armstrong jumped he felt a tendency to tip over backwards as soon as he landed. They had to take care in turning and drawing to a halt. "I noticed immediately," Aldrin recounted, "that my inertia seemed much greater. Earthbound, I would have stopped my run in just one step ... an abrupt halt. I immediately sensed that if I tried this on the Moon, I'd be face-down in the lunar dust. I had to use two or three steps and sort of wind down. The same applied to turning around ... on Earth, it's simple, but on the Moon, it's done in stages."

Whilst performing these mobility evaluations, Aldrin set his own record – by becoming the first man to urinate on the Moon. "My kidneys, which have never been of the strongest," he explained later, "sent me a message of distress. Neil might have been the first man to step on the Moon, but I was the first to pee in his pants on the Moon." Fortunately, Aldrin was linked up to his suit's urine-collection device, but with the whole world watching, he was amused by the fact that "*I* was the only one who knew what they were *really* witnessing!"

Having assured themselves of a more-or-less solid footing on alien soil, the astronauts' next task was to unveil a commemorative plaque on the strut of the lander that held the ladder. At 11:24 pm EST, less than half an hour after setting foot on the surface, Armstrong described the plaque to his television audience: "First, there's two hemispheres, one showing each of the two hemispheres of the Earth. Underneath it says, 'Here men from the Planet Earth first set foot upon the Moon, July 1969 AD. We came in peace for all mankind.' It has the crew members' signatures and the signature of the President of the United States." Other objects left behind on the surface included a small silicon disk, bearing statements from Dwight Eisenhower, John Kennedy, Lyndon Johnson and Richard Nixon, together with tidings of goodwill from more than 70 other heads of state. Pope Paul VI had quoted the Eighth Psalm and, touchingly, the astronauts left medals and shoulder patches in memory of fallen comrades and adversaries in the exploration of the heavens: Yuri Gagarin, Vladimir Komarov, Gus Grissom, Ed White and Roger Chaffee.

As Apollo 11 was an *American* venture, and paid for by the *American* public, but one undertaken in the name 'of all mankind', the problem of what kind of flag to plant on the Moon arose with frequency in the months before launch. Some felt that the flag of the United Nations was appropriate as an endorsement that the United States was neither directly or indirectly laying any sort of claim to the Sea of Tranquillity, but others argued with equal vigour for the Stars and Stripes. At President Nixon's inauguration six months earlier, he had spoken of going "to new worlds together ... not as new worlds to be conquered, but as a new adventure to be shared". Was he hinting that a United Nations flag should be raised on Apollo 11? Some spectators believed so, and it was perhaps with this in mind that in February 1969 newly appointed NASA Administrator Tom Paine formed a Committee on Symbolic Activities for the First Lunar Landing to determine how one of the most historic events in human history *should* be marked. The committee heard convincing arguments in favour of a UN flag and in favour of depositing a collection of miniature flags of *all* nations, but finally it decided that a Stars and Stripes would be erected. To be fair, many asked the inevitable question: Why not? The American public, after all, had footed the $25 billion bill for Apollo and *why shouldn't they* mark the triumphant landing with a symbol of their national pride? The problem was that 1967's Outer Space Treaty had expressly forbidden its signatories from making territorial claims on the heavens or any celestial bodies, including the Moon. Any attempt to impose sovereignty or any form of 'occupation' was out the question. The Stars and Stripes *would* be raised on the Moon as nothing more than a symbolic gesture.

Certainly, opinion within United States senior politics seemed to underline that many congressmen and senators supported the notion of planting the Stars and Stripes on the lunar surface ... not only on this mission but also on those that would follow. When NASA formally notified members of Congress on 10 June 1969 that it intended to raise the national flag on the Moon, its appropriations bill for the next fiscal year was immediately approved. When the final version of the $3.7 billion bill was agreed by a House and Senate conference committee on 4 November, one provision stated that "the flag of the United States, and no other flag, shall be implanted or otherwise placed on the surface of the Moon, or on the surface of any planet, by the members of the crew of any spacecraft ... as part of any mission ... the funds of which are provided entirely by the Government of the United States ... " It was, indeed, a symbol of national pride.

For something which, in essence, was relatively simple, several months of work and planning went into preparing the Apollo 11 flag for its mission. Bob Gilruth, head of NASA's Manned Spacecraft Center, approached his technical services chief Jack Kinzler for advice; the latter proposed a full-sized flag which could be deployed using a specially designed pole. Together with his colleague Dave McCraw, Kinzler added a horizontal crossbar to create the illusion that the flag was flying in the 'breeze' of an airless world. The pole would be a two-part, fold-out affair, complete with telescoping crossbar, easily erectable by two men clad in bulky space suits and cumbersome pressurised gloves.

The flag itself measured 1.5 m wide by 0.9 m high and is said to have been purchased from the Government Stock Catalogue for the grand sum of $5.50. It was altered by having a hem sewn across its top to secure the crossbar and a loop sewn around the bottom of the flag would hold it onto the pole. Armstrong would then deploy it on the lunar surface by extending the crossbar and raising it firstly to a position just above 90 degrees, then lowering it such that it was perpendicular to the pole, where a catch prevented the hinge from moving. The upper portion would then slip into the top of the pole, which the men had driven into the regolith with a geological hammer. A red ring, painted near the base of the pole, allowed the astronauts to judge that it had been hammered to a sufficient depth into the soil. During its journey to the Moon, the flag, pole and crossbar were attached to the left side of Eagle's ladder, covered by a shroud to guard against high temperatures from the descent engine.

Armstrong's perspective was that he did not want it to stand out or be put into a rigid framework. "I thought the flag should just be draped down, that it should fall down the flagpole like it would here on Earth," he told James Hansen. "I soon decided that this had gotten to be such a big issue, outside of my realm and point of view, that it didn't pay for me to even worry about it." In truth, when it *was* finally assembled on the surface, the flag looked somewhat odd – 'crinkled', even – and the artificial steps which had been taken to make it appear to 'fly' on a world with no breeze certainly provided plenty of ammunition for those cynics who had convinced themselves that the Moon landing had been faked. When the time came to set up the flag, in some eyes it hardly seemed worth the effort: firstly, the crossbar would not telescope out properly, then the flag refused to lie flat or fully stretched and finally the pole would not penetrate far enough into the soil. For a relatively simple act, the problem was magnified because, by this point, the men had set up a portable television camera and more than a fifth of the world's population was watching their every move! Surely there could be nothing worse than the American flag collapsing into the lunar dust in front of a live audience of six hundred million people ...

Thankfully, the flag remained upright and Armstrong was able to shoot a picture of Aldrin rendering a smart military salute, against the desolate backdrop of the mare and an incomprehensibly black, totally starless sky. *This* picture, and a few others, proved to be among the most memorable from the mission ... but one crucial photo-opportunity had been missed: for when the images from the Moon's surface were processed, there were *plenty* of Aldrin, but only a *handful* of Armstrong himself. It is an omission which James Hansen has described as "one of the minor tragedies of Apollo 11". In his biography of Armstrong, Hansen noted that, aside from one famous image of Aldrin with the mission commander reflected in his visor, there are only a few frames in which show the First Man on the Moon: some of them underexposed, others showing him from the back or in shadow ... and one quite remarkable shot from a movie camera inside Eagle, released to much public fanfare in June 2009.

This latter image had actually been found by Andrew Chaikin more than two decades ago, whilst conducting research for his seminal work *A Man on the Moon*. It had been captured shortly before Aldrin ventured outside and shows Armstrong in the midst of collecting his contingency sample; therefore, it must have been very early, perhaps only ten minutes or so, into humanity's first Moonwalk. Not only does it show a full-body view of Armstrong, with the collapsible handle of the contingency sample collection tool at his feet, but the observer can actually see the First Man's *face*. Facing the shadow of the lunar module, he had temporarily raised the faceplate of his visor … and, within, it is just barely possible to see a flushed, somewhat puffy countenance, a pair of microphones to either side of his mouth and the black-and-white communications skullcap known as the 'Snoopy hat'. In Chaikin's mind, *this* was by far the best frontal image of Armstrong on the Moon. "There were other pictures," he explained, "taken with the Hasselblad camera … but none of those photos shows him from the front in very good light. Even though the Hasselblad photography is inherently higher resolution than [the] 16 mm movie footage still, in order to get the best view of Armstrong on the Moon, we have to go to [that film]."

Over the years, outrageous claims have been made that Aldrin 'intentionally' avoided taking direct photographs of his commander on the Moon, with some even pointing to continued bitterness over the First Man decision. In reality, of course, both men were outside Eagle for little more than two and a half hours and virtually every minute of that time was spent on assigned tasks: getting the contingency

In one of very few shots of the First Man on the lunar surface, Neil Armstrong works in Eagle's shadow. Note the United States flag and the solar wind experiment, set up earlier in the Moonwalk.

sample, unveiling the plaque, erecting the flag, deploying a pair of instruments for the Early Apollo Scientific Experiments Package (EASEP) and conducting geological inspections and taking specimens. They were *not* there to 'smell the roses', said Aldrin. Rather, they had a *job* to do.

Having said this, the primary reason that there were so few images of the First Man was because Armstrong had possession of the Hasselblad for most of the time. "As the sequence of lunar operations evolved," Aldrin wrote later, "Neil had the camera ... and the majority of the pictures taken on the Moon that include an astronaut are of *me*. It wasn't until we were back on Earth and in the Lunar Receiving Laboratory, looking over the pictures that we realised there were few pictures of Neil. My fault, perhaps, but we had never simulated this during our training." For his part, Armstrong agreed: he cared little about who took pictures of whom, as long as those pictures were *good* ... and they *were*. "I don't think Buzz had any reason to take my picture," he said, "and it never occurred to me that he should." To this author, the episode perfectly illustrates the gulf which existed between the businesslike pragmatism of the astronauts and the acute need for millions of people back on Earth to have the symbolic facets of the flight – an image of the First Man, the flag, the plaque – recorded for posterity.

The historic nature of the mission, indeed, made it inevitable that there would be a live telephone conversation with the astronauts' head of state ... and *this* was the event that both men blamed for their inability to get a good photograph of the First Man. According to Aldrin, seconds after Armstrong had taken the picture of him saluting the Stars and Stripes, Houston came on the line to say that President Nixon wished to talk to them. Apparently, Aldrin explained, the men were just about to swap the Hasselblad at that point, with the intention of taking some images of Armstrong, but were distracted by the request and the subject was later forgotten in the hurry to get everything done. In his study of the events on the surface, however, James Hansen notes that a full *five minutes* actually elapsed between the 'salute photo' (taken at 11:42 pm EST) and the call from Houston that Nixon was on the line (at 11:47 pm). In those five minutes, the two men would have had *plenty* of time to exchange the Hasselblad, but did not do so: both headed off in separate directions to tend to their respective tasks of collecting samples and making observations: Armstrong back towards Eagle and Aldrin further west, to a distance of some metres from the lander.

None of this, of course, even implies that the failure of either man to suggest taking a posed photograph of Armstrong was anything less than an oversight and something neither man thought important at the time. "I was *intimidated* by the enormity of the situation," Aldrin recalled later. "At the time, there was certainly a gun-barrel vision of focusing in on what you were supposed to be doing, rather than being innovative and creative." To be fair, almost all of the pictures that Aldrin *did* take on the few occasions that he had possession of the Hasselblad *were* pictures which the flight plan called for him to take. A picture of Neil Armstrong was *not* on the list. Other astronauts, including Al Bean and Gene Cernan, have stopped short of openly speculating that Aldrin deliberately excluded his commander from the photography, but have certainly been vocal in their 'surprise' that on such a historic occasion, inserting an item into the flight plan to get good-quality pictures of both men *should* have been a NASA priority. Whatever the reality, at 11:47:47 pm, Bruce McCandless called both men from their respective work.

"We'd like to get both of you in the field of view of the camera for a minute." McCandless paused for a second, then continued: "Neil and Buzz, the President of the United States is in his office now and would like to say a few words to you. Over."

"That would be an honour," replied Armstrong.

"All right. Go ahead, Mr President. This is Houston. Out."

"Hello, Neil and Buzz," Nixon began. "I'm talking to you by telephone from the Oval Room at the White House and this certainly has to be the most historic telephone call ever made. I just can't tell you how proud we all are of what you have done. For every American, this *has* to be the proudest day of our lives. And for all people all over the world, I am sure they, too, join with Americans in recognising what an immense feat this is. Because of what you have done, the heavens have become a part of man's world. And as you talk to us from the Sea of Tranquillity, it inspires us to redouble our efforts to bring peace and tranquillity to Earth." Then, Nixon added the words which would bring a lump to many a throat and reinforce the reality that the human race had never been as unified as it was now: "For one priceless moment," he intoned quietly, "in the whole history of man, all the people on this Earth are truly one ... one in their pride in what you have done and one in our prayers that you will return safely to Earth."

Although he had been in office for scarcely six months, and the decision to land an astronaut on the Moon had been proposed by a man against whom he had vigorously competed for the presidency a decade earlier, Nixon *was* entirely sincere in his praise of Armstrong and Aldrin; for although he had no real love for space exploration, he *did* understand that it was good for a nation to have heroes. The comment about redoubling efforts to achieve peace and tranquillity on Earth may well have seemed hypocritical if it had been spoken by his predecessor, Lyndon Johnson, but only weeks after declaring the first formal troop withdrawals from Vietnam, Nixon was, for the moment at least, riding a tidal wave of popularity.

Closing his remarks with wishes for a smooth return to Earth and anticipation of a happy reunion aboard the aircraft carrier *Hornet*, Nixon doubtless left the two astronauts swelling with pride and allowed them to calm down, too. For Aldrin, the announcement that his commander-in-chief wanted to speak to them generated stage fright. "My heart rate," he explained in his autobiography, "which had been low throughout the entire flight, suddenly jumped. Later, Neil said he had known the president might be speaking with us while we were on the Moon, but no one had told me. I hadn't even considered the possibility. The conversation was short and, for me, awkward. I felt it somehow incumbent on me to make some profound statement ... for which I had made no preparation whatsoever ..."

Fortunately for Aldrin, it was his commander who spoke and responded to Nixon's remarks. Armstrong had been told by Deke Slayton in the days before launch that there was a likelihood of some form of 'special communication', but, in his discussions with James Hansen, it would seem that he too had little idea *who* it might be. Judging from his response to the president – a polite "thank you", a couple of instances of "it's an honour" and a brief note about his desire for "peace for all nations" – the brevity of Armstrong's words would seem to suggest that both men felt unprepared, nervous and decidedly ill at ease. His mother, Viola, who listened to the conversation, could tell from her son's voice that he was "emotionally shaken" and detected an unmistakable "tremor" in his tone. Years later, Aldrin would express mild annoyance at not being given the courtesy of a heads-up on what might occur, but Armstrong could only state to Hansen that all he knew was that it was "going to be a surprise" and "maybe it wasn't even going to happen".

After their awkward conversation with Nixon, the two men had little time to ponder, for they were now almost an hour into the Moonwalk, closing in on its halfway mark, and still had so much to do. One of Armstrong's key tasks was operating the Apollo Lunar Surface Close-up Camera (ALSCC), designed by Thomas Gold, the scientist who had earlier confidently predicted that the men might sink up to their knees in dust. Nicknamed 'the Gold camera', its purpose was to acquire extreme close-up stereoscopic shots of lunar rocks and soils, but the lateness of its addition to Apollo 11's science payload irritated Armstrong, who saw it as an "inconvenience". In his recollections to Hansen, he could only remark that he took some pictures "that I thought would be of particular interest to him".

If Gold was interested in the close-up imagery of the lunar surface, then one other image taken by Aldrin during one of his brief spells with the Hasselblad camera grabbed the attention of the average man in the street: a photograph of his own footprint in the lunar soil which has since become world-famous. Since taking his first steps, Armstrong had commented on the quite remarkable clarity of the imprints of his boots, which sank barely a few millimetres into the terrain of this airless, waterless, lifeless wilderness. Aldrin noticed the same thing and made a conscious decision to photograph the soil, then photograph his bootprint, then photograph the print *and* his boot off to one side. He also took a series of 360-degree panoramas of the landing site, a few shots of Earth, hanging like a lone Christmas decoration in the black sky, and pictures to document the lander's grey and gold-coated bulk.

With the unveiling of the plaque and the raising of the flag and the words with Nixon now behind them, the astronauts could set to work on the scientific side of their mission. Armstrong's role during this time would be to collect samples of lunar material. "The geology community had hoped we would provide what they called 'documented samples'," he explained to Hansen, "that is, samples whose emplacement was photographed prior to and after lifting the samples. Time did not permit our doing as much of that as we had hoped." Even so, he obtained 21.7 kg of rocks and soil. Most of the samples were basalts – a dense, dark-grey-coloured, finely grained igneous rock, composed mainly of calcium-rich plagioclase feldspar and pyroxene – and the oldest of the specimens were later pegged at 3.7 billion years old. This would offer further evidence against the cold-Mooners' theory that *all* traces of a molten past ceased in excess of 4.5 billion years ago.

As Armstrong laboured with the samples, it was Aldrin's responsibility to take the lead in setting up an automated research station on the surface. This Early Apollo

Scientific Experiments Package (EASEP) was a forerunner of the more sophisticated Apollo Lunar Surface Experiments Packages (ALSEPs) which would be deployed by subsequent landing crews. It came in two parts. Firstly, there was a passive seismometer to detect subsurface quakes. This had a detector affixed to it to measure the rate at which lunar dust accumulated on its upper surface. Secondly, there was a device known as 'LR-cubed', which was a flat plane of corner-cube reflectors and essentially functioned as a special mirror to reflect an incoming beam of laser light aimed toward Mare Tranquillitatis from a terrestrial telescope. There were several such telescopes, including one at the University of California's Lick Observatory, near San Jose. Its objective was to precisely measure the distance between Earth and the Moon and, together with follow-on devices set up by two subsequent crews and one on a Soviet roving vehicle, this research has produced some of the most scientifically meaningful results from the Apollo landings. It has permitted more precise computations of the Moon's orbit, on variations in its rotation and established that it is receding from Earth at the rate of 3.8 cm per year. Furthermore, the data has provided a direct measurement of the rate at which the tectonic plates are travelling across the Earth's surface.

The seismometer, too, more than proved its worth, sensing and transmitting the tremors induced by Aldrin as he hammered core tubes into the soil several metres away and later the 'thud' as the astronauts dumped their now-unwanted backpacks, food packets and other junk onto the surface shortly before liftoff.

A third instrument, the Solar Wind Collection Experiment, was also erected – in fact, there is a photograph taken by Armstrong which shows Aldrin setting it up. This sheet of highly pure aluminium foil was suspended from a staff so that it could collect solar wind particles, including argon, krypton, xenon, neon and helium. This was left exposed for 77 minutes, then rolled up and stowed in one of the rock boxes for return to Earth and subsequent analysis.

The decision to push ahead with a relatively modest set of experiments caused some raised eyebrows, not least those of Wilmot 'Bill' Hess, director of science and applications at the Manned Spacecraft Center for the Apollo lunar missions. Hess had long lobbied for an ALSEP to fly aboard Apollo 11, but many NASA managers felt that weight limitations and timeline issues made it impossible. In fact, on the Apollo 12 and 14 missions, which each featured two Moonwalks, assembly and activation of the ALSEP required the better part of three or even four *hours*. Getting NASA managers to approve a *single* excursion of *that* length for Armstrong and Aldrin on the *very first* landing was virtually impossible and, as a result, no ALSEP flew on Apollo 11.

Two others intimately involved with the development of the ALSEP were Bill Anders and NASA's only geologist-astronaut, Jack Schmitt. At first, both men were astounded by what they perceived to be an experiments package which *maximised*, rather than minimised, the astronauts' workload on the lunar surface. Something needed to be done. "Bill Anders immediately introduced what ... I call 'the Anders big red button concept'," Schmitt told a NASA oral historian in July 1999, "in that ... the ideal was for the astronaut to go over and kick a big red button and the ALSEP would deploy itself. Well, by then it was too late, because they had received

Buzz Aldrin walks towards the solar wind experiment.

guidance some time before 'to give us something to do on the surface' (and these were from non-geological astronauts who apparently felt that they were going to be bored on the surface of the Moon, which is a little bit hard to believe). So when Anders and I got up there, we had to do an awful lot of work to get this thing down to the point of where it was not going to occupy all the time that the astronauts would have on the lunar surface. For example: they had a particular type of fastener holding this whole thing together, called a calphax fastener, that required a one-and-three-quarter turn to release; and there were 50 of these fasteners. Doing that kind of work in a suit was going to be time-consuming and would take a lot out of the muscles of the hand ... We finally got that down to a new kind of fastener that required only about a quarter turn to release and just flick your wrist and it's released. That's the best we could do. We, of course, wanted to have no fasteners, but there was a limit on what we could do."

As the Apollo 11 Moonwalkers' time outside drew towards its close, one of the few changes in the plan came when Armstrong took it upon himself to go and photograph a yawning bowl-shaped crater about 60 m east of Eagle which has since become known as 'East Crater'. To get there as quickly as possible, he adopted a loping, foot-to-foot stride. He took half a dozen Hasselblad images, including outcroppings in the crater. By the time he returned to Eagle, his adventure had lasted a little over three minutes. It was now 12:45 am EST on 21 July and Aldrin had been advised that they had only a few minutes left before packing their equipment away.

"There was just far too little time to do the variety of things that we would have liked to have done," Armstrong explained in a press conference after the mission. "When you are in a new environment, *everything* around you is different and you have the tendency to look a little more carefully at 'What is this? and 'Is this important?' or 'Let me look at it from a different angle', which you would never do in a simulation. In a simulation, you just picked up the rock and threw it into the pot!" Similarly, both men had seen rocks through Eagle's cabin windows before they set foot on the surface – rocks which may have been pieces of lunar bedrock, potentially priceless geological specimens – which they did not have time to inspect, photograph or collect. President Nixon's telephone call had eaten more time out of their excursion, as had the assembly of the flag and the reading of the plaque.

However, it was always to be expected that, as the *first* manned lunar landing, Apollo 11 could never have been much more than a 'flags-and-footprints' affair. It would be the task of successive crews to explore and attempt to comprehend its mysteries more thoroughly. In retrospect, Buzz Aldrin lamented the fact that he was on the first landing mission – he would rather have been on the second or even the third, where scientific exploration of the Moon would have taken precedence. Both men knew that they would only visit this barren wilderness once and both dearly wished for more time; but as astronaut Dick Gordon would later remark, "Time is relentless," and, indeed, the passage of time on the Moon was agonisingly rapid.

As Aldrin headed up the ladder at 12:56 am EST, Armstrong sealed the last rock box. Then, working together, the two men used the LEC to haul up the film magazines from the ALSCC and the Hasselblad and the two rock boxes – one

Aldrin gazes across the "magnificent desolation" of Tranquillity Base, towards the lunar module Eagle. The flag and Early Apollo Scientific Experiments Package are clearly visible.

containing the solar wind experiment. At 1:09 am, as Aldrin stowed the cargo, the First Man on the Moon jumped with both feet into Eagle's footpad and set his gloved hands on the ladder; after a little more than two full hours, this was his last direct contact with lunar soil. He then crouched into a kind of deep-knee bend, getting his torso as close to the footpad as possible ... and sprang upwards, easily reaching the third rung. It wasn't a case of showing off; it was sheer curiosity to see how high he could jump on the Moon. Two minutes later, he was back inside the ascent stage and Aldrin had pushed shut and sealed the hatch. All in all, the world's first excursion on alien soil had lasted two hours and 31 minutes from depressurisation to repressurisation of the cabin, of which Armstrong had actually been on the surface for two hours and 14 minutes and Aldrin for one hour and 46 minutes.

Removing their helmets, the first scent which greeted the men was something not dissimilar to burnt gunpowder. It was the odour of the lunar dust ... the *smell of the Moon itself* ... and its effects would give later Apollo astronauts swollen sinuses upon their return to Earth. As well as the surprising smell, Armstrong and Aldrin were doubtless relieved that one prediction had not come true: for some doomsday scientific prophets had predicted that large quantities of lunar dust, which had never been exposed to oxygen, might spontaneously burst into flame as soon as it was exposed to the oxygen-richness of Eagle's cabin. It was fortuitous that such a prediction proved entirely unfounded, for both astronauts were literally blackened from the knees down in the abrasive, gritty stuff. "The dust was so fine," Armstrong said later, "that you *couldn't* have got rid of all of it." Their suits, once snow-white, would never be truly clean again.

Despite having not slept for the better part of a full day, housekeeping would occupy the two men for the next few hours: a well-earned meal, photography through Eagle's windows to use up the last of the film, since there was no reason to return it *unexposed* ... and, for Aldrin, the chance to make good on what he had not done whilst outside. Turning the camera in Armstrong's direction, he snapped a picture of his commander, grinning through whiskers and eyes which did little to disguise their exhaustion. A brief chat with their boss, Deke Slayton, sitting at the capcom's mike in Houston, was followed by refitting their helmets, depressurising the cabin for a second time and reopening the hatch to throw out unwanted equipment.

First were the massive PLSS backpacks, which the television camera on the surface recorded bouncing off the porch and thumping heavily onto the ground. When capcom Bruce McCandless told him that the seismometer had detected the shock of the two impacts, Armstrong quipped, "You can't get away with anything anymore, can you?" Next came the dust-covered lunar overshoes, a load of empty food packages, urine and 'solid waste' bags, an expired lithium hydroxide canister and a spare Hasselblad. Sealing the hatch for the final time, Armstrong and Aldrin certainly had more volume in which to move within the broom-cupboard-sized cabin, but the place in which they would sleep for the next few hours was hardly clean. Everything was coated with a fine layer of lunar dust.

Sleep was virtually impossible and very uncomfortable. As Armstrong later told James Hansen, "The floor was adequate for one person, not to stretch out, but to lay halfway between a foetal position and a stretched-out position. That was where Buzz slept. The only other place to rest was the engine cover, which was a circular table some two and a half feet in diameter. To support my legs from it, we configured a sling from one of our waist tethers. We attached that to a pipe structure that was hanging down." Armstrong stuck his legs in there, kept the bulk of his body atop the engine cover and laid his head on a small shelf at the rear of the cabin. "It was a jerry-rigged operation," he added, "and not very comfortable."

Their lack of comfort was compounded by the fact that they were sleeping with their helmets and gloves on, in order to protect their eyes and lungs from the lunar dust. Indeed, the filtered oxygen entering their suits from Eagle's environmental control system was significantly 'cleaner' than the dust-laden air circulating through the cabin. Lights which could not be dimmed disturbed them frequently and the ever-present hum of life-support systems and water coolant pumps irritated them throughout the night. It was also cold, particularly after they had put window shades up to darken the cabin and blot out the glare from the sun-drenched surface. Resting for maybe an hour or two, at most, they shivered their way through a decidedly unpleasant night on the Moon in which temperatures inside Eagle did not climb much higher than 17°C.

It must have been with some relief that they were awakened by capcom Ron Evans at 10:32 am EST on 21 July and plunged into their efforts to prepare for liftoff from the Moon and a return to Mike Collins aboard the command and service module Columbia. During their last couple of hours on the surface, Armstrong and Aldrin ploughed their way methodically through their checklists, taking star sightings, establishing the proper state vector for the ascent, inputting computer data and tracking Columbia in an effort to gain a more accurate fix on their landing spot. The rendezvous radar, Evans directed, should be switched off for ascent to hopefully prevent them from suffering a recurrence of any of the program alarms that they had gone through the previous afternoon.

At length, at 1:36:04 pm, Evans cleared them for takeoff. The ascent engine that would get them away from the Moon was a single-chambered design whose operation was kept as simple as possible to minimise the risk of a malfunction. "The tanks and the propellants and the oxidiser were what they were," Armstrong recalled. "We did have various means of controlling the circuitry to the valves – opening the flow of propellants to the engine." He had actually proposed that a big manual valve should also be added as an additional form of redundancy, but ultimately his suggestion was not taken. "It wasn't really a problem," he said, "because if we fired the engine and it didn't fire, we weren't out of time. We had a lot of time to think about the problem to figure out what else we could do. When pilots *really* get worried is when they run out of options *and* run out of time."

Thankfully, Eagle was charmed and at 1:53 pm, Aldrin counted down their last seconds on the Moon, then executed the final sequence that he had choreographed so many times back on Earth: he hit the Abort Stage button, armed the engine for ascent and hit Proceed. With a muffled bang of pyrotechnics and a splintering shower of brilliant insulation particles, the engine lit instantaneously and the ascent stage literally punched its way directly upwards with all the unwavering accuracy and precision of an express elevator. Three subsequent landing crews would benefit from a lunar rover, whose onboard television camera would record these impressive liftoffs ... and *impressive* they were.

Neither man had much time to sightsee or gaze wistfully outside – in fact, during Aldrin's only chance to look up from his computer, he caught just a fleeting glimpse of the Stars and Stripes toppling over. Seconds after rising from Mare Tranquillitatis, the ascent stage pitched forward by about 45 degrees to begin the long climb to lunar orbit. The engine, which was to fire for seven minutes, looked good, but the three-hour celestial dance of rendezvous and docking with Collins was challenging and incredibly difficult. Armstrong and Aldrin performed three burns of Eagle's reaction control thrusters to raise their orbit and gradually close the distance between themselves and the mother ship: the first, at 2:53 pm EST, placed them into a higher orbit which was barely 20 km 'below' Columbia, then a second established them 'in-plane' with their quarry and finally the third brought them into what Collins would later, light-heartedly, describe as "a collision course".

The last day of isolation in lunar orbit had been a strange time for Mike Collins and, although he would admit in his autobiography to having felt the pangs of solitude, there certainly was no feeling of fear and the crackle of the capcom's voice from Houston was frequently in his ear and *that* kept him from getting lonely. He had turned the lights up in the command module, removed the centre couch to give himself ample room in which to move around and the overall effect was "a happy place". The only times at which he felt an unusual sensation were the spells when Columbia drifted behind the limb of the Moon and began its passage across the far side; *then*, communication with Earth was promptly cut and he was truly on his own. "I am it," he wrote. "If a count were taken, the score would be three billion, plus two over the other side of the Moon and one-plus-God-knows-what on *this* side." The sensation was both powerful and intensely mystifying, but, conversely, it was also a sensation that Collins *liked*.

One of his greatest regrets, though, had been his inability to spot Eagle on the surface. The 28-power sextant in the command module's lower equipment bay provided superb magnification, but its narrow field of view gave him limited chance to actually find the landing spot. On more than one occasion, Collins had asked Houston for any topographical cues which might guide him in the right direction. There were a few vague craters, but nothing of substance to help him. He was also frustrated (owing to rescheduling) to have been on the far side of the Moon when Armstrong took his historic first steps and, upon re-establishing contact with Houston, his crewmates had already set up the flag. Collins was not surprised, though, by his commander's entirely appropriate choice of words. "Neil doesn't *waste* words," he wrote, "but that doesn't mean he can't *use* them." Now, a little less than a full day after seeing them last, they drew steadily into view, twirling and pirouetting and finally presenting their docking adaptor to him.

Physical docking was eventually achieved by Collins at 5:38 pm and a little under an hour later, the hatches were opened and a very dusty Armstrong and Aldrin floated through the connecting tunnel to be received by their first human contact since leaving the Moon. So overjoyed was Collins to see them that he grabbed Aldrin's balding head, "a hand on each temple" and made to give him a big, welcome-home smooch on the forehead. Remembering, however, that he was a no-nonsense astronaut, a test pilot and a possessor of 'the right stuff', the CMP quickly changed his mind. A couple of hearty handshakes would have to suffice.

There were other things to bring over from Eagle, aside from themselves. "Get ready for these million-dollar boxes!" Aldrin yelled. "Got a lot of weight. Now watch it!" The two silvery boxes, each measuring about 60 cm long and capable of sealing the samples in an uncontaminated, lunar-type vacuum, were quickly shoved into white fibreglass fabric containers and zipped. Later that evening, at 7:42 pm, after closing the hatch for the last time, Eagle was jettisoned. Unlike several subsequent Apollo ascent stages, it was not purposely targeted to impact the surface and set off a reading on the seismometer. Rather, it would linger in lunar orbit for some time, before the mascon perturbations caused its orbit to decay and it hit the Moon. Since it was not tracked, and the precise mechanics of its orbit were not known with sufficient accuracy, it remains unclear exactly *when* it made landfall and, indeed, *where*. As a result, no one can say what remains of Eagle now: obviously, the descent stage – lacking at least some of its gold-coloured insulation – is still in one piece at Tranquillity Base, but the remnants of the cabin which kept Armstrong and Aldrin alive are probably minuscule and buried somewhere in the walls or floor of a small, man-made crater. Perhaps, one day in the far future, some shards will be recovered and put on permanent display in a museum. Perhaps.

The late John Kennedy's pledge, though, would not be entirely fulfilled until all three men had been returned "safely to the Earth" and of absolutely critical importance in achieving this final leg of the expedition was the two-and-a-half-minute transearth injection (TEI) burn of Columbia's Service Propulsion System (SPS) engine. In Mike Collins' mind, TEI was merely a euphemism for what he called "the-get-us-out-of-here-we-don't-want-to-be-a-permanent-Moon-satellite" manoeuvre; thankfully, as with so many events which could conceivably have gone wrong in the last five days, the burn was perfect, increasing their velocity by 3,600 km/h, causing them to barrel out of lunar orbit and onto a welcoming path homeward. "Beautiful burn, SPS," Collins exulted from the left-side commander's seat, "I love you. You are a jewel! Whoosh!"

Aside from their joy at having pulled off a mission whose chances of success Collins had earlier placed at no more than fifty-fifty, the journey home was hardly one of comfort. The command module's cabin was literally packed with equipment, food packets and some rather unpleasant waste. "The right side of the lower equipment bay," Collins wrote, "wherein are located old launch-day urine bags, discarded washcloths and worse, is now a place to be avoided. The drinking water is laced with hydrogen bubbles – a consequence of fuel-cell technology which demonstrates that H_2 and O join imperfectly to form H_2O. These bubbles produce gross flatulence in the lower bowel, resulting in a not-so-subtle and pervasive aroma which reminds me of wet dog and marsh gas ... "

As they headed back to Earth, the astronauts were granted plenty of time to 'rest' in this decidedly unpleasant setting and to fool around with their colour television camera. All three spoke their own words of wisdom and reflection to terrestrial audiences, but little could they have realised that a potential disaster of mission-fouling proportions awaited them. A violent thunderstorm brewing in the prime recovery location had been identified by an officer named Hank Brandli at Hickam Air Force Base in Hawaii. Brandli was a meteorologist working with then-classified data from the National Reconnaissance Office's Corona military satellites and quickly recognised that the situation could cause a catastrophe by tearing Columbia's parachutes to ribbons. The secret nature of the project prevented Brandli from contacting NASA directly, so he proceeded through the chain of command and, fortunately, the naval recovery group, centred around the aircraft carrier *Hornet*, was re-routed and Apollo 11's re-entry trajectory was modified to produce a splashdown some 345 km further downrange, just off Hawaii. The Brandli story was eventually revealed in the mid-Nineties, nearly three decades later ...

Mike Collins referred to the trajectory adjustment in his autobiography. "The weather in our recovery area is full of thunderstorms," he wrote, "and they are going to move our splashdown point ... to the east in search for clearer skies and calmer seas." His worry was not so much that the computer could not handle the revised trajectory, but that if the automatic system failed and *he* had to assume manual control, the circumstances could become hairy. "To get that extra range will require a great soaring arc after our initial penetration into the atmosphere and the difference between soaring an extra [345 km] and skipping out of the atmosphere altogether is slim indeed." He went to sleep for his final night in space with a head filled with two things: the speeches he would need to have memorised in time for landing ... and his own heightened sense of mental preparedness to make a manual re-entry.

Just to make absolutely certain that no fundamental oversights were made by the crew, astronaut Jim Lovell radioed them on the morning of 24 July with a word of helpful advice: to make sure that they re-entered the atmosphere *blunt end forward*! Collins politely acknowledged the call. Entry interface occurred at 12:35 pm EST, just to the north-east of Australia. "We are scheduled to hit our entry corridor," wrote Collins, "at an angle of six and a half degrees below the horizon, at a speed of nearly [40,000 km/h]. We are aimed at a spot south-west of Hawaii." Deceleration at such blistering lunar-return velocities produced a spectacular light show as the infinite blackness of space was steadily replaced by "a wispy tunnel of colours: subtle lavenders, light blue-greens, little touches of violet, all surrounding a central core of orange-yellow".

Dropping further into the atmosphere, the three men beheld their first direct Earthly sight in more than eight days – a vast bank of stratocumulus clouds – and were then jolted sharply by the perfect deployment of the three orange-and-white main parachutes. Six minutes after entry interface, 'Air Boss 1', callsign of the head of the inter-service rescue team, reported that the command module was in sight. In Florida, of course, it was early afternoon, but in the central Pacific, dawn was just breaking. Eight minutes after Air Boss 1's call, at 12:51 pm EST, Columbia and its crew of three lunar heroes splashed down hard, like a ton of bricks. Armstrong radioed Air Boss 1 to say they were safe, but in reality the swells off the Hawaiian coast were much stronger and considerably higher than he had anticipated and quickly dragged Columbia into a so-called 'stable 2' position, with the apex underwater and the men hanging from their harnesses.

Quickly, the astronauts started motorised pumps to inflate three brown and white airbags on the apex which, after several minutes, turned Columbia into a 'stable 1' configuration, bobbing, base-first, in the ocean. Armstrong had told his comrades to take an anti-motion sickness pill just before re-entry – only to discover that they should actually have taken *two* – and Aldrin, the Air Force officer, appears to have suffered the worst of the nausea from the unpleasant landing. Mike Collins, meanwhile, found that he now owed Armstrong a beer, having earlier bet him that the command module would *not* capsize. All three waited in silence for the frogmen to arrive, each willing himself and the others *not* to be sick. "It was one thing to land upside down," Aldrin recalled later, "it would be quite another to scramble out of the spacecraft in front of television cameras, tossing our cookies all over the place!"

Notwithstanding a decidedly unsettling return to Earth, Columbia had actually splashed down right on target and the recovery ship *Hornet* – with President Nixon aboard – was only about 20 km away and steaming in their direction. Navy helicopters buzzed in the area and at 1:20 pm, the pararescue swimmers arrived in the water near the command module. The first person they saw was 25-year-old Lieutenant Clancy Hatleberg, but from his appearance the astronauts could have been forgiven for believing themselves to have landed on an alien planet and just met one of the natives: for he was clad, head-to-toe, in a peculiar, greyish-green rubberised, hooded and visored ensemble called a Biological Isolation Garment (BIG). Without saying a word, Hatleberg swung open Columbia's hatch and tossed three sets of BIGs into the cabin.

It was nothing personal, of course. No one knew if they had brought any potentially lethal bacteria back with them. Some doomsday prophets had speculated

Clad in their Biological Isolation Garments (BIGs), the three heroes of Apollo 11 relax in a life raft as the hatch of the command module Columbia is sealed.

that 'Moon germs' could be as virulent as bubonic plague was to medieval Europe. Others, including Chris Kraft, scoffed at the notion. "It was ... politically mandatory," he wrote. "We went along with the game because we had to. The same thing will happen, I'm sure, when men and women first return from Mars. Hysteria cows common sense every time." Whatever the reasoning on both sides of the fence, for three weeks, Armstrong, Collins and Aldrin – the first ambassadors to travel from the surface of our world to another – would be isolated, poked and prodded and viewed only from behind multi-layered glass. The saving grace was that they would have plenty of quiet time to themselves, to watch the Moonwalk, to write up their reports, to rest, to eat and to contemplate the enormity of what they had done in a few short days. Hopefully, the three 'amiable strangers' made the most of what private time they had ... for on 10 August, they would be released into the full glare of worldwide attention: onto an unending merry-go-round of public engagements and foreign goodwill tours ... and their lives would never be the same again.

"THIS IS *THEIR* FLAG, TOO"

As successful, epochal and transcendental as Apollo 11 had been, the completion of one of humanity's oldest and most enduring dreams was met with a mixture of joyful optimism and cynical disbelief. "I was very proud when I saw that space ship and the men with the flags on their sleeves," said John Furst, a student at the University of Pennsylvania, "but I must confess that I also thought of all the people who live in the ghettos. This is *their* flag, too. The flag may be flying on the Moon, but it is also flying in their neighbourhoods, where there are poverty, disease and rats." The violent and bloody decade known as the Sixties was drawing to its close and, indeed, enormous advances had been made: improvements in education, in civil rights, in the treatment of minorities, in tackling homelessness and poverty ... but, clearly not enough had been done and the billions of dollars spent on Apollo raised many questions and many more eyebrows.

Why couldn't such sums be spent on the poor, many asked, or on other, more pressing needs on the ground? It was a valid point. "The poor cannot be blamed," *Time* told its readers on 18 July, "for being indifferent or even bitter when they watch the shining and vastly expensive rocket travel into the sky on a mission that does not improve their immediate future." In Algiers, Eldridge Cleaver of the Black Panther Party, a fugitive from justice in California, denounced Apollo 11 as "a circus" and accused it of having been engineered to distract attention from the *real* issues of intolerance and inequality that were rotting away the heart and soul of America. Others disagreed. The Nobel Prize laureate Felix Bloch cautioned that progress in science could *not* be measured in dollar signs and that the long-term benefit of Apollo was impossible to guess. Others pointed out that the discovery of the Americas had opened Europe up to a wider world and the voyage of Armstrong, Collins and Aldrin, likewise, could mark the start of a new chapter in the human story. Of course some directed their anger elsewhere: at the Pentagon, whose years spent pouring billions of dollars into the bottomless pit of Vietnam dwarfed the expenditure of NASA.

The outrage and the admiration and the criticism and the counter-criticism could not derail the fact that a fifth of the human race had just witnessed a truly historic event, the magnitude of which almost certainly would not be repeated in their lifetimes. For NASA employees, the 'splashdown parties' and celebrations went on for two whole weeks. It was hard for some to believe how much had been achieved in such a short span of time; for less than 3,000 days earlier, the United States had struggled to put a man aloft for just 15 minutes of suborbital flight. "It was only eight years," wrote Chris Kraft in his autobiography, "since Kennedy stood before Congress and proclaimed the Moon as an American goal ... *How did we do it? ...* and the only answer I had was that we did it because we *had* to do it."

In Houston, bars began selling 'Moonshots' – fearsome concoctions of orange juice, laced with healthy helpings of cognac and champagne – together with 'Lunar Cocktails' and 'Armstrong Benders'. Elsewhere, the city's swanky Marriott Motor Hotel hosted a $20,000 bash for the brass of the space industry. Bob Gilruth was there, together with Kraft and a couple of dozen astronauts and their wives. Jan Armstrong, Pat Collins and Joan Aldrin finally got some much-deserved recognition and respect for the enormous emotional investments they had made to 'the mission'. Wernher von Braun, whose fearsome wartime V-2 had helped pave the way for the mighty Saturn V, was raised onto the shoulders of sheriffs and local councillors and paraded around in triumph.

Feelings across the world were also apparent in the reactions of their mass media. Notwithstanding the predictable responses from the Soviet Union and China – the former merely sending a terse congratulatory telegram to the White House, the latter unashamedly trying to jam all five Voice of America broadcasts – even communist nations of eastern Europe expressed their support and admiration. Radio Zagreb emphasised the differences between American honesty and Russian secrecy in space matters, Czechoslovakia issued commemorative stamps, Hungarian television described the "amazing tasks" done by the astronauts and Poland unveiled a huge statue in their honour. Newspapers in New Delhi and Montgomery in Alabama printed large footsteps on their front pages, whilst São Paulo's *O Estado de São Paulo* ran Armstrong's famous words in no fewer than *nine* languages.

Elsewhere, the effects were just as dramatic. Television sets were virtually sold out in Japan, as people sought every opportunity to witness history in the making, and more than 50,000 South Koreans watched a giant screen in Seoul, awestruck. A 26-year-old Londoner scooped a massive bookies payout, after placing a bet at thousand-to-one odds back in 1964 that an American *would* walk on the Moon before the end of the decade. A Beirut woman is said to have given birth to her 11th child and named him 'Apollo Eleven Salim' ... the list of anecdotes and stories from this quite remarkable time in 1969 is endless and everyone who was there has their own tale to tell. In spite of the subsequent dishonour which fell upon him, maybe Richard Nixon had been right when he spoke to Neil Armstrong and Buzz Aldrin on the Moon. Maybe, despite all the criticism and the bitterness, the most important legacy of the first manned mission to the lunar surface was *not* its vast price tag ...

but the fact that, for just one priceless moment, the people on Planet Earth *were* truly one.

GOLDFISH BOWL

When Jim Lovell, sitting in Mission Control, spoke to Armstrong's crew shortly before re-entry, he had a few words of advice for them. "I just want to remind you," he said, "that the most difficult part of your mission is going to be *after* recovery!" His words came true sooner than anyone could have expected. It began when Clancy Hatleberg threw the otherworldly-looking BIGs into the cabin and the astronauts struggled into them – *struggled*, indeed, for eight days of weightlessness had left them light-headed and scarcely able to move – then tumbled clumsily through the hatch into the life rafts which were now bobbing alongside Columbia.

They had come from *another world* and no one really knew if that world harboured bacteria which might be harmful to life on Earth. Several doomsday prophets, of whom a large number emerged from the woodwork in the months before the mission, speculated that 'lunar germs' – to which mankind had never been exposed – could trigger pandemics on a scale not unlike the one which wiped out a third of Europe in the 14th century. In response to this threat, in mid-1966, an Inter-Agency Committee on Back Contamination (ICBC) had been formed and was charged with assisting NASA in the development of a programme to prevent contamination by any lunar organisms or materials. Sitting on its various panels were representatives of the Public Health Service, the Department of Agriculture, the Department of the Interior, NASA itself and the National Academy of Sciences; proof, if ever it were needed, that such threats were taken with the most solemn seriousness.

The facilities and procedures which arose from this committee were, in the words of the official Apollo 11 press kit, "well beyond the current state-of-the-art ... and the overall effort has resulted in a laboratory with capabilities which have *never* previously existed". Steps to minimise the possible harm to the environment began immediately after splashdown: all three astronauts were scrubbed with a decontaminating iodine solution by one of the rescue divers – and *themselves* then scrubbed *him* down – and the hull of the command module around the hatch was thoroughly cleaned. Armstrong, Collins and Aldrin were then flown aboard one of the Sea King rescue helicopters to the deck of the aircraft carrier *Hornet*, setting down next to a trailer-sized Mobile Quarantine Facility (MQF). Waving briefly to the cheering sailors, the three men tramped out of the helicopter and disappeared into the MQF. This was promptly sealed and there they would remain, with a lounge, galley and sleep/bathing area to themselves, until the carrier docked in Pearl Harbour.

From behind a picture window in the rear end of the trailer and speaking through an intercom, the astronauts' first official welcome came from President Nixon himself, his eyes brimming with pride and excitement. "I want you to know that I

President Richard Nixon greets the Apollo 11 crew, screened away in the Mobile Quarantine Facility aboard the *Hornet*.

think I'm the luckiest man in the world," he told them. "I say this not only because I have the honour of being the President of the United States, but particularly because I have the privilege of speaking for so many in welcoming you back to Earth." He concluded his remarks by inviting all three men and their wives to a state dinner in Los Angeles on 13 August – following their release from quarantine ... then referred to the "long week" of Apollo 11 as "the greatest week in the history of the world since the Creation".

In terms of completely isolating any lethal lunar germs, the whole procedure of donning the BIGs and crossing over to the MQF was full of holes and it is fortunate that nothing living *was* found in the samples, for there was realistically little prospect of *completely* preventing cross-contamination. Original plans had called for the men to remain aboard Columbia, which would be hauled from the ocean and placed next to a sealed corridor on the *Hornet*, through which they would transfer into the isolation cabin. Simulations of this plan before launch were compounded by the realisation that the command module was top-heavy, by no means seaworthy and it was unwise to leave Armstrong, Collins and Aldrin inside for too long as it pitched and rolled in the waves.

Central to this effort was the $15.8 million Lunar Receiving Laboratory (LRL) at Houston's Manned Spacecraft Center, into which the three men and a handful of technicians and medics would spend the remainder of the post-flight portion of the quarantine. After docking in Pearl Harbour, the MQF and the astronauts were loaded aboard the cavernous belly of a C-141 Starlifter transport aircraft and flown six hours east to Houston, touching down at Ellington Air Force Base around midnight on 26 July. Despite the late hour, crowds thronged the streets as a flatbed truck bore the MQF and its three heroes to the Manned Spacecraft Center.

For all the sophistication about which the press kit had spoken, the LRL, into which Armstrong, Collins and Aldrin were ensconced in the small hours of the 27th, had a few home comforts, too, including a lounge for cards, a games room with pool table, a film library and a miniature gym with exercise equipment. The meals were pretty good, too, with the men typically served enormous T-bone steaks and a steady supply of 1964 Chateau Lafite Bordeaux. They were not alone. In addition to their two chefs, there were a couple of doctors, a NASA public relations officer, a lab specialist, a janitor and even a journalist, who kept up a stream of daily reports. The astronauts spent their time relaxing and writing up their post-mission reports, in which they analysed the flight in the minutest detail. When complete, the report ran to more than five hundred pages of single-spaced type.

It was not just the crew of Apollo 11 who were being examined in the LRL; the samples from the Moon were there as well and when the containers were opened for the first time, there was full television coverage. Subsequent analysis would show approximately half of the rocks and soil specimens to be forms of basalt, originating from the lava flows which, the hot-Mooners had long argued, were responsible for laying down the lunar maria. Most of them were dated to around 3.6 billion years old and a new titanium-bearing mineral was actually called 'Armalcolite', deriving its name from the surnames of the men who first found it: ARMstrong, ALdrin and COLlins. What was *not* found, in any of the lunar material, however, was the slightest hint of anything biological; its utter deadness was quite profound.

"Some of the things we saw in them," geologist-astronaut Jack Schmitt told a NASA oral historian in July 1999, "were very, very unusual, relative to terrestrial rocks. One of the first things that catches your eye [is that] there's no evidence at all of *any* water activity; and there isn't any water in these rocks ... and because of that, there are mineral assemblages that are very, very clearly identified just in hand specimens, which you'd normally have trouble doing here on Earth with the volcanic rocks. It was exciting to be able to sit and work on the actual rocks that you had dedicated a small part of your life to."

The Moon, said Schmitt, is diverse, though not nearly as diverse as Earth. "The absence of water and water-driven processes on the Moon reduce that diversity significantly," he explained, "but still, in its own way, it's not only very different [from] Earth, but it has a diversity as a result of that ... a diversity we don't see here on Earth, because that part of Earth history has been obscured or destroyed. Earth history beyond three billion years ago is *very* difficult for us to look at; in fact, Earth history beyond half a billion years ago is somewhat difficult. You have to go to very special places. On the Moon, we just *start* looking at history at three billion years and go *back from there*! It's a pitted and dusty window into our own past. No question about that."

WHERE NEXT?

The joy at being *uncontaminated* by Moon germs was overwhelmed by another feeling of joy on the evening of 10 August ... when the doors of the LRL were opened and Armstrong, Collins and Aldrin were finally allowed to go home and see their families. For all three men, their astronauting days were over. Their lives as three of the most famous men on the planet were about to begin.

Mike Collins had already declined Deke Slayton's invitation to take backup command of Apollo 14 and maybe one day set foot on the Moon himself; the stress and constant travel had affected his family too much. Neil Armstrong, for whom aviation was his life, had long since desired to enter academia and write a textbook and his next steps beyond NASA would take him to the University of Cincinnati to teach engineering. For his part, Buzz Aldrin would remain with his parent service for several years, receiving a coveted command within the Air Force, but for a man attuned to overachieving and being the best at whatever he set his mind to, his *inability* to ever surpass what he had done that midsummer's evening in 1969 would drive him for a time into the depths of alcoholism and severe depression. Now recovered, he is a vocal and enthusiastic proponent of returning to the Moon.

All three men would 'survive' the effects of Apollo 11 and would be seen together on various occasions over the years: with President George H.W. Bush in 1989 when he announced the ill-fated Space Exploration Initiative to return to the Moon, with President Bill Clinton on the 25th anniversary of Eagle's triumphant landing and, most recently, with President Barack Obama on the 40th anniversary. All three will turn 80 in 2010 and it seems unlikely – though we can continue to hope – that they will still be alive when the *next* generation of lunar explorers follows in their esteemed footsteps. *That* is perhaps one of the depressing truths to come from Apollo; that after four decades, we are no closer to *ever* going back.

The decision to *not* continue exploring the Moon came, ironically, from the very man who welcomed the astronauts back to Earth: Richard Nixon himself. Soon after entering office, he had directed Vice-President Ted Agnew to head up a new Space Task Group, whose focus would be upon charting a course for America's future in the heavens in the Seventies and beyond. Agnew proved a strong supporter of ambitious and energetic NASA Administrator Tom Paine, who was already working on an audacious plan for a manned mission to Mars. "Paine," wrote Andrew Chaikin, "believed that there was something implicit in Kennedy's challenge beyond

Chicago welcomes the Apollo 11 astronauts with a tickertape parade.

its words; that it was a call for the United States to become a spacefaring nation." The price of a Martian venture, Agnew had told Nixon, would be high, but not *that much* higher than the $25 billion which had been invested in Apollo.

The timescale for achieving a manned Mars landing, Agnew explained, would be entirely dependent upon the level of funding that America was prepared to allocate. The options were these: if Congress would spend between eight and ten billion dollars per year on space throughout the Seventies, then a landing on the Red Planet was possible as early as 1983. At the other extreme, a reduction in funding below, say, six billion dollars annually would postpone the mission for almost a full decade; but finding some sort of 'middle ground' – perhaps a compromise of seven and a half billion dollars – could see an astronaut setting foot on Mars around 1986. Agnew himself backed the compromise and felt sure Nixon would do the same. NASA would not, after all, be requesting a dramatic budget increase at a time when America was under increasing pressure to meet urgent social needs. It would also allow Nixon to defer a firm decision on Mars until 1976 (the last year of a possible second presidential term) *without* harming the 1986 target.

Others felt quite differently. George Romney, the secretary for housing and urban development, felt that the time was instead ripe for refocusing attention on "problems on Earth". Many doubted Agnew's price tag for the Mars mission as being anywhere near realistic, with estimates as high as $40 billion being made in some quarters. At the same time, there was a very real fear that the *social* and *technological* cost of simply allowing the space programme to wither would be equally profound: Wernher von Braun felt that the United States stood to lose "a national asset" and senior NASA managers predicted the dramatic shrinkage of the space industry workforce and a loss of talent. Even as Apollo 11 flew home, thousands of North American personnel were being laid off and towns like Huntsville, Alabama, where the Saturn V was assembled, were expected to be hit the hardest.

The huge cost of a Mars effort would eventually prove its undoing in the eyes of Nixon and, in the following years the focus settled on a means of astronauts reaching space more cheaply and more frequently, aboard the reusable Space Shuttle. In April 1972, as the penultimate Apollo crew were working on the lunar surface, approval was finally given for the Shuttle to go ahead. It was to be used, among other things, to build a permanent space station, its supporters said, but they could scarcely have realised that this vision would not become a reality for almost three decades. As the Shuttle consumed ever more funding, plans for a Mars expedition steadily slipped off NASA's radar, as did parallel efforts to build permanent lunar bases as a fitting legacy of Apollo. It is a bitter irony, therefore, that the very man who had so welcomed the first ambassadors to the Moon on the deck of the *Hornet* would be the one who would effectively destroy the last realistic hope of returning there for the next half century.

As Neil Armstrong, Mike Collins and Buzz Aldrin emerged from the LRL in the scorching month of August 1969, there was still much to do. Nine further lunar landings were planned, beginning with the ten-day voyage of Apollo 12 in November. On *that* flight, astronauts Pete Conrad and Al Bean would leave crewmate Dick Gordon in lunar orbit and spend more than a day on the surface at a place called the Ocean of Storms. Conrad and Bean would make not one, but *two* excursions, each lasting four hours. They would deploy a full Apollo Lunar Surface Experiments Package and would attempt a pinpoint touchdown, just a few hundred metres from an unmanned Surveyor lander. Pete Conrad knew that it would be an exciting mission, a logical extension beyond Armstrong's giant leap. But what could *he* say to mark his first steps on alien soil? How could Conrad possibly compete with the profound, meaningful, thought-provoking words of the First Man? Then, completely out of the blue, he had a chance encounter with an Italian journalist ... and from that moment, he *knew* what he was going to say.

A BET TO WIN

"That's a vagina," quipped Charles 'Pete' Conrad Jr. "Definitely a vagina." The psychiatrist noted his response without a word, maybe realising, maybe not, that he was the victim of yet another wisecrack from the gap-toothed, balding Navy lieutenant. Yet despite his comments about each of the Rorschach cards which he was shown, Conrad was not entirely obsessive about the female genitalia. He had actually been tipped-off the night before by another astronaut candidate, Al Shepard: what the NASA psychiatrists were really looking for was *male virility*. "I got the dope on the psych test," Shepard had assured him. "No matter what it looks like, make sure you see something sexual." So Conrad did.

His key concern, though, that spring in 1959, had been the impact that this crazy 'Project Mercury' idea might have on his career. Instead of logging hours in the Navy's new F-4 Phantom fighter, he spent a week at the Lovelace Clinic in Albuquerque, New Mexico, serving up stool, semen and blood samples and collecting 24-hour bagfuls of urine. On the evening before a major stomach X-ray, told not to drink alcohol after midnight, Conrad had sat up until 11:57 pm draining a bottle with Shepard and another naval aviator called Wally Schirra to loosen themselves up for the next day. Conrad doubted that Lovelace's invasive tests had anything remotely to do with flying in space: the physicians, he told Shepard, seemed far more interested in "what's up our ass" than in their flying abilities, which was typical of military flight surgeons. Shepard warned him to be careful, to give the right answers to questions and to remember that the staff were watching their every move.

In spite of his frustration, Conrad persevered. He followed Shepard's advice, saw the female anatomy in every Rorschach card, deadpanned to a psychologist that one blank card was upside down, pedalled a stationary bicycle for hours, sat in a hot room for an age, then dunked his feet into ice-cold water and argued with one of the physicians that he considered it pointless to have electricity zapped into his hand through a needle. However, all this torture, Conrad felt, would at least give him the opportunity to lay his entire naval career on the line for one chance to fly something even faster: to ride a rocket, outside Earth's atmosphere, "at a hell of a lot more Machs than anything he was flying right now". Flying higher and faster, and pushing his own boundaries, had been the story of Conrad's life.

Born in Philadelphia, Pennsylvania, on 2 June 1930, the offspring of a wealthy family which made its fortune in real estate and investment banking, Conrad's father insisted that he be named 'Charles Jr' – "no middle name" – although his strong-willed mother, Frances, felt that this tradition of Charleses should be broken. Frances liked the name 'Peter', wrote Nancy Conrad in her 2005 biography of her late husband and although it never became his official middle name, Charles Conrad Jr would become known as 'Peter' or 'Pete' for the rest of his life. His fascination with anything mechanical reared its head at the age of four, when he found the ignition key to his father's Chrysler and reversed it off the drive. Later, in his teens, he worked summers at Paoli Airfield, mowing lawns, sweeping and doing odd jobs for free flights. Aged 16, he even repaired a small aircraft single-handedly.

"Most people I meet are doing eight things," fellow astronaut and Apollo 12 crewmate Al Bean once said. "Some people can do that. Pete Conrad, my best friend and my mentor, could do that. We could be talking about space and he'd get a phone call about something like the transmission on his race car. 'We've gotta get the transmission changed because the gear ratio's not working on this track.' Suddenly, all his effort would be on that. The next day, someone would call about motorcycle racing. He could do *all* of them well." At heart, Conrad was an engineer and tinkerer.

Education-wise, he would partly follow in his father's footsteps: the private Haverford School, from which he was expelled, then the Darrow School in New York, where Conrad's dyslexia was identified and where he shone. Although his father intended him to go to Yale University, he actually enrolled at Princeton in 1949 with a Reserve Officers Training Corps scholarship from the Navy to pay for studies in aeronautical engineering. Graduation in 1953 brought him not only his bachelor's degree, but also a pilot's licence with an instrument rating, marriage to Jane DuBose and entrance into naval service.

He breezed through flight training, earning the callsign 'Squarewave' as a carrier pilot. In *Rocketman*, his second wife and widow Nancy wrote that Al Teddeo, executive officer of Fighter Squadron VF-43 at Naval Air Station Jacksonville, Florida, had doubts on first meeting the young, seemingly-wet-behind-the-ears ensign one day in 1955. Those doubts were soon laid to rest when Teddeo discovered that Conrad could handle with ease any manoeuvre asked of him. Tactical runs, strafing runs, spin-recovery tests; Ensign Conrad did it all. "Hell, we refuelled three times till I just had to get back to my desk," Teddeo recalled years later. "It was like telling a kid at the fair that it was time to go home."

Next came gunnery training at El Centro, California, and transition from jet trainers to the F-9 Cougar fighter. Then, in 1958, he reported to Pax River to qualify as a test pilot. Later that same year, he received, along with over a hundred others, classified instructions to attend a briefing in Washington, DC. Conrad was told to check into the Rice Hotel under the cover name of 'Max Peck'. Only when he got there did he find that another 35 'Max Pecks' were also there – including an old naval buddy, Jim Lovell. Neither he nor Lovell would make the final cut for the Mercury selection, but their day would come three years later.

Whereas Lovell was cast aside for a minor liver ailment, Conrad's cause for failure proved a little ironic. "Unsuitable for long-duration flight," asserted the explanatory note. He had, it seemed, shown a little too much cockiness and independence during testing; characteristics which went against the panel's notion of a good, all-rounded, level-headed astronaut. During his career, Conrad would fly four space missions in total, two of which would set new world records ... for long-duration flight!

When it came to training, Conrad was a genius in the simulator and could handle with ease virtually any malfunction that the instructors threw at him. He knew the lunar module like the back of his hand and had played a significant role in helping to design its interior and displays – in fact, his only regret was that *he* had not been first to set the spidery craft down onto the Moon's surface. In terms of the words that came from his mouth, Conrad could be just as much a maestro as he was in the simulator: he could speak appropriately when protocol demanded, but he also swore like a sailor. Moreover, he was irritated by people who had convinced themselves that what astronauts *did* say was what they had been *told* to say. One afternoon, in the summer of 1969, he and his then-wife, Jane, were entertaining Oriana Fallaci. This petite, fiery Italian journalist had come to Houston several years earlier to write a book about the astronauts and she was particularly fascinated by the psychology behind this strange band of supermen. During the course of their conversation, Fallaci said she was certain that NASA's senior management had *told* Neil Armstrong what to say when he stepped off Eagle's ladder.

Fallaci had already gained a reputation as a fearsome interrogator; in later years, she would lambast Ayatollah Khomeini over the "medieval" regime he had imposed on Iran and Henry Kissinger would suffer the most "disastrous" interview of his career at her hands. In October 1968, she had narrowly escaped death in Mexico City, caught up in the infamous Tlatelolco massacre and had a well-known dislike for what she regarded as the "oppression" of the people by any political elite. Sitting with Conrad that summer's afternoon in 1969, Fallaci now turned her fervour onto NASA's elite and the 'control' that she perceived they had over the astronauts. There was no way, she reasoned, that Armstrong could possibly have dreamed up such poetic, figurative words on the spot.

Conrad tried to persuade her otherwise; that, although he could not *prove* that Armstrong had dreamed them up, they certainly had not been imposed on him. "Pity the twit who would *try* to script a bunch of test pilots and fighter jocks, egos fully intact, riding a rocket to the Moon," reflected Conrad's widow in 2005. For her part, though, Fallaci scoffed at the idea.

"Okay," Conrad said. "I'll make up my first words on the Moon right here and now." Being one of the shortest astronauts in the corps, an idea had already popped into his head.

"Impossible," Fallaci retorted. "They'll *never* let you get away with it."

"They won't have anything to say about it. They won't *know* about it until I'm on the Moon."

They agreed on five hundred dollars and shook hands. A few months later, true to his word, Conrad hopped off the ladder, onto the lunar surface ... and said it. Fallaci never paid up, though!

TO THE OCEAN OF STORMS

The mission of Apollo 12 should be viewed through the prism of NASA's race with the decade; for at the beginning of January 1969, it might very well have wound up as humanity's first manned lunar landing. Certainly, Pete Conrad made no secret of his personal wish to command the maiden touchdown on alien soil. Less than a month before the launch of Jim McDivitt's crew, NASA Headquarters had issued a forecast of its long-range planning for the year ahead. In the aftermath of Tom Stafford's dress rehearsal in May, no fewer than *three* opportunities for lunar landings were timetabled: Neil Armstrong's mission in July, followed by Apollos 12 and 13 in September and November, respectively. All three were to be 'open-ended' flights, perhaps running as much as 11 days apiece. Conrad's crew had been directed at Apollo 12 ever since they finished backing up McDivitt's team and a new trio of Jim Lovell, Ken Mattingly and Fred Haise were assigned to Apollo 13.

Of course, when one considers NASA Administrator Tom Paine's promise to

Armstrong in the days preceding his launch – that they would be assigned to the *very next mission* if something went wrong on Apollo 11 – it is difficult to see how these plans might otherwise have played out. So it was that by June 1969, Apollo 12 had two quite different profiles: if Apollo 11 did not succeed, then Apollo 12 would inherit that same mission, but, if Apollo 11 succeeded, its scope would be significantly expanded: by performing a *precision touchdown* in a place called the Ocean of Storms, just a couple of hundred metres from an old NASA spacecraft named Surveyor 3. During their two Moonwalks, Conrad and Al Bean would visit that craft, retrieve a couple of instruments and return them to Earth. By the end of July, the race had been won, John Kennedy's goal had been met and Conrad's crew began training feverishly for the precision landing at a point near the equator and some 1,300 km west of the Sea of Tranquillity.

Oceanus Procellarum, the Ocean of Storms, lies in the western portion of the lunar nearside and gained its name from an old superstition that its appearance during the 'second quarter' heralded bad weather. It is the largest of the lunar maria, covering some four million square kilometres and stretching 2,500 km from north to south. Like Tranquillitatis, it is a thick, almost flat plain of lava. Around its periphery are a number of other, smaller maria, including the Sea of Clouds and the Sea of Moisture. Prior to the visit of Apollo 12, Procellarum had received no fewer than four emissaries from Earth: NASA's Surveyor 1 and 3 and the Soviet Union's Luna 9 and 13. What Pete Conrad and Al Bean discovered when they strode its crater-pocked terrain in November 1969 was a place of mystery which would generate more questions than answers.

The need to make pinpoint landings was entirely understandable, for there seemed little point in training crews to achieve specific geological objectives during traverses across the surface if they had no guarantees *where* their lunar modules would set down. Through no fault of their own, Armstrong and Aldrin had landed six kilometres downrange of the intended position. The trajectory specialists had several explanations, including the 'lumpiness' of the Moon's gravitational field due to the mascons, but it was mathematician Emil Schiesser who made the final breakthrough. The key was the Doppler effect: an apparent 'shifting' in frequency of light or sound waves emitted by a moving object when viewed by a stationary observer.

Radio signals from lunar modules in orbit around the Moon had a predictable pattern of Doppler effect, explained Andrew Chaikin. That effect was at its most acute when the craft was flying over the limb of the Moon and at its weakest when it was over the geometrical centre of the near side. "If planners could predict the pattern of Doppler shifts," Chaikin wrote, "they could compare that information with the actual shifts they detected. The differences would in turn reveal whether the lunar module was off course and by how much." This was the mathematical solution, but the problem remained of how to provide it in a form which could be easily be fed into the lander's guidance computer. It was decided essentially to 'fool' the computer into thinking that the landing point had moved. It was an elegant ruse because, as Chaikin observed, it "required entering only a single number".

Touching down with precision and so close to another spacecraft on only the *second* manned lunar landing was both daring and audacious. In planning the first landing, the site selectors had avoided craters. However, Surveyor 3 had landed amongst a nest of craters and was actually on the inner wall of a crater about 200 metres in diameter. Nevertheless, the geologists were delighted, because they wanted the astronauts to inspect and sample those craters. Assuming that Conrad and Bean managed to touch down within walking distance of Surveyor 3, they would examine the lander's condition after more than two years on the Moon and remove its television camera and various other components. They would also assemble the first Apollo Lunar Surface Experiments Package (ALSEP) and perform extensive geological exploration of their landing site.

When a tentative schedule for future flights was published by NASA on 29 July 1969, it showed that Apollo 12 had now moved back a couple of months to mid-November to accommodate planning for the precision landing. Three subsequent missions would follow every four months up to Apollo 15 in November 1970, each featuring two Moonwalks and 34 hours on the surface. These 'H' missions would be followed by as many as *five* 'J' missions flown at five-monthly intervals between April 1971 and December 1972, utilising long-duration lunar modules for three-day stays on the surface and a battery-powered car called the Lunar Roving Vehicle (LRV) to extend to as much as 10 km the distance astronauts could venture from their lander. The later missions would also be more ambitious in terms of *where* they landed – with options including the Littrow and Marius Hills volcanic areas, the snaking, river-like channel of Schröter's Valley and the enormous craters Copernicus and Tycho. Even at this time, with the report of Vice-President Ted Agnew's Space Task Group hot off the press, there remained some optimism that after this expeditionary phase of Apollo, exploration might intensify with the construction of a permanent lunar base.

Such dreams, of course, would be soon quashed and by January 1970 one of the later Apollo missions would have been scrapped in order to use its Saturn V to launch Skylab and in September of that year two more would be lost due to lack of money. In the eyes of the public, and some of the media, the excitement of seeing men walking on the Moon – just a few months after Apollo 11 – was already beginning to wear thin. Not everyone was totally blinkered to the reality of how *difficult* these mission were, however. "Though the flight of Apollo 12 may seem like history relived," *Time* told its readers on 24 October, "the second American effort to land men on the Moon should be almost as dramatic as its predecessor. It will demand every bit as much daring from its all-Navy crew."

'ANIMAL' AND 'BEANO'

With Pete Conrad in command, the remainder of that all-Navy crew included Dick Gordon as the command module pilot and Al Bean, the man who would become the fourth person to set foot on the Moon. At the time of Apollo 12's launch, *all three* wore the silver leaf of a commander and, after splashdown, *all three* would be promoted to captains. In Bean's case, that would make him one of the service's youngest-ever captains, at the age of just 37. Yet, as recently as a couple of years earlier, Bean could not have imagined that he would be in this exalted position. Born in Wheeler, Texas, on 15 March 1932, Alan LaVern Bean's father worked for the Soil Conservation Service and his mother ran her own ice cream shop. A keen athlete, he received his degree in aeronautical engineering from the University of Texas at Austin in 1955 and joined the United States Navy through the Reserve Officers Training Corps. Initial flight instruction was followed by assignment to a jet attack squadron, based at the Naval Air Station in Jacksonville, Florida.

From an early age, Bean was fascinated by painting and took night classes in oils whilst in Florida; indeed, after leaving NASA, he would devote his life to conveying the story of his lunar adventure on canvas. Today, his paintings routinely fetch thousands of dollars and he always adds a few unique finishing touches to each one: a smearing of real Moon dust, salvaged from his own space suit patches, or a print made by a real lunar boot or a scratch made by a real Apollo geological hammer or a groove etched by a core-tube sampler. Bean has been interviewed many times over the years and has always described himself as "an artist, creating paintings that record for future generations mankind's first exploration of another world". Some astronauts have said that going to the Moon did not change them. For Bean, walking the dusty plains of the Ocean of Storms was a profound part of his life and has guided him ever since.

In his history of the Apollo missions, Andrew Chaikin described how 'different' Bean was compared to most other astronauts; for he exhibited few of their macho qualities and their overarching desire to be first. Instead, he had followed a simple motto: Work hard and *someone* will notice. Since his days in Jacksonville, he had been alternately nicknamed 'Sarsaparilla', because he never touched a drink, or simply 'Beano'. As a squadron pilot and, later, as a test pilot at Patuxent River in Maryland, his sheer determination had enabled him to master weapons delivery systems and turned him into an outstanding aviator. Whilst at test pilot school, he met the one instructor who would most significantly change his life ... Charles 'Pete' Conrad.

All three Apollo 12 crewmen responded to NASA's call for astronaut candidates in the spring and summer of 1962; Conrad made it, but both Bean and Gordon would have to wait another year before their time came. So too would another pilot, a Marine named Clifton 'C.C.' Williams, who, although killed in an air accident long before the second manned landing on the Moon, would be associated with it to the extent of being honoured by a star on that crew's mission patch.

Selection to the world's most elite flying fraternity in October 1963 would bring both frustration and triumph for Bean, Williams and Gordon. Within months, like the other members of the Fourteen, they had received their individual technical assignments. Gordon would be named as head of the Apollo branch of the astronaut corps, whilst Bean and Williams were given duties pertaining to spacecraft recovery systems and range operations, respectively. Unsurprisingly, Gordon was first to draw a flight opportunity: he served as Conrad's pilot on Gemini XI in September 1966 and performed a spacewalk. However, Bean and Williams wound up as the backup crew for Gemini X. That made Bean the first of his class to receive a command position (albeit in a backup capacity) on his *very first* assignment. The downside was that Deke Slayton's three-flight crew rotation system should have next pointed the Bean/Williams pairing at the prime crew slot for Gemini XIII ... except that the project *ended* with Gemini XII. Nevertheless, the fact that he had been assigned a backup command implied to Bean that he was highly regarded by Slayton and Chief Astronaut Al Shepard and hence a firm flight assignment *must* be around the corner. His hopes subsided when, in late 1966, he was detailed instead to work a project known as 'Apollo Applications' – the main feature of which was to be a space station that would be launched in the early Seventies – and any chance of a lunar mission seemed gone forever. In his autobiography, however, Slayton endorsed Bean's initial supposition: that he *was* highly regarded. "Al was just a victim of the numbers game," Slayton wrote. "I would only point to the fact that he was the first guy from his group assigned as a crew commander. I was confident he could do the job."

So it was that in the autumn of 1966, when Pete Conrad was asked to pick an astronaut to someday serve as his lunar module pilot, Al Bean was his first choice. Slayton turned him down, saying that Bean was currently working on Apollo Applications, and Conrad opted instead for the burly and jovial C.C. Williams. He "brought a gentle, self-effacing presence to the astronaut corps," wrote Andrew Chaikin. "He would set his jaw and utter with mock seriousness: 'I'm a marine. I'm a trained killer.'" Less than a year later, on 5 October 1967, Williams was flying his T-38 from Cape Canaveral to Mobile, Alabama, to visit his cancer-stricken father. Near Tallahassee, the jet suddenly went into an uncontrolled aileron roll. It "lost all its lift and became a ballistic projectile, accelerating ... at a horrendous rate," wrote Chaikin. Although Williams followed the prescribed ejection procedures, he was too low and too fast for his parachute to save him ...

Within weeks of receiving the devastating news of Williams' death, Conrad caught up with Bean with some good news, at last. Conrad had gone back to Slayton and asked him to reconsider Bean as his replacement LMP. Slayton had relented. After a year in the relative purgatory of Apollo Applications, Bean's future now looked more promising. If the schedule ran as planned, he might – just might – end up flying one of the early lunar landing missions. As 1969 dawned, it could easily have been the first one.

The other member of the Apollo 12 crew, and a man who deeply wished to walk on the Moon himself, was the CMP, veteran astronaut Dick Gordon. In fact, in the weeks after returning from their lunar expedition, Conrad would gather his men together and advise them to move over to the Skylab space station project, where the flight opportunities lay. Conrad saw that the writing was already on the wall for lunar landings, even at the end of euphoric 1969. Bean followed his old instructor's advice and both men would go on to fly Skylab missions; but Gordon accepted backup duties on Apollo 15 in the hope of rotating to Apollo 18 and walking the lunar surface as a commander of his *own* crew. When NASA budget cuts deleted *that* mission, Gordon diverted his energies into a push to lead Apollo 17 ... but without success.

This acute disappointment has long coloured many opinions of Richard Francis Gordon Jr, a man who would ultimately chalk up two space flights to his credit and

The Apollo 12 crew listen to a briefing prior to an emergency water-egress training exercise. Left to right are Dick Gordon, Pete Conrad and Al Bean.

come so close, but so far, from actually reaching the Moon's surface. He was an almost perfect match with Pete Conrad and the two had shared a friendship which long pre-dated their NASA days, back to the late Fifties when they were roommates aboard the aircraft carrier *Ranger*. A decade later, as astronauts, they earned a reputation for being cocky and fun-loving – Gordon, indeed, was such a ladies' man that Conrad came to nickname him 'Animal' – yet both were intensely focused on their jobs.

The respect in which Gordon was held as a test pilot and naval aviator also preceded his time at NASA. In fact, when he unsuccessfully applied to join the 1962 astronaut group, he was already on first-name terms with some of the *illuminati* of the Mercury Seven, including Al Shepard, Wally Schirra and Deke Slayton. In 1971, Slayton would consider it one of his most difficult tasks trying to choose between Gordon and Gene Cernan to command the final Apollo lunar mission. Gordon would lose, but by barely a whisker.

He had been born in Seattle on 5 October 1929 and attended high school in Washington State with dreams of the priesthood, rather than any aspiration to fly. Upon receiving his bachelor's degree in chemistry from the University of Washington in 1951, his focus shifted to professional baseball or a career in dentistry. He had settled firmly on the latter when the Korean War broke out and, in 1953, joined the Navy and discovered his life's true calling: aviation.

He would win top honours for his precise aerial manoeuvres, which guided him through All-Weather Flight School to jet transitional training to the all-weather fighter squadron at the Naval Air Station in Jacksonville, Florida. It was in 1957, shortly after being selected to join the Navy's test pilot school at Patuxent River, Maryland, that he and Conrad met and became lifelong friends. The pair would frequently while away raucous nights in bars and nightspots, knocking back beers and shots, then show up at the flight line six hours later, models of sobriety. "They were not only good pilots," wrote Deke Slayton, "but a good time."

After graduation, Gordon test-flew F-8U Crusaders, F-11F Tigers, F-J4 Furies and A-4 Skyhawks and served as the first project pilot for the F-4H Phantom II. Later, he moved on to become a Phantom flight instructor and helped introduce the aircraft to both the Atlantic and Pacific Fleets. His expertise and reputation as one of the hottest F-4H fighter jocks reached its zenith when Gordon used it to win the Bendix transcontinental race from Los Angeles to New York in May 1961. By doing so, he established a new speed record of almost 1,400 km/h and completed the epic coast-to-coast journey in barely two hours and 47 minutes.

In light of such professional accomplishments, it came as a surprise to many – not least Pete Conrad – when Gordon did not make the final cut for NASA's 1962 astronaut intake. The intensely competitive Gordon would describe his reaction as "pretty pissed off", but he plunged straight into applying for jobs in commercial aviation, intending to retire from naval service. One night in the bar he was met by the just-selected Conrad. As Conrad's widow Nancy would later describe, this encounter changed Gordon's career.

"Still crying in your beer, Dickie-Dickie?"

"Just crying for you, Pete, ya poor dumb sumbitch. Stuck in a garbage can in space with some Air Force puke while I'm out smoking the field in my Phantom."

"So, Dick. They're gonna fill out this Gemini program now that Apollo's approved. At least ten more slots. I think you oughtta apply again."

"And why would I do that?"

"Because you miss me."

A few months later, in October 1963, Gordon and 13 others, including Al Bean and the ill-fated C.C. Williams, were picked as astronauts. Three years after that, to his surprise and great joy, he would fly right-seat alongside his long-time buddy. And three years after *that*, they would fly to the Moon together. It would bring back memories of a picture of a flight-suited Conrad that he had sent Gordon in 1962, just after his own selection, on the back of which was written: 'To Dick: Until we serve together again'.

ROCKY START ...

If the Apollo 11 crew had been amiable strangers, with their conversations encompassing little more than the technical details of their mission, then the men of Apollo 12 were quite different and a multitude of tales exist of their time together. Andrew Chaikin tells the story of a friendship cultivated with Jim Rathmann, a car dealer in Cocoa Beach, whose contacts within General Motors allowed him to get them three matching gold Corvettes, the licence plates of which were emblazoned with their respective crew positions: CDR for Conrad, CMP for Gordon and LMP for Bean. Another anecdote is that Conrad – a long-time collector of baseball caps – tried to get a huge blue-and-white one made that would fit over the helmet of his space suit; he then intended to bounce in front of the television camera on the lunar surface to give his audiences a chuckle. Unfortunately, he could not think of a way of sneaking it aboard the spacecraft. Even the studious Al Bean was not exempt from the pranks: he planned to carry a camera-shutter self-timer to the Moon, with the intention of taking a photograph of himself and Conrad, standing in front of Surveyor 3. The point of it all? After landing, the inevitable question from everyone would have been: *Who* took the picture?

The humour also did not detract from the respect in which Conrad, Gordon and Bean were held as one of the sharpest crews in the simulator. Moreover, on a serious note, the crew patch paid touching homage to the man who might otherwise have become the fourth person to set foot on the Moon: the late C.C. Williams. Their naval backgrounds (Williams was, after all, a Marine) had already led them to choose the name 'Yankee Clipper' for the command and service modules and 'Intrepid' for the lunar module, from a selection of names submitted by workers at North American and Grumman respectively.

The Yankee Clipper name, in fact, had been submitted by George Glacken, a senior flight test engineer at North American; he felt that such ships of old had "majestically sailed the high seas with pride and prestige for a new America". Intrepid came from Grumman planner Robert Lambert, who felt that it denoted "this nation's resolute determination for continued exploration of space, stressing our astronauts' fortitude and endurance of hardship in man's continuing experiences for enlarging his Universe." Having picked the names, the crew's patch showed an old clipper ship, its sails set, in orbit around the Moon, surrounded by the naval colours of blue and gold. Bean dug out several pictures of American clippers – "we felt that the clipper ship was definitely an American symbol" – for incorporation into the patch and four stars were added, one for each crewman, plus Williams. The over-indulgence of the naval theme, however, may have proven a little too much for the Apollo 12 backup crew, Dave Scott, Al Worden and Jim Irwin ... *all of whom* happened to be Air Force officers!

November the fourteenth – launch day – dawned cold, cloudy and drizzly at the Kennedy Space Center. Weather reconnaissance had already identified a front of rain showers 130 km to the north and moving southwards; coupled with broken, low clouds and overcast conditions at 3,000 m, it seemed inevitable to some that the launch would be postponed. As Conrad, Gordon and Bean ate breakfast, the storm clouds rolled overhead and, later, as they lay in their couches aboard Yankee Clipper, they could see trickles of rainwater on the window. Somehow, it had worked its way underneath the spacecraft's boost protective cover. "The weather was erratic," wrote Andrew Chaikin. "The skies would seem to clear for a time and then gloom over again." Still, there were no predictions of thunderstorms or severe turbulence in the area and all of the conditions were 'better' than the minimum requirements specified in launch safety rules: cloud ceilings were acceptable, wind speeds within limits and no lightning for 30 km. As Launch Director Walter Kapryan deliberated whether or not to proceed, Pete Conrad lightened the mood by announcing that the Navy was always ready to do NASA's all-weather testing for it. It was a statement he would live to regret.

Barely an hour before the scheduled 11:22 am EST launch, a liquid oxygen replenishment pump failed, but the countdown continued. With more than 3,000 invited guests in attendance, including Richard Nixon, watching his first manned launch since assuming the presidency, Apollo 12 took flight precisely on schedule and quickly disappeared into the low deck of murky clouds. Pete Conrad reported the completion of the 'roll program' manoeuvre, as the rocket's guidance system placed it onto the correct heading for insertion into Earth orbit, then added "this baby's *really* going!" From their seats inside the darkened command module, the three men were astonished when, 30 seconds into the ascent, a bright flash illuminated the cabin, accompanied by a loud roar of static in their headsets … and then the wail of the master alarm. Glancing at the instrument panel, Conrad was shocked to behold more red and yellow warning lights than he had ever seen in his life. He had seen maybe three or four warning lights glow during simulations, but this looked like a Christmas tree. Even the worst training run had never shown up *this* many failures.

It seemed that something had gone horribly wrong with the spacecraft's electrical system: momentarily, all three fuel cells went down, the AC power buses died and the ship's gyroscopic platform tumbled out of control. "Okay," Conrad announced calmly to Mission Control, "we just lost the platform, gang. I don't know what happened here; we had everything in the world drop out." But what *had* happened? The crew was mystified. Al Bean, seated in the right-hand couch, guessed that *something* must have severed the electrical connections between the command module and the service module, which contained the fuel cells … but the gauges were showing that Yankee Clipper was still drawing power, albeit at a greatly reduced rate.

What the astronauts did *not* know – at least, not yet – was that the Saturn V had been twice hit by lightning. In fact, the first strike, which came 36.5 seconds after liftoff, was clearly visible to spectators on the ground. By that point, the giant rocket was hurtling towards orbit and had reached an altitude of a couple of kilometres: the strike had hit the vehicle and travelled down the long plume of flame and ionised gases of its exhaust all the way to the launch pad! At two kilometres long, it had set a new record for the Saturn V: making it the world's longest *lightning rod*. "Apollo 12 had created its own lightning," wrote Tom Stafford, who was watching the launch in his current capacity as chief astronaut, "when this huge, ionised gas plume from the first-stage engines opened an electrical path to the ground." Yankee Clipper's systems shut themselves down in response to this massive electrical surge and a *second* strike, some 52 seconds into the ascent, knocked out the gyroscopes. Automatic cameras close to Pad 39A recorded both strikes.

Immediately after the shutdowns, the command module had automatically transitioned to backup battery power. Almost straightaway, Conrad – his hand instinctively closing around the abort handle – began to suspect that lightning was to blame. As the mission commander, the decision was essentially his to make: he could abort several hundred million dollars of hardware and splash into the ocean a few minutes later ... or he could hold out and wind up in orbit with a dead spacecraft. Neither option particularly appealed. With these considerations in mind, it is unsurprising that Conrad opted to ride it out as long as possible.

Fortunately, the Saturn's guidance system was working perfectly and kept them on a smooth track into orbit. At this point, Conrad reported that he suspected a lightning strike to have caused the power dropout. Meanwhile, Gerry Griffin, seated in the Mission Operations Control Room (MOCR) in Houston, was undertaking his *first* stint as an ascent flight director and was almost certain that he would soon be forced to call an abort. If Apollo 12 wound up in Earth orbit with a lifeless spacecraft, then the crew would be as good as dead. Before doing so, however, he asked John Aaron, the 24-year-old electrical, environmental and communications officer, callsigned 'EECOM', for his recommendation.

Aaron's computer screen showed a jumble of virtually indecipherable numbers ... but he had encountered a similar problem during a training run a year earlier in which power had been inadvertently removed from the spacecraft and thought he knew how to resolve it. "Flight," he called over the intercom loop, "try SCE to Aux." Neither Griffin nor capcom Gerry Carr had the foggiest notion what this switch instruction meant and when it was radioed to Apollo 12, neither did Pete Conrad. In fact, it was Al Bean who recognised the switch for the Signal Conditioning Equipment and moved it to the Auxiliary position. Immediately, data reappeared on screens in Mission Control. The crew was instructed to bring the fuel cells back online by activating their reset switches. "The whole thing," concluded Nancy Conrad, "had taken less than 30 seconds."

The SCE converted raw signals from the instrumentation into data which was usable by Yankee Clipper's telemetry and Aaron had correctly deduced that it had gone offline following a major electrical surge. In Conrad's mind, two men had effectively saved the mission: Al Bean, by finding and acting on the SCE-to-Aux instruction, and John Aaron for making the call which restored control. Gradually, as Bean brought the fuel cells and electrical buses back online, the warning lights blinked off. When Apollo 12 reached orbit, Dick Gordon set to work taking star sightings and punching numbers into the guidance computer, recovering and realigning the internal navigational platform with just moments to spare before the spacecraft emerged from Earth's shadow.

However, there were still no guarantees that the mission was out of the woods. No one knew if the lunar module – Intrepid – had been damaged by the electrical surge and there would be no means of accessing it or checking its systems until *after* transposition and docking ... which itself had to occur *after* the make-or-break translunar injection burn for the Moon. "I listened while Griffin was briefed by his experts," wrote Chris Kraft of those critical minutes. "The lunar module was probably unscathed, they told him. But nobody knew for sure. Go or no-go? It was a decision that only Flight could make. Gerry Griffin made it ... one of the gutsiest decisions in all of Apollo and I was proud of it."

President Richard Nixon (centre) and NASA Administrator Tom Paine (right, with umbrella) brave the rain to watch the launch of Apollo 12. This was the first occasion on which a serving president had witnessed a manned launch in person; only months later, Nixon would respond by slashing NASA's budget and obliging the agency to cancel its last few Moon landings.

Halfway through their second orbit, with everything returned to normal, the S-IVB was reignited to commence the three-day journey to the Moon. Shortly thereafter, Gordon uncoupled Yankee Clipper and performed the transposition and docking with Intrepid, extracting the spider-like lunar module from the Saturn's final stage. "Everything's tickety-boo," Conrad reported, but to make sure, later that afternoon he and Bean opened the tunnel and completed a quick inspection of their lander. None of its electrical equipment had been damaged by the power surge and, by now, the astronauts were in such high spirits that they even asked Mission Control to replay the voice communications from those first few, adrenaline-fed seconds of launch.

The flight across the 370,000 km gulf of cislunar space was as uneventful and 'routine' as a mission to the Moon could possibly be, with long periods of dullness broken by the sound of country music cassettes carried by Conrad and Bean and bubblegum hit 'Sugar, Sugar', which had been sneaked aboard by Gordon. A nine-second mid-course correction was performed on 15 November to place Apollo 12 onto a 'hybrid' trajectory. Previous lunar crews – those of Borman, Stafford and Armstrong – had utilised a trajectory known as the 'free-return', in which, if they took no action, they would fly around the far side of the Moon and return to the vicinity of Earth with a minimal amount of mid-course corrections or manoeuvres. In effect, they utilised lunar gravity to slingshot them back home, essentially getting a 'free' return ticket. The main benefit of this trajectory was safety: even if a major malfunction occurred – such as a complete failure of the SPS engine – the spacecraft would return to Earth and to an ocean splashdown.

However, wrote Robin Wheeler in the *Apollo Flight Journal*, it also "severely limited the area on the Moon that Apollo missions could reach ... [and] was costly in terms of spacecraft performance requirements". Consequently, parallel investigations of other techniques were conducted and these led to the hybrid design. "The spacecraft," wrote Wheeler, "was injected into a highly eccentric elliptical orbit which had the free-return characteristics; that is, a return to the [re]-entry 'corridor' without any further manoeuvres. The launch vehicle energy requirements were reduced and a greater payload could be carried. Some three to five hours after TLI ... a mid-course manoeuvre would be performed by the spacecraft to place it on a lunar approach trajectory. This would not be free-return and hence would not be subject to the same limitations in trajectory geometry. This would therefore open the opportunity to reach landing sites at higher latitudes, with little or no plane change, by approaching the Moon on a highly inclined trajectory."

The hybrid plan offered large improvements in performance, but still retained many of the safety features of the earlier free-return. In fact, Apollo 12 would not depart from the free-return until after transposition and docking, after it had separated from the S-IVB and after the SPS had been tested. However, if the SPS failed to make the LOI burn, the spacecraft would emerge from behind the Moon on a trajectory that would not return it to Earth – in which case the lunar module's descent engine would be used to ensure that the crew could get home.

Four days after launch, at 8:47 am EST on 18 November, Yankee Clipper and Intrepid entered an initial lunar orbit of 312 x 115 km, which was later refined into a near-circular path of 122 x 100.6 km. Conrad and Bean floated through the connecting tunnel to begin preparing the lander for the descent to the Ocean of Storms the following day. Everything seemed to be going exceptionally well – the astronauts even found time to photograph the hilly Fra Mauro site, where Apollo 13 was scheduled to land in March 1970 – although Conrad and Bean's biomedical sensors seemed unwilling to co-operate: Conrad's were causing his skin to blister and those of Bean were producing erratic signals. Both men removed, cleaned and reattached their electrodes without further incident and finished donning their space suits in readiness for undocking and Powered Descent. Gordon gave them a few words of last-minute advice. "Let's go over this again, Pete," he grinned. "The *gas* is on the right; the *brake* is on the left!" With this last spell of banter behind them, Intrepid and Yankee Clipper parted company at 11:16:03 pm EST on 18 November, beginning a two-and-a-half-hour-long sweeping curve to the Ocean of Storms.

Understanding the nature of the terrain upon which Conrad and Bean would walk was aided by the presence of Surveyor 3, which, after bouncing several times, had touched down on the inner slope of a crater on 19 April 1967 and then transmitted a series of remarkable photographs of its surroundings. The landing site was located on photographs taken from orbit by one of the Lunar Orbiter spacecraft flown between 1966–67 to extensively map the Moon and reconnoitre potential landing sites for Apollo.

"Ewen Whitaker ... was a member of the Surveyor team," wrote Eric Jones in the *Apollo Lunar Surface Journal*, "and had the responsibility of identifying the landing sites. As the first pictures came in from Surveyor 3, it was immediately apparent that the spacecraft had landed in a crater. It was a relatively featureless crater, but there were a number of good-sized rocks scattered around, particularly to the north of the spacecraft. One pair of large rocks looked as though they were almost touching each other and it seemed to Whitaker that he might be able to find them when he started to examine the appropriate Orbiter pictures through the microscope. Within a couple of days, he was sure he had them; the rocks looked like mere pinpricks through the microscope, but there were other rocks visible as well and they made a pattern which matched up nicely with the Surveyor 3 pictures. Whitaker had found the crater."

The 200 m-wide crater would come to be known as 'Surveyor Crater' and it formed part of a somewhat distinctive cluster which, when viewed from Conrad and Bean's angle of approach, closely resembled a fat and jolly snowman with the unmanned probe sitting squarely within its belly. By using Lunar Orbiter images and photographs taken by the Borman, Stafford and Armstrong crews, it was possible for trajectory planners to construct a fairly realistic topographical model of the region in which Intrepid would land. Indeed, during training, the televised view through Conrad's left-hand window would be so realistic that, on landing day, 19 November, he would be astounded at how well the planners had done their jobs. "He would have a comfortable sense of déjà vu," wrote Eric Jones.

To better familiarise themselves with the spot, the astronauts had assigned nicknames to the craters which outlined other parts of the snowman's body: Head, Left Foot and Right Foot, to identify just a handful. Conrad hoped to land in what looked to be a relatively flat, smooth place close to Surveyor Crater – he called it 'Pete's Parking Lot' – but, either way, he knew that he had to touch down relatively close to target to allow them to walk over to the probe without difficulty. In one of his last conversations with Conrad, Deke Slayton had advised him not to say too much to the press about landing close to Surveyor 3, because if things went awry the newspapers were sure to portray the mission as a failure. Indeed, in the official Apollo 12 press kit, released by NASA a few days before launch, it is noted only that the crew would "develop techniques for a point landing capability". Having said this, the two geological traverses planned for Conrad and Bean *were* centred around the snowman, so if they could not land close to Surveyor Crater, much of the exploration phase of their mission would be lost.

It was with some scepticism, therefore, that Intrepid and Yankee Clipper parted company, for none of the astronauts were entirely convinced that the trajectory planners would indeed bring them directly down towards Surveyor Crater. Earlier that day, as he tucked into a breakfast of Canadian bacon, Conrad had told Bean: "I don't know what I'm gonna see when I pitch over. You know, I'm either gonna say 'Aaaaaa! *There* it is!' or I'm gonna say 'Freeze it, I don't recognise nothin'!'"

After undocking, Gordon had tracked them against an endless backdrop of craters with the command module's 28-power sextant until they disappeared from view. On the far side of the Moon, during their 13th orbit, Intrepid's descent engine roared to life for 29 seconds to reduce their perilune altitude to just 15 km. As they approached perilune, Intrepid was flying on its 'back', with the descent engine pointing along the flight path. At the appropriate time, the computer initiated the Powered Descent. After the braking phase, the vehicle pitched over until it was almost vertical and the two men were granted their first glimpse of the lunar landscape beneath them. They were astonished. Despite all their doubts, *there* was the snowman, right ahead of them.

"I think I see my crater," Conrad told Bean. "I'm not sure." At first, it looked like a maze – thousands of shadows inside thousands of craters – but then he caught sight of it and blurted out: "*There it is!* Son-of-a-gun, right down the middle of the road!" A conversation a few weeks earlier with trajectory specialist Dave Reed had led him to request a touchdown *in* Surveyor Crater; but *now*, as he beheld the landscape for real, Conrad decided to shift his aim-point a little shorter and to the north; in other words, directly into the patch of ground he had called 'Pete's Parking Lot'. A couple of hundred metres above the Moon, though, this area looked more like a battlefield and Conrad began making adjustments with his hand controller; pitching Intrepid forwards slightly to accelerate their forward velocity, passing around Surveyor Crater's northern rim and eyeballing a smooth spot to the north-west, close to Head Crater.

Moving lower now, with barely 30 metres to go, the descent engine began to kick up so much dust that it obscured the landing site. In the strange airless environment that he and Bean were preparing to visit, the dust did not billow around, but shot radially outwards, in bright streaks. As he 'felt' his way down using only rocks sticking up through the sheet of dust, he was relying on both eyeballs through the window and Bean reading the instruments. Sometime around 1:54 am EST on 19 November, four and a half days since leaving Earth, one of Intrepid's footpads found alien soil and the Lunar Contact light glowed blue. Instantly, Conrad's hand went to the Engine Stop button and the lander dropped the last half-metre or so to the surface.

"Good landing, Pete!" yelled Bean. "Out-*standing*, man!"

They had landed on an undulating plain, literally peppered with hundreds of craters, ranging from a few centimetres across to several hundred metres in diameter; the larger ones rimmed by large blocks of rock. The area was littered with six-metre-wide boulders, many of them angular, rather than rounded, and Bean noticed a peculiar 'patterning' to the surface: there were parallel cracks in the soil, maybe several millimetres deep, and the dominant colour seemed an intense white. Despite their enthusiasm and banter, neither man was unaware of the enormity of what they had done ... and where they were. "Those rocks," Conrad gushed, "have been waiting four billion years for us to come and grab them!"

Conrad knew that he was close to Surveyor Crater, but since there was no window in the back of Intrepid's cabin, he had no way of knowing exactly how far away they were from its rim. The landmarks which had seemed so obvious during the final minutes of descent were now less conspicuous. They knew that Head Crater should have been directly in front of them – and, indeed, it was – but it took them some time digesting the scene to finally realise that, yes, *there it was*. They were close to its eastern rim and were looking directly away from the sunlight of the early lunar morning. With a lack of colour variation, the crater was hard to see.

It was actually Dick Gordon, in orbit aboard Yankee Clipper, who managed to nail down their precise co-ordinates. During his second overhead pass, he spotted the snowman ... and then saw Intrepid's long shadow. "He's on the Surveyor Crater!" Gordon jubilantly told Mission Control. "He's about a fourth of the Surveyor Crater diameter to the north-west ... I'll tell you, he's the only thing that casts a shadow down there." A few seconds later, he added, with clear excitement in his voice, that he could see Surveyor 3 itself. The eastern wall of the crater was in shadow, but the body of the unmanned lander was catching the Sun. In fact, Conrad and Bean had set Intrepid down a mere 163 m from the old craft, which by any measure was a precision landing.

Now, Conrad could relax. "Holy cow," he breathed, "it's beautiful out there!"

... AND A ROCKY MOON

The two astronauts could not help but chuckle when capcom Gerry Carr advised them to rest before their first excursion onto the lunar surface. "After all the training, preparation, the dreams and visions of humankind for thousands of years," wrote Conrad's widow, Nancy, "and here they were, the third and fourth in the history of the species to set foot on this thing ... *and you expect us to sit down for a smoke break?*" As with Armstrong on Apollo 11, the chance to rest had been conservatively built into their schedule to allow for the possibility of the descent to the surface being delayed by one orbit; but now that they were *here*, Conrad, the mission's commander, elected to begin EVA-1 at the earliest opportunity. Like a couple of children eager to go outside and play in the snow, he and Bean proceeded crisply through the two-hour effort to don their lunar overshoes, snap on their gold-tinted visors, put on their life-support backpacks and connect the umbilicals, chest-mounted control units and gloves.

Four and a half hours after landing, they made one final check and Conrad radioed Houston for permission to depressurise Intrepid's cabin.

"Houston, are we Go for EVA?"

"Stand by, Intrepid. We'll be right with you," replied capcom Ed Gibson.

"*Stand by?*" asked Conrad, incredulously. "You guys oughtta be spring-loaded!"

Seconds later, after checking the status boards, Gibson gave them the go-ahead and as the final wisps of oxygen left the cabin, the astronauts stood in a near-pure vacuum. Within minutes, Conrad had squeezed himself through the hatch and onto the porch. He drew the lanyard deploy the Modular Equipment Storage Assembly (MESA) on the side of the descent stage, on which was the television camera – this time a colour one – and then dropped lightly down the ladder and onto the surface.

Without doubt, Oriana Fallaci's comment about astronauts being unable to speak for themselves popped into his mind. Conrad was the second-shortest astronaut in the corps and he had a five-hundred-dollar bet to win, so he spoke: "Whoopie! That may have been a *small* one for Neil, but it's a *long* one for me!"

In the wake of the mission, Conrad would relate that he never did catch up with Fallaci to collect his winnings. For now, though, the prize before him was beyond anything of monetary value. His first step onto the Moon came at 6:44 am EST and his initial moments were spent learning to walk in the peculiar environment. In fact, the timeline provided them each with *five minutes* to gain their bearings and a sure footing on lunar soil. Both would report that they never seemed to get tired in the bouncy lunar gravity, although the suit's limited flexibility meant that 'walking' often took the form of a sort of stiff-legged lope: running with straight legs, landing flat-footed, then pushing off with the toes. There were other aspects of Moonwalking which were quite different to their pre-flight simulations in the Houston 'sandpit': the fine, deep lunar dust quickly covered everything and, before long, the astronauts' pure-white suits were black from the knees down. Each time they moved, small clouds of dust kicked up around their feet and they grew more than a little nervous about the effect of this charcoal-like stuff on the working parts of their suits and Intrepid's systems.

Among Conrad's earliest tasks was collecting a contingency sample of lunar material from the Ocean of Storms. Even this was easier said than done. "We learned things that we could never have found out in a simulation," he recalled later. "A simple thing like shovelling soil into a sample bag, for instance, was an entirely new experience. First, you had to handle the shovel differently, stopping it before you would have on Earth and tilting it to dump the load much more steeply, after which the whole sample would slide off suddenly."

Bean joined him on the surface within half an hour. One of his first tasks was to retrieve the television camera from the MESA, mount it on a tripod and then position it so that the audience could observe their activities. However, as he carried the camera away from Intrepid, Houston announced that the camera seemed to have failed. When initial attempts to rectify the problem proved fruitless, Bean first shook the camera and then tapped it with his hammer. Nothing worked. It subsequently became clear that in carrying the camera on its tripod, Bean had inadvertently pointed it directly towards the Sun, burning out the light-sensitive coating on its vidicon tube. Audiences back on Earth had seen a bobbing image of the lunar landscape, then a brief glimpse of the Sun and then ... a meaningless pattern that remained fixed and unchanging. Although the television camera was not critical for the mission, its loss was a public-relations disaster for NASA on its second lunar landing. Apollo 12 became a radio show.

The early part of the excursion was akin, in some ways, to Apollo 11: the men set up the Stars and Stripes, finding the process of hammering the flagstaff into the soil nowhere near as difficult – although this time the hinge of the horizontal rod failed, leaving the flag draped around the staff – and unveiled a plaque on Intrepid's ladder leg. Unlike the plaque aboard Eagle and, indeed, the plaques aboard subsequent landers, that of Apollo 12 did not have a depiction of Earth and was textured

Al Bean manoeuvres himself down Intrepid's ladder. Note the square hatch, left partly open to reduce thermal stress on interior systems.

differently; instead of black letters on polished stainless steel, its *letters* were in polished stainless steel against a brushed-flat background.

Much of their time would be spent gathering samples, photography and setting up the first ALSEP scientific station. To ensure that they did not stray from the timeline, each man had a checklist attached to the cuff of his space suit and *this* ran through, sometimes minute by minute, what he was supposed to do. There were also a few 'additions' to Conrad and Bean's spring-bound cuff checklists, thanks to the antics of their backups, Dave Scott and Jim Irwin. "Part of my job," wrote Scott in his autobiography, "was to keep some levity in the game, keep things light and loose, relieve the tension when I could. In the last days before Pete and the crew were due to launch we got a cartoonist to draw some sketches to stick round their flight plans. We

stuck some pictures of 'Playboy' bunnies in there, too, which brought a few laughs."
Each of the images was accompanied by a lewd and suggestive comment, mostly about
inspection of geological features: 'Don't forget to describe the *protuberances*' or 'Seen
any hills and valleys?' for example. Conrad would occasionally ask Bean to flip to a
certain procedure, revealing only that he "might need your help on this". Controllers
who knew Pete Conrad were aware that he had a tendency to chuckle and hum to
himself whilst working, but even *they* were surprised when, on occasion, he broke into
a hearty cackle for no obvious reason. Only Scott and Irwin knew the truth ...

Conrad's humming and laughter and Bean's boylike wonder at being able to jump
impossibly high in the gentle gravity did not distract either of them from their packed
timeline. Within minutes of arriving on the surface, Conrad began erecting the S-
band communications antenna, but this was rendered redundant when Bean ruined
the television camera. Bean's major task was to remove the two pallets of ALSEP
equipment from the rear of Intrepid's descent stage. If the television camera had
been working, he would have relocated it to provide the audience with a clear view of
this activity. After connecting the two pallets to a horizontal bar, he would lug them
a hundred or so metres to the site chosen for their deployment.

Powered by a plutonium dioxide nuclear generator – whose rod-shaped fuel
cartridge was hot enough to melt their suit gloves – the ALSEP consisted of a central
station and a number of experiments: a seismometer to record quakes and tremors, a
suprathermal ion detector to characterise the low-energy positive ions of a near-
surface ionosphere, a spectrometer to study the electrons and protons of the solar
wind which impinge on the lunar surface, a magnetometer to conduct detailed
magnetic-field analyses, an instrument to measure the density of any ambient
atmosphere and any variations of a random character or correlated with local lunar
time or solar activity, and a dust detector, mounted on the central station, to provide
engineering data on the rate at which dust accumulated on the ALSEP as a measure
of the degradation of its thermal surface. Although the generator relied on energy
from plutonium, it was not a 'nuclear reactor'; the heat liberated by the fission
process was converted into electricity by a thermocouple, which is why it was
formally called a radioisotope thermal generator.

The central station was switched on at 9:21 am EST on 19 November, a couple of
hours into the Moonwalk, and would be shut down, along with the other ALSEPs in
September 1977. During its eight-year operational lifetime, it returned a vast wealth
of data about conditions on Earth's only natural satellite. In fact, when Intrepid's
ascent stage was sent crashing down to the Moon in a couple of days' time, the
seismometer would record no fewer than 55 minutes' worth of bell-like oscillations.
The 'strangeness' of these reverberations, said the instrument's principal investi-
gator, Gary Latham of Columbia University, was totally unlike terrestrial quakes
and might have been caused by a layer of fractured rock sandwiched between
bedrock in the floor of the Ocean of Storms and a more solid cover of finer materials
above it. Latham speculated that in the absence of dampening fluids or gases, this
rubble may have acted as a gigantic echo chamber. As later missions deployed a
network of seismometers, the true nature of the Moon's interior became evident,
which is that the pulverised layer extends to a considerable depth.

Pete Conrad sets up the Stars and Stripes on the Ocean of Storms.

Al Bean's hammer, earlier used in a futile attempt to get the television camera to work, was also employed by Conrad to hammer out the ALSEP's plutonium fuel rod, which had gotten stuck in its protective casket. "Never come to the Moon without a hammer!" was one of the astronauts' key pieces of advice. The two lunar-atmosphere experiments were placed into standby mode to enable internal gases to bake themselves out in two weeks of fierce lunar sunlight. In total, the men spent more than an hour setting up the ALSEP – undoing bolts, setting out the strangely shaped experiments, flattening the soil to ensure a level surface, attaching its bright-orange ribbon cables that carried power, command and data – and then moved on. Time, as Dick Gordon would later comment, was relentless.

Having finished the ALSEP on time, and been awarded an extension to the excursion, they inspected a pair of metre-tall conical mounds which Conrad likened to miniature volcanoes – undoubtedly prompting the geologists in Mission Control to wince somewhat. They quickly convinced themselves that these were large clods of compacted soil, most likely ejected by whatever impact had created the nearby Head Crater. Finally, they ran to the rim of Middle Crescent Crater, some 70 m west, took pictures of its shallow 300-m-diameter interior and then ran home, where Bean acquired a core sample.

Back inside Intrepid after a walk which, measured from cabin depressurisation to repressurisation, had lasted three hours and 56 minutes, the astronauts set to work stowing samples and recharging their PLSS backpacks with fresh reserves of oxygen and water for tomorrow's excursion. Their worries about how lunar dust might damage the lander's delicate systems and their own breathing meant that they kept their suits on throughout the 'night'. In reality, it was not 'night'; it was early in the lunar morning and 'morning' lasts a full week on the Moon. Consequently, Conrad and Bean's first excursion started at 6:30 am local time and by the time their second outing began – 13 hours later – only *half an hour* of lunar time had elapsed.

Sleeping in their suits, Conrad found, was about as snug as sleeping in football pads and he caught himself glancing several times out of Intrepid's triangular windows. The sky was blacker than anything he had ever seen, punctuated only by the blue and white marble of Earth, hanging there like a Christmas decoration. "The stars weren't brilliant," Nancy Conrad related. "He could hardly make them out at all in the harsh white light bouncing off the [surface]. It was all so cold . . . and as silent as silent got." Despite having a pair of light, beta-cloth hammocks, they were both particularly uncomfortable: some cooling water had gotten into one of Conrad's boots and the right leg of his suit had been misadjusted before launch and was too short. It was now pulling on his shoulder, wrote Andrew Chaikin, "like a vice". Conrad woke Bean at one stage and the two men set about undoing cords around the leg of the suit and retying them. Bean was not overly put out to have his rest period disturbed; he had tried taking a sleeping pill, but had slept little. Neither man got much sleep in the cold, cramped cabin and they ended up radioing Houston and starting preparations for their second Moonwalk a good two hours earlier than planned.

Conrad and Bean gulped down a quick breakfast, completed their suiting-up and were back outside by 11:00 pm EST on 19 November. After checking that the ALSEP was healthy, they set about their early exploration, trudging to Head Crater, then Bench Crater, then the relatively fresh Sharp Crater – a 12 m-wide and 3 m-deep bowl whose bright rim and ejecta implied that it was only a few million years old – and later Halo Crater. The men collected, documented and photographed rock and soil specimens, dug trenches, took core-tube samples and remarked on the strange colour of the surface: grey in some areas, brown in others, depending on the angle of the Sun. The lightness of the soil in places drew an excited response from the geologists, particularly when, on rounding Head Crater's western rim, Bean noticed that Conrad's bootprints had uncovered lighter textures beneath the dark upper coating. It had been theorised that such light soil could represent 'ray' material ejected by the impact which created the vast Copernicus crater, more than 300 km to the north. In fact, analysis of Conrad and Bean's samples helped to peg the age of the Copernicus impact to about 800 million years. Moreover, potassium-argon dating would reveal the Apollo 12 basalts to be around half a billion years *younger* than those from the Apollo 11 site, indicating the Moon was volcanically active for a substantial fraction of its early history.

Two hours after leaving Intrepid, they moved over to the southern rim of Surveyor Crater and could clearly see the three-legged, spindly craft, sitting on a 12-degree slope, some 45 m inside the pit. The pristine white surfaces that it had displayed on the day of its launch were gone – its light-tan discolouration was probably caused by more than two years of exposure to harsh sunlight and more than a little lunar dust. Descending slowly into the crater – they had been provided with a rope for safety – the men set to work on their respective tasks. Bean photographed the probe whilst Conrad removed samples for return to Earth: first, a

Pete Conrad inspects the Surveyor 3 probe in its crater. The lunar module Intrepid is
clearly visible on the horizon, just beyond the rim of the crater.

piece of insulated cable, then its television camera, a few other fragments and finally
its mechanical scoop.

Whilst alongside the robotic lander, the time seemed ripe for a touch of banter.
Conrad had been unable to smuggle his giant baseball cap aboard Apollo 12, but
they *had* managed to sneak a little chrome Hasselblad timer into one of their space
suit pockets to get a shot of themselves standing in front of Surveyor 3. They
dropped it into the tool carrier at the start of the Moonwalk ... but as *that* steadily
filled with rocks and soil, they couldn't find it! Bean rummaged around for a while,
but all he could see was ubiquitous dust on everything. The glint of chrome was
nowhere to be seen. In order not to give the game away to Mission Control, they
tried using hand signals, but in the end gave up, disappointed.

An hour or so later, back at Intrepid, Conrad emptied the tool carrier into a rock box and out popped the timer! "I've got something for you," he called to Bean. Exasperated, Bean grabbed the timer and threw it as far as he could into the distance. It is a pity that the two-man photograph could not be taken. When they finally re-entered the lander each man had at least 40 percent of his oxygen remaining. Before launch, Conrad had agreed with mission planners that if he and Bean were granted one extension to their Moonwalk, he would not ask for another. In total, they had spent around eight hours outside during their two excursions and neither man was exhausted, having expended less energy than anticipated. They felt, it is true, some pain in their forearms: the fingers of their gloves were fully pressurised and this had made them stiff and difficult to move. During the second period on the surface, Conrad and Bean had spent much time carrying tools and other equipment and *this* required them to keep their hands clenched almost constantly. After their return to Earth, both men would remark that this stiffness had impaired their efficiency on the Moon.

Still, there was plenty of pride when Intrepid's ascent stage lifted off at 9:25 am EST on 20 November. The astronauts had spent 31 hours on the Moon and the climb back to Yankee Clipper was picture-perfect – so perfect, in fact, that on the far side of the Moon during the early stages of the rendezvous, Conrad handed control to Bean for a few minutes. It was normally the commander's prerogative to fly the vehicle, and most presumed this prerogative without a second thought, but yielding control was also a moment of pure Pete Conrad. "Al would never forget," wrote Nancy Conrad, "the simplest, most natural gesture Pete offered, the only time it happened in the Apollo programme ... the commander let the rookie fly."

The reunion with Dick Gordon was both joyful and more than a little embarrassing. No sooner had the hatches been opened between Intrepid and Yankee Clipper, the CMP grinned, took one look at his two filthy crewmates – literally blackened from exposure to lunar dust – and refused to let them come aboard.

"You're not coming in my ship like that, Pete. Strip down."

"Say what?"

"You heard me. Get out of those suits and you can come in."

Despite his naval background, Gordon was not simply being finicky. No one knew what effect the abrasive lunar dust might have on the systems of the ship which would keep them alive for the three-day return to Earth; it might clog filters and hamper airflow in the command module. Gordon was not about to take the risk. As a result, Pete Conrad and Al Bean – the third and fourth humans in history to walk on the Moon – crossed from ship to ship in their birthday suits. Years later, Conrad would chuckle at the picture: if *something bad* had happened at *that* precise moment, and a thousand years later someone found them, what *would* they think?

"That I'm a sick and lonely man," Gordon deadpanned, "and I went to a *lot* of trouble and expense for some privacy!"

The ascent stage of Intrepid was sent to crash into the Moon to stimulate the ALSEP seismometer. Yankee Clipper spent another day in lunar orbit, its crew – now clothed – shooting photographs of the Fra Mauro foothills, targeted for a visit by Apollo 13 in March 1970, and possible future landing sites near the Descartes and Lalande craters. They also talked about the prospects for geologists on an alien world. It would be difficult, they admitted, to carry out efficient fieldwork on the Moon. Certainly, Al Bean felt that 'on the spot' geologising was hard and astronauts would benefit from selecting and documenting as many different kinds of lunar specimens as possible. Indeed, this practice of 'finding the suite' – capturing the *variety* of a geological site, from the mundane to the exotic, in just a dozen or so rock samples – was already being drilled into future landing crews, including Jim Lovell and Fred Haise, whose Apollo 13 mission promised to be the most dramatic yet flown ... and which would more than live up to predictions, although not quite in the manner intended.

THE LUCK OF THE DRAW

When Yankee Clipper splashed down into the somewhat choppy waves of the Pacific at 3:58 pm EST on 24 November, several kilometres from the recovery ship *Hornet*, the historic year of 1969 ended on a high, with eight further lunar landings seemingly on the cards. Behind the scenes, however, the chances of all of those missions taking place were in doubt and in January 1970 one of them – Apollo 20 – was cancelled so that its Saturn V could instead launch the Skylab space station. Some senior NASA managers, including Bob Gilruth, director of the Manned Spacecraft Center, felt that too many chances were being taken, too many risks accepted and were concerned that, sooner or later, an unknown or undetected problem would catch a crew out. If such a problem reared its head on the lunar surface or during the journey to or from the Moon, it could spell disaster and prove an absolute catastrophe for America's future in space.

Disaster, indeed, could quite easily have befallen Apollo 12. In early February 1970, North American, in its analysis of the two lightning strikes, recommended that when atmospheric conditions exhibited electrostatic gradients in excess of several thousand volts with severe fluctuations or when heavy cloud conditions associated with frontal passages existed – even in the absence of rain showers – a launch delay should be considered. Also, one of the command module's three fuel cells was placed onto a separate circuit to prevent total power loss in the event of a future electrical emergency. In the meantime, the Apollo 13 crew of Lovell, Haise and command module pilot Ken Mattingly were wrapping up their final months of training for what would be a true extravaganza in lunar geology. Like Conrad, Gordon and Bean, they would also spent ten days aloft, a little more than a day of which would be on the surface and Lovell and Haise would perform two Moonwalks, each running for perhaps four or even five hours. However, this was one of the earliest opportunities for the geological community to actually get their hands on the astronauts, metaphorically and literally, to help prepare them for the experience. Key to this effort was a young scientist – in fact, the *only* professional geologist in the astronaut corps – named Harrison 'Jack' Schmitt; a man who would someday walk the lunar surface himself.

Schmitt had already found an ally in Jim Lovell, who was, at the time, the world's most experienced space traveller. By now a veteran of three previous missions, Lovell had jointly secured the record for the longest manned flight in December 1965 and, three years later, had been a member of the first crew to circumnavigate the Moon. His career seemed blessed. In fact, Mike Collins' spinal injury in the summer of 1968 had catapulted Lovell from the backup to the prime slot on Apollo 8 and, having then backed up Neil Armstrong on Apollo 11, he was now primed to plant his own bootprints in the lunar soil, commanding his own crew. Yet he found it hard to see how the world would possibly remember the *third* manned Moon landing: Armstrong had made history and Conrad had perfected it with a pinpoint touchdown near Surveyor 3. The focus of Lovell's crew would be pure science.

It is no coincidence that the Apollo 13 patch proudly displayed the Latin words 'Ex Luna, Scientia' – 'From the Moon, Knowledge' – and that the mission's two spacecraft would be dubbed 'Aquarius' (for the lander) and 'Odyssey' (for the command and service modules). The former honoured the ancient Egyptian god who brought life, fertility and knowledge to the Nile Valley, whilst the latter paid homage to Homer's epic poem, denoting a long and arduous voyage with many changes in fortune. Both would prove highly appropriate for the mission they were about to undertake. The choice of 'Ex Luna, Scientia' harked back to all three men's naval backgrounds: for Lovell and Mattingly were both active Navy officers and Haise, though now a civilian, had served in the Marine Corps. Lovell borrowed the naval motto 'Ex Tridens, Scientia' ('From the Seas, Knowledge') and altered it slightly to fit Apollo 13's destination.

"I started out the design of this patch with the idea of the god Apollo, driving his chariot across the sky and dragging the Sun with it," Lovell said in a 1995 interview with the magazine *Quest*. "We eventually gave this idea to an artist in New York City, named Lumen Winter, and he came up with [a] three-horse design which symbolised the Apollo, but also included Earth and the Moon. The funny thing is that Winter, prior to making this patch for us, made a large wall mural of horses crossing the sky with Earth below, which is prominently displayed at the St Regis Hotel in New York City ... Anyway, we said, 'why put names on it?' We decided to eliminate the names and instead put in the Latin 'Ex Luna, Scientia'."

By the early spring of 1970, Jack Schmitt's involvement with future Apollo crews had long since shifted into high gear and many of his contacts in the geological community were playing an increasingly important role in preparing test pilots to become scientific observers on another world. Already, Schmitt and his colleague Gordon Swann had helped educate Pete Conrad and Al Bean in specific geological terminology and observation techniques, both of whom gave impressive performances on the Moon. Now Schmitt turned his focus to getting Lovell and Haise ready. To do this, he enlisted the help of his mentor, Gene Shoemaker, to sign up two of the professors at the California Institute of Technology who had inspired him while he was a student: Bob Sharp – who, in 1969, had famously investigated the mysterious 'moving stones' of Racetrack Playa in Death Valley – and the exceptional geochemical analyst Lee Silver. Also summoned for their advice were a handful of Schmitt's old Harvard colleagues, Jim Hayes, Gene Simmons and Jim Thompson.

Jim Lovell (left) and Fred Haise during a geological training exercise in the Quitman Mountains of far-western Texas.

Lee Silver's first encounter with the Apollo 13 crew was not with Jim Lovell, but with his lunar module pilot, Fred Haise. Silver offered to take Lovell and Haise and their backups, John Young and Charlie Duke, to a site in Southern California's Orocopia Mountains for a few days of impromptu training in late September 1969. Both Silver and Schmitt knew the pilots were deep in mission preparations and had to convince them that it would not be a waste of their time and, indeed, the eight-day trip was judged a great success. First, amidst a wide range of rocks and soils and a virtual absence of vegetation, Silver encouraged the quartet to describe their setting, then steadily honed their descriptions into more refined geological language. Lovell asked Silver to schedule further trips and subsequently toured Meteor Crater in Arizona and the volcanic terrains of Iceland.

Elsewhere, two other men were granted their own scientific teacher. Apollo 13's command module pilot, Ken Mattingly, and his backup, Jack Swigert, were aided by the Egyptian-born geologist Farouk el-Baz (nicknamed 'the Pharaoh') in preparing for their solo programmes of study in lunar orbit. Certainly, by the time Swigert flew with Young and Duke on Apollo 16, NASA was anticipating that the spacecraft's service module would carry a fully fledged observatory of cameras and instruments. "We don't know very much about the Moon," el-Baz told the astronauts. "*You* have a chance to help us know more." Even specific photography objectives were set: as well as looking out for 'unexpected' items of geological interest, the astronauts would keep track of glassy features, rock-soil junctions, undisturbed areas, rock surfaces and the morphology of craters, bumps and furrows. It was with much anticipation, therefore, that el-Baz, Silver, Schmitt and every other budding lunar geologist must have awaited the launch of Apollo 13. However, those of a superstitious disposition had long predicted that anything bearing the number 13 was bound to encounter ill-fortune ...

For Apollo 13, that brush with ill-fortune and that run of cruel luck did not even wait for the launch. In fact, it came *several days early*.

VETERAN, ENTHUSIAST, BACHELOR ... AND SPECTATOR

Thirteen has for hundreds of years been regarded by many cultures as an lucky number. Some have seen its origins in Judas Iscariot – Jesus' 13th disciple – who betrayed him in the Garden of Gethsemane and caused his arrest. Others have traced it all the way back to the great Babylonian lawgiver Hammurabi, whose legal stelae of the 18th century BC is supposed to have omitted a 13th law. In reality, the Code of Hammurabi was not numbered and, indeed, 13 is not a universally bad number in Judeo-Christian belief. Still more have drawn links with the mischievous Loki, 13th god of the Norse pantheon, and the notion that if 13 people gather, one will die within the year. Some Iranians still leave their homes on the 13th day of the Persian Calendar to avoid bad luck, the Knights Templar were arrested on Friday the 13th in 1307, King Harold decided to commit his troops to battle on Friday the 13th in 1066 (and they were heavily defeated at Hastings the following day) ... and, worst of all, this author's mother-in-law celebrates her birthday on October 13th!

Seriously, there are many who scoff at the idea. Statistics have suggested that, for whatever reason, driving is somehow 'safer' on Friday the 13th, that there are fewer accidents on Friday the 13th and Thirteen Clubs have sprung up all over the United States with the explicit purpose of disproving its fatalistic connotations.

For their part, Lovell, Mattingly and Haise were not immune to superstitious concern for their mission. In August 1969, when Lovell told his wife, Marilyn, about his next assignment, she experienced a twinge of superstition and in Ron Howard's 1995 movie *Apollo 13*, he has Ken Mattingly (played by Gary Sinise) telling journalists that the crew "had a black cat walk over a broken mirror under the lunar module ladder ... didn't seem to be a problem". Mattingly is said to have wished that the mission *could* have launched on Friday the 13th, so that he could include a black cat in the crew patch. Lovell, however, has scorned the notion that his crew was superstitious. His immortal words, again taken from the movie, but perhaps rooted in real events, sum up the extent of his thinking about the mystical number: "It comes after twelve!"

In truth, Apollo 13 would have absolutely no relationship with *Friday* the 13th: it would launch on a Tuesday and, although it was due to enter the field of lunar gravitational influence on the 13th, that day would be a Thursday, not a Friday. However, this date, together with the number of the mission, together with the launch time – at 13:13 Houston time – did little to dissuade the superstitious that something bad was inevitable. Already, in January, the press had read much into NASA's announcement that Apollo 13 would be postponed by a month from March 12 to April 11. Coupled with the decision to move Apollo 14 from July to October 1970, there had been speculation that budget problems were to blame or that difficulties had been encountered with the spacecraft or the Saturn V. In reality, it simply represented a 'stretching-out' of the remaining lunar missions and gave Lovell's crew more time to plan their surface science and orbital photography activities.

It also gave the astronauts a little more free time to spend with their families and, in Charlie Duke's case, to visit a friend a couple of weeks before launch. Unfortunately, one of his friend's children, three-year-old Paul Hause, had fallen ill with German measles and Duke succumbed. As the mission's backup lunar module pilot, he had routine contact with Lovell, Mattingly and Haise, had eaten meals with them and attended meetings with them ... and there was a very real possibility that he had infected them. Flight surgeons quickly verified that Duke was not contagious, but the prime crew found themselves providing daily blood tests. By 6 April, it was concluded that Lovell and Haise were probably immune, but Mattingly's case was uncertain – he might already be fighting the illness, the two-week incubation period of which meant that he might develop full-blown symptoms *during the flight*. His prospects deteriorated rapidly. On the 7th, as launch drew ominously closer, NASA physician Chuck Berry was already arguing vocally that Mattingly should be grounded for at least ten days, enough to accommodate the incubation period for measles. However, doing *that* would effectively remove him from the mission. Lovell complained about the recommendation. "Ken ... was one of the most conscientious, hardest-working of all the astronauts," he remarked years later. Lovell even spoke directly to NASA Administrator Tom Paine, telling him that, surely, measles were not *that* bad – a mild illness in adults – and if Mattingly fell ill, it would probably be during the journey back to Earth, a relatively quite time. Furthermore, Lovell knew from his experience on Apollo 8 that he and Haise could handle the command module by themselves, if necessary.

Paine was unmoved. In a way, his response made sense and can be compared to Jim Webb's vehement opposition to the C-prime decision in August 1968: any problem during the mission would be hungrily leapt upon by the press, by the public and by an increasingly fickle Congress, some of whose members were keenly waiting for the slightest opportunity to slash NASA's budget. No, concluded Paine, a case of German measles in space could prove dangerous – blurred vision, perhaps, or swollen joints in the hands – and could turn into a public relations disaster. Other options included postponing Apollo 13 until the next available lunar launch window on 9 May ... but *that* also ran the risk of allowing some Saturn V components to degrade and would cost the agency in excess of $800,000.

In Ron Howard's movie, the backup command module pilot, Jack Swigert, is presented as something of an inexperienced novice and, at one stage, actor Kevin Bacon describes his duties as having to "set up the guest list and book the hotel room" for the VIPs. This was far from the truth. For the last eight months, John

Air Force Instructor Chester Bohart (second left) with astronauts Charlie Duke (right), Ken Mattingly (second right) and Jack Swigert (left), pictured during desert survival training in August 1967. The careers of all three astronauts would figuratively collide and bring them both fortune and misfortune less than three years later with the unlucky voyage of Apollo 13.

Young, Charlie Duke and Swigert had virtually shadowed Lovell's team and were not only *ready* to step into their shoes if needed, but were also able to integrate their individual members into the prime crew. "Jack was a real good command module pilot," Duke recalled in a 1999 interview. "We were ready to go as a crew and it showed the beauty of the synergy in all of our training, that you could take somebody, a week before liftoff, stick him in and everyone felt comfortable."

That is not to say that there were no concerns. In his autobiography, Deke Slayton expressed reluctance to swap the *entire* backup crew of Young, Swigert and

Duke, feeling that Lovell's team were better prepared, but bringing even three highly motivated pilots together at three days' notice and moulding them into a lunargoing unit posed challenges. "Jim and Fred and Ken had developed their own language and their own way of working together," Slayton recalled. "There had been some cross-training and certainly some of these guys knew each other fairly well, but it was not the same thing." It was, he concluded, like dropping Glenn Miller into Tommy Dorsey's band; both were great musicians, but each had individual styles. In Slayton's mind, the deciding factor was that the CMP would spend a lot of time operating alone – particularly in lunar orbit – so it was reasonable to send Swigert in Mattingly's place. Had it been one of the landing crew who had fallen ill, the outcome may have been quite different and Apollo 13 would probably have been delayed until the next lunar launch window in May.

It also made the final days of training increasingly hectic ones. "Normally in that period," Haise recalled years later, "you would literally quit training, go off to a beach house ... and be isolated and sit around. Generally, you might read your checklist and go a little fishing on the beach, but get rested up for the launch and the subsequent mission. With *that* change though, we did go back into the simulation mode through most of the critical phases of the flight, like launch or lunar orbit insertion or transearth injection, leaving lunar orbit [and] rendezvous. It was more an exercise ... to make sure that the backup crew hadn't developed something different ... than we had."

Lovell had nothing against Jack Swigert, but certainly put up something of a fight to keep his crew together. "During the course of the previous year," he wrote in his autobiography, co-authored with Jeffrey Kluger, "the original crew members had become so accustomed to working with one another that Lovell and Haise could even interpret the nuances and inflections in Mattingly's voice – a valuable skill at those moments in the flight when the two pilots would have to rely on the [CMP's] radioed commands alone to steer their lander to a safe rendezvous." In a 12-hour integrated test in the command module simulator on 9 April with Lovell and Haise, Swigert proved himself to be a maestro with the systems and an expert at rendezvous, transposition and docking. Lovell agreed to the crew switch.

Poor Ken Mattingly, on the other hand, was left very much in limbo. As Swigert took all of the simulator time in those last few days, Apollo 13's former CMP could only do so much running and so much flying to occupy his time. "He considered breaking the pre-mission medical quarantine," wrote Andrew Chaikin, "to cheer himself up with some Dunkin' Donuts – those and the barbecue sandwiches at Fat Boy's were his main weaknesses at the Cape – but decided against it." At length, driving home to the crew quarters, he heard the formal announcement on the radio: that Swigert was taking his place on the mission. It was with intense disappointment, therefore, that Mattingly sat in Mission Control and watched his two former crewmates spear for the heavens on the afternoon of 11 April.

James Arthur Lovell Jr – nicknamed 'Shaky' for his bubbling-over stores of nervous energy – was undergoing his seventh stint on either a prime or backup crew and was now getting ready for a record-breaking fourth trip into space. He was born on 25 March 1928 in Cleveland, Ohio, and his fascination with rockets manifested itself at a young age. In his book with Kluger, he recalled sheepishly visiting a supplier in Chicago one day in the spring of 1945 to buy chemicals to enable himself and two school friends to build a rocket for their science project. Despite the boys' admonitions over wanting to create a liquid-fuelled rocket, like those flown by Robert Goddard and Hermann Oberth, their teacher guided them instead toward a solid-propelled option, loaded with potassium nitrate, sulphur and some charcoal. Days later, after packing the gunpowder ingredients into a shell made from cardboard tubes, a wooden nose cone and a set of fins, they took their rocket into a field, lit the fuse and ran like hell.

"Crouching with his friends," Lovell wrote in third-person narrative about his exploits, "he watched agape as the rocket he had just ignited smouldered for an instant, hissed promisingly and, to the astonishment of the three boys, leapt from the ground. Trailing smoke, it zigzagged into the air, climbing about 80 feet before it wobbled ominously, took a sharp and surprising turn and with a loud crack exploded in a splendid suicide."

Lovell's interest in the workings and possibilities of such projectiles eclipsed that of his friends, who regarded this as little more than a lark, but his family situation made it seem unrealistic to hope that a career in rocketry was within his grasp. His family had moved to Milwaukee, Wisconsin, when he was a young boy and his father's death in a 1940 car accident had placed enormous pressure on his mother to make ends meet. The military and, in particular, the Navy, seemed an attractive alternative. (His uncle, in fact, had been one of world's earliest naval aviators during the First World War.) He was accepted into the Navy, which offered to pay for two years of an undergraduate degree, provide initial flight training and six months of active sea duty. Lovell jumped at the chance and, within months, was registered as an engineering student at the University of Wisconsin at Madison. He completed his studies in 1952 with a bachelor's degree from the Naval Academy at Annapolis. Whilst attending the academy, he met Marilyn Gerlach, whom he married barely three hours after his graduation ceremony ... and who had typed up his carefully prepared thesis on liquid-fuelled rocketry.

Flight training consumed much of the next two years, after which he was attached to Composite Squadron Three, based in San Francisco, whose speciality included nighttime takeoffs and landings on aircraft carriers at the height of the Korean conflict. Several months later, Lovell was flying F-2H Banshee jets from the carrier *Shangri-La* over the Sea of Japan. It was during this time that a routine flight went seriously wrong. It was his first mission in darkness. The only means of determining where the carrier was at night, he wrote in his book with Kluger, was a beamed, 518-kilocycle signal from the *Shangri-La*, designed to allow the aircraft's automatic direction finder to guide him home. However, poor weather forced the ship to cancel the mission of Lovell, his teammates Bill Knutson and Daren Hillery and their group leader Dan Klinger; in fact, Klinger had not even left the deck of the *Shangri-La* when the flight was terminated. Unfortunately for Lovell, his direction finder had picked up the signal of a tracking station on the Japanese coast – which also happened to be transmitting at 518 kilocycles – and, far from guiding him back to the *Shangri-La*, it was actually leading him away. Around him, he saw nothing but a "bowl of blackness".

Perhaps the homing frequencies had changed, Lovell thought. At once, he turned to the list of frequencies on his kneeboard, but switching on his small, jury-rigged reading light, "there was a brilliant flicker – the unmistakable sign of an overloaded circuit shorting itself out – and instantly, every bulb on the instrument panel and in the cockpit went dead". His options seemed dire: ask the *Shangri-La* to switch its lights on, which was hardly advisable and would prove hugely embarrassing, or ditch in the icy sea. Then, in a story repeated by Tom Hanks, who played Lovell in the movie *Apollo 13*, his dark-adapted eyes perceived a faint greenish glow, like a vast 'carpet', stretching out below and ahead of him. It was the phosphorescent algae churned up in the *Shangri-La*'s wake and it served as his guiding light to bring him back to the company of his two wingmen, Knutson and Hillery, and a safe, though hard, landing which he later described as "a spine-compressing thud".

For his efforts, the sweat-drenched Lovell was given a small bottle of brandy, which he downed in a single gulp, and the opportunity to fly his next nocturnal mission ... the *very next night*. This time, his automatic direction finder behaved flawlessly. Eventually, he accumulated no fewer than 107 carrier landings and became an instructor in the F-J4 Fury, F-8U Crusader and F-3H Demon jets, before moving to the Navy's Test Pilot School at Pax River. Less than two years later, in early 1959, he was one of 110 military test pilots ordered to attend a classified briefing in Washington, DC. Like Pete Conrad, he would be turned down as a candidate for Project Mercury (a minor liver ailment was to blame), but secured admission into the exalted ranks of NASA's astronaut corps in September 1962.

By the time Neil Armstrong set foot on the Moon, Lovell had flown in space three times and would later recall that he raptly watched history in the making and yearned for the chance to do it himself. His wife Marilyn, on the other hand, was far more cautious, particularly when they went to the cinema one evening in November 1969 to watch a Gregory Peck movie called *Marooned*. During the course of the movie, a team of astronauts were stuck in orbit, unable to return to the space station which they had just departed and unable to return to Earth. It was, she noted, the *commander* of the crew who died at the end. Lovell was the commander of Apollo 13; the parallels seemed inescapable. Despite her fears, Marilyn accepted the risks as best she could, doubtless glad that her husband had promised that *this* would be his final space mission.

His two colleagues, John Leonard Swigert Jr and Fred Wallace Haise Jr, were at the opposite end of the astronauts' pecking order; for Apollo 13 would mark their first flights into space. Both had been selected four years earlier, in April 1966, as members of NASA's fifth astronaut intake. Ironically, this 19-strong class had nicknamed themselves 'the Original Nineteen', in a play on words referring to the Original Seven astronauts chosen for Project Mercury, with an expectation that each of them might wait for years before getting a chance to fly.

As well as being new to the crew, Jack Swigert was also a total opposite of Lovell and Haise; like Mattingly, he was unmarried, childless and on Apollo 13 he would become the first bachelor astronaut to fly into space. Born in Denver, Colorado, on

Jim Lovell trains for his ALSEP tasks in the spring of 1970.

30 August 1931, the son of an ophthalmologist, he earned a degree in mechanical engineering from the University of Colorado and entered the Air Force. Initial training at Nellis Air Force Base in Las Vegas was followed by three years' operational service in Japan and Korea. Swigert then resigned his commission, but continued as a National Guardsman in Massachusetts and Connecticut from 1957–60. As a civilian, he worked for Pratt & Whitney and North American Aviation as a test pilot, then completed a master's degree in aerospace science from Rensselaer Polytechnic Institute in 1965. Selection by NASA came the following year and he finished *another* master's credential (in business administration) from the University of Hartford in 1967.

Swigert gained a measure of fame amongst the astronauts, and has garnered more popular appeal since the *Apollo 13* movie, for his reputation as something of a ladies' man. Stories abounded in the days leading up to the mission that he presented himself as a "rambunctious bachelor" with a girl in every port. "[He] did what he could to perpetuate the image," wrote Lovell and Kluger. "His Houston apartment included a fur-covered recliner, a beer spigot in the kitchen, wine-making equipment and a state-of-the-art stereo system." NASA's good-natured acceptance of Swigert's lifestyle was very much at odds with its efforts in the early Sixties to present the astronauts as being squeaky clean; perhaps, speculated Lovell and Kluger, as the nation's attitudes loosened up later in the decade, so did those of the space agency itself. Some astronauts joked about Swigert's flat-footedness and his penchant for wearing white socks with a suit and tie ... but all jaws dropped when he *always* turned up to parties with a drop-dead-gorgeous date on his arm.

Yet Swigert's competence, his fierce confidence and his composure under stress were more than proven. Years earlier, whilst flying combat missions in Korea, he had landed in a driving squall, crashed into a road-grading machine mistakenly left on the runway ... and walked away unscathed from a fire which totally destroyed his aircraft. On another occasion, at Buckley Air Field, near Denver, his brakes failed and the aircraft slammed into an arresting net, but he escaped unhurt. Shortly after selection by NASA in April 1966, he went to Deke Slayton and volunteered to be a command module pilot, knowing that this would rule him out of someday walking on the Moon, but the pilot in him considered the CMP's job to be the more challenging option. After all, the CMP was the driver, the navigator, the rendezvous expert ... and the man who had to capable of flying home alone if necessary. In Swigert's mind, it was the perfect job.

If Swigert knew the command module like the back of his hand, then Fred Haise lived and breathed the lunar lander. Much of his time at NASA was spent at Grumman's Bethpage facility in New York, supervising the development of the spidery, bug-like machine, to such an extent that he would tell stories of actually catnapping on the ascent stage floor whilst waiting out another delay to another test run. He knew the vehicle as if it were an extension of himself and his knowledge of its inner workings was unparalleled. He was born in Biloxi, Mississippi, on 14 November 1933, attended high school in the area and joined the Marine Corps. Haise received initial flight training at the Naval Air Station in Pensacola, Florida, and served as a Marine Corps fighter pilot at Cherry Point, North Carolina, and as an interceptor pilot with the Oklahoma National Guard. As a consequence, he saw active service in *three* branches of the military and it was aviation and the outbreak of the Korean War which drew him away from his first love of journalism and into a quite different world. Haise cemented his academic credentials in 1959 with a degree in aeronautical engineering from the University of Oklahoma, then moved to the famed Edwards Air Force Base in California to attend test pilot school. Graduation in 1964 (he was honoured as the Outstanding Graduate of Class 64A) led to a series of assignments as a civilian NASA research pilot.

Selected as an astronaut a couple of years later, Haise would be described by Deke Slayton as "very capable" and "one of the best people in his group". It has been remarked that it was no accident that Haise ended up as the first of his class to receive a crew assignment: when Mike Collins' spinal injury meant Jim Lovell advanced to the prime crew for Apollo 8, Haise was inserted into the backup crew along with Neil Armstrong and Buzz Aldrin. After that, he backed up Aldrin on Apollo 11. Haise therefore owed his position on Apollo 13 to the injury suffered by Collins.

By the spring of 1970, Haise and his wife Mary had three children, and were expecting their fourth in June. Although generally referred to as Freddo, Lovell and Kluger related that as a result of playing a woodpecker in a first-grade play he also gained "the excruciatingly youthful nickname 'Pecky'".

Slayton's confidence in Haise was reinforced by his former crewmate, Ken Mattingly, in a 2001 NASA oral history. "Fred was the fastest switch-thrower, button-pusher in the west," he recalled. "He could sit there and rattle this stuff off. It eventually became a big joke with us that Freddo had this characteristic of being abrupt in everything he did. If he opened the door, he never used the doorknob; he *explosively* opened it. With Fred in the area, you always were cautious about walking by a door, because that was just his style! He knew every acronym. You mentioned a procedure and the instructors couldn't believe that Freddo knew all these things. They were astounded."

The expertise of both Swigert and Haise, and that of their commander, Jim Lovell, would prove vital when, late on the evening of 13 April 1970, the hands of fate turned on Apollo 13.

"BORED TO TEARS"

None of this could have been further from the astronauts' minds as they suited up on the morning of 11 April for the third manned landing on the Moon. Lovell and Haise had trained extensively to make a pinpoint touchdown in a patch of lunar highland near the crater Fra Mauro. It is named after a 15th-century Venetian cartographer-monk who completed one of the earliest, relatively accurate maps of the Old World. Unlike the open mare upon which Armstrong had walked, or even the cluster of craters on which Conrad landed, this patch was much more rugged, looking very much like a low 'island' in the Ocean of Storms, about 170 km east of where Intrepid landed. Many geologists were convinced that the lunar highlands had remained essentially unchanged, geochemically and morphologically, since the Moon formed some four and half billion years ago, so this pioneering expedition into the older and more heavily cratered highlands promised to offer some of the most ancient rocks on the Moon.

Several specimens returned by the earlier crews included chippings and fragments which differed quite markedly in their composition from 'ordinary' mare materials and they were thought to have been violently ejected, in some cases across hundreds of kilometres, by vast impacts in the distant highlands. The most obvious example of such an impact was the one which produced the 1,200 km-wide Imbrium basin, whose southern rim was 480 km to the north of the selected landing site, and the rugged terrain near Fra Mauro, known as the 'Fra Mauro Formation', was believed to be some of the ejecta from this cataclysmic event. Much of its material was expected to shed significant new light on the composition of the original, pre-

Imbrium lunar crust and help to establish an absolute date for when the impact occurred, for it would tie down an analysis of lunar history based on the stratigraphic order of deposits.

Of specific interest within this patch of the Fra Mauro Formation was a crater known as Cone, whose impact, lunar geologists suspected, had dug into a ridge of Imbrium ejecta. The crater was 300 m in diameter and Lunar Orbiter pictures showed its rim to be littered with large rocks drawn from deep in the ridge. Clearly, the samples obtained by Lovell and Haise as they approached Cone Crater were expected to reveal tantalising clues about the nature of the Imbrium impact.

Reaching Fra Mauro and Cone Crater involved a new fuel-conservation plan. Previously, Apollo spacecraft had entered more-or-less circular orbits, around 110 km above the Moon, after which the lunar module undocked to commence its descent. In this case, however, Odyssey would be inserted into an elliptical orbit with a high point of 110 km and a low point of only about 15 km. This would essentially relieve Aquarius of the need to perform the Descent Orbit Insertion (DOI) manoeuvre and provide Lovell with 15 seconds of additional hovering time to select a landing site. During his approach, he was expected to clear the 100 m ridge in which Cone was embedded and then find a safe patch on which to land on the hummocky plain beyond, somewhere in between two groups of craters, nicknamed 'Doublet' and 'Triplet'.

If reaching this hilly site was difficult, then rising from Earth and achieving a satisfactory translunar injection was by no means a walk in the park. However, by April 1970, with two successful Moon landings already achieved, the public had come to believe that such missions were *routine* ... and popular enthusiasm and interest declined precipitously. That Apollo 13 might be something out of the ordinary became apparent just five and a half minutes after its 2:13 pm EST launch, when the centre J-2 engine on the second stage prematurely shut down. Fortunately, the remaining four outboard engines automatically compensated for the loss by firing an additional 33 seconds longer than programmed, but even this produced a shortfall in the order of 68 m/sec. Consequently, the S-IVB third stage had to burn for nine seconds longer to achieve satisfactory insertion into Earth orbit.

Post-flight analysis would reveal that pogo oscillations, of which the Saturn V already had a long history, had subjected the second-stage engine to 68 G vibrations at 16 hertz, flexing its thrust frame by several centimetres. These oscillations, however, caused a sensor to register excessively low average pressures and the engine was shut down by computer command. Subsequent Saturns would benefit from helium gas reservoirs in the liquid oxygen line to dampen out such oscillations, together with automatic cut-off sensors and simplified propellant valves on all engines. In fact, such anti-pogo measures were already in the works before Apollo 13 set off.

The glitch provided a momentary fright in Mission Control, but the spacecraft was soon established in a near-circular parking orbit of 185 km. The standard checks were completed satisfactorily and the S-IVB fired for the second time at 4:54 pm to commence the journey to the Moon. Within an hour, Jack Swigert had uncoupled

Odyssey from the top of the third stage, swung it around and extracted the spider-like Aquarius. A mid-course correction burn later removed Apollo 13 from a free-return trajectory and placed it onto a hybrid trajectory. So far, the post-launch scare had not detracted from what seemed to be a smooth, near-perfect flight.

As Lovell, Swigert and Haise settled down for their first night's sleep in cislunar space, the now-discarded S-IVB had another mission ahead of it. At 8:13 pm, it fired its auxiliary propulsion system for three and a half minutes to deliberately hit the lunar surface and stimulate Apollo 12's seismometer. That quake turned out to be far more severe than anticipated. The spent rocket stage hit the Moon over 130 km from Pete's Parking Lot ... but the seismic signal it triggered lasted no less than *three hours and 20 minutes* and was so intense that ground controllers were forced to reduce the seismometer's gain to prevent its readings from going off-scale! Additionally, the suprathermal ion detector noted a jump in the number of ions from zero to 2,500 and then back to zero. Scientists would later theorise that the ionisation had either been produced by impact temperatures as high as 10,000°C or that particles from the impact had reached an altitude of 60 km above the surface and been ionised by sunlight. As events would transpire, this would be the only scientific result from the mission ...

For the first two days, everything ran exceptionally smoothly. Forty-six hours after launch, capcom Joe Kerwin even told them that Odyssey and Aquarius were behaving themselves so well that mission controllers were "bored to tears". From the spacecraft itself, Swigert's only concern was that, in the feverish rush of the last few days, he had forgotten to file his federal income tax forms and as a result of being in space he would not be home in time for the 15 April deadline. Kerwin, good-naturedly, rubbed salt in the wound by asking all three of them if they had completed their returns.

"How do I apply for an extension?" Swigert asked, in all seriousness.

In Mission Control, Kerwin laughed. But Swigert was not to be put off.

"Joe, it ain't too funny. Things happened kinda fast down there and I do need an extension."

Now more of the mission controllers were echoing Kerwin's laughter.

"I'm *really* serious," Swigert protested. "I didn't get mine filed."

"You're breaking the room up down here," Kerwin said.

"Well," Swigert acquiesced, "I may be spending time in *another quarantine* when we get back, besides the medical one they're planning for us!"

"We'll see what we can do, Jack," replied Kerwin, before pressing on with other matters. Thankfully, a request for an extension was filed on Swigert's behalf.

Fifty-five hours into the flight, in one of his last acts before going to bed, Jim Lovell guided his terrestrial audience through a tour of the ship: firstly showing them the machine which he and Haise would shortly use to touch down in the Fra Mauro hills, then drifting back up the connecting tunnel into the command module. He showed the helmet visors, other equipment and the hammocks they would string across Aquarius' cabin whilst on the Moon, then proudly held up a tape recorder, pounding out the thunderous theme from *2001: A Space Odyssey* and paid homage to the craft which was bearing them gracefully towards the Moon. A minute before

Crew-cutted Gene Kranz (back to camera) watches a television transmission from the Moon-bound Apollo 13, shortly before the explosion. Fred Haise's face is visible on the display in Mission Control.

10:00 pm EST – an hour earlier in Houston – he drew the telecast to a conclusion by wishing his audience a pleasant night's sleep. He could not possibly have known that, by now, going to the Moon had lost the sheen of the previous year and *The Doris Day Show*, NBC's *Rowan & Martin's Laugh-In*, the movie *Where Bullets Fly* and CBS' *Here's Lucy* were deemed to be more interesting to audiences on Earth. In fact, Lovell, Swigert and Haise's audience that evening were mainly flight controllers in Houston and, seated in the glassed-in gallery at the back of the Mission Operations Control Room, the astronauts' families. When Lovell signed off, all were relieved that things were going so well.

Seven minutes later, disaster struck.

STRUGGLE FOR SURVIVAL

Deep inside Odyssey's service module were two cryogenic tanks filled with liquid oxygen and two with liquid hydrogen to feed three fuel cells. "Flowing into the cells and reacting with the [catalysing] electrodes," wrote Lovell and Kluger, "the two gases would combine and, in a happy coincidence of chemistry and technology, produce a trio of byproducts: electricity, water and heat. From just two gases, the cells would produce three consumables no life-sustaining spacecraft could do without." During a routine countdown demonstration a few weeks earlier, these tanks were test-filled with cryogenic propellants and then drained. However, technicians had been unable to properly remove the liquid oxygen from the No. 2 tank; the quantity dropped to 92 percent and refused to budge further. "Gaseous oxygen at 80 psi was applied through the vent line to expel the liquid oxygen," Jim Lovell remarked later, "but to no avail." On 27 March, after round-table discussions and with the input of engineers, managers, technicians and even Ken Mattingly himself, additional attempts were made to remove the oxidiser.

At length, following a conference between NASA personnel and contractors, it was decided to 'boil off' the remaining liquid oxygen in the No. 2 tank with the aid of its electrical heater. (In space, this heater would be used to warm and expand the oxygen if the tank pressure dropped too low.) The process of boiling off the oxygen took eight hours of 65-volt power from the ground equipment, but was successful. Lovell and Mattingly were approached for their opinion and both men considered it safe to fly. "With the wisdom of hindsight," Lovell recalled, "I should have said 'Hold it. Wait a second. I'm flying on this spacecraft. Just go out and *replace* that tank.'" However, no one really wanted to face the prospect of a month-long delay costing $800,000 to replace a tank which *seemed* to be fit for purpose. In fact, the No. 2 tank had originally been earmarked for Tom Stafford's Apollo 10 mission, but was removed and replaced when it exhibited trouble. Shortly afterwards, in October 1968, it was accidentally *dropped* – admittedly only by a couple of centimetres, but enough to damage some of the tubing needed to fill and drain its propellants. *That* problem with the tubing was now at the root of why the tank could not be properly emptied. Unknown to everyone, including the astronauts, contractors and NASA managers, there was a deeper problem – a potentially catastrophic design flaw, the effect of which would not be felt until Apollo 13 was 55 hours into its mission . . . and 320,000 km from home.

In 1965, the Apollo spacecraft, which was originally intended to operate at 28 volts of electrical power, was upgraded to accept 65 volts from ground support equipment. Everything was duly modified to accept this change – except, that is, for a small thermostatic switch inside the oxygen tank, which was still rated at only 28 volts. This thermostat was meant to shut off the tank's heaters when the internal temperature reached about 26°C. No one picked up the error. When the residual oxygen in Apollo 13's tank was boiled off by the heater in late March 1970, the thermostat was activated and the excess voltage caused an arc which welded its electrical contacts shut. The technician monitoring the test had no idea of what was happening, since the gauges on his instrument panel went no higher than about 30°C. By the time the liquid oxygen had fully boiled off, the temperature in the tank had risen to over 500°C! The result was dangerous: the intense heat had baked and cracked the Teflon insulation covering wires for an electric fan which would be used to stir the tank's contents in space. When the highly volatile liquid oxygen was loaded, shortly before launch, there was still Teflon debris and now-bare electrical wiring inside the tank. Although Teflon itself is usually not flammable, the fragments were immersed in a pressurised oxygen atmosphere, in which almost *anything* would burn. A spark or a short-circuit would trigger an explosion. The No. 2 tank was nothing short of a bomb, waiting to go off. At 10:06 pm EST on 13 April 1970, a little more than two days into the voyage of Apollo 13, Mission Control issued a routine request to Jack Swigert, floating in the left-hand couch of Odyssey. He was to flip several switches to 'stir' the cryogenic tanks in the service module. In weightlessness, the slushy fluids, which were held at 'supercritical densities' at -206°C, tended to 'stratify' and make it hard for engineers to obtain precise quantity readings. Swigert complied, reaching over and flipping the switches marked H_2 FANS and O_2 FANS. All was silent for an instant. Then, without warning, Apollo 13 shuddered with a dull, metallic bang ... and fell quiet again.

Jim Lovell was floating back into the command module from Aquarius when the bang occurred and he instantly assumed it was caused by his mischievous crewmate, Fred Haise, who enjoyed setting off a noisy cabin repress valve to scare them. Glancing back down the tunnel towards the LMP, however, it was clear that Haise had no idea what had caused the bang ... and Swigert's eyes were as wide as saucers. Both men's expressions, wrote Lovell and Kluger, showed that they were "truly, wholly, profoundly frightened". No one could possibly have known at this stage that when power was applied to the electric fans, the stripped-bare wiring had shorted and the spark had ignited the Teflon fragments and caused a violent explosion that blew the top off the No. 2 tank, punctured the No. 1 tank and shed an entire side panel of the service module.

Instantly, Lovell and Swigert's attention was arrested by the blaring sound of the master alarm and the unnerving red glow of a warning light on Odyssey's instrument panel. It was the MAIN BUS B UNDERVOLT light. The very fact that it was illuminated suggested that there had been a loss of power from one of the command module's two main electrical buses. If the warning was for real, and was *not* simply a glitch, it signified *very* bad news. Apollo 13 was 320,000 km from Earth, at least four days from returning home ... and its very lifeblood was draining away at a rate which might kill the entire ship in a matter of minutes. The crew were in dire peril.

"Hey," Swigert shouted to Houston. "We've got a problem here."

"This is Houston," replied the duty capcom, astronaut Jack Lousma, in a perplexed tone. "Say again, please."

With more urgency, Lovell floated to his couch and keyed his mike. "Houston, we've had a problem. We've had a main B bus undervolt."

"Roger, main B undervolt," verified Lousma. "Okay, stand by, 13, we're looking at it."

Seymour 'Sy' Liebergot, the electrical, environmental and communications (EECOM) flight controller in Mission Control, was responsible for monitoring the power and life-support utilities of the spacecraft from launch until landing. Ordinarily, he and his fellow EECOMs requested 'cryo-stirs' of the oxygen and hydrogen tanks at least daily. As Jim Lovell concluded his telecast, Liebergot had noted from the data on his screen that a low-pressure reading was coming from one of the two hydrogen tanks. He recommended giving all four tanks a stir in order to stabilise their pressure readings.

Like all flight controllers, Liebergot was attuned to looking for seemingly minor instrumentation glitches, for he knew that something which at first *might* be nothing could expand into something beyond anyone's control. He had been bitten on several occasions by misinterpreting problems in training simulations. A few weeks before Apollo 13 set off, Mission Control and the astronauts had been midway through a fully integrated sim. Lovell and Haise had just landed on the lunar surface, whilst Mattingly orbited overhead in Odyssey. Just before the command ship disappeared behind the Moon and contact was lost, Liebergot noticed some peculiar readings in his data: a tiny, almost imperceptible drop in cabin pressure. He discussed the issue with his support team and concluded it was nothing more than 'ratty data'. They were wrong. The simulation team had thrown Mattingly a *real* pressure-loss emergency and when he emerged from behind the Moon·he reported that after a sudden, near-total depressurisation of Odyssey's cabin he was now wholly reliant on his space suit for life-support.

Liebergot had been stunned. He had misread the data, wasted time and failed to respond appropriately to an emergency. Gene Kranz was also being wrung through the mill that day, as the lead flight director, and in the next few frantic minutes the controllers assembled a makeshift plan for Lovell and Haise to immediately liftoff from the Moon, dock with Mattingly and use their limited fuel and life-support reserves to offer a 'lifeboat' capability. The astronauts would then crowd into Aquarius and essentially *exist* until just before re-entry, when they would crawl back into the command module, jettison the lunar and service modules and return safely home by the skin of their teeth. In truth, such lifeboat scenarios had been in NASA's mind since at least 1964, and during Apollo 9 Jim McDivitt's crew had rehearsed a few options. However, realistically, there were many flight controllers who wondered if it would really work.

Now something *very* bad had afflicted the command and service modules *for real*. Any possibility that it was merely ratty data was quickly eliminated by the astronauts themselves: for they had not just *seen* lights flash onto the instrument panel ... they had also *heard* the bang and they had *felt* the spacecraft shudder in response. Lovell could only compare it to a reverberating thunderclap and Haise had seen the walls of the tunnel visibly flex around him. For his part, Swigert was perplexed. Voltage and current levels seemed normal, as did the fuel cell data ... for now. At first, they guessed that Aquarius might have been hit by a micrometeoroid: as soon as Haise was back inside the command module, they moved to close the hatch and minimise any further pressure loss. Before they could do so, however, it became clear that nothing was wrong with the lander – if its hull had been compromised, the two craft, sharing an atmosphere, would have quickly depressurised. In fact, it was *Odyssey* which was crippled.

Sy Liebergot and his back room support team were trying to interpret the data. Although he tried excluding the most likely options first, it soon became clear that this was definitely no instrumentation error. Yet, for a while, Haise's report of a bang seemed to pass unnoticed. Whatever had happened, its effects were rapid. Before their very eyes, controllers could only watch in horror as the quantity and pressure readings for Odyssey's No. 2 oxygen tank crept to zero. From their perches in the command module, Lovell, Swigert and Haise were looking at the same thing.

More warning lights came on: now *two* of the three fuel cells were showing up as dead and the No. 1 tank's oxygen level was also steadily falling.

How *both* tanks could have been so affected was almost beyond belief. "The oxygen tanks were built with the fewest number of parts possible," wrote Lovell and Kluger, "making the likelihood of a breakdown as slim as possible. Even if one tank did fail, the other tank would still be more than adequate to power all three [fuel] cells. And as long as all three fuel cells were operating, both buses [A and B] would continue operating too. The probability of any one of these components failing was down in the multi-multi-decimal places. The probability of one tank, two fuel cells *and* one bus failing at the same time was off the numerical charts."

Ken Mattingly, too, was stunned. He was convinced at first that it *had* to be an instrumentation failure. Nothing like this had *ever* cropped up during simulations. "All our procedures," he told a NASA oral historian, "were based on two practical rules. One of them is: structural things *don't break*. Actually, that drove *everything*. Fluid lines and joints can leak, shorts can happen to wires ... but *physical structure* doesn't break. The reason we had that rule is, if you admitted to that, then the number of things that you could have to prepare for is infinite. It's big anyhow, but it was a practical matter. We had done a lot of testing of this, a lot of margin of safety in the hardware, so we never looked at those kinds of implications."

Yet, despite being so unlikely, a potential catastrophe was *happening*.

So too were other anomalies. The shuddering bang and the myriad problems had caused the computer to restart itself and the high-gain antenna ceased functioning and switched communications and tracking over to four smaller, omni-directional antennas, which could not carry the same bandwidth. To the men in Mission Control, these responses seemed to support the notion of an instrumentation problem. Neither Liebergot or his back room specialist Larry Sheaks believed that the oxygen tank pressure data was reliable – the manifolds were good, as was the environmental control system – but the readings on *their* screens were not mirrored by the situation aboard Apollo 13.

In the spacecraft, Haise, aware that Bus B was now effectively dead, set about reconnecting Odyssey's systems to Bus A ... only to discover that it, too, was starving of current. "We've got a main bus A undervolt showing," Swigert told Lousma. It seemed that the dead Bus B was dragging its counterpart along with it towards failure. Only fuel cell No. 2 was now producing electrical current. A mission rule forbade a lunar landing with only one operating fuel cell, so it was clear that exploring Fra Mauro was now out of the question.

Realisation that this emergency was for real came 13 minutes after the bang. Lovell noticed that as the spacecraft shuddered, the service module's thruster quads were automatically firing to compensate for the unwanted motions. Under normal conditions, during the cislunar voyage, Apollo spun like a top in order to provide passive thermal control to all its surfaces and prevent any section from becoming either too hot or too cold. The incessant vibrations and thruster firings, wrote Lovell and Kluger, "had shot the graceful choreography all to hell". For a while, Lovell operated his hand controller in an attempt to steady the spacecraft manually, but without success. Irritated, he floated to the left-hand window and craned his neck to look outside, hoping to see what was causing the problem … and what he saw sent a chill down his spine.

Outside, he beheld a thin, whitish cloud of gas emanating from the side of the service module. It crystallised as soon as it entered space, then formed a huge halo which extended in all directions. Instrumentation this was *not*. The ship was leaking.

Lovell keyed his mike. "Houston, we are venting something out into space. It's a gas of some sort."

"Roger," replied Jack Lousma, "we copy you're venting."

"Then, abruptly, all the pieces of the puzzle came together," wrote Gene Kranz in his autobiography. For this mission, Kranz was Flight Director of the White Team and the most senior of its four directors. "A shock rippled through the room as we recognised that an explosion somewhere in the service module had taken out our cryogenics and fuel cells. The controllers felt they were toppling into an abyss." Anger at having wasted precious minutes was now replaced by a steely determination to do whatever was needed to resolve the problem.

It was gradually becoming apparent that the explosion and the venting of oxygen from the No. 2 tank had led directly to the rolling torrent of anomalies that Apollo 13 was now experiencing. The jolt of the blast had snapped the fuel cells' reactant valves shut, effectively cutting off their supply of oxygen and hydrogen and starving the electrical system. Fuel valves to the service module's thrusters had also closed, making it difficult for Lovell or Swigert to regain control, and the propulsive effect of the still-venting oxygen caused the excessive vibrations and prevented the S-band antenna from locking onto Earth. An hour and a half after the explosion, at 10:38 pm EST – 9:38 pm in Houston – the situation had changed from one of dealing with an instrumentation problem to orchestrating a coherent means of getting Lovell, Swigert and Haise back home alive.

From the EECOM's console, watching as the remaining oxygen steadily drained away, Sy Liebergot and his back room colleague George Bliss were keenly aware that the final fuel cell was close to death. However, if the reactant valves for the other cells were closed, it might allow Mission Control to isolate the problem and perhaps save whatever was left in the final cell. "If the leak that was killing Tank 1 could not be found in the body of the tank or in the gas lines that ran from it," wrote Lovell and Kluger, "perhaps it was located downstream in one or both of the dead cells. Shutting off the valves either would stop the [oxygen] bleeding, allowing Odyssey to stabilise itself and power back up, or it would do nothing at all, allowing the controllers to give up on the ship altogether." By shutting down the reactant valves to the fuel cells, there would be no turning back: for they could not be restarted in space. It would be nothing short of a formal acknowledgement that Apollo 13 was now an aborted mission.

Gene Kranz's calm response belied his comprehension of the seriousness of what Liebergot was now advocating. In truth, ever since Lovell had seen the venting gas and everyone in Mission Control had linked this to the dropping oxygen gauges and the failing fuel cells, he knew that the prospect of a lunar landing was now gone. Getting the astronauts home alive was the new mission. It did not lessen the disappointment for the astronauts at losing their chance to walk on the Moon; in fact, a clipped directive from Lousma – "We want you to close to close the reac valve on fuel cell 3. You copy?" – stopped them in their tracks. Their long-awaited prize of being fifth and sixth men to walk on the Moon would instead belong to someone else.

The obvious course of action at this stage might have been to use Odyssey's huge SPS engine to swing them in a great U-turn and bring them back to Earth; a so-called 'direct abort'. It was an option initially favoured by Milt Windler, flight director of the Maroon Team, but Kranz and Glynn Lunney, in charge of the Black Team, disagreed. They felt that there was still no clear understanding of exactly *what* had caused the explosion *and* how severely damaged the service module really was. This precluded such a dangerous move as attempting a direct abort. Having Aquarius attached, Lunney chimed in, would buy them additional time and keep another option open. Seated at Kranz' console, Chris Kraft – the newly-appointed deputy director of the Manned Spacecraft Center, who had been called at home and pulled out of the shower by his wife to rush into Mission Control – agreed: a direct abort could totally close off all remaining options. It was too risky.

Other controllers felt the same way; there was some inexplicable sense of dread warning them that they should *not* even attempt to use the SPS. The decision quickly became a moot one, anyway. The damage to the service module could quite easily have wrecked the engine and, even if everything was still fine with it, they now lacked electricity to open valves to its combustion chamber, swivel its nozzle and effectively steer the five-minute-plus burn. The only real choice was to continue on their path to the Moon, loop around it, somehow re-establish a free-return trajectory and then have the astronauts simply *exist* for long enough to make it back home.

If this option were to stand the slightest chance of working, it depended on one thing: the lunar module, Aquarius.

"We are starting to think about the LM lifeboat," Jack Lousma radioed.

"That's what we've been thinking about, too," Swigert responded. In fact, Lovell and Haise were already scurrying down the tunnel into the lander, in an effort to bring it alive in around 30 minutes instead of the two hours or more on the checklist. Aquarius' descent engine did not have sufficient power to return them directly to Earth, but it *did* have enough oomph to place them back onto a free-return trajectory. In the meantime, Houston told Swigert to begin shutting down a fifth of the command module's systems in order to conserve what little power was left and everyone knew that very soon he would have to switch off the rest, including the guidance platform and the computer. Only batteries in the command module, needed for re-entry, parachute deployment and splashdown and offering just ten hours of life, would sustain the crew after the lunar module was finally jettisoned.

Before Swigert could shut down the command module, guidance and control had to be transferred over to Aquarius' computer, and this was no simple task. It was compounded by the fact that Apollo 13 was now surrounded by a cloud of debris, maybe 12 km wide, which caught the sunlight and glimmered, making it virtually impossible to acquire reliable star sightings for the lander's navigation platform. The only solution was that Lovell and Haise had to copy, by hand, all the navigational and alignment data from Odyssey's computer, calculate the different frames of

A group of astronauts gather around Jack Lousma's capcom console at the height of the crisis. Seated left to right are Deke Slayton, Lousma and pipe-smoking backup commander John Young. Standing are Ken Mattingly (left) and Vance Brand.

reference between the two ships and input it into Aquarius' computer ... and do so with haste, for the command module was very close to death.

As Lousma passed up instructions from back-room lunar module experts and Haise worked through the activation procedures, Lovell struggled with a fill-in-the-blanks conversion sheet to realign their platform of navigational reference from that of Odyssey to the lander. On the ground, during worst-case, disaster-scenario simulations, he had made errors when doing this and Haise urged him to take his time and do it right. Aware of his tiredness, Lovell asked Lousma to double-check his arithmetic before entering it into Aquarius' computer. Once this had been done, the platform was aligned and Lovell yelled up the tunnel to Swigert, telling him to shut down the remainder of Odyssey's systems. At 1:51 am EST on 14 April, a little under four hours since the explosion, Odyssey fell silent and Swigert floated through the tunnel into the lunar module.

"It's up to you now," he said to Lovell and Haise. Swigert was undestandably glum, for he had lived and breathed the command and service modules for over two years. The bug-like lander, on the other hand, was an unknown quantity to him. As command module pilot, he had just lost his ship.

However, even for two experts on the lander's systems, it was difficult for Lovell and Haise to control the Apollo 'stack' from inside Aquarius; it was strange and awkward, much like steering a wheelbarrow down the street with a long broom handle. They had to learn many of the facets of flying exceptionally quickly, for very soon they would burn the descent engine to put Apollo 13 back onto a free-return trajectory. If the engine failed, they would swing around the Moon and 'miss' Earth by thousands of kilometres. Thankfully, at 3:43 am EST, five and a half hours after the explosion, the engine lit successfully and burned perfectly for 30.7 seconds, committing Apollo 13 to a splashdown somewhere in the Indian Ocean.

"How do you like this sim?" Jack Lousma asked with a twinkle in his eye.

"It's a beauty!" Lovell shot back.

Splashing down in the Indian Ocean, though, would bring Lovell, Swigert and Haise home a significant distance from any of the United States Navy's recovery vessels ... as well as taking almost four days to get back to Earth. Such a length of time might be greater than Aquarius' systems could sustain three men. Other options included *trying* to burn the SPS as they swung around the far side of the Moon, which would bring them down into the Atlantic, off the coast of Brazil, in barely 38 hours. This idea was quickly abandoned for the same reason as not allowing a direct U-turn: because no one knew to what extent the big engine had been affected by the explosion. Could the 24,000 kg service module itself be jettisoned and Aquarius made to complete a 'long' burn of its descent engine, thereby shaving off some more time? This alternative could achieve a South Atlantic splashdown within 40 hours. On the other hand, without the protection afforded by the service module, there were real fears that the base of Odyssey – whose ablative covering would shield the astronauts from the furnace-like heat of re-entry – could be impaired by solar ultraviolet radiation and temperature fluctuations during the return journey to Earth. This option, too, was rejected.

The burn by Aquarius had, in fact, been so good that Lovell and Haise would probably not need to make any further manoeuvring burns: they were back on the free-return trajectory and *would* now make it back to Earth. Whether they would still be alive when Odyssey hit the ocean was another matter.

Their predicament and their chances of survival had by now reached the ears of the media, whose disinterest in seeing men landing on the Moon was now overtaken by a renewed enthusiasm in covering the unfolding drama. At first, reports were issued by the television networks, informing audiences that there had been an accident aboard Apollo 13, but that the crew "were in no immediate danger". The true severity of what had occurred came in Chris Kraft's comment to a journalist during the course of the night. When asked for his assessment of the magnitude of the situation, Kraft had responded: "I would say that this is about as serious a situation as we've ever had in manned space flight." Even after burning Aquarius' engine and setting themselves back onto the free-return trajectory, the chances of survival were slim. Years later, Lovell remarked that, if someone had given him Apollo 13's list of problems before launch and asked for his judgement on chances, "I'd have said they were virtually nil".

Perversely, the explosion could not really have happened at a *better* time, in terms of getting the men back to Earth alive. If the stir of the cryogenic tanks in the service module and the resultant blast had occurred *earlier* in the mission, there would have been insufficient stores of electrical power and water to even sustain them for the swing around the Moon ... and if it had occurred *later*, after Lovell and Haise's landing, there would have been no descent engine to press into service at all. "We could probably have gotten up and rendezvoused with the command module," Lovell said, "but we wouldn't have had any fuel to go home."

The main worries were four-fold: oxygen, carbon dioxide, electrical power and water. Aquarius had plenty of oxygen because two Moonwalks had been planned – both of which would have involved totally venting the cabin, then repressurising it from tanks in the descent stage – and there were reserves in the oxygen tanks of the ascent stage and in the PLSS backpacks. Haise had calculated that the lander's oxygen stores could keep the three of them alive for eight to ten days. Carbon dioxide from the astronauts' exhaled breath, though, was more difficult to overcome. Normally, it was removed by a canister of lithium hydroxide, which 'scrubbed' it from the air to ensure that it did not accumulate to toxic levels. Unchecked, it could kill the crew: firstly by giving them severe headaches and then, in turn, racing hearts, drowsiness and finally ... quiet oblivion. The lunar lander was designed to accommodate a crew for a maximum of 45 hours and therefore had five lithium hydroxide canisters, but the journey back to Earth would take 90 hours. More canisters were needed, but the two vehicles had been designed by different companies and the lithium hydroxide canisters in the command module were square, whereas those of the lander had been designed to fit a round hole; they simply could not be plugged into Aquarius' environmental control system. After a day and a half aboard the lunar module, a warning light illuminated to advise that the carbon dioxide was climbing toward toxic levels. Enter Ed Smylie, head of NASA's crew systems division, who had been directed to develop a means of overcoming this hurdle. In a day and a half, his team had conceived a routine to build something which resembled a mailbox from materials that Lovell, Swigert and Haise had aboard: sections of duct tape, socks, the ripped-off cover of a flight plan, the oxygen hose from a space suit and a few plastic bags.

In Ron Howard's movie, Smylie and his team were portrayed dumping an assortment of bric-a-brac onto a table and clumsily sorting through it, like a car-boot sale, during a coffee-fuelled, round-the-clock effort to cobble something – *anything* – together to save the astronauts. The reality was quite different, and better organised. "If you saw the movie, it wasn't like that," Smylie said later. "Everything is pretty calm, cool and collected in our business." It was also surprisingly elegant: encased and taped within plastic bags, the box boasted little fashioned 'archways' at the top to keep it from being sucked against the inlet screens. This improvisation *was* critical in keeping the crew alive long enough to return to Earth and Fred Haise would later praise its creators with words to this effect. After the instructions had been radioed up, the astronauts dug out the bits and pieces and spent about an hour building a pair of boxes. Within six hours of fitting the first one into place, the carbon dioxide level in the lunar module began to fall and, at length, became unmeasurable. Smylie's team had saved the day. The astronauts photographed their work for posterity and when they returned to Earth they found that their boxes were remarkably similar to what was intended.

The next overarching issue was power, since Aquarius, unlike Odyssey, had no fuel cells and ran exclusively on batteries. A lander of the H-series mission design – capable of a 33-hour stay on the surface and two Moonwalks – could support an operational lifetime of less than two days and stretching that further would require turning off most of its systems: its cabin lights, its gauges and even its computer. During quiet periods, Haise thumbed through his procedures book to add up the requirements for individual components and determined that they could operate on 18-20 amps at bare-bones minimum power, switching virtually everything off, apart from the air circulation fan and glycol coolant system to keep themselves and the lander alive and, of course, the radio to talk to Houston. That *should* be enough, he figured, to see them through to the time when they would have to re-enter the command module for the return to Earth.

Cooling water for the lander's systems, finally, was the resource in real short supply; Haise anticipated that it would run out five or six hours before re-entry. However, when Neil Armstrong and Buzz Aldrin jettisoned the ascent stage of their lander in July 1969, for an engineering test they deliberately left all of its systems running and turned off its water supply ... and it continued transmitting valuable data for no less than eight hours, until it finally overheated. Haise was certain that Aquarius could do at least as well. His certainty was shared by many experts in Mission Control and by the fifth day of the mission, it seemed likely that – barring any further problems – they would probably make it.

"SOLD DOWN THE RIVER"

When Jim Lovell was named to command Apollo 13 in August 1969, despite the prospect of performing humanity's third landing on the lunar surface, he struggled to find a way for his flight to stand out. It is a poor reflection on the culture and society of the time that so little popular interest was given to one of our species' most remarkable achievements ... if not *the* most remarkable. Little could he have suspected that the mission he thought would be virtually ignored by the press and public would, in fact, seize front-page headlines all over the world, for entirely the wrong reasons. A few years ago, in 2001, I happened to interview Rick Husband, the commander of Space Shuttle Columbia on what would prove to be its final flight. He was virtually unknown in an era where astronaut numbers had risen from a few dozen into the hundreds; yet as soon as his ship plummeted tragically to Earth, I began receiving calls from newspapers asking me about him. I found myself musing over the curious *need* for the media and public at large to adopt such a strange, almost perverse fascination with disasters in space, whilst paying pitifully little heed to the triumphs and the successes.

Would Apollo 13 have been so extensively covered if it had landed on the Moon successfully? Perhaps. Certainly, images of Lovell and Haise bouncing across the hummocky Fra Mauro in one-sixth of terrestrial gravity would have drawn some attention ... particularly since, hopefully, their colour television camera would have worked better than that of Al Bean and Pete Conrad. What actually happened was that it again drew people together, quite remarkably, in a time still dominated by the endless fighting in Vietnam and in a society haunted by the aftermath of the Tate-LaBianca murders, the subsequent trials of the 'Manson Family' and increased distrust of the hippy movement. For that single week in mid-April 1970, everyone's focus was on the welfare of three men, far from home. A Chicago taxi driver, taking a fare to O'Hare Airport, suddenly turned off the expressway and drove to the nearest tavern to watch the command module splash down into the Pacific. In Rome, Pope Paul VI led 50,000 people to share the "universal trepidation" about the fate of the astronauts and prayers were uttered at Jerusalem's Wailing Wall, echoing Psalm 19: "Their line is gone out through all the earth, and their words to the end of the world". In India, a hundred thousand Jain pilgrims said prayers. Thirteen nations, including the Soviets, offered ships or aircraft to assist in the recovery of the Apollo 13 crew. The United States' embassy in London announced that it had not received such an outpouring of concern since JFK's assassination ...

Yet at the same time, the near-disaster of the $380 million mission offered some the perfect piece of ammunition to reinforce their arguments that the Moon landing programme, together with planned robotic ventures to Mars, should be terminated. "I cannot justify approving monies to find out whether or not there is some microbe on Mars," complained Edward Koch, a member of the House Committee on Science and Astronautics, "when in fact I know there are rats in Harlem apartments." Others were more supportive of continued space exploration, and many congressmen voted against recommendations from the White House not to increase NASA's budget. As George Miller, chair of the science committee, put it, "The sympathies and interest generated by this flight, both in this country and around the world, cannot be sold down the river."

Notwithstanding the support, and an extra $265 million tacked onto NASA's 1971 budget, the future after Apollo seemed bleak. Richard Nixon had already issued a statement in March 1970 to the effect that space exploration would no longer retain its lofty position in national priorities and that there would be no more expensive space projects, especially *manned* space projects. Ted Agnew's Space Task Group had failed miserably in most of its efforts to create a roadmap for the future; only the Space Shuttle, in fact, would bear fruit. Dreams of human beings walking on Mars by the Eighties and the establishment of a 50-man space station in Earth orbit were out of the question in an era dominated by an increasing economy drive; the budget simply *could not* accommodate the kind of enormous spending that space had enjoyed under John Kennedy and Lyndon Johnson. Even the Shuttle would come under increasing fire and only narrowly survive cancellation before being finally approved in the spring of 1972. It is ironic, therefore, that in one of the bleakest periods in the United States' space programme, the next four lunar voyages after Apollo 13 would prove the most spectacular and brilliant missions ever flown in the annals of space science.

RETURN TO EARTH

As the Moon loomed in Aquarius' triangular windows, Jim Lovell and Fred Haise took turns monitoring the ship's performance and executing hourly thruster firings to roll the Apollo stack and maintain passive thermal control. In normal circumstances, Odyssey's computer would automatically conduct this 'barbecue' roll, but with the command module out of action and the lander's computer unable to perform the delicate manoeuvre, it had to be done manually, by hand controller. The next major task, according to the trajectory specialists in Mission Control, was to perform a long burn of Aquarius' descent engine, two hours after 'pericynthion' –

The LMP's station inside the lunar module. This image was recorded during Apollo 11, but offers a perspective of the kind of environment in which Lovell, Swigert and Haise lived during their long journey back to Earth.

their closest approach to the Moon. The burn was known as 'PC + 2' and it would shave Apollo 13's return journey from four days down to just two and a half.

Lovell had his own worries, because by the time of the PC + 2 burn more than 20 hours would have elapsed since Aquarius' navigational platform was aligned and he was concerned that it would have significantly drifted. Due to the glow of the debris surrounding them there was no way of checking the platform by a star sighting unless it was done while the spacecraft was in the Moon's shadow and time would not be on his side. Thankfully, capcom Charlie Duke, now fully recovered from the measles, radioed up instructions for a solution: Lovell and Haise were to use the only star that they *could* see clearly – the Sun – to make an alignment sighting. Using co-ordinates radioed up to them, Lovell and Haise steered the spacecraft into position and performed the sighting with perfection.

Shortly thereafter, as Apollo 13 made its closest approach to the Moon, both Swigert and Haise were glued to the windows, shooting off photographs. For the first time since the explosion, they were granted a few fleeting moments simply to *look* ... and to forget the life-and-death drama in which they had found themselves embroiled. Lovell was impatient, telling them that none of their photographs would ever get developed if they failed to complete the PC + 2 burn. Yet the two rookies' enthusiasm was entirely understandable. Lovell, after all, had been here before. It must have been exhilarating for Swigert, who had not even expected to behold this vista a little more than a week ago, and profoundly saddening for Haise, who had anticipated walking on that crater-pocked surface for many months.

Haise was an explorer at heart. In spite of his military and experimental flying background, he wore the 'Ex Luna, Scientia' patch on his space suit with pride and unlike many of the other pilot astronauts, he did not pay simple lip-service to the scientific aspects of his mission. Rather, he had been genuinely bright-eyed and enthused by the geology briefings of Jack Schmitt and Lee Silver and Farouk el-Baz. He had been looking forward to planting his own bootprints into alien soil ... and, indeed, one of his assigned experiments on the surface allowed him to do just that, then photograph their imprints to study the 'clinging power' of lunar dust. So excited was Haise at the variety of equipment that he would take to the Moon – sampling hammers, core tubes, scoops, tongs, a battery-powered lunar drill – that his mentors nicknamed him 'Drilling Fool'. Now he could only gaze down those last few hundred kilometres in sorrow at his lost goal.

The PC + 2 burn duly took place at 9:40 pm EST on 14 April and lasted four and a half minutes, effectively cutting a sizeable chunk of time off Apollo 13's return journey and aiming the command module for a splashdown squarely in the mid-Pacific. The burn was perfect and even Lovell was impressed by the performance of Aquarius, which had fulfilled roles for which it had never been designed. However, conditions inside the broom-cupboard-sized cabin were far from good. Most of the lander's systems were shut down after the PC + 2 burn and the temperature fell rapidly. Since it was not possible to close the hatch leading into Odyssey, the total volume was too big to keep warm. "We got clammy," Lovell recalled years later. "Moisture started to form on the metal pieces of the spacecraft and all the couches and the windows were foamed up and running with water ... It was *not* a very pleasant journey." Sleep inside the command module was virtually impossible, with each man grabbing perhaps a couple of hours at best, unable to rest in an environment which resembled a dark, damp cellar. Eating was difficult, too. On one occasion, Swigert floated into Odyssey to fetch some foil-wrapped hot dogs from the food locker ... and found them frozen solid. The crew drank very little water, partly due to limited supplies and partly because they did not wish to perform urine dumps, which they feared might nudge Apollo 13 off-course. All three would return to Earth severely dehydrated and Haise was later found to be suffering from a kidney infection.

"Time and again during training," wrote Lovell and Kluger, "the flight surgeons had cautioned all astronauts that if they did not consume and pass enough water in space, their bodies could not excrete toxins ... and if they did not excrete toxins, the noxious substances would accumulate in their kidneys, leading to an infection." That infection would remain with Haise for the rest of the mission, manifesting itself in a burning sensation whenever he urinated; after splashdown, he was quickly whisked into the infirmary and would be conspicuously absent from many of the post-flight tours, press conferences and a lavish tickertape parade in Chicago.

Nevertheless, home was drawing closer with each passing day and by Friday 17th, nearly a week since leaving Cape Kennedy, the blues and whites of Earth filled Aquarius' windows. It also filled Mission Control with a juxtaposed sense of fear and jubilation: for, against all the odds, Lovell, Swigert and Haise were *almost home* – but there was no way of knowing if Odyssey's systems, critical for re-entry and splashdown, would come back to life after more than three and a half days of frigid cold and with moisture dripping from every surface. Indeed, it was very likely that moisture existed *behind* the instrument panels, too. For his part, Lovell considered it inevitable that some electrical gear would short out as soon as it was reactivated.

Jack Swigert, the command module pilot, was the man upon whom fell the responsibility for bringing Odyssey back to life. He had been worrying about how to do so ever since shutting her down. He knew that reactivation *had* to occur in a specific order of events that would have to work correctly, the *first time* – since Earth was looming and re-entry could not be postponed. Normally, it took three months to write a detailed, step-by-step, circuit-breaker-by-circuit-breaker checklist for re-entry ... but the maestros in Mission Control had devised the procedures for Swigert in just *two days*!

Key to this effort was Gene Kranz and his White Team. Ordinarily, an Apollo lunar mission had four shifts, but since the accident Kranz' men had become a 'tiger team' charged with planning the return to Earth – the burns of Aquarius' descent engine, the platform alignments, the mid-course corrections and finally the procedures for the critical last four hours of the mission. Within Kranz' team were three focal groups, the first of which wrote instructions for each and every procedure, the second translated those procedures into checklists for Lovell's crew and the third ensured that none of the steps would overwhelm the meagre power levels in either Aquarius or Odyssey.

For Apollo 13 to enter the atmosphere at the appropriate time and angle, its trajectory had to be near-perfect and another mid-course correction burn was needed to ensure that it did not drift off-course. Late on the evening of 15 April, Lovell was instructed to orient the spacecraft such that the Sun shone directly through the lander's overhead rendezvous window; he would then rotate the stack until the crescent Earth appeared in his forward window. When the 'horns' of the crescent touched the crosshairs on the window, he would perform the 14-second burn. With all three men involved – Lovell in control, Haise maintaining orientation in pitch and Swigert tracking time – the burn was made as perfectly as perfect could be. Apollo 13 was on target and heading home.

Before 'home' arrived, of course, there was the small matter of a re-entry checklist, which was still being completed. Whilst the astronauts waited, their conversation within the chilly confines of Aquarius turned to their comrades on the ground. Backup crew members John Young and Charlie Duke and Apollo 14 backups Gene Cernan and Joe Engle had been rehearsing procedures in the lunar module simulator, but it was chiefly Ken Mattingly who was sweating the details of the re-entry checklist for Odyssey, making sure that switches were thrown in precisely the right order and systems reactivated in exactly the right way to avoid overstressing the meagre battery power. In the last couple of days before launch, Lovell arranged a series of code words with the shift capcoms to discover if Mattingly had indeed fallen ill with the measles and capcom Vance Brand, a member of Gerry Griffin's Gold Team, knew the routine.

"Are the flowers in bloom in Houston?"

"Nope, not yet," said Brand. "Still must be winter."

"Suspicions confirmed," Lovell replied.

"I doubt if they will be blooming even Saturday, when you return." Like Mattingly and Swigert and Haise, Brand was a member of the 1966 astronaut class and doubtless was also keen to seize any opportunity to take a dig at the flight surgeon.

Humour aside, the uncomfortably long wait for the re-entry procedures began to irritate Lovell, to the extent that he pressed Brand for an accurate prediction of *when* they would be ready. The crew was tired and needed time to study them, after which they would require a good night's rest. The following day, 17 April, would be the real test of whether Apollo 13 would end as a triumph or a disaster. At this point, Deke Slayton came on the line, his reassuring demeanour calming the situation down.

At 8:30 pm EST, Ken Mattingly finally walked into Mission Control with a sheaf of power-up and re-entry procedures in hand. He sat down next to Brand and plugged his headset into the the capcom's loop. "Hello, Aquarius; Houston. How do you read?"

"Okay," replied a happy Swigert. "Very good, Ken."

"Let me take it from the top here." For the next two hours, he and Swigert delved into a 39-page, 400-step checklist which established the order of every switch throw and every keystroke needed to bring Odyssey back to life within power and time constraints. First, though, came the reactivation of Aquarius at 3:15 am on 17 April. Four hours later a 21-second firing of the lunar module's thrusters made a final refinement to their trajectory, and a few minutes after *that*, the crippled service module was jettisoned. This provided the astronauts with their first glimpse of the damage caused by the explosion.

Tiredness and dehydration had already caused the astronauts to make mistakes – Lovell had accidentally called up the computer program for Aquarius' descent engine, instead of the tiny manoeuvring thrusters, when preparing for the final burn – and Jack Swigert was acutely sensitive to the risk of his jettisoning the *wrong ship*, by setting the lander loose with Lovell and Haise still inside. It might sound foolish and unlikely, but it should be borne in mind that all three had done little but *exist* for nearly four days, with little sleep, little sustenance, little comfort and an awareness that they had little chance of survival. To make sure he did *not* screw up, Swigert taped a bit of paper with the word 'NO' over the toggle switch for LM JETTISON. He tripled-checked it and then invited a bemused Fred Haise into Odyssey to verify that it was correct. When the time came to release the service module, right on cue, at 8:16 am EST, he pushed the SM JETTISON button and heard a bang of pyrotechnics.

Within seconds, the massive cylinder was suddenly and clearly visible in the glint of sunlight. All three men could do little but gasp ... and shoot photograph after photograph of how close they had come to death. "There's one whole side of that

The crippled service module reveals the extent of the damage from the oxygen tank explosion. One entire side panel was torn away in the blast, effectively disabling Apollo 13, eliminating any chance of a Moon landing ... and placing all three astronauts in dire peril.

spacecraft missing!" Lovell reported in astonishment. Joe Kerwin tried to make light of the matter by joking that Lovell had not taken very good care of his spacecraft, but his gallows humour fell on deaf, and rather shocked, ears. In fact, a panel some four metres long and covering one sixth of the service module's hull had been completely blown away, leaving a jumble of torn and shredded wiring and remnants of wrecked plumbing. Lovell had expected to see a small puncture wound, but *nothing* as destructive as this. The S-band assembly, deployed from the side of the service module, had been bent out of position and the damage extended from the bell of the SPS engine ... right up to the ring that supported the base of the command module. The base of the command module, of course, carried something else of vital importance to the crew's survival: the *heat shield*. Had the explosion damaged *that*? If so, then Lovell, Swigert and Haise were still in deadly danger. After braving and overcoming all the odds of an emergency on the way to the Moon, a damaged heat shield could now render them vulnerable to a fiery death during atmospheric re-entry.

There was precious little time to ponder such dire possibilities, because Earth's gravity was accelerating their fall and, whether they liked it or not, they *would* be landing there today, 17 April. By ten in the morning, with much relief, Swigert was back inside the command module, running through the power-up procedures with Haise's support. As they pushed in each new circuit breaker and flipped each new switch, their eyes watched for sparks and their noses keenly sniffed the chill air for any trace of smoke, but there were none. The specifications said that Odyssey could not be restored after so many days in hibernation. Thankfully, those specifications were wrong.

At length, the time came to jettison Aquarius itself. With Lovell and Haise now having stored a few bags of unneeded gear inside the lander's tiny cabin, they returned to the command module and wrestled its bulky hatch into position. Swigert pushed the LM JETTISON button and the machine which might have ferried James Arthur Lovell Jr and Fred Wallace Haise Jr down to the Moon – but which actually accomplished so much more – drifted away into the inky blackness. In years to come, Haise would reflect with sadness that this machine had no heat shield and could not be brought back to be enshrined in a museum. Instead, it would end its days in a blaze of glory: with a true and fiery Viking funeral, plunging back to Earth in a shower of burning debris.

Alone now, the cone-shaped command module plummeted Earthward and hit the tenuous upper atmosphere at 12:53 pm EST, a little more than an hour shy of six full days since leaving Cape Kennedy. The sheath of plasma surrounding Odyssey during its plunge blocked communications with Mission Control. This was expected to last around three minutes and its drama was played out to great effect in the movie *Apollo 13*. The unspoken fear that the heat shield had been compromised, or that the parachutes would fail to deploy, or that for some other reason the command module would not survive, was pervasive.

"Odyssey, Houston," radioed Joe Kerwin. "Standing by."

He was greeted by silence. For perhaps the first time in almost four days, there was absolute quiet and stillness in Mission Control, with all eyes riveted on the

"A good ship": the lunar module Aquarius, its job done, is finally jettisoned shortly before re-entry.

fuzzy television images of a desolate stretch of Pacific Ocean. Recovery forces from the amphibious assault ship *Iwo Jima* – including its detachment of Marine Corps helicopters – had already been scrambled as part of a massive search-and-rescue effort called Task Force 130 to recover Apollo 13. The long seconds of silence continued. There was, recalled Gene Kranz, a sickening feeling of dread in Mission Control.

After what seemed like an eternity of waiting, a Boeing EC-135 modified to provide telemetry and tracking as an Apollo Range Instrumentation Aircraft reported that it had acquired a signal from the rapidly descending capsule.

"Odyssey, Houston," Kerwin repeated. "Standing by."

At length, there came a reply. It was Jack Swigert's voice. "Okay, Joe." Seconds later, a television camera from the recovery ship picked up the tiny black cone slung beneath three perfect parachutes, in amongst the clouds a few kilometres away. It was *then* that everyone applauded. In the movie *Apollo 13*, the controllers stand up and cheer at the first word from Odyssey; in reality, wrote Kranz, if one of his team *ever* did that before the mission had ended it "would be the last time he sat at a console". Not until the parachutes deployed, and deployed successfully, could everyone breathe an enormous, collective sigh of relief.

"Odyssey, Houston, we show you on the mains," yelled Joe Kerwin, as the television screen in Mission Control clearly picked out the orange-and-white canopies of the command module's main parachutes. "It *really* looks great!"

Splashdown at 1:07 pm EST was smooth and graceful. Within 45 minutes, Lovell, Swigert and Haise were safely aboard a helicopter. Almost exactly six days, to the minute, since they launched, the three bearded, pale and exhausted astronauts stood on the deck of the *Iwo Jima* and, in bewilderment, listened to the cheers of the ship's crew and a rousing rendition of 'Aquarius' by its band. Bewilderment, indeed, for none of them could have imagined that millions of people around the world had been watching and praying for their safe return. Before Apollo 13 set off, Jim Lovell had expected his mission to receive little attention from the press or the public. By the time he and his crew splashed down, they would be feted for the heroes that they were.

LOST MISSION

It was not just Lovell, Swigert and Haise who had demonstrated heroism and courage during the short, ill-fated week that took them to the Moon and back; it was also the four shifts of Mission Control staff – Gene Kranz's White Team, Milt Windler's Maroon Team, Glynn Lunney's Black Team and Gerry Griffin's Gold Team – which had operated, near-constantly, to bring the three men home safely. On 18 April 1970, the day after Odyssey splashed down, Richard Nixon visited Houston and awarded each of the flight directors the Presidential Medal of Freedom to honour the achievements of themselves and their teams. Several weeks later, they joined the astronauts in a tickertape parade through the streets of Chicago. Jim Lovell was awarded the Medal of Merit and the flight directors received ceremonial keys to the city.

The events of that week would forever be described, in a rather oxymoronic way, as "a successful failure", since the men returned safely to Earth, despite losing the lunar mission; and many would come to see Apollo 13 as NASA's finest hour. Controllers, engineers, technicians, astronauts – in fact, *everyone* with anything to do with Apollo, from Grumman to North American – had of their own volition gone to their offices and stayed there until the job was done. Astronauts had worked through procedures for Lovell's exhausted crew in the simulator, weeding out problems or hidden flaws which might spell doom. By re-entry morning, most of the flight controllers had not seen their families since the night of the explosion, almost four days earlier, catching a couple of hours' sleep on the floors of their offices and surviving on cigarettes and black coffee. Gene Kranz had much to say about his team. "I remember their eyes," he wrote, "dull with fatigue and shadowed by anxiety. But their confidence and focus never wavered."

Yet, now that it was all over, the very real question remained: Would Apollo continue? Would there be further manned expeditions to the Moon? Many observers, even within NASA, harboured doubts and in the wake of the events of April 1970 became more vocal in their belief that the programme should be

Command module Odyssey, scorched from the intense heat of re-entry, is gently
lowered onto the deck of the *Iwo Jima* on 17 April 1970.

terminated before a crew did indeed die in space. The Nixon administration had
made no secret of the fact that space exploration was low on its list of national
priorities. Even Jim Lovell, speaking from aboard the chilly Aquarius during the
long flight home, had told Jack Swigert that he expected Apollo 13 to be the last
mission to the Moon for some time. NASA Administrator Tom Paine assured
journalists of the opposite: that missions *would* resume as soon as the problems were
ironed out, but as April wore into May, uncertainty still remained over what had
happened.

A review board, chaired by Ed Cortright, director of the agency's Langley
Research Center in Hampton, Virginia, set to work on the very day that Odyssey hit
the waves of the Pacific Ocean. Its initial focus was a theory that a defect in the
oxygen tank walls was responsible for the accident. Maybe a rivet or a piece of one of
the cryo-stir fans had broken away and created a spark that caused the liquid oxygen
to expand and burst the tank with explosive force. This possibility was advocated by

Max Faget himself in late April. By the time the board's final report was issued on 15 June, however, more extensive analysis had identified the 28-volt thermostatic switch and a loosely fitting fill tube assembly as the primary culprits. A combination of human mistakes and deficiencies in design, it seemed, were to blame.

None of the contractors or technicians involved with boiling off the liquid oxygen from the No. 2 tank in late March had apparently recognised "the possibility of damage from overheating" and "many were not aware of the extended heater operation". Beech Aircraft Corporation, which built the tank, had not changed the thermostat switch specifications to be compatible with 65-volt power. None of NASA or North American's quality assurance overseers had spotted the error. Nor did subsequent testing, including acceptance runs conducted by Beech itself, uncover any problems. Following the loading of liquid propellants, and whilst on the launch pad, atop the Saturn V, the No. 2 tank "was in a hazardous condition when filled with oxygen and electrically powered". Combustion within the tank "probably overheated and failed the wiring conduit where it entered the tank and possibly a portion of the tank itself".

Cortright's panel, which included Neil Armstrong, representing the astronaut corps, made a number of key recommendations, including modifying the liquid oxygen storage system to remove from direct contact *all* wiring and unsealed motors which could short-circuit or ignite adjacent materials. Furthermore, all Teflon, aluminium and other relatively combustible materials in the presence of the oxygen and other ignition sources should be minimised and the tanks and caution-and-warning systems should be overhauled. Given that Lovell's crew had only narrowly survived, the report added that consumables and emergency equipment in both the command module and the lander should be reviewed in light of possible future 'lifeboat' scenarios.

One of Cortright's recommendations which was *not* expected was described by Chris Kraft in his autobiography.

"I'm recommending that we design and build a new tank," Cortright told Kraft and Bob Gilruth.

"Why would you do that?" Gilruth asked in amazement. Kraft agreed; the *tank* itself was sound and had been certified by hundreds of hours of test and actual flight experience. The fault lay squarely with the *thermostatic switch*, so why not just replace that? Cortright remained fixed on his recommendation and, although Gilruth and Kraft eventually fought him to the higher echelons of NASA Headquarters, the decision was ultimately taken to redesign the tank. "That new design," wrote Kraft, "cost us $40 million. Cortright faded into the woodwork and left us to work out the details. It was a difficult engineering job and ground tests couldn't possibly provide the kind of testing and flight experience we were throwing away."

Late in November 1970, Bob Gilruth confirmed that fan motors would be removed from future oxygen tanks; the electrical leads would be encased in stainless-steel sheaths with hermetically sealed headers and kept from any contact with remaining Teflon parts. Consumables and survival equipment had been "reviewed", he said, to allow for a safe return from lunar orbit in the event of an oxygen supply

The Apollo 13 crew during a post-flight debriefing on 20 April. From left to right are Jim Lovell, Jack Swigert and Fred Haise.

loss. Three design changes were made with relation to the oxygen tanks and a descent battery was added to the lander and a water storage system installed in the command module. The final irony is that, had the No. 2 tank not been *dropped*, and thus damaged, NASA might have escaped the near-disaster of Apollo 13 and the flaw in the thermostatic switch might have gone completely unnoticed.

One of the greatest disappointments was that Lovell, Swigert and Haise never got their opportunity to put their geological training to the test at Fra Mauro. Without doubt, Jack Schmitt and Lee Silver and Farouk el-Baz lamented the mission that was lost in April 1970. It is interesting, though, to speculate on what might have been achieved. Assuming a satisfactory insertion into lunar orbit, Lovell and Haise would have set Aquarius down at 9:55 pm EST on 15 April, a little over four days into the mission. Plans for a 'rest' period after landing had long since been abandoned and their first Moonwalk would have begun at 2:13 am, followed by a meal and a rest, then a second excursion at 9:58 pm on 16 April.

Lovell and Haise's first outing would have been very similar to that of Pete Conrad and Al Bean, lasting some four hours (extendable to five hours) and encompassing the collection of a contingency sample, erecting the Stars and Stripes and an S-band antenna, unveiling a plaque on Aquarius' leg and deploying a solar wind experiment and the second ALSEP. Their second walk, however, would focus on real scientific inspection of the site. Indeed, the Apollo 13 press kit, published shortly before launch, noted that "sample locations will be carefully photographed before and after sampling" and that Lovell and Haise would "carefully describe the setting from which the sample is collected". The two men would put their geological training to good use in collecting several core tube samples, digging a 60 cm-deep trench to evaluate soil mechanics and gas dynamics, using a battery-powered drill to emplace probes for a heat-flow investigation and collecting a variety of rock samples.

Liftoff from the Moon would have occurred at 7:22 am on 17 April, followed by docking with Odyssey. During their time apart, Swigert would have pursued his own programme, using a large-format topographical camera known as the 'Hycon'. It was a huge device, with a lens that completely filled the window of the command module's access hatch. Swigert would have acquired numerous high-resolution black-and-white images in overlapping sequences for use as mosaics or single frames; key surface targets would have included candidate landing sites for subsequent Apollo missions, including Censorinus, the crater chain of Davy Rille and the Descartes highlands. So important was it that, after the return of Lovell and Haise, the crew were to spend another full day in lunar orbit using the Hycon and their other complement of cameras. Barrelling away from the Moon on 18 April, they were scheduled to splash down in the Pacific Ocean on the 21st, completing a mission of almost exactly the same duration as Apollo 12.

Sadly, it was not to be ... at least, not for Lovell and Haise. By the end of the year, Apollo 14 had slipped until January 1971 and the sites being considered for it were discarded in favour of Fra Mauro, whose scientific importance took precedence.

The astronauts who would finally plant their bootprints at the Fra Mauro site were Al Shepard – who had achieved worldwide renown in May 1961 as America's first man in space – and Ed Mitchell. Joining them in the command module was sandy-haired Stu Roosa. As a crew, they had much to prove ... and not simply in terms of demonstrating that Apollo was again spaceworthy. Shepard, the only veteran amongst them, had flown only once, on a suborbital hop almost a full decade earlier and on a flight lasting *only* 15 minutes. Even within NASA and the astronaut corps, there were many who wondered if he was really ready to lead a make-or-break expedition to the Moon. An inner-ear disorder, a lengthy period of medical grounding and a ruling preventing him from flying solo for several years had intensified such worries. On Apollo 14, Shepard, Roosa and Mitchell would do much to allay the fears of the naysayers. They would re-establish an American toehold on the Moon, bring home a haul of priceless samples, show both interest and indifference to the nuances of lunar geology ... and even find time for a spot of golf in one-sixth gravity.

5

End of the beginning

ROAD TO RECOVERY

Alan Bartlett Shepard Jr was not a man to mince words. As America's first man in space and head of its astronaut team, he rarely had to. Yet even his old friend Deke Slayton was shocked one morning in the spring of 1969, when Shepard strode into his Houston office and *told him* that he wanted a flight to the Moon. It wasn't a request; it was an announcement. For any other astronaut, such an outburst would have been utterly objectionable and Slayton – the man in charge of picking crews for space missions – would ordinarily have delivered a severe verbal roasting and most likely ensured that no flight assignment came his way for some considerable time. Shepard, though, was different. It wasn't just that they were old friends or that they had both been chosen as members of the very first group of American astronauts. It was because they shared a bond which few others would have wanted: Shepard and Slayton had been astronauts for longer than almost anyone else at NASA and yet both had been barred from flying into space. In fact, for the last five years, neither of them could even fly a jet aircraft on their own.

Slayton's heart problem and its effect on his career has already been discussed, but at least Shepard had been granted the opportunity to fly into space. Eight years earlier, on 5 May 1961, he had become the most famous man in the United States, riding a converted Redstone ballistic missile on a suborbital mission into the history books. His 15-minute, up-and-down adventure had been a mere shadow of the orbital flight undertaken by Russian cosmonaut Yuri Gagarin a month earlier, but for America – still reeling from an embarrassing diplomatic fiasco in Cuba and fearful of a perceived technology gap with the Soviet Union – it was an unqualified success. In Shepard's mind, becoming the first American in space had been a deeply personal achievement, particularly as he had spent most of his life searching and striving for the next challenge . . . but the one thing he continued to yearn for was the opportunity to do it again.

By 1963, his chances of doing so seemed bright. Then, his career and health, figuratively and literally, took a spin. Years earlier, just after being selected as one of the Mercury Seven, Shepard had complained about feeling light-headed during a game of golf; every time he tried to swing the club, he felt that he was about to fall over. It was an isolated, peculiar incident, which did not resurface again until the summer of 1963. It came with a vengeance, usually striking him in the mornings and taking the form of a loud metallic ringing in his ears, coupled with feelings of intense dizziness and nausea. At first, Shepard dealt with the problem himself: he saw a private physician, who prescribed diuretics and vitamins such as niacin, which had little effect. It did not prevent Slayton from assigning him to command the first Gemini mission and, indeed, Shepard and his co-pilot, Tom Stafford, completed the first six weeks of training, visiting McDonnell's plant in St Louis, Missouri, to watch their spacecraft being built.

Although Shepard told no one in the astronaut corps of the problem, it very soon became impossible to hide. An episode of dizziness whilst delivering a lecture in Houston forced him to admit his concerns to Slayton, who sent him to Chuck Berry for tests. In May 1963, unknown to everyone else in the corps, Shepard was temporarily grounded. The diagnosis was that fluids were regularly building up in the semicircular canals of his inner ears, affecting his sense of balance and causing vertigo, nausea, hearing loss and intense aural ringing. The incidents were intermittent, but proved sufficiently unpredictable and severe to render him ineligible to fly into space.

Known as Ménière's Disease, the ailment was a recognised but rather vague condition. Indeed, formal criteria to define it would not be established by the American Academy of Otolaryngology-Head and Neck Surgery until 1972. The academy's criteria would describe exactly the conditions suffered by Shepard: fluctuating, progressive deafness – he would be virtually deaf in one ear by 1968 – together with episodic spells of vertigo, tinnitus and periodic swings of remission and exacerbation. Nowadays, it can be treated by vestibular training, stress reduction, hearing aids, low-sodium diets and medication for the nausea, vertigo and inner-ear pressure, such as antihistamines, anticholinergics, steroids and diuretics. In mid-1963, however, the physicians who examined Shepard had next to no idea what caused the ailment, with some even speculating that it was a 'psychosomatic' affliction. Moreover, there was absolutely no cure.

His removal from flight status was temporarily revoked in August, with the prescription of diuretics and pills to increase blood circulation, in the 20 percent hope that the condition would clear up on its own. This allowed Shepard to be internally assigned back onto his Gemini mission, but when the early diagnosis was confirmed and no sign of improvement was forthcoming, he was formally grounded in October. Not only was Shepard barred from space flight, but, like Deke Slayton, he also could not fly NASA jets unless accompanied by another pilot. Subsequent examinations revealed that he also suffered from mild glaucoma – a symptom of chronic hyperactivity – and a small lump was discovered on his thyroid. It was surgically removed in January 1964 and, the press statement announced, "would have no impact on his status in the space programme". In reality, Shepard had been effectively grounded for months by that point.

One evening, at the Rice Hotel in downtown Houston, Shepard pulled Stafford aside and asked him if Slayton had mentioned anything about the Gemini assignment. No, Stafford replied, and could only listen, open-mouthed, as his former crewmate told him about the dizziness, the vertigo, the Ménière's diagnosis ... and the bombshell that Shepard was grounded.

By this time, Deke Slayton was chief of flight crew operations and needed someone to run the day-to-day activities of the Astronaut Office. In spite of his illness, Shepard was not ready to retire – he would stick around, as long as there was the slimmest of chances that he might fly again – and Slayton invited him to lead the corps. There were times during Shepard's five-year tenure as chief, from early 1964 until the summer of 1969, that Slayton despaired: for his friend was a less-than-diplomatic administrator, a fearsome and malevolent figure for the younger, newer astronauts and an often-ineffectual public relations manager. On one occasion, whilst being interviewed by the fiercely pro-feminist journalist Oriana Fallaci, he had offered to sell her a cow from his ranch, telling her "You need a cow!"

If Fallaci needed a cow, then many astronauts needed God's help whenever they were summoned into Big Al's office at the Manned Spacecraft Center. Stu Roosa, who would later fly with him on Apollo 14, always made sure that his thoughts were in good order and his hair was combed – a little like going to see the General – and to a man, virtually every one of them would describe Shepard as frequently ice-cold, remote and utterly mysterious. They would avoid him in corridors and be perplexed by the motivations of an individual who would buy them drinks in the bar one night and then nail them to the wall the next morning over their misdemeanours. Some historians over the years have seen Shepard's attitude as an indicator of his frustration at watching these "young pups" flying into space whilst he could not, but the man himself would cogently defend his position: he was treating them, he said, just as *he* would expect to be treated. He had no tolerance for mistakes and his military background had drilled into him the skills necessary to run a tight and efficient ship.

At times, Shepard could be genuinely warm and approachable, sharing jokes and good humour, but at others his character and his stare would freeze the uninitiated like ice. For this reason, his secretary placed a picture of a smiling or scowling face on her desk each morning to pre-warn astronauts of which personality they could expect from Big Al that day. However, a glowing testimony of him has come from Deke Slayton's second wife, Bobbie. "If you were a friend of Al's and you needed something," she once said, "you could call him and he'd break his neck trying to get it for you. If you were in, you were in. It was just tough to get in."

The man who provoked these swinging extremes of opinion had been born in East Derry, New Hampshire, on 18 November 1923, the son of an Army colonel-turned-banker father and Christian Scientist mother and the progeny of a close-knit, fiercely loyal and wealthy family. His key qualities – bravery, a spirit of adventure and an absolute determination to be the best – emerged at a young age: as a boy, chores around the home and a paper round enabled him to buy a bicycle, which he rode to the local airport, cleaning hangars and checking out aircraft. At school, his boundless energy led teachers to advise that he skip ahead two grades, making him the youngest in each class he attended. After spending a year at Admiral Farragut Academy in New Jersey, Shepard entered the Naval Academy in Annapolis, Maryland, receiving his degree in 1944 and serving as an ensign aboard the destroyer *Cogswell* in the Pacific during the closing months of the Second World War.

He subsequently trained as a naval aviator, taking additional flying lessons at a civilian school, and in 1947 received his wings from Corpus Christi, Texas, and Pensacola, Florida. Shepard served several tours aboard aircraft carriers in the Mediterranean Sea and was chosen in 1950 to join the Navy's Test Pilot School at Patuxent River – the famed 'Pax River' – in Maryland; whilst there, he established a reputation as one of the most conscientious, meticulous and hard-working fliers. On more than one occasion, he was hand-picked to wring out the intricacies of a new aircraft, purely on the basis of his intense technical skill and precision. His test work included missions to obtain data on flight conditions at different altitudes, together with demonstrations of in-flight refuelling systems, suitability trials of the F-2H Banshee jet and evaluations of the first angled carrier deck.

Later, as operations officer for the Banshee, attached to a fighter squadron at Moffett Field, California, Shepard made two tours of the western Pacific aboard the *Oriskany*. A return to Pax River brought further flight testing: this time of the F-3H Demon, F-8U Crusader, F-4D Skyray and F-11F Tiger jets, together with posts as a project officer for the F-5D Skylancer and as an instructor at the school. Graduation from the Naval War College in Rhode Island in 1957 led to his assignment to the staff of the commander-in-chief of the Atlantic Fleet as an aircraft readiness officer. By the time he was selected as an astronaut candidate by NASA in April 1959, Shepard had accumulated 8,000 hours of flying time, almost half of it in high-performance jets. His flight-test experience surpassed that of the other members of the Mercury Seven, although he was alone among them in having never flown in combat.

Reputation-wise, he was quick-witted, a top-notch aviator and, as a leader, possessed all the characteristics of a future admiral – a rank which, even having never commanded a ship, he attained in 1971. In fact, when he told his father of his selection as an astronaut candidate, the older Shepard expressed grave misgivings that he was abandoning a promising naval career for what was at that time perceived by many as an ill-defined programme with limited prospects, run by a newly-established civilian agency. For Shepard, though, Project Mercury was a logical extension to a life spent looking for the next challenge. His competitive nature had become the stuff of legend years before Mercury and had gotten him into hot water with superiors on more than one occasion. After several illicit, close-to-the-ground flying stunts, known as 'flat-hats' – one over a crowded naval parade ground, another looping under and over the half-built Chesapeake Bay Bridge in Maryland and a third blowing the bikini tops off sunbathing women on Ocean City beach – he came dangerously close to court-martial.

Undoubtedly, Shepard's less-than-reputable exploits had come to the attention of the NASA selection board, but his biographer Neal Thompson speculated that it was interpreted as an aspect of his fearless and competitive personality, rather than as an excuse to discard his application. He indulged in other hobbies, too. After taking up water-skiing, he progressed rapidly from two skis to one and later experimented on the soles of his bare feet. His wife, Louise, whom he married whilst at Annapolis, would remark that it was "characteristic" of Shepard to always be restless for new challenges. His biggest feat – and, he would say later, his proudest professional accomplishment – was selection to fly the first American manned space mission. "That was competition at its best," he said, "not because of the fame or the recognition that went with it, but because of the fact that America's best test pilots went through this selection process, down to seven guys, and of those seven, I was the one to go. That will always be the most satisfying thing for me."

Also satisfying were the many other perks which came from being a Mercury astronaut and a national hero. Despite outcry from the press and, on occasion, from NASA's then-Administrator Jim Webb, Shepard would dabble in dozens of lucrative business ventures and by the middle of the Sixties he was well on his way to becoming a millionaire, owning a couple of expensive homes in the Houston area. He once told Oriana Fallaci that "all of it" – the wealth, the business activities, the flying in space, the fame – was important to him, but it was the tiny chance that he might again ride a rocket which kept him at NASA. By the spring of 1968, that chance seemed to be drifting further and further away ... and then, one day, in a conversation with Tom Stafford, he learned of an ear, nose and throat surgeon in Los Angeles by the name of Bill House who was experimenting with techniques to cure Ménière's Disease. Shepard and his wife were Christian Scientists and had a strong aversion to surgery. Nevertheless, as the years went by without an improvement in his condition, a visit to House seemed the only realistic option. He checked into St Vincent's Hospital in Los Angeles under the assumed name of 'Victor Poulos' – suggested to him by House's Greek nurse – and underwent the procedure, which entailed an incision into his mastoid bone and into part of his inner ear known as the 'sacculus', where endolymph fluid resides. It was excessive pressure and inflammation of the sacculus which had disrupted the delicate balance of Shepard's inner ear. After cutting into the sacculus, House inserted a thin rubber tube which, if the operation succeeded, would drain excessive endolymph fluid into the spinal column. By removing the source of pressure and causing the fluid to dissipate over a wider area, it was hoped that the surgery – known as an 'endolymphatic shunt' – would do the trick.

With absolutely no guarantees that the surgery would work, Shepard returned to Houston in late May 1968, bandaged, and continued his daily duties in charge of the astronaut corps. Over the next eight months, his hearing steadily returned and he regained his balance. Privately, he was elated and had already approached Deke Slayton with a view to securing command of a future flight. Other astronauts began to notice a difference in his personality, for in the spring of the following year, Shepard began showing up at simulator sessions more often and taking a greater than normal interest in the systems of the Apollo spacecraft. Yet his efforts to regain a seat on a mission culminated in a very bitter, acrimonious and public confrontation with the only other remaining member of the Mercury Seven, a veteran astronaut named Gordo Cooper.

Cooper was several years younger than Shepard and already had two flights to his credit, one of which had lasted eight days in August 1965 and claimed the world endurance record from the Soviets. Since then, however, Cooper's star had steadily dimmed. Senior managers had long seen him as a maverick, a complainer, somewhat unpredictable and his stance was often completely at odds with the public image that NASA wanted its astronauts to display. He once landed on a runway that was deemed too short for his type of aircraft and asked to be refuelled; when ground staff objected that it was too dangerous for him to take off again, Cooper shrugged, took off regardless and touched down at a nearby air base with fumes in his tanks. On another occasion, just two days before his first space flight, he had buzzed Cape Canaveral in an F-106 Delta Dart, flying so low that second-floor office workers could actually look *down* onto his screaming jet. To borrow an expression from Project Mercury's operations director, Walt Williams, many superiors wanted Cooper's "ass on a plate".

In his autobiography, Deke Slayton would recall his own doubts about assigning Cooper to missions, but perhaps his own grounding made him feel somewhat sentimental towards his colleague. To be fair, Cooper was a smart and exceptional pilot and his Mercury mission had been spectacular, but on his Gemini flight he had done little to hide his irritation at several decisions made by senior managers. Moreover, his 'strap-it-on-and-go' attitude towards flying, though legendary, went against the grain and he did not always apply the same enthusiasm to the hard grind of training; in fact, he frequently had to be goaded into the simulator. Slayton had assigned him as backup commander to Tom Stafford on Apollo 10, hoping that if Cooper turned in a good performance, he might be eligible to rotate to the prime slot on Apollo 13.

Unfortunately, Cooper's efforts did not improve and he certainly did not help his cause by entering a 24-hour road race in Daytona Beach, whilst in dedicated mission training. When Slayton pulled him from the race at the last minute, Cooper had sarcastically told journalists that NASA "ought to hire tiddlywinks players as astronauts!" The final months of Cooper's training coincided with Shepard's recovery from surgery and in March 1969 America's first man in space passed NASA's physical, requalified for a Class I pilot's licence and on 7 May became eligible for a future space assignment. As soon as he heard the news, Shepard had gone to Slayton and asked for command of the *next* available mission – which happened to be Apollo 13. It should be stressed at this point that the three-flight rotation normally applied by Slayton was *not* a rule, but many astronauts confidently expected that backup duties on a specific mission would lead to a prime slot three flights later, and in most cases this was what happened. Cooper cannot be blamed, therefore, for *assuming* that his backup work on Apollo 10 would guarantee him the Apollo 13 command. When Shepard asked for Apollo 13, it made Slayton's task a lot easier. "I didn't feel any obligation," he wrote, "technical or moral, to keep [the rotation system] in place from that point on. Thirteen looked as though it would be the third lunar landing, sometime in the spring of 1970. We had a year to get ready for it, as opposed to six months." For his part, Cooper felt cheated and would later admit that it took him years to forgive Shepard for what he saw as a stab in the back. In reality, as Andrew Chaikin noted, Cooper had *allowed* the Moon to slip from his grasp and Slayton would later admit that, even if Shepard had *not* returned to active duty, his eyes were on Jim Lovell to lead Apollo 13.

Slayton now felt that the additional time would be enough for Shepard to thoroughly prepare for a return to space and become familiar with the systems and the hardware. However, at least two of his fellow astronauts and a whole gaggle of senior managers at NASA Headquarters in Washington disagreed. The first, not surprisingly, was Gordo Cooper himself, who learned that Apollo 13 was now Shepard's mission at a very awkward meeting in Slayton's office in May 1969 . . . and complained bitterly about it. The second was Jim McDivitt, a veteran astronaut who had led Apollo 9. He felt that Shepard was not yet ready to lead a lunar crew. His judgement, however, may have been impaired by Slayton's offer to assign him as LMP on Shepard's flight, rather than as commander. Since McDivitt had already commanded two missions, he objected to flying as a subordinate, even to Al Shepard, and even if that subordinate position gave him the chance to walk on the Moon. After turning down an offer to succeed Shepard as chief astronaut, McDivitt ended up taking a senior management role as the Apollo spacecraft programme manager.

Other astronauts were equally miffed that they had to slog away in backup duties and now Shepard was using his clout to bag himself first pickings by jumping directly onto a prime crew. Even his CMP, Stu Roosa, and LMP, Ed Mitchell, were floored when Shepard told them, point-blank, that they were moving directly onto a prime crew with him. Mitchell had backed up Apollo 10, but Roosa was one of very few rookies to have not yet served in a backup role. (He had, though, done an outstanding support job in Mission Control on Apollo 9 and his performance on *that* flight had caught several managers' attention.) Still, since Shepard had only 15 minutes of space time – and even *that* had been suborbital – other astronauts snickered behind his back that his team was an 'all-rookie' crew. It was, wrote Gene Cernan, "a needle to Al's ego for having stomped on the rest of us for so many years". Some even argued that he had deliberately chosen two rookies for his crew to "lock in his own seniority among the group and guarantee command of the mission". To be fair, the CMP and LMP roles were already being increasingly assigned to rookies. The Apollo 12 backup team, named in March 1969, were the first such example of this and subsequent crews would also follow suit.

Regardless, the media quickly got in on the act. William Loeb, publisher of the *Manchester Guardian* in Shepard's home state of New Hampshire, wrote a series of letters directly to the White House, expressing concern about the assignment. Loeb pointed out that Shepard had been grounded for more than half a decade and questioned his medical fitness to fly, as well as his relatively advanced age. Nor was Loeb just 'any' journalist: a staunch Republican, with a keen interest in politics and an opinion of himself as something of a 'president-maker', he had been persuaded by some ear, nose and throat surgeons that allowing Shepard to fly would damage Richard Nixon's administration. "Poor Chuck Berry," Deke Slayton wrote, "had the White House climbing all over him looking for a way to answer Loeb. I think Chuck had to write about a ten-page letter justifying Al's assignment on medical grounds alone. It was a real pain in the ass and totally unjustified."

Al Shepard (right) and a bearded Ed Mitchell inspect some of the rocks which they had collected from Fra Mauro.

Notwithstanding the criticism, Shepard had been the *first* of any of them to fly and his Gemini mission had been snatched away from him: through *this* training and his perseverance, he *had* paid his dues to the programme and therefore was fully deserving of another flight into space ... but not yet. NASA Headquarters normally rubber-stamped Slayton's recommendations, but when he submitted the names of Shepard, Roosa and Mitchell for Apollo 13, they were rejected. George Mueller, for one, felt that after so much time off active status, Shepard needed more time to train. Slayton's solution, therefore, was to assign the Apollo 11 backup team (Jim Lovell, Bill Anders and Fred Haise) to Apollo 13 and hold Shepard's crew back for Apollo 14. Anders, though, had already announced his retirement from NASA and was replaced in the simulators by Ken Mattingly. This time, the response from Headquarters was a positive one.

One other astronaut who had some cause for concern was Gene Cernan. In May 1969, he had returned from orbiting the Moon with a clear vision of where he wanted *his* future to carry him ... but not at all sure if it would actually come true. As a veteran of two space missions, Cernan wanted, more than anything, for the opportunity to lead his own lunar landing flight. Slayton *had* factored Cernan into his future plans, but had not envisaged him as a commander; in fact, together with John Young and Jack Swigert, he would have backed up Apollo 13 and then flown Apollo 16. Cernan would have walked on the Moon ... but as the LMP, the same position he had held on Apollo 10. When Slayton put the assignment to him, astonishingly, Cernan turned it down, explaining that he wanted to hold out for a command of his own.

In his autobiography, Cernan explained that he might have been less forthright about turning down the chance to walk on the Moon with Young if he had known about Slayton's discussion with another astronaut, Mike Collins, who was only days away from flying Apollo 11. Collins had greatly impressed Slayton through his hard work and his speedy recovery from neck surgery and had been his first choice for the backup command of Apollo 14. Only when Collins turned it down did Cernan's own chances brighten. Had Collins accepted the Apollo 14 slot, it is almost certain that he would have gone on to command Apollo 17, the final landing mission, and Cernan would have been out of luck.

Years later, Cernan would firmly believe that Slayton's next move was a kind of 'acid test' to prove he was capable of leading a lunar crew: he was assigned, along with rookie astronauts Ron Evans and Joe Engle, to back up Shepard's team on Apollo 14. On the day that Cernan learned the news of his assignment and went into Big Al's office, he noticed that the secretary had put the scowling face onto her desk. Ignoring this bad omen, Cernan presented himself well and promised to do all he could to ensure the success of Apollo 14. Over the next two years of dedicated training for the mission, he would win Shepard's respect ... and go a long way towards securing command of his own lunar flight.

DEEPENING CRISIS

One day during the traumatic week of Apollo 13, a professor from Duke University encountered one of his students and asked him pointedly: "Do you think they'll get them back?" The student automatically responded by talking about the likelihood of withdrawing American troops from Vietnam. This response underlined an increased concern in the United States that the conflict in south-east Asia was far from being won and even Richard Nixon's promises of bringing 150,000 more troops home would not come to pass. Indeed, at the end of April 1970, a major series of more than a dozen US-supported military incursions – the 'Cambodian Campaign' – began in an effort to defeat the combined forces of the People's Army of Vietnam and the Vietcong, then based in pro-communist sanctuaries in eastern Cambodia. The response from within the United States had been both immediate and bloody. On 1 May, the day after Nixon announced the incursion into Cambodia as a "necessary means" of ending North Vietnamese aggression, more than 500 students assembled on the commons at Kent State University in Ohio to protest. Dramatic scenes showed one student burying his copy of the Constitution and another burning his draft card. The situation turned decidedly uglier in the following days, with Kent Mayor Leroy Satrom declaring a state of emergency on 2 May and ordering the National Guard onto the streets to restore order. Despite calls for calm, a protest on the Kent State campus on 4 May drew a crowd of more than 2,000 and attempts by the National Guard to disperse them with tear gas led to a barrage of rocks and bottles being thrown. At 12:22 pm, the nervous National Guardsmen had opened fire – some rounds directed toward the ground, others into the air – but during the next few minutes, four students were killed and nine seriously wounded. Two of the dead were not even involved in the protest, but were simply walking across campus between lectures ...

Throughout the nation, the reaction was one of outrage. Hundreds of thousands gathered in Washington, DC, and photojournalist John Filo won a Pulitzer Prize for his iconic image of teenage runaway Mary Ann Vecchio screaming over the corpse of Kent State student Jeffrey Miller, who had been shot in the mouth. In the following weeks, massive demonstrations, involving perhaps four million students, were organised and hundreds of schools and college campuses were closed in an effort to counter unrest. Students at New York University hung a banner declaring 'They Can't Kill Us All'.

Against this backdrop of a nation bitterly at war with itself, and the seeming indifference of the Nixon administration to what was happening, it is hardly surprising that America's space programme and its drive to return to the Moon had a low priority. Workers at Cape Kennedy knew that Apollo would probably only survive for a few more missions and bumper stickers began appearing in Florida, reading 'Apollo 14: One Giant Leap for Unemployment'. In July, the *Washington Post* speculated that as many as *four* more Apollo landings might be cancelled. By the late summer, George Low, who had assumed the role of acting administrator following the resignation of Tom Paine, formally announced the cancellation of Apollos 18 and 19. These flights, involving long-duration lunar modules and rovers for three-day extended stays on the surface, would have been the most ambitious geological extravaganzas ever planned. In fact, the Apollo 15 backup crew, pointed towards Apollo 18, included geologist Jack Schmitt. Although Schmitt *would* ultimately fly to the Moon, in 1970 that was by no means certain.

The cancellations were the result of lengthy meetings in August, organised by Paine and Low, which included representatives of NASA's Lunar and Planetary Missions Board and the Space Science Board of the National Academy of Sciences. The choices had been stark. The first option was to fly four of the six remaining lunar missions, as planned, at six-monthly intervals, beginning with Apollo 14 in January 1971, then taking a break to conduct three lengthy flights to the Skylab space station in 1972–73 and finally performing the last two lunar missions in 1974. The second option was to cancel the last two Apollo landings – 18 and 19 – and make their Saturn V launch vehicles "available for possible future uses". One such use would, NASA announced, probably be for placing a future space station into orbit, but no one could have guessed at this stage that the other gargantuan Saturn V would end its days as little more than a museum exhibit. The cancellation allowed the agency to cut its Apollo budget by $42.1 million to a total of $914.4 million and enabled it to fit more snugly within the $3.27 billion budget allocation for 1971.

To be fair, many had known for some time that it was unlikely that missions through Apollo 20 would fly. Internally, Apollo 20 had been scrubbed from the books as long ago as July 1969 – and officially since January 1970 – so that its Saturn V could launch Skylab. Pete Conrad had seen the writing on the wall in advising his former crewmate Dick Gordon to be careful and not hold out too much hope that there would *be* an Apollo 18 for him. Nonetheless, since March 1970, Gordon had persevered with training as Dave Scott's backup on Apollo 15, together with his crew of CMP Vance Brand and LMP Schmitt, hoping that he would get his chance to walk on the Moon. Also hopeful was Fred Haise, who had been denied his own opportunity on Apollo 13 and was now assigned as backup commander of Apollo 16. Together with CMP Bill Pogue and LMP Gerry Carr, *that* assignment might have rotated all three of them into the Apollo 19 prime crew slot. The dreams of all six men to one day visit Earth's closest celestial neighbour were abruptly shattered in September 1970. "We had lost our opportunity to go to the Moon," Carr related. "We moped around for quite a few weeks."

Reading between the lines of Bill Pogue's words, spoken to a NASA oral historian in July 2000, many astronauts feared that the last few Apollos would disappear from the funding manifest; but that did not make the actual hammer blow any less devastating. "I was ... put on a 'phantom' backup crew, they called it, for [Apollo] 16," Pogue recalled. "They did not want to announce us, and for good reason, because it looked pretty bad in Washington, as far as the budget was concerned. We [went] ... on geology field trips and we were out on one in Arizona. We came back to Flagstaff [and] stayed at a motel. The next morning ... I walked out the door of the motel and Fred was holding this newspaper and it said 'Apollos 18 [and] 19 Cancelled'. That's how *we* found out about it."

From his perspective, Chris Kraft would later write that everyone in NASA would have fought much harder to keep Apollos 18 and 19 if they had known that

The 'final' Apollo 16 prime and backup landing crews undergo geological training with Lee Silver, near Taos in New Mexico in September 1971. By this time, a year after the cancellation of Apollo 19, Gerry Carr had moved toward Skylab, but Fred Haise chose to remain as backup commander. From left to right are Charlie Duke, Lee Silver (pointing), Fred Haise (partially obscured behind Silver), a bearded Ed Mitchell – who replaced Carr as Haise's LMP – and John Young.

humanity would be waiting another half-century before bootprints again dotted the lunar surface. "None of us," he wrote, "thought that America would turn into a nation of quitters and lose its will to lead an outward-bound manned exploration of our Solar System. That just wasn't possible." Kraft felt that even Bob Gilruth, who feared another failure, would have supported the additional few missions, had he known that low-Earth orbit would be the only domain of astronauts for another half century. Gene Kranz was similarly disgusted. "It was as if Congress was ripping our heart out," he wrote, "gutting the programme we had fought so hard to build."

As the months ticked down to Apollo 14's scheduled January 1971 launch, therefore, it seemed increasingly likely that there would be no more than four manned landing missions. The J-series had been eagerly awaited: its long-duration lunar module could not only support three-day stays on the surface, but could also facilitate the oxygen, water and power requirements for a trio of Moonwalks – each lasting perhaps seven hours or more – and accommodate a huge battery of scientific gear, including advanced ALSEP stations, the rover and a fully-fledged caddy of geological tools for the astronauts. As things stood, Apollos 14 and 15 were classed as the final H-series missions, staying on the Moon for 33 hours and supporting two Moonwalks, after which Apollo 16 onwards would be J-missions. For that reason, John Young – as backup commander of Apollo 13 – must have looked forward to rotating with his crew to Apollo 16 and leading the first such mission.

The cancellation of Apollos 18 and 19 changed the situation markedly, particularly in terms of *where* the remaining missions would land. Two touchdowns on relatively flat plains of lunar mare had been conducted by the Armstrong and Conrad crews and the importance of sampling Imbrium ejecta had already prompted NASA to reassign Apollo 13's Fra Mauro site to Apollo 14. In September 1970, a conference in Houston on the structure, composition and history of the lunar surface considered a number of options for the final landings. The options included Descartes in the central highlands, which was deemed to be volcanic terrain, the very prominent crater Tycho in the south, the crater chain Davy Rille in the north-eastern corner of Mare Nubium, the low domes of the Marius Hills and finally the thousand-metre-high central peak complex of the large crater Copernicus.

Descartes, firstly, looked as if it would offer geologists a chance to sample 'typical' terrain from the *central* highlands, rather than their periphery, and many believed that the fissures, grooves and hills of such regions had been formed through ancient volcanic activity and had remained essentially unchanged since shortly after the Moon's formation. "Samples from the Descartes site," read a NASA press release, issued on 1 October 1970, "would be important in determining whether or not highlands were formed by a very early differentiation of the Moon or whether they represent a primitive, undifferentiated planetary surface."

Davy Rille, too, would have offered ancient highland material, with craters that some geologists thought might have been formed by explosive eruptions which ejected material from 100 km beneath the surface, although limited photographic coverage rendered it inadequate for detailed mission planning. "It does not appear likely," the NASA release concluded, "that adequate photography of Davy will be obtained on Apollo 14 or 15." The collection of domes and cones known as Marius Hills *had already* been extensively documented by the Lunar Orbiters and, located near the centre of the Ocean of Storms, would have been the most westerly Apollo landing site. On the other hand, the Marius Hills were thought to offer little hope of yielding any highland material. Tycho had been visited by the Surveyor 7 probe in January 1968. Considered one of the last major impacts in lunar history, it was hoped that Tycho would turn up primordial material from 10 km beneath the highlands ... but its drawbacks were twofold: firstly, it would be the most difficult terrain yet negotiated by a landing crew and secondly, *reaching it* would demand a trajectory "far removed from the free-return path". In recognition of the difficulties encountered on Apollo 13, it is not surprising that Tycho was not selected. Lastly, the 95 km-wide Copernicus, with its yawning terraced walls, central peaks and brightly-rayed ejecta, appeared to mark where an impact blasted a hole through the mare surface of the Ocean of Storms into the underlying highland material. A violent impact propagates a shockwave into the ground and, upon being reflected back from the denser subsurface, causes the surface to essentially 'rebound' and uplift material from deep underground to form a central peak complex. The peak of Tycho was beyond reach, but Copernicus, being nearer the lunar equator, was attainable. A landing in Copernicus was expected to reveal clues about the differentiation of the Moon.

Marius Hills and another site, called Hadley, were nailed down as candidates for the Apollo 15 mission. Unlike the Marius region, Hadley, which lay on the very edge of one of the Moon's great mountain chains – the Apennines, whose peaks rose to 6 km – offered the chance to sample both Mare Imbrium and the primordial lunar material thrust upwards by the enormous shock of the Imbrium impact. "The chunks of basalt from Tranquillity Base and the Ocean of Storms," wrote Andrew Chaikin, "had taken geologists back to the era of mare volcanism. The Apennines promised to open a window on an even earlier time, perhaps all the way back to the Moon's birth." The choice of Hadley was aided by the presence of lava plains as a ready-made landing strip for a lunar module. Moreover, the relatively gentle topography of the area would more than likely allow the astronauts to drive in a battery-powered rover partway up the slopes of a 4 km mountain called Hadley Delta. Also within reach was the long, winding channel of Hadley Rille, which geologists suspected had once been a sinuous 'river' of lava. As a target for the first J-mission, it was enticing. If it discovered a single fragment of primordial, almost unchanged crust, the mission would make the entire Apollo effort worthwhile. Fittingly, it was Apollo 15's commander, Dave Scott, who was offered the final decision. Privately, he felt that he could land at either Marius or Hadley, but he favoured the latter, not just on the basis of its scientific promise, but because of its sheer *grandeur*. Scott felt that it was good for the human spirit to explore beautiful places.

As exuberant as Scott and his crew must have been, their backups – Dick Gordon, Vance Brand and Jack Schmitt – had already been dealt a hammer blow that month of September 1970, when *their* mission, Apollo 18, was scrubbed from the books. John Young and his team of Ken Mattingly and Charlie Duke seemed firmly pointed towards Apollo 16 and the crew of the final mission were not expected to be announced for several more months, though Gene Cernan, Ron Evans and Joe Engle seemed to have the edge. Regardless, Gordon decided, he and his team would sweat it out, in the hope that Deke Slayton would break the rotation system and assign them instead. After all, surely the presence of Schmitt – a professional *geologist* – on his crew would make Slayton's decision easier.

For his part, Cernan had the same future goal in mind. Yet there were mutterings, even in the astronaut offices, that Apollo 14 might be the *last* lunar landing; that Congress might pull the plug entirely on the hugely expensive project. After all, John Kennedy had promised to send "a man" to the Moon ... he had *not* specified that the feat then had to be repeated five or six times. *If* Apollo 17 survived the budget axe – and it *was* a big 'if' – there were no guarantees that Cernan, Evans and Engle would be aboard. Consequently, Cernan took it upon himself to ensure that his crew did the best backup job possible on Apollo 14. With Al Shepard and, by extension, also Deke Slayton, on their side, they surely had a better-than-average chance to scoop the mission, ahead of Gordon's team, Schmitt notwithstanding.

In his autobiography, though, Cernan noted that he felt Engle was not as knowledgeable about the lunar module's quirky systems as Cernan would have liked. However, Engle was one of the most gifted pilots in the corps, having flown the X-15 rocket-propelled aircraft before he was even selected by NASA, and Cernan felt that his deficiencies would not preclude them from forming an outstanding crew. None of them could ever have guessed that by the end of 1971 Engle would have forever lost his chance to walk on the Moon … and not through any fault of his own. It is therefore ironic that, in December 1970, Cernan attended a meeting with Shepard and Slayton to discuss a concern about the Apollo 14 LMP, Ed Mitchell.

"He was fed up with Mitchell's penchant for playing around with experiments in extrasensory perception," Cernan wrote of Slayton's worries, "even wanting to take some ESP tests along to the Moon. Ed just wouldn't let it go and Deke said he was uncomfortable with the possibility that Mitchell's full attention might not be on the mission." Moreover, Mitchell had earlier refused to take on 'dead-end' backup duties for Apollo 16 and an annoyed Slayton had given him a choice: he either fulfilled these important duties or lost his place on Apollo 14. Mitchell had complied, but the doubts remained. In an exchange of opinions he would later regret, Cernan was asked for his input over whether to drop Mitchell from Shepard's crew and replace him with Engle. Both Shepard and Cernan felt that Mitchell was more than qualified, in terms of his knowledge about the lander's systems. However, the very fact that Engle was not quite at the same proficiency level led both commanders to stand by Mitchell. When Engle lost his chance to fly Apollo 17 later in 1971, Cernan would lament not fighting harder to get his former crewmate a seat on a landing mission.

"BEEP-BEEP"

A special, often unspoken fear exists between prime and backup crews; the former never knows whether the latter might someday step into their shoes, through a mistake in training or an illness or an accident or some other unforeseen event. Ken Mattingly is the most obvious victim of such a crew swap, when the fear of German measles caused him to be dropped from Apollo 13 in favour of his backup, Jack Swigert. Other victims included Deke Slayton, dropped from Delta 7; Al Shepard, dropped from the first Gemini mission; and Elliot See and Charlie Bassett, killed in an aircraft crash just weeks before their own space flight.

Throughout the 19 months that Cernan, Evans and Engle acted as understudies for Shepard, Roosa and Mitchell, they never once missed the chance for some good-natured ribbing of the prime crew. At 47 years old, Shepard would be one of the oldest men yet to fly into space and the younger Cernan's team designed their own *backup* crew patch to exploit this fact. It showed a grey-bearded, red-furred and pot-bellied Wile E. Coyote – representing the aged Shepard, the sandy-haired Roosa and the plump Mitchell – rising from Earth, only to find that the backup crew, illustrated by the Roadrunner, had beaten them to it and were already on the Moon. The image was emblazoned with the Roadrunner legend: 'Beep-Beep'. Many astronauts had already called the Apollo 14 astronauts 'The Three Rookies', since even Shepard only had 15 minutes of suborbital space time, and Cernan mercilessly pressed the advantage; after all, *he* had two missions to his credit, one in lunar orbit, *and* a spacewalk. The backups consequently dubbed themselves 'First Team' and presented themselves as more than willing and capable of replacing Shepard's crew. Every time Shepard found a sticker or a patch with the Roadrunner or First Team emblem on it – and they even floated out of command module lockers during the journey to the Moon – he would growl: "Tell Cernan: Beep-Beep – *your ass!*"

Perhaps Cernan pressed his advantage a little too far, because on 23 January 1971, whilst flying a helicopter at Cape Kennedy, he nearly killed himself by crashing into the Indian River. The tiny Bell H-13 chopper, which was little more than a bubble canopy with a set of rotors and featured in the Korean War movie *M*A*S*H*, was routinely used by Apollo commanders to practice lunar landings. On that day, Cernan flew down the Atlantic side of Cocoa Beach, over Melbourne and back up the Indian River towards the Cape. Mischievously, he decided to 'flat-hat' the river ... but as he looked at the reflective bottom, his eyes lost touch with the water. One of his skids touched the calm surface of the river and the H-13 crashed in a spectacular explosion.

"Spinning rotor blades shredded the water, then ripped apart and cartwheeled away in jagged fragments," he wrote later. "The big transmission behind me tore free and bounced like a steel ball for a hundred yards before going down. The lattice-like tail boom broke off and skittered away in ever-smaller pieces, the plexiglas canopy surrounding me disintegrated, one of the gas tanks blew up and what remained of the demolished chopper, with me strapped inside, sank like a rock." Miraculously, Cernan survived and swam to safety through water coated with burning fuel. Boaters on the river hurried to his aid. After being patched up at nearby Patrick Air Force Base, Cernan – his eyebrows singed, his backside charbroiled – strode into the crew quarters to see an astonished Al Shepard having breakfast. In true 'right stuff' fashion, Cernan told him that since things were so boring at the Cape, he had to do *something* to get some publicity for Apollo 14! To reinforce his point that there would be no backup interference, he told Shepard that Apollo 14 was now well and truly *his* flight.

"Right!" Shepard grinned, his bullshit detector shifting into overdrive.

Deke Slayton was in no mood for humour. At first, he tried to give Cernan an easy way out before talking to the press – offering to tell them that the helicopter itself was to blame, that its engine had failed. No, Cernan told him, "it didn't fail. I just screwed up". When the investigation board, chaired by Jim Lovell, published its report on the accident in October 1971, it concluded that "misjudgement in estimating altitude ... [was] the primary cause". In admitting blame and telling the truth, Cernan knew that he may have just screwed his chances of someday commanding Apollo 17, but was aware that honesty went a long way with Deke Slayton.

'SMOKY' AND 'THE BRAIN'

Ever since returning to active flight status in the spring of 1969, Shepard had enjoyed the confidence of Slayton in being able to train for and execute a complicated lunar landing mission. Even Gene Cernan would laud his ability to mop up knowledge like a sponge and admitted that Shepard had more than paid his dues to the programme. So too did his crewmates, Stuart Allen Roosa and Edgar Dean Mitchell. The former was one of few astronauts to move directly onto a prime crew, without having first served a detail as a backup. For this reason, Roosa had been stunned when Shepard summoned him into his office and told him that he was going straight onto a prime crew. Roosa had to ask him twice to ensure that he had even heard correctly.

From Slayton's perspective, though, Roosa was an ideal choice. Ever since joining NASA in April 1966, he had worked the 'trenches' of Mission Control on the last few Gemini missions and had served six months in a support capacity for Apollo 9, impressing that flight's commander, Jim McDivitt, with his efforts. In fact, Roosa had been at the capcom's mike for virtually every hour that the crew was awake. When the mission was over, the flight directors had unanimously voted Roosa as their 'most valuable player'; he had the honour of hanging a ceremonial Apollo 9 patch on the wall of the control room. With someone like McDivitt – a highly-regarded astronaut and a leading candidate to replace Shepard as chief of the corps – fighting his corner, it is little surprise that Roosa quickly established a flight assignment for himself.

Roosa had been born in Durango, Colorado, on 16 August 1933, but grew up in Oklahoma. After serving as a smoke jumper for the US Forest Service in the early Fifties and earning the nickname 'Smoky', he joined the Air Force. He completed gunnery instruction and aviation cadet school at bases in Arizona, then attended the University of Colorado and received his degree in aeronautical engineering in 1960. The following two years of his military career were spent as chief of service engineering at Tachikawa Air Base in Japan. His next steps, from 1962–64, were as a maintenance flight test pilot at Olmsted Air Force Base in Pennsylvania, focusing on the F-101 Voodoo, and then as a fighter pilot at Langley Air Force Base in Virginia, flying the F-84F Thunderstreak and F-100 Super Sabre. Shortly before his selection as an astronaut candidate, in early 1966, Roosa graduated as an experimental test pilot from Edwards Air Force Base in California.

In his history of Project Apollo, Andrew Chaikin would note that Roosa's conservative, soldier-like devotion to duty and to country led him to treat the Moon project as literally a 'peaceful war' with the Soviet Union. His military career before NASA, though, had been spent preparing for anything but peaceful conclusions. Assigned to the 'special-weapons' fighter-bomber squadron at Langley, he trained extensively for a full-scale conflict with the communist bloc: a conflict which might someday require him to fly F-100s, laden with nuclear weapons, toward Russia, drop to an altitude of just a few tens of metres to avoid enemy radar, then unleash destruction on a massive scale. By the autumn of 1968, as a would-be astronaut, Roosa had spent two years with Charlie Duke, covering Saturn V issues, when Al Shepard approached him with a job on the Apollo 9 support team. It hardly registered in Roosa's mind when Shepard added, matter-of-factly, "Be patient. I've got something in the works."

Now, at the start of a new decade and with Apollo 14 drawing inexorably closer, Roosa had become part of a fierce drive to make *this* flight a so-called 'full-up' mission: in other words, they wanted to accomplish *every one* of their major objectives. Full stop. It had nothing to do with wanting to show Cernan's Roadrunner team who was boss or even to prove that the Three Rookies could indeed do the job, but more that Apollo 14 had to show a visible recovery from the near-disaster of Jim Lovell's mission. The urgency of this was summarised by a NASA official, quoted by *Time*, in the days leading up to the launch: "If anything goes wrong *this* time," he said, "you'll *really* hear the hounds baying at the Moon – literally." Shepard trusted Roosa to do an outstanding job: as chief of the astronaut corps, he had already seen the younger man perform. Roosa also had a sly sense of humour which appealed to Shepard. In recognition of Roosa's early association with the Forestry Service, Ed Cliff, its then-chief, contacted him with a view to carrying some seeds to the Moon. The astronaut complied and packed several hundred seeds from loblolly pine, sycamore, sweetgum, redwood and Douglas fir in his personal belongings. Despite contamination of the specimens by space radiation, nearly all of the seeds would successfully germinate after their return to Earth and in 1975 many were given to various forestry organisations as part of the United States' bicentennial celebration. One of the loblolly pines was planted at the White House, others in Brazil and Switzerland, one was given to the Emperor of Japan and dozens of others flourished throughout the United States. Some of these 'Moon Trees', indeed, are still alive today.

The man who would accompany Shepard down to the hummocky Fra Mauro site was the soft-spoken Ed Mitchell, one of the few pilot-astronauts to possess a doctorate. He was quiet, studious and nicknamed 'The Brain'. Mitchell came from Hereford, Texas, where he was born on 17 September 1930, the son of a rancher father and a fundamentalist Baptist mother. As a boy, he was active in the Scout movement and achieved its second-highest rank. He was torn between the religious convictions of his mother and the pragmatism of his father and by the time he entered high school he had concluded that the story of the Creation was allegorical, not literally true. "As much as he respected his parents and the church he was raised in," wrote Andrew Chaikin, "as much as he sensed a great body of truth in its teachings, he could not follow that path." Mitchell chose a path into mathematics and engineering, balanced with an insatiable desire to better understand the workings of the Universe. Upon receiving a degree in industrial management from the Carnegie Institute of Technology in 1952, he joined the Navy. After initial flight instruction, he was detailed to Patrol Squadron 29 in Okinawa and subsequently flew missions from the aircraft carriers *Ticonderoga* and *Bon Homme Richard*. Later, he taught at the Navy's research pilot school and studied for a master's degree in aeronautical engineering and a doctorate in aeronautics and astronautics. His DSc thesis, today enshrined in the Astronaut Hall of Fame, focused upon the design of space vehicle guidance systems and interplanetary low-thrust navigation.

"There were programmes set up in the late Fifties," Mitchell told a NASA oral historian. "Once we had launched the space era, it was realised [that] we didn't have any academic career path at the PhD level having to do with space exploration, so these programmes were initiated at MIT, Caltech and Princeton for people who wanted to do that. It was an eclectic programme of study: orbital mechanics, star formation, exotic fuels. I was privileged to work under Stark Draper ... who invented inertial platforms ... and [I] specialised in the navigation phases of it."

Mitchell was chosen as an astronaut candidate in April 1966 and spent his first few years becoming a master of the lunar module. In fact, Stu Roosa has opined that Mitchell's wizardry with the lander's systems was so great that he actually carried some of the load for Al Shepard, enabling the commander to concentrate his energies on piloting them to a safe landing at Fra Mauro. Mitchell's request to focus on the lunar module would offer him a chance to reach the Moon, but his own ambitions ranged further afield ... to the Red Planet: Mars itself. "I wrote my thesis ... by a navigational program that would go to Mars with low-thrust engines," he recalled later, "and there was no reason why Mars, in its nearest conjunction after that period, that we couldn't have launched a mission to Mars in 1982. If we'd continued the progression and interest and putting the funds into it that we did in the Sixties, had we made a similar commitment to going to Mars during that period, we *could* have done so." It is not difficult to discern the frustration in Mitchell's words and such frustrations were obviously shared by many others, too.

As well as frustration, there was also some disappointment. As backup LMP for Apollo 10, teamed with Gordo Cooper and CMP Donn Eisele, Mitchell had not only anticipated assignment to Apollo 13, but also the prospect of becoming the *first* of the 1966 astronauts to fly into space. When Lovell's and Shepard's crews swapped places, that honour instead went jointly to Ken Mattingly (later replaced by Jack Swigert) and Fred Haise. "We were the dearest of friends," Mitchell told the NASA oral historian of his relationship with Mattingly and Haise, "but we were in continuous personal competition. By being assigned on Apollo 10 and then going to 13, it looked like I would get to fly first ... I was the 'top dog' and then we switched."

The Apollo 14 trio did not share the same camaraderie as Pete Conrad's crew, but a few episodes of note during their training were highlighted by Shepard's biographer, Neal Thompson. One geological exercise took Shepard and Mitchell and their backups, Cernan and Engle, to a remote part of Bavaria. Each evening, the men would throw back huge tankards of beer and on one occasion they drunkenly climbed an old bell tower, just outside Munich, and then had to pound on the doors of their dormitory which the proprietor had locked at 10:00 pm. On another expedition in southern Arizona, Cernan had arranged to cross the border into Mexico to visit a friend's restaurant for dinner. When they arrived, they were shown into a motel room and on a dresser stood four bottles of scotch and four room keys. Expecting dinner, the astronauts were shocked when four Mexican woman entered the room, each hopeful of sex with an astronaut that night.

"For the next 45 minutes," wrote Neal Thompson, "in a mélange of tortured Spanish and charades, the spacemen tried to explain why they couldn't stay. They each tried what little Spanish they knew: *el presidente ... no es possible ...* we can't stay out late ... we can't drink ... have to train for mission in the morning ... going to Moon ... *la luna* ... During the ride back, the astronauts joked that they didn't even get dinner." Nor did any of them get chance to surreptitiously grab the bottles of scotch!

Shepard had changed beyond recognition from an icy head of the astronaut corps to a fully-fledged commander, flying the unforgiving Lunar Landing Training Vehicle (LLTV) more times than Cernan could remember, teaching himself to pilot helicopters and sweating every detail of the mission he was about to undertake. On the evening of 30 January 1971, just a few hours before the scheduled launch of Apollo 14, Shepard grabbed Cernan, dumped him in a car outside the Cape Kennedy crew quarters and drove to the pad to see the Saturn V undergoing its final preparations. Standing in the glare of million-candlepower floodlights, the two men gazed at the vehicle which would propel Shepard, Roosa and Mitchell toward the Moon ... and Cernan could not help but admit that, against all the odds and all the criticism, Big Al *was* the right man for the job.

JUICING IT

Launch day dawned cloudy and dreary. A persistent drizzle turned into a torrential Sunday afternoon downpour and the 3:23 pm EST launch was postponed by 40 minutes in the hope that it would subside. The 'window' for that day actually extended until 7:12 pm, after which Apollo 14 would have to be postponed to 1 March. From within the command module, which they had named 'Kitty Hawk' in honour of the place in North Carolina where the Wright Brothers made the first sustained and controlled heavier-than-air flight in 1903, Shepard, Roosa and Mitchell felt supremely confident that they would not suffer a problem like that which had befallen Apollo 13. A third oxygen tank, isolated from the others, was aboard their service module, together with a spare 400-amp battery, capable of supplying them with enough electrical power to handle all of their needs from *any* point in the mission.

All three men were keenly aware of the Nixon administration's stance on manned space flight; the axe on Apollos 18 and 19 had fallen barely four months earlier and, Slayton and Shepard would later write, "the doom merchants were already aiming at more Apollo scalps to hang from their office walls". Congress had slashed NASA's budget to its lowest level in a decade and even Bob Gilruth was now openly suggesting that Apollo 14 should be the final lunar mission. Shepard, Roosa and Mitchell's voyage, therefore, encompassed far more than 'just' going to the Moon: they had to restore a dimming sense of national pride and restore confidence in the Apollo programme.

Before they could do that, they had to wait for the forces of nature to take their course. Deke Slayton radioed the astronauts to give them a status update and received an impatient growl from Shepard. Thankfully, the storm lashed the launch pad only briefly, then headed out to sea and the countdown resumed and proceeded without further incident. The access arm to the command module's hatch was swung away and the call of "Initiate firing command" signified the transition of the remainder of the countdown to the computers. With 50 seconds to go, the Saturn V transferred its systems onto full internal power; and from the astronauts' perspective, they could *hear* and *feel* the behemoth coming to life, far below them. "He felt the

The Apollo 14 crew enters the white room at Pad 39A on launch day. Commander Al
Shepard is clearly identified by the stripes around the elbows and knees of his space suit.
In the background is Tom Stafford, who had assumed Shepard's mantle as chief of the
astronaut corps since the summer of 1969.

first distant whispers of the Saturn V flexing its sinews," Shepard later wrote in
third-person narrative of the experience, "the rush of thousands of gallons of
propellants hurtling downward through their lines, turbopumps spinning ... He had
the wild thought that the giant rocket was ten inches shorter than before fuelling.
How could they get to the Moon with a booster that had *shrunk*? The fuel, of course.
Millions of pounds of cold fuel contracting the rocket, bending metal ... "

Nine seconds before launch, the Ignition Sequence Start command was given and
kerosene and liquid oxygen poured into the combustion chambers of the five F-1
engines. Louise Shepard braced herself with her friend, Dorel Abbot, against a
hurricane fence, 5 km from the pad, and her recollection of those final seconds was

of little more than a steadily increasing brightness of flame. The *sound* – the agonising staccato crackle of over three million kilograms of thrust – would not reach the assembled VIPs for 15 long seconds. "The Saturn V roared and screamed," wrote Shepard, "anchored to its launch pad by huge, hold-down arms chaining it to Earth until computers judged the giant was howling with full energy." That energy was unleashed at 4:03 pm, when countdown clocks in Mission Control touched zero and 'Big Al' – the man who had waited almost a decade for this day to come – headed for space once more.

When one watches film footage of the Saturn V launches – and for a relative youngster like myself, that is all I can do – it is impossible not to be touched by the grandeur and majesty and sheer, raw power that this immense beast produced. Merely watching one of these boosters thunder skyward, the casual observer could be forgiven for thinking that the ride would be a hard, jackhammering slog. Shepard, Roosa and Mitchell had these same thoughts and could hardly believe their ears when other astronauts assured them that, no, most of the uphill journey was exceptionally smooth. As Apollo 14 left Earth, they were proven right: for the sound of the F-1 engines seemed like nothing more than distant, muted thunder, followed by a very gentle, almost jerky, motion as they rose from Pad 39A. The spectators saw something quite different: a terrifying, shrieking, earthquake-like cataclysm of fire and thunder which sent flocks of birds fleeing in all directions and which pummelled chests and the soles of feet with intense shockwaves. Nor was the impact only felt at Cape Kennedy. Nearly two thousand kilometres to the north, at Palisades Park in New Jersey, the roar of the Saturn shook atmospheric instruments at the Lamont-Doherty Geological Observatory ...

The smooth ride gave way to a somewhat bumpier one when the first stage was jettisoned and the J-2 engines of the second stage ignited. "Without constant acceleration and with the sudden cut-off of stage one," wrote Shepard in third-person narrative, "the three men jerked forward in their seats. The accordion stretched out and then compressed again; the fuels sloshed and the astronauts felt a series of bumps, just like a train wreck." Next came the jerky jettison of the escape tower, which uncovered the command module's windows, but through which they could see virtually nothing, save the profound blackness of space. Thirteen minutes since leaving Florida, and with the S-II and the first burn of the S-IVB finally behind them, Shepard, Roosa and Mitchell were in orbit.

"That alone," exulted Big Al, "was *almost* worth the entire trip!"

So far, the mission of Apollo 14 seemed charmed and the crew was given the customary go-ahead for TLI – the translunar burn which would propel them towards the Moon. "There was no sensation of movement," wrote Shepard, "only the delicious freedom of floating without weight." Next on the agenda was a time-honoured task assigned to the command module pilot: for Roosa was now to apply 19 months of training to the intricate transposition and docking manoeuvre to extract the lunar module (named 'Antares', after the brightest star in the constellation Scorpio) from the top of the spent S-IVB.

With pinpoint precision, Roosa guided Kitty Hawk's docking probe into the cone atop the roof of Antares' ascent stage and all three men waited for three capture latches to signify a 'soft-dock', after which they would retract the probe and pull the spacecraft together in a metallic embrace. The soft dock, however, never came. Perplexed, Roosa pulled away, informed the Houston capcom, then tried a second time. Again, there was no success. In Mission Control, there was similar confusion: for the mechanical latches needed no electrical power to operate, no pneumatic pressure or drive – they were supposed to simply *come together*, click into place and lock. Huddled around a console, flight controllers and managers speculated that maybe a piece of debris, or even dirt, could be lodged inside the mechanism. Their consensus was that Roosa should try again.

The CMP did try again ... and again, *and again*. No fewer than five attempts were made to 'soft dock' with Antares and all failed. More worryingly, every time Roosa manoeuvred Kitty Hawk towards the lander, he was consuming more and more of the precious fuel which would be needed later in the mission. Shepard was becoming increasingly irritated, reporting "No joy" after each attempt. At length, he offered to depressurise the command module's cabin, open the apex hatch and withdraw the probe into the cabin to determine what was afoot. When Roosa had moved Kitty Hawk up to Antares with the tunnel open, Shepard would then have poked out his suited arms to align the two spacecraft for an attempt at hard docking. Mission Control turned him down flat, declaring that such an exercise was too risky. Instead, they advocated having Roosa come in *harder* and *faster*, ramming the probe into the docking cone and hopefully allowing the capture latches to 'telescope' into position from the impact. Backup crewman Gene Cernan also advised them to push the retract switch a split-second before the two craft touched, thereby using the probe only to align the two collars. "That would drive Kitty Hawk hard up against Antares," wrote Shepard, "and hold it there long enough to engage the 12 latches of the docking rings directly. If the smaller latches, which normally made the first connection, were faulty, then they could be bypassed and a hard dock achieved."

Aware of the dwindling supplies of propellant, Roosa asked for a fuel status reading from Mission Control and was advised to make one more docking attempt before *that* was "re-evaluated". From his seat, Al Shepard had by this point had enough of gentle efforts to capture the lander and conserve fuel. It was time to push the throttle full forward. Turning to his command module pilot, he told Roosa: "Just forget about conserving fuel. This time – *juice it!*"

Minutes later, as some cautious managers in Mission Control continued to debate whether or not to cancel the flight, Roosa juiced it and the two craft came together. A welcoming cacophony of clacks sounded through Kitty Hawk's cabin and when Shepard announced success, applause broke out in Houston. However, they were still not out of the woods. It would later be determined that dust or debris had indeed prevented a soft docking, but the fear remained that a similar problem could arise as Shepard and Mitchell rose from the Moon to redock in a few days' time. Until the problem could be properly understood, and assurances received that disaster would not befall Apollo 14 in lunar orbit, the landing remained very much in question. The new head of flight operations, Sig Sjoberg, who had replaced Chris Kraft the previous year, told his troops that he wanted to be sure "that this thing is indeed satisfactory for docking, again, before we commit to the Moon landing".

By now, the astronauts had been awake for 19 hours and were directed to get some sleep. That did not, however, inhibit Mitchell from conducting an unusual task during what he already expected would be his one and only space flight. For some time, he had been fascinated by the mysteries of extrasensory perception – ESP, for which he felt neither religion or science provided a satisfactory explanation – and a few weeks before launch, he and some acquaintances had agreed to perform an experiment. Forty-five minutes into each sleep period, during the flight to and from the Moon, he would attempt to transmit thoughts from space. He started on the very first night. From within his sleeping bag, by the glimmer of a flashlight, Mitchell pulled out a clipboard on which were written a series of random numbers; each designating a typical ESP symbol – for example, a circle, a square, a set of wavy lines, a cross or a star.

"Mitchell chose a number," wrote Andrew Chaikin, "and then, with intense concentration, imagined the corresponding symbol for several seconds. He repeated the process several times, with different numbers, knowing that on Earth, four men were sitting in silence, trying to see the pictures in their minds." In fact, Stu Roosa saw Mitchell's flashlight during the first night's sleep period and did not think to ask him about it the following morning. Not until after the mission, whilst reading the newspaper, would Shepard learn of Mitchell's experiment.

With the exception of the lingering worry about the docking mechanism, the remainder of Apollo 14's cruise to the Moon was uneventful. As it drew closer, the astronauts began to discern traces of greys and browns; totally different to the bright object that they had known for all of their lives. Eighty-two hours after launch, Roosa fired the big Service Propulsion System (SPS) engine to drop them into an elliptical orbit with a low point of just 15 km above the surface. This manoeuvre essentially eliminated the need for Antares to perform the early stages of descent and represented a refinement of mission techniques since Apollo 12. The propellant savings would then give Shepard additional time to hover to find a suitable landing spot in the hummocky Fra Mauro.

By this time, experiments on Earth had satisfied mission controllers that the astronauts could repeat their 'juice-it' manoeuvre to bring Kitty Hawk and Antares together if a similar problem arose in lunar orbit. The landing was officially back on the schedule. On their 12th circuit of the Moon, early on 5 February, the lunar module undocked and began its descent. Despite their inherent knowledge of the landing site from months spent studying Lunar Orbiter photographs, the view was profound. Shepard called it a "wild place" and Mitchell considered it to be "the most stark and desolate-looking piece of country I've ever seen".

Apollo 14 was not out of the woods yet. In the centre of Antares' control panel, bordered with black and yellow tape, was a red circular push button labelled 'Abort'. Its purpose does not require a huge amount of explanation, except that pressing it would set in motion a chain of events to terminate the lunar landing, throttling up the descent engine and boosting Shepard and Mitchell back towards Roosa and Kitty Hawk. "The switch," wrote Gene Kranz, one of the mission's four flight directors, "had electrical contacts to issue signals to the LM engines, computer and abort electronics. When the abort switch for Apollo 14's LM had been manufactured, a small piece of metal had been left in the switch. Now, in zero gravity, and with both crew and ground oblivious, this piece of metal was floating among the contacts of the switch, randomly making intermittent connections."

Since the drama of Apollo 13, more than $15 million-worth of modifications had been incorporated into the Mission Operations Control Room, one of which included adjustments to help a controller rapidly identify any change in status in 'critical' spacecraft systems. From his chair as Antares' control engineer for descent and landing, Dick Thorson glanced at his monitor and noticed a red light blink on; it seemed to imply that either Shepard or Mitchell had pushed the Abort button. Thorson was perplexed. Why would they do that? They hadn't even begun the Powered Descent. Maybe there had been a telemetry patching error to the light panel on his console; a quick check, however, confirmed that everything was as it should be. As the engineer's eyes widened, it became increasingly clear that, if this was for *real*, it signalled bad news for the landing at Fra Mauro. In the back room, two of Thorson's colleagues, Hal Loden and Bob Carlton, also noticed the problem and suggested that one of the astronauts should tap the panel on which the Abort switch was located, in an effort to resolve the indication.

"Gerry," Thorson called Flight Director Gerry Griffin on the intercom loop, "I'm seeing an abort indication in the lunar module. Have the crew verify that the button is not depressed."

Capcom Fred Haise duly passed the request up to Antares and Mitchell tapped the panel with a flashlight. The abort light blinked off, then came back on again a few minutes later. "What's wrong with this ship?" Shepard wondered. They were barely 90 minutes away from the initiation of Powered Descent and the landing was temporarily waved off until a solution could be found. "Thorson's dilemma was a thorny one," explained Kranz. "To land, we needed to bypass the switch, but if we had problems during landing, we needed the switch to abort. It was a hell of a risk-gain trade."

Thorson's team identified a software 'patch' for Antares' computer, which would lock out both the Abort and Abort Stage switches, thereby allowing the mission to continue. Unlike the Abort option, Abort Stage was intended to support an abort if problems arose later in the Powered Descent. However, in an emergency situation, should Shepard and Mitchell need to perform such an abort close to the surface, they would need to use the keyboard to manually initiate the abort program. Gerry Griffin was willing to accept the risk, confident that Al Shepard would probably do the same. He rescheduled the landing attempt for two hours' time, on the next pass.

Key to this effort was the Draper Laboratory at Massachusetts Institute of Technology, which had developed the guidance and navigation systems for the Apollo spacecraft. Their engineers now shifted into high gear to wring out the software patch and make it work. Within the hour, a procedure had been devised, whereby the Abort switch could be bypassed when the descent engine was ignited. Amidst ratty communications with Antares, Fred Haise radioed up instructions to Ed Mitchell. First, the astronauts would start the descent engine at low power, using the acceleration to move the contaminating metal – probably a bit of solder – away from the switch contacts. As soon as Shepard fired the engine, Mitchell would input a string of 16 keystrokes into the computer to enable steering and guidance, then another string of 16 more keystrokes to disable the Abort program and finally *another* 14 keystrokes to lock into the landing radar and the descent software.

"This entire sequence," wrote Kranz, "would occur as the crew was descending to the Moon. The mission now rested on an emergency patch to the flight software that was less than two hours old, had been simulated only once and was being performed by a crew that had never practiced it."

Nevertheless, when the engine lit, Kranz was astounded that Shepard had lost nothing of his sharpness and marvellous calmness as the instructions were entered into Antares' computer. As the engine climbed steadily towards 10 percent, Thorson monitored his display and saw no evidence that the Abort switch had been activated. So far, the mission was back on track.

"Thank you, Houston," radioed Shepard. "Nice job down there!"

In his biography of Shepard, Neal Thompson related the singular contribution of one young MIT programmer, Don Eyles, who had helped to design Antares' software. In an almost-*Men In Black* kind of way, Eyles recalled being shocked from sleep as an Air Force car screeched to a halt outside his apartment at two in the morning and a uniformed officer hammered on his door. He was told that he had 90 minutes to come up with a solution for Apollo 14's problems. Eyles threw a jacket over his pyjamas and was driven to his nearby lab to create, virtually from scratch, a substitute program to eliminate Antares' faulty abort signal.

A few minutes into the Powered Descent, another problem reared its head. Both astronauts were now feeling positively snakebitten. This time, it was the landing radar – the device upon which Shepard and Mitchell would depend to feed them accurate altitude and rate of descent data as they headed for Fra Mauro. (The trajectory data from Mission Control could include errors as great as a thousand metres, making the radar indispensable.) From his seat in the MOCR, guidance officer Will Presley began to wonder why the radar had not locked-on. At first, he advised Griffin to continue the descent, but knew that he would need to call an abort if the radar did not kick in within the next 60 seconds.

From aboard the lander, Mitchell was irritated by the long delay in the radar acquiring the surface. "C'mon, radar," he repeated quietly to himself. At length, Dick Thorson, upon whose shoulders fell the responsibility for the radar, suggested cycling its circuit breaker. When Fred Haise passed up the request, Antares was less than 7 km above the surface – if the radar did not come on by 3 km or so, flight rules demanded that the landing be aborted – and Mitchell could scarcely hide the urgency in his tone: "*Come on!*" Shepard plucked out the circuit breaker, cutting its power, then jammed it back into place. "Hell, it works with my toaster," Mitchell quipped. Thankfully, this did the trick and a stream of radar data began flowing into the computer. By the time they reached an altitude of a couple of thousand metres, they could see Cone Crater off to the right, embedded in the crest of a ridge. They passed low over the ridge, heading for their target near Doublet.

"Fat as a goose!" Shepard exulted as he guided Antares towards a perfect touchdown, just 53 m from the intended spot; closer to target than Armstrong or even Conrad had achieved. The Lunar Contact light glimmered blue at 4:17 am EST on 5 February. It was the third time humans had landed on an alien body and Shepard and Mitchell were electrified, primed and ready to begin the expansive programme of exploration so cruelly denied to Jim Lovell and Fred Haise.

PERSONAL TRIUMPH

In the adrenaline-charged minutes after touchdown, Ed Mitchell had turned to Big Al and asked him, in all honesty, *would* he really have aborted the mission if the radar had not locked-on in time?

Shepard merely grinned. "Ed," he said, "you'll *never* know!"

Earlier, when Haise advised him to consider looking over the abort procedures, Shepard responded, with clear sarcasm in his voice, that he was *well aware* of the rules. Knowing Big Al as everyone did, it seems more than likely that after ten years waiting for this opportunity, he would have rewritten the mission rules on the spot: as long as the vehicle continued to fly beautifully and everything else seemed to be going well, he would have attempted to land, even without the aid of the radar. In fact, in a joint autobiography, co-authored with Deke Slayton, he wrote that, with the future of Apollo hanging by a thread, by hook or crook, there was no alternative *but* to land ...

Others were more sceptical. Gene Kranz recalled the feelings of Gerry Griffin and his team. "On Apollo 14," he wrote, "the error in the LM computer's knowledge of the actual altitude was almost 4,000 feet [more than 1,200 m] before the landing radar data update. With an error *this* great in the computer, Griffin and the [flight controllers] were convinced Shepard would have run out of fuel before landing. But everyone who knew Al never doubted he would have given it a shot. We also never doubted he would have had to abort. The fuel budget was just too tight." Now, it was immaterial; they had landed.

Glancing at the bright, undulating terrain of the Fra Mauro Formation beyond Antares' windows, both men reported that it was considerably more rugged than Tranquillity Base or the Ocean of Storms. There was certainly more topographic relief, Mitchell added, than they had anticipated by looking at their maps. Even before they ventured outside, a key obstacle of operating on the Moon was apparent: the complete absence of any recognisable features made it incredibly difficult to judge distances. Indeed, Shepard and Mitchell would come face to face with this problem during their second Moonwalk. The lack of a sensible atmosphere lent the same unreal clarity to the scene as it had to Armstrong and Aldrin and Conrad and Bean and the horizon seemed to slope noticeably away from them, framed against a totally black sky. Unlike Earth, they could actually *sense* that they were on a spherical body.

Just under five hours later, at 8:54 am, after abandoning plans to take an early rest period, Shepard dropped down onto the surface of that spherical body and planted his boots into alien soil. The importance of 'first words' had been on everyone's mind, it seemed, even as Armstrong prepared for his mission and the need to prove a point and win a bet had been Conrad's primary motivation for what *he*

A stunning view of the lunar module Antares on the undulating plain of Fra Mauro.

said on Apollo 12. In Shepard's case, however, a dozen years since joining the astronaut corps and after the better part of a decade spent chained to a desk, his words were important and intensely personal for a different reason: for they reflected the end of a long and difficult journey. Not only had he beaten Ménière's Disease, not only would he become the *only one* of the original seven Mercury astronauts to set foot on the Moon ... but he had overcome all the odds in being able to *fly* again. Clinging tentatively onto Antares' ladder, he turned and looked out onto the barren landscape and an eerie silence of ages, all around him.

"It's been a long way," he said, wonder evident in his voice, "but we're here."

Then, in a display of emotion so unlike the normal Alan Bartlett Shepard Jr, he wept. Not so many years ago, he had been grounded, with the chance of even returning to the pilot's seat in an aircraft considered a virtual impossibility. Now, he was standing on the surface of the Moon. His eyes welled up with tears; tears which quickly dried in the confines of his air-conditioned suit, but whose emotional impact remained. The impact of seeing the blue and white jewel of Earth, hanging in the black lunar sky, was equally emotional, if not more so. It was, he later wrote in third-person autobiographical narrative, "as though he were sent here, he and the others, so they might look back at that lovely, sensitive sphere and then carry home the message that everyone there must learn to live on this planet together".

Moving around the lander, he gazed to the east and could clearly see the ridge which he and Mitchell would attempt to negotiate the following day in search of Imbrium-era ejecta at Cone Crater.

For now, though, Mitchell himself was itching to get outside. With both men on the surface, they began assembling the customary Stars and Stripes, the erectable S-band antenna, the solar wind composition experiment, a rickshaw-like tool carrier known as the Modular Equipment Transporter (MET) and the ALSEP. For the latter, they picked a spot a couple of hundred metres to the west of Antares, setting up the nuclear-powered central station, from which radiated, in a star-like formation, its five main research instruments: the passive and active seismic experiments, designed to acquire data about the physical properties of the lunar crust, together with a cold cathode gauge to monitor the density of the Moon's atmosphere, a suprathermal ion detector to measure the composition of its ionosphere and a charged-particle detector to study the solar wind impinging on the lunar surface. In addition, they were to set up another laser reflector.

Of these, the active seismic device is particularly notable, since it required the implantation of three 'geophones' onto the surface, in a straight line at distances of 3 m, 50 m and 100 m from the ALSEP's central station. These provided a sort of mechanical 'ear', which Mitchell activated using a 'thumper' to produce a series of artificial quakes. "Earlier lunar seismic experiments," *Time* told its readers on 1 February, "have been largely passive; that is, the seismometers have usually depended on the occurrence of Moonquakes or other natural rumblings to make readings." Now, with the help of the thumper, not dissimilar in shape to a heavily-weighted walking stick, Mitchell was to detonate a series of 21 charges as he walked down the line. The intention was that the geophones would determine the travelling time of seismic waves created by the charges and measure their velocity through the lunar surface material. However, eight charges failed to fire and Mitchell had a fit over the others. "Houston," he radioed, "this thing's got a pretty good kick to it." Far from having the desired effect of a moderate firecracker, they actually resembled both barrels of a 12-gauge shotgun going off. The scismologists back on Earth were pleased, but the job of the astronauts had not been an easy one. As Shepard later wrote: "It was quite a trick to take the Moon's pulse when you're on another world." They also set up a mortar unit to lob grenade-like charges several hundred metres after the crew had left the Moon to extend the seismic survey, but it was never actually used.

Enabling them to transport a wide range of equipment – including cameras and lenses, sample bags, trenching tools, tongs, scoops, core tubes, hammers, brushes, maps and a gnomon to be placed next to samples in order to indicate local vertical and illumination conditions – across the hilly terrain was the 8.2 kg MET, a sort of two-wheeled, two-legged caddy which the astronauts pulled along by hand. Despite a number of puns from jokers that it would enable 'Old Man Shepard' to take a breather when he got tired, it was capable of holding 160 kg of equipment, but would only be used on this single mission. It tended to become bogged down in lunar dust, which seemed thicker and deeper than the soils encountered by Armstrong and Conrad's crews and was like dragging a golf buggy through deep sand. Moreover, being lightweight, it tended to bounce in the low gravity and one man had to follow behind to pick up items that fell off! A battery-powered Lunar Rover was already in the works for Apollo 15.

However, a common misconception over the years was that the MET was scheduled to have been aboard Jim Lovell's ill-fated Apollo 13 to Fra Mauro. This was not, in fact, the case. Originally, the MET was manifested on Apollos 14, 15 and 16, after which the final flights (17–20) would carry the Rover. When Apollos 18, 19 and 20 were scrubbed from the books during the course of 1970, the remaining missions were retasked: the Rover was reassigned to Apollos 15, 16 and 17 and Shepard's flight wound up as the only expedition to feature an MET.

In between tugging the MET, they found that the easiest form of locomotion was a sort of one-two, one-two, one-two gait, akin to a horse's trot, rather than the standard left-right moves of a jogger. With every step, they involuntarily kicked up

Al Shepard works with the Modular Equipment Transporter, a tool caddy which proved more trouble than it was worth in the difficult terrain of Fra Mauro.

clouds of dust, whose consistency they likened to talcum powder, and this quickly clung to their boots and worked its way up their legs. Ahead of them, the harsh glare of lunar sunlight made shadows unreliable, distance difficult to judge and, with no visible points of reference, craters seemed to appear at their feet, as if from nowhere. Boulders which looked to be some way off were suddenly within spitting distance. Unlike Earth-based walking, they had to 'plan' each step four or five paces in advance. It was, said Shepard, like trying to find one's way across the featureless expanse of the Sahara Desert.

Four hours and 47 minutes after the last wisps of air had been vented from Antares, the two men lugged the last of their rock and soil samples inside the lander and repressurised the cabin. Their sleep that night was far from comfortable, particularly since Shepard had set the lander down with one footpad in a shallow crater, meaning their whole point of reference was noticeably tilted to one side. Within the darkened cabin, shades having been put up against the triangular windows and the small rendezvous window in the roof, the sensation was one of silent, eerie stillness in the desolate landscape of the Fra Mauro Formation. With his burgeoning interest in psychic phenomena, it is not surprising that Mitchell acutely sensed the spiritual 'weight' of being the only two living creatures on this otherwise lifeless world. Every so often, he rose from his hammock to lift the shades and gaze outside at the desolate, alien scene.

On more than one occasion, both men were awakened from their light slumbers by the mild groaning of Antares' systems, the rustling of its paper-thin walls and the gentle hiss of its life-support machinery; Shepard even wondered if the lander was tipping over. One conversation between them is particularly comical:

"Ed? Did you hear that?" Shepard whispered.

"Hell, yes, I heard that."

"What the hell *was* that?"

"I don't know."

A few seconds passed. Then:

"Ed?"

"What?"

"Why the *hell* are we *whispering*?"

CLIMB TO CONE

After convincing themselves that Antares *might* be tipping over, and decidedly uncomfortable in their suits, Shepard and Mitchell clambered out of their hammocks and pulled down the window shades ... only to discover that they were still firmly planted on the undulating plain of Fra Mauro. By 3:30 am EST on 6 February, they were back outside, loading the MET with their tools to begin the traverse to Cone Crater, from where it was expected they would photograph the interior and gather samples from the rim. Perhaps, lunar geologists hoped, those samples would reveal clues about the Imbrium impact.

The hike across the hummocky terrain proved tougher than anticipated. With

Shepard pulling the MET – which, thanks to Gene Cernan and the First Team, was also emblazoned with a 'Beep Beep' sticker; even on the lunar surface, it seemed the Old Man could not escape – and Mitchell studying the map, it became clear after an hour that the rim of Cone Crater was further away than expected. 'Checkpoint' craters which seemed obvious on the map were now hard to find and the whole region seemed like an endless 'sea' of sand dunes, with yawning depressions, several metres deep, everywhere. "That next crater," they would think, "*ought* to be 100 m away" … but invariably, it was nowhere to be found. Sometimes, Shepard later wrote, they "would appear to walk along flat ground when their legs disappeared and reappeared, like small ships on a heaving sea. In reality, they strode through great shallows in the plains of Fra Mauro".

With difficulty, they found their first sampling stop, then their second, and managed to linger for just a few minutes to grab some rock and soil specimens, take a few photographs and acquire readings with their portable magnetometer. Then it was time to tackle the ridge itself. Here, Shepard and Mitchell found surer, firmer footing, but as they threaded their way upwards the presence of rocks everywhere slowed their progress considerably. Whenever the MET hit a rock, it jumped off the ground and the astronauts were concerned that it might topple over. At length, Shepard grabbed the back of the rickshaw and the two men carried it.

Watching from Mission Control, in addition to capcom Fred Haise – who had trained to perform just this traverse as LMP of Apollo 13 – were the backup landing team of Gene Cernan and Joe Engle. Before launch, they had bet Shepard and Mitchell a case of scotch that they would be unable to make it up the ridge whilst dragging the bulky MET behind them.

"There are two guys sitting next to me who kinda figured you'd end up carrying it up," Haise observed.

Yet the climb was difficult and their bulky space suits fought their every move. Every so often, Shepard and Mitchell paused for a breather and took time to look back down the slopes toward Antares; clearly visible in the slanting lunar sunlight were the tracks from the MET's two rubber wheels, interspersed with their own bootprints. It was becoming clear, though, that they were still far from their objective. The lack of reliable landmarks and the unreal clarity of the scene made it impossible to determine exactly where they were; at one stage, convinced that they were approaching the rim of Cone Crater, both men were disappointed that they had only crested an inflection in the flank of the ridge and that their climb was not yet over. After the mission, they would liken the optically illusive effect to looking at a mountain in clear air conditions on Earth; although a peak might *seem* to be quite close, in reality it was some distance away.

Many of the geologists listening to Shepard and Mitchell's efforts were not surprised that they were having trouble navigating their way around. It was one thing to recognise a lunar feature from orbit, but something quite different to view it on the surface. The climb was steadily taking it out of the astronauts – at one point, Shepard's heart rate reached 150 beats per minute – and the flight surgeon asked for a pause, then queried the geologists: How important *was* it to reach the rim of Cone Crater? From the perspective of better understanding the Moon's early history, it *was* important, but not vital. It was believed that rocks near the crater's edge would yield some of the deepest and oldest material. In effect, Cone Crater was a drill hole that would enable them to sample material which was otherwise deeply buried. By finding such material, Shepard and Mitchell would be travelling back in time, to only a few hundred million years after the Moon's formation.

The men's thoughts seemed to be diverging at this stage, more than an hour into the traverse, with Shepard keen to gather samples where they were at that time and then turn back and Mitchell eager to press on. Mitchell felt that the mission would be a failure if they did not reach the rim of Cone Crater, if only to have the awe-inspiring opportunity to look into its 300 m diameter bowl. On the other hand, geologists doubted that they would *see* anything of significance, because photographs from orbit showed no signs of exposed layering or other structural features. The message to call a halt was passed to Fred Haise. They should, he told Shepard and Mitchell, consider where they *were* to be the edge of the crater.

Shortly after abandoning their search for Cone Crater, the astronauts briefly explored this field of boulders ... little realising that the rim of the elusive crater was just a few tens of metres away.

Mitchell was bitterly disappointed. "Think you're finks!" he said. However, Haise gave them some good news: if they thought that they could reach the rim soon, they could press on a little further. Shepard decided to give it a shot. On and on they climbed, again stopping for a breather a few minutes past 5:00 am EST, gazing in bewilderment at the enormous boulders all around them, ejected by the impact which formed Cone Crater. From Mission Control, Deke Slayton offered to cover Cernan and Engle's bet if they left the MET and carried on without it. Much as Shepard and Mitchell appreciated the thought, neither of them wanted to leave their stash of geology tools behind. As far as they knew, they might be almost at the rim.

Another breather. Both men were drenched in perspiration, gulping oxygen and the internal temperatures of their suits were rising dramatically. The search was proving fruitless. At length, Shepard radioed: "We're right in the middle of the boulder field on the west rim. We haven't quite reached the rim yet." Mitchell took this as an admission that the search was over. In fact, both astronauts believed themselves to be at different places: Shepard thought they were close to the crater's western edge, but Mitchell was sure that they were somewhere to the south. If they headed north, Mitchell argued, they would reach the rim. They continued walking and, studying his map, Mitchell was convinced that they *should* be able to see a prominent, bright boulder. They could not ... and by this point, nearly two hours since leaving Antares, they were running out of time. Haise told them to stop searching for the crater and begin taking samples.

The return downhill towards the lander, though disappointing, was exhilarating and both men could take full advantage of one-sixth terrestrial gravity, bounding in wide, slow-motion jumps. Back at Antares, they loaded the rock and soil boxes – totalling 45 kg of lunar material – inside and took a few moments to catch a last glimpse of Fra Mauro. Before returning inside, Shepard indulged in one final chance to have a spot of fun on the Moon, taking a golf ball from his pocket and dropping it onto the surface. (He had gotten the idea when he and Deke Slayton showed comedian Bob Hope, golf club in hand, around Mission Control a few months earlier.) "Houston," he radioed, "I have in my hand the handle for the contingency sample return and it just so happens I have a genuine six-iron on the bottom of it. In my left hand, I have a little white pellet that's familiar to millions of Americans. I drop it down."

Unfortunately, with the stiffness of the suit, Shepard could only operate one-handed and his first swipe missed totally. Mitchell told him that he got more dirt than ball. A second swipe shanked the ball and sent it into a pathetic dribble. He dropped a second ball into the dust. *This* time, he was more successful and the ball sailed in impossibly slow motion into the distance. "Miles and miles!" he cried. In truth, it flew a few hundred metres. Before launch, he had told Deke Slayton, who made him promise only to do it if everything else was going well, but Mitchell had not been made privy to the caper. In fact, Shepard had sneaked out of crew quarters on a number of occasions, donned his suit and practiced his swing. The *last* thing he wanted, with a worldwide audience tuning in, was to fall flat on his face ...

Al Shepard makes his way back towards the lunar module Antares.

By the time Antares' cabin was repressurised, the second Moonwalk had lasted four hours and 35 minutes, bringing the Apollo 14 experience to over nine hours. Subsequent analysis of Shepard and Mitchell's photographs, correlated with images from orbit, would show that they came within 20 m or so of the rim of Cone Crater. In fact, a large saddle-shaped rock they had sampled was the bright one on Mitchell's map! A few steps further north and they would have seen the pit of Cone Crater open up before them! Less than eight hours later, Antares' ascent stage was barrelling its way back into orbit in order to rejoin Kitty Hawk.

During his time alone, Stu Roosa had undertaken much of the detailed observation work originally planned for Apollo 13. Key to this effort was the huge Hycon Lunar Topographic Camera, which he had installed in the command module's hatch window. The device, with its motorised film transport, exposure controls and timer, was capable of resolving surface features just a couple of metres across, although it had proven somewhat sluggish to get up and running. It took 140 frames, then began making peculiar clanking noises. Roosa had unplugged all of its cables and reconnected them, to no avail, and despite troubleshooting advice from Houston, he ran out of time and had to press on with his other work. One task was to photograph the Descartes site, being considered for a future landing, which he did using a hand-held Hasselblad fitted with a 500-mm lens.

Docking at 3:35 pm EST was followed by the sound of Shepard knocking from his side of the hatch. "Who's there?" asked Roosa, before admitting them. A perfect SPS burn later that day set them on course for home. The return journey was a calm one, although a few experiments were performed in electrophoresis, liquid transfer between containers, heat transfer and the casting of materials from a molten state. These would later be developed further aboard the Skylab space station and, in the Eighties, using the Shuttle. Kitty Hawk splashed down in the South Pacific, less than a kilometre from its intended point and within view of the aircraft carrier *New Orleans* at 4:05 pm EST on 9 February.

With the exception of narrowly missing the rim of Cone Crater and the close shaves during the journey to the Moon, Apollo 14 had proven hugely successful. In the eyes of the space workforce, Shepard, Roosa and Mitchell had saved the lunar programme and recovered from the ordeal which befell Jim Lovell's crew. Yet the public at large still seemed peculiarly disenchanted by the prospects of further Moon landings. Only three more expeditions remained on NASA's books, although *those* were expected to be the grandest of all. All three would be designated as 'J-series' lunar voyages and would spend three days on the surface, feature three Moonwalks as long as seven hours apiece and conduct advanced scientific research from orbit. The first J-mission, Apollo 15, scheduled for July 1971, was targeted for the rugged Hadley region of the Apennine Mountains. Commanded by Dave Scott, it would be one of the most brilliant missions ever undertaken in the annals of space science. Then, in March 1972, John Young would lead Apollo 16 to Descartes in the central highlands and the programme would conclude with Apollo 17 later that year. In the spring of 1971, the crew for Apollo 17 had yet to be announced, but as time wore on, the scientific community was pressuring NASA and hammering Deke Slayton to put a professional geologist aboard. There was only one man in the astronaut corps who fitted the bill.

GOLDEN CREW, GOLDEN MISSION

The rocks and soils returned by Apollo 14 turned out to be quite distinct from those brought back by the Armstrong and Conrad crews from the lunar mare. Whereas the samples from the mare were primarily basaltic, most of the rocks from Fra Mauro were breccias – mechanically assembled rocks comprising a number of rough fragments of earlier rocks which had been broken up by the shock of a violent impact, then bound together in a fine grained matrix of powdered rock. Some of the inclusions, called clasts, were basaltic, but many were themselves breccias and hence evidence of several cycles of brecciation. Late in June 1971, Arch Reid of the Geochemistry Branch of NASA's Manned Spacecraft Center told the Committee on Space Research in Seattle that the Fra Mauro material was considerably lower in ratios of calcium to aluminium and iron to magnesium and was higher in several minor and trace elements, such as potassium, barium and the rare-earth elements.

"Detailed studies of soil returned by Apollo 14," read a NASA press release, dated 25 June, "reveal that the Fra Mauro Formation contains a series of similar basaltic materials which are quite different from the iron- and titanium-rich mare basalts." According to Reid, the mare samples and the Fra Mauro samples had enabled geologists to examine two different types of lunar material and their results suggested that the Moon was "a complex, heterogeneous body with marked differences in composition between the interior and outer portions".

Yet humanity's knowledge of the Moon was still far from complete. The *real* lunar highlands remained largely unknown and it was here that Dave Scott and Jim Irwin were destined to visit. The precise landing site for Apollo 15 had been chosen in October 1970: a place called 'Hadley-Apennine', about 700 km north of the lunar equator. It is characterised by a small patch of Mare Imbrium at the base of the spectacular Apennine Mountains, some of which stand 6,000 m tall, and a long, meandering gorge known as Hadley Rille. The rille, whose precise nature is still unknown, but which may be an ancient lava channel, is almost a kilometre wide, 200 m deep and 40 km long and would set the stage for a true geological extravaganza.

The Apennines themselves, named in honour of the Italian mountain range, form the south-eastern border of Imbrium and the north-western border of the Terra Nivium ('Land of Snows') highland region. They arise to the west of the vast Eratosthenes crater, which touches the southern face of the range, and then sweep in a steady, north-easterly direction, forming an arc-like 'chain', which terminates at Promontorium Fresnel. In total, they cover 600 km of the surface. Mount Hadley Delta, the slopes of which Scott and Irwin would negotiate during their three days on the Moon, is one of its most easterly peaks, rising to 5,000 m.

"A mission to the Hadley-Apennine region," read the NASA news release of 1 October, "presents a unique scientific opportunity. On the lunar surface, the astronauts will obtain samples and make observations relating to three key problems. First, they will collect materials from the base of the Apennine Mountains ... Such samples are expected to contain a mix of the old lunar crust existing before the formation – possibly by impact – of the Imbrium basin, perhaps more than four billion years ago, and of rocks from deep within the Moon which were ejected during the impact. Second, the astronauts will make trips to the rille area in an attempt to obtain evidence bearing on the origin of those strange lunar features, resembling dry river beds on Earth. Third, sampling of the fresh-looking mare and volcanic-looking features at this location is expected to extend the age scale established on Apollo 11 and 12 to younger ages ... "

The importance of such areas was further underlined in June 1971, when NASA decided to send Apollo 16 to the mountainous Descartes region in the central highlands. This site featured two primary geological objectives for surface sampling. "The first," read a 17 June press release, "is the highlands basin fill. This is a volcanically-appearing material, flooding many of the large old highland craters. The geological evidence indicates that this material is older than the old mare sampled on Apollo 11 and 12, but younger than the Imbrium basin ejecta sampled on Apollo 14. When combined with the expected very old rocks from ... the Hadley-

When Dave Scott gave his recommendation to land at Hadley-Apennine, one of his reasons for doing so was a desire to explore somewhere visually and geologically spectacular. This image of the lunar module Falcon, backdropped by the bulk of the Apennine front, reveals that Apollo 15's landing site was precisely that. Note the softened nature of the mountaintops and the peculiar, unreal clarity of the scene, thanks to the complete absence of atmospheric haze. Judging distance on this alien land was notoriously difficult; in fact, one of the few measures of distance available to the astronauts were footprints, tyre tracks and the apparent size of the lander itself.

Apennine site, these samples should help scientists develop the story of lunar evolution. The second sampling objective is topographically hilly, grooved and furrowed terrain again thought to be volcanic. Called 'uplands volcanics', the area is thought to be of a similar age but of a different composition than the basin fill. Much as the mare basalts are giving clues to the lunar interior composition in mare regions, it is thought that the upland volcanics will yield data on the interior composition of the thick highland crust."

Sampling such diverse terrain would require a means of transport several orders of magnitude beyond the simple bootprints of Armstrong and Aldrin and Conrad and Bean and even the rickshaw-like caddy dragged along by Shepard and Mitchell. Since at least the summer of 1966, plans had been afoot to develop a battery-powered car, known as the Lunar Roving Vehicle (LRV). Nicknamed 'the Rover', it would see its debut on the first J-series mission. Weighing just 208 kg, it was a collapsible, four-wheeled buggy and could carry a pair of astronauts, together with up to 55 kg of scientific gear and 27 kg of lunar samples across a total distance of 92 km. It could overcome obstacles as high as 30.5 cm and cross crevasses as wide as 70 cm, descend slopes as steep as 25 degrees and park on 35-degree inclines. When one adds the weight of the astronauts themselves, and their suits and associated life-support gear, the Rover was remarkable in that on the Moon it could carry more than *twice* its own weight. Conversely, though, it also posed its own risks. A major failure could leave the astronauts stranded and a conservative 'walk-back limit' was implemented to ensure that they always had enough oxygen, water coolant and power reserves in their backpacks to enable them to walk back to the lunar module from any point on their excursion.

Six weeks before the launch of Apollo 11, NASA's Marshall Space Flight Center of Huntsville, Alabama, was given the go-ahead to begin developing the Rover. In his initial summing-up of requirements for the machine on 20 June 1969, Chris Kraft had stressed two major needs: a reliable and continuous system of voice communications between the astronauts and Mission Control and a simple 'dead-reckoning' system for navigating across the surface and safely returning the men to the lander. On 28 October, a formal agreement was made with Boeing to develop, build and test the Rover; they would be responsible for the electronics and navigational systems and for overall testing and subcontracted the construction of wheels, motors and general suspension to General Motors. The initial contract amounted to $19 million, but ultimately $38 million was spent on the development effort.

Weight was a primary concern, since the Rover would need to be housed in a folded configuration on the descent stage of the lunar module and unfolded as it was lowered to the ground by a system of pulleys and lanyards. Original plans called for an automatic deployment system to bring the Rover onto the surface, but problems encountered during its genesis prompted NASA to direct Boeing in March 1971 to terminate work on this in favour of a manual method.

Fabricated from aluminium, the front and back sections of the Rover folded over the middle for storage inside a prism-shaped compartment in the descent stage. It was driven by a T-shaped hand controller, located on a control and display 'post' between the two astronauts' seats, which permitted driving forwards and reversing, as well as left and right. Its wheels, which were topped off with bright-orange fenders, had spun-aluminium hubs and titanium bump stops inside tyres made from a woven 'mesh' of zinc-coated piano wire and a chevron-like patterning of riveted titanium treads. A traction drive affixed to each wheel had a motor harmonic drive gear unit and brake assembly which reduced motor speed at an 80-to-one rate for continuous operation at all speeds without gear shifting. An odometer monitored the distances travelled and sent data to the dead-reckoning navigational system. Seatbelts were provided to keep the fully-suited astronauts aboard the vehicle as it rolled and tumbled its way, like a bucking bronco, across the undulating lunar terrain. Power came from a pair of 36-volt silver-zinc potassium-hydroxide, non-rechargeable batteries that not only supported the drive and steering motors but also the communications relay unit and colour television camera.

"The cooling system," wrote Jim Irwin, "had to be rather unusual. They tried to radiate the heat with a radiator, but they couldn't get rid of it all, so they put seven pounds of beeswax in one of the electronics units to absorb the heat. The electronics in the unit generated heat and we had the solar heat coming in, so we had mirrored surfaces on top of heat-generating devices to radiate it and we frequently had to brush [lunar dust] off these surfaces to help the process of radiation."

Working in one-sixth of terrestrial gravity had, of course, already been done by three previous landing crews and some of the procedures to prepare the astronauts for the experience of moving from zero-gravity to conditions of lunar gravity had been rehearsed exhaustively aboard a modified KC-135 aircraft that was appropriately nicknamed 'the Vomit Comet' for its stomach-churning aerial manoeuvres to produce such low conditions. "Zero gravity lasted only 15–20 seconds," wrote Scott, "as the plane went over the top of a parabola. One-sixth gravity lasted about the same time, but was flown as a separate profile by the KC-135. As soon as the buzzer sounded, we started performing a small section of whichever manoeuvre it was we could practice ... such as climbing onto the Rover and putting our seatbelts on. The buzzer sounded again to signal that the period of one-sixth gravity was over and we had to wait for the next period to carry on with the activity." On their final such flight before launch, Scott and Irwin performed no fewer than 130 parabolas.

It would be something of an understatement, therefore, to label Apollo 15 as the most adventurous, ambitious and exciting lunar expedition yet undertaken. With this in mind, it is ironic that the triumph of Apollo 15 actually arose from the ashes of disappointment. Original plans called for four H-series missions – Pete Conrad's Apollo 12, Jim Lovell's Apollo 13, Al Shepard's Apollo 14 and Dave Scott's Apollo 15 – each of which would spend around 33 hours on the Moon and perform two excursions onto the surface. In September 1970 these plans changed with the cancellation of one H-mission and one J-mission; as a result, the schedule shifted to maximise the scientific return from the last few expeditions. Upgraded to 'J' status, Apollo 15 would be extended from ten to 12 days in duration, spend nearly 70 hours on the lunar surface, feature three long Moonwalks and benefit not only from the Rover, but also from an advanced suite of cameras and other scientific instrumentation aboard the service module. Its heavyweight lander – which, at 16,300 kg, was some 1,800 kg heavier than previous models – would be equipped with a bigger exhaust nozzle, larger fuel tanks and extra batteries and storage space to support a longer stay on the Moon. At the same time, weight remained the overriding concern and deciding what could and could not be taken to the lunar surface was a painful process. Years later, Dave Scott remembered a number of awkward conversations with Deke Slayton, Jim McDivitt and Rocco Petrone, arguing over adding simple tools like telephoto lenses or surface rakes aboard the lunar module.

During their Moonwalks, the astronauts would be clad in upgraded suits. Previously, the life-support, liquid-cooling and communications connections had been arranged on the chest in two parallel rows of three. From Apollo 15 onwards, these connectors were set in triangular pairs; together with a relocated entry zipper, which now ran from the right shoulder down to the hip, rather than a straightforward up-and-down motion, the new suit allowed the astronauts to bend completely and even 'sit' aboard the Rover.

"I understood [the] concerns about the extra weight the items would add to the LM," Scott wrote, "which was already heavier than any of the previous missions. Weight was a crucial issue on every spacecraft and was closely monitored by NASA's Configuration Control Board. Early on in the Apollo programme, technicians had been asked to review every single item carried aboard and, in an effort to rid the spacecraft of even the smallest unnecessary weight, they reduced the number of Band-Aids in the first-aid kit! But I was confident that the two extra items I wanted to take *could* be accommodated. Deke eventually came around to my way of thinking, but I still had to spend time arguing my case at higher levels." In this way, Scott was not only a strong advocate for his own crew, but also a fierce advocate of the *scientists*, too.

Despite the frustrations, Scott and his team – Al Worden as the command module pilot and Jim Irwin as the lunar module pilot – were more than delighted with their lot on Apollo 15. At the same time, he and Irwin knew that they could not operate on the Moon in the same manner that previous crews had done. "We would ... truly have to adapt to living on the Moon," he wrote. "We decided to plan our schedule in rough accordance with the circadian rhythms of a working day in Houston. We knew we would have to perform at our peak and so we would *have* to get some sleep. Previous missions had lasted 36 hours, so the crew could just about get away with not sleeping. But for three days that would be impossible." As a result, Scott decided that they could not possibly sleep in their space suits; they would have to remove them and use the light, beta-cloth hammocks to their maximum effect. One night at Cape Kennedy, they tried the hammocks out. In normal gravity, the men were too heavy; under lunar conditions, they would be lighter, but the experience was still far from a comfortable one.

Still, the Apollo 15 crew was one of the best-suited to accept such a significant science mission. Dave Scott had long harboured an interest in archaeology and his duties as backup commander of Apollo 12 had introduced him to geology and fired his fascination with it. At length, his home in the Houston suburb of El Lago boasted its own collection of rocks in a specially-made wooden cabinet and in April 1970 he had arranged a first meeting with Lee Silver. A month later, Scott and Irwin and their backups, Dick Gordon and Jack Schmitt, hiked into the Orocopia Mountains of southern California on their first geology expedition with Silver. Scott's focus on the mission was very clear and almost dictatorial. When Al Shepard told him in November 1969 that he would be commanding Apollo 15, Scott was happy with both the prime and backup crews, but he had a very coherent idea of what he wanted, telling his brood that the mission would return to Earth with the *maximum* amount of scientific and geological data possible. The pivotal role that Apollo 15 would play became yet more acute in the autumn of 1970, when it was promoted to 'J' status. Scott pushed Silver to accelerate their impromptu geology trips into something more serious and closer to a 'real' mission simulation. At length, they were taking one or even two geology trips a month, in their own time, with Silver. Scott and Irwin, clad in sunglasses, jeans and lumberjack shirts, toted mock backpacks, radios and cameras and by November a model of the Rover had been added for them to use on expeditions into the San Gabriel Mountains. They trekked across Hawaii, into the Rio Grande, the Coso Hills and the inhospitable Mojave Desert. They pored over aerial maps – deliberately blurred to match as closely as possible the best-quality images of Hadley-Apennine – and laid out geological traverses on them. Even astronaut Joe Allen, who, as the mission scientist, would serve as the capcom during the Moonwalks, was there, sitting in a tent with a bunch of geologists, talking to the astronauts by two-way radio. It is testament to Scott's determination that not only he and his crew, but also the flight controllers and geologists, would be as ready as possible for Apollo 15.

"Dave would select the rocks by placing a gnomon in such a way that it gave an indication of the vertical," wrote Irwin in his 1973 autobiography, *To Rule the Night*, "and he'd check the chart for colour classification and we'd take a total of five pictures for each rock. I supported Dave as a rock-carrier. We worked our routine out carefully and practiced it over and over." One picture was a wide shot called the locator, which included a recognisable feature. The pre-sampling documentation would include a view looking down-Sun and a pair taken from the side to facilitate stereoscopic analysis of the sample *in-situ*. Finally, a picture would be taken of the site after the sample had been lifted.

As the backup crew, Gordon and Schmitt were equally enthusiastic about being ready to fly if necessary and a good-natured competition arose during these trips. "Jack was the first trained geologist recruited to the astronaut corps," Scott noted in his autobiography, "and his expertise spurred us to learn. It also meant he and Dick were a tough team to beat. On a trip to Hawaii, Jim and I got caught in a thunderstorm on a pretty remote mountainside and approached our geology task with a little less enthusiasm than usual; consequently, we missed something important. Jack and Dick hadn't missed it, of course, and said mockingly that we needn't think we could get out of a task because of a little freezing rain. It kept us on our toes!"

At the same time, family life suffered. Andrew Chaikin related in his book that Scott's wife, Lurton, had taken introductory classes in geology at the University of Houston ... not just to fill the time whilst her husband was away, but also to give them something to talk about when he was home. Often, during geology trips with Silver, Scott would comment that he was not spending enough time with his children. In spite of their own excitement about the impending mission, similar feelings and emotions had undoubtedly marked his two crewmates.

"Dave was a slave-driver," Worden recalled with a hint of humour in a May 2000 NASA oral history, "and he made us do a lot of things. He was a very professional, very no-nonsense kind of commander ... but he urged us [to do more and more during training], so by the time we got to be prime crew, flying the spacecraft, we were already there. We could have gone anytime ... so we spent a big percentage of our time learning lunar geology, learning astronomy, learning all the other things, all the things we're going to see and we're going to look at when we get out there. I thank Dave for that. We could handle *anything* that happened on the flight." Having said that, in 1969 Worden's marriage had broken down under the incessant grind of training and lengthy absences from home and, a year later, that of Irwin also came under severe strain.

Alfred Merrill Worden was born into a farming family in Jackson, Michigan, on 7 February 1932 and completed his high school education in his hometown. "Aviation was not really something that was foremost in my mind," he told the NASA oral historian. "From the age of 12 on, I basically ran the farm, did all the fieldwork, milked the cows; did all that until I left for college." When he went to the Military Academy at West Point, Worden came into contact with academic classes that were predominantly focused on leadership skills and he won a degree in military science in 1955. At first, he aspired to become an army leader, but a couple of comrades talked him into the Air Force; partly, Worden believed (mistakenly, as it turned out) that the promotions there would be faster than in the army. Nevertheless, Worden turned out to be a gifted pilot. One of his flying instructors was ominously nicknamed 'Bendix', because he 'washed-out' so many students, but Worden was one of the few who shone in his classes.

Following initial flight instruction at Moore and Laredo Air Force Bases, both in Texas, he transferred to Tyndall Air Force Base in Florida as a member of an all-weather interceptor squadron. Subsequent work as a pilot and armaments officer guided him back towards education and he received a master's degree in astronautical, aeronautical and instrumental engineering from the University of Michigan in 1963. Next came assignment to test pilot school: in 1965, the year before he was picked by NASA, Worden graduated from both the Aerospace Research Pilots School at Edwards Air Force Base *and* from the Empire Test Pilot School in Farnborough as part of an exchange programme with the Royal Air Force.

Even in the early Sixties, Worden would confess to only a passing interest in the manned space programme. When Al Shepard flew Freedom 7 in May 1961, Worden and the remainder of his fighter squadron were far more interested in one of their comrades, a Major Henderson, nursing his ailing jet towards an emergency crash-landing than in the live television coverage of Shepard riding as 'spam in a can' a couple of hundred kilometres into space. By the spring of 1966, however, he faced a choice: to apply for the NASA astronaut programme or the Air Force's Manned Orbiting Laboratory effort. Worden was convinced that the latter would never reach fruition (and, indeed, it was cancelled in 1969) and opted for NASA instead.

It is interesting that the 19 astronaut candidates selected in April 1966 very quickly headed off in quite different directions, with some (like Jim Irwin and Charlie Duke) mastering the intricacies of the lunar module and others (like Worden and Jack Swigert) focusing exclusively on the command and service modules. By the summer of 1968, Worden and Swigert were spending virtually every week at North American's facility in Downey, California, monitoring changes to wiring and subsystems and developing from scratch many of the malfunction procedures that their comrades would someday need to know. His confidence was steadily growing, although even Worden would admit that as Dick Gordon's backup on Apollo 12, he "probably could have flown if we had to, but probably not totally comfortable with it either".

Backup duties were, for Worden, little more than a prerequisite for what he considered to be the 'real deal' of a prime crew assignment. When that prime assignment arose, it was for a H-series lunar expedition, but was expanded into a J-series mission in September 1970. For Worden, that decision placed an added burden on the crew, because while Scott and Irwin handled three days of surface duties, he would be tasked with operating a complex battery of cameras and instruments mounted in Apollo 15's service module and collectively known as the

The command module pilots of the J-series missions carried out their own detailed scientific research programmes in lunar orbit, using a battery of instrumentation aboard the SIMbay. During their transearth coast, each CMP would then perform a spacewalk to retrieve film canisters from the bay. Here, Apollo 16 backup CMP Stu Roosa practices the exercise during a few seconds of weightlessness aboard a KC-135 parabolic aircraft.

Scientific Instrument Module Bay (SIMbay). "Our purpose," Worden told the NASA oral historian, "changed from getting there and getting back to going out there and collecting all this science. There was an *end game* here; there was an end purpose to going. It wasn't just to go and come back. It was to go out there and really do something scientific that was worthwhile."

Worden's scientific payload was centred on Sector One – an originally empty portion of the service module which had been modified to carry an additional cryogenic oxygen tank in the aftermath of Apollo 13. It consisted of no fewer than eight individual experiments: an X-ray fluorescence detector, a gamma-ray spectrometer, an alpha-particle spectrometer, panoramic and mapping cameras, a laser altimeter, a dual-beam mass spectrometer and a subsatellite which would be deployed into lunar orbit before Apollo 15 headed home. These experiments would acquire data on the chemical composition of the lunar terrain, detect traces of its tenuous 'atmosphere' and precisely measure the spacecraft's altitude above the surface. The cameras would also prove something of a highlight for Worden: capable of resolving surface features as small as a few metres across, their film cassettes would be retrieved by him during a dramatic spacewalk on the way home to Earth. As such, it would be the first excursion ever performed in cislunar space.

Planning for lunar orbital science began in May 1968, when Wilmot 'Bill' Hess, head of science and applications at the Manned Spacecraft Center, asked George Low to consider the placement of instrumentation into the service module's empty bay. Indeed, when the Block 2 service module was defined in 1964, one bay had deliberately been left empty for precisely this purpose. North American perceived no major difficulties and in March 1969 a Steering Committee drew up a tentative list of possible instruments. Funding for the formal development of the SIMbay was released by George Mueller in May, with an original expectation that it might make its maiden voyage on Apollo 14 in mid-1970. However, in order to work out how to best integrate the experiments and finalise their operation, it was concluded that the first SIMbay would fly aboard Apollo 16 in mid-1971. Later, following the cancellation of several J-series missions, the first SIMbay was advanced to Apollo 15.

Testing the experiments of the SIMbay on the ground, though, had turned up problems right from the start. Since the instruments were designed to operate in zero-gravity, they had to be tested under normal terrestrial conditions. The extendable, deployable booms for the mass and gamma-ray spectrometers, for instance, could only be wrung out using railings which mimicked the space environment as closely as possible, although this was far from satisfactory and they never worked out particularly well on the ground. Then, when technicians tried to integrate the whole bay into the Apollo 15 spacecraft, data streams failed to synchronise properly and last-minute changes and adjustments were made by principal investigators until shortly before launch.

The hexagonal subsatellite was the first of its kind and carried three experiments of its own. Housed in a small container not dissimilar to a mailbox, it would be spring-ejected from the SIMbay, after which a pair of booms would deploy to stabilise its spin rate at around a dozen revolutions per minute. A third boom housed a fluxgate magnetometer. Measuring 78 cm long and coated in solar cells for electrical power, the 36.3 kg subsatellite was intended to examine the plasma, particle and magnetic-field environment of the Moon, as well as mapping its gravitational field for up to a year. Two such satellites were planned – one aboard Scott's mission, the other aboard John Young's Apollo 16 in the spring of 1972 – and both would prove hugely successful.

In this way, Worden's duties during his solo time in lunar orbit were far from simply being a 'caretaker' of the command and service modules; in fact, Dave Scott later wrote, he would be "performing alone all the manoeuvres carried out by some of the earlier *three-man* Apollo missions". Together with the experiments' principal investigators, he actively planned many of the SIMbay activities and integrated them into his flight schedule. Years later, he would take justifiable pride in having mapped around a quarter of the Moon's surface with the high-resolution imaging gear, although he did note that one of the cameras was a modified version of an obsolete military reconnaissance camera. "When the Russians found out about it," Worden told the NASA oral historian, "they had a fit. As a result of that, we were restricted from turning the camera on until we got to the Moon. We couldn't take pictures of the Earth with it. There was a lot of uncertainty in somebody's mind that maybe we could take pictures they didn't want us to see. Of course, what's silly about it is that those cameras probably flew all over Russia in U-2s and *they* knew it and *we* knew it! What are they going to *do* about it?" Of course, the camera could not be used anyway until its protective side panel on the service module had been jettisoned, so there was no scope for performing such 'sneaky' observations in Earth orbit. Nevertheless, at the time of Apollo 15, relations between the two superpowers were steadily thawing and the first glimmer of a joint manned mission was already on the horizon. With this in mind, NASA felt an acute need to establish guidelines to avoid unwanted political fallout.

Just as Scott and Irwin and Gordon and Schmitt were guided by their own dedicated scientific mentor, so Worden and his backup, Vance Brand, had the outstanding Farouk el-Baz to sharpen their eyes in the skills of lunar orbital geology. "He made me memorise the name of every crater there was on the surface of the Moon," Worden recalled, "and where it was and how it got there and what was happening to it. Farouk and I spent endless hours training on the lunar geology. My kind of geology on the Moon was different from surface geology. Lee Silver taught all of us how to describe a scene. There's a difference. Dave and Jim were very good at [analysing rocks] ... [but] I had to look at *major* features. I'm looking at a volcanic crater or a meteor impact crater from 60 [nautical] miles away, but there's no way I can look at individual rocks. So I looked at 'macro' features and Dave and Jim looked at 'micro' features."

It was actually el-Baz who proposed the name 'Endeavour' for Apollo 15's command and service modules. In addition to the appropriateness of the name as a word to denote doing one's duty or making an effort, it was also fitting for a venture of scientific discovery. "By coincidence," Scott later wrote, "on 12 July 1771, almost exactly 200 years before Apollo 15 was due to soar into the Florida skies ... Captain Cook had dropped anchor in Deal harbour at the end of the first truly scientific

The all-Air Force crew of Apollo 15. Left to right are Dave Scott, Al Worden and Jim Irwin. During their mission, they completed one of the greatest expeditions of scientific endeavour in history ... but all three paid a price: Worden lost his marriage, Irwin would lose his health and Scott – as the flight's commander – would lose a great deal of respect and unintentionally tarnish the reputation of the astronaut corps.

expedition by sea." During the past three years, Cook and his crew had carried out a detailed exploration of the South Pacific. In honour of that expedition, Scott's crew arranged with the Marine Museum at Newport in Rhode Island to carry with them a small block of wood from the sternpost of Cook's historic ship. As the only all-Air Force crew to fly to the Moon, Dave Scott had already decided that the lunar module would be named 'Falcon', in honour of the mascot of the Air Force Academy.

Rounding out the crew of Apollo 15 was James Benson Irwin, the man who would share Scott's experience of walking the slopes of the lunar Apennines and exploring the mysteries of Hadley Rille. As his name implies, he was of Irish descent; his grandparents having emigrated to the United States from Pomeroy in County Tyrone sometime in the mid-19th century. For him, an expedition to the mountains of the Moon was entirely appropriate; as a boy, he had derived great joy from exploring the hills near his home in Pittsburgh, Pennsylvania, and savouring their solitude. "It is very exciting country," Irwin wrote, "and the complexity of the land features always makes me feel I would like to fly over and explore from above." This fascination with hills and mountains and aerial exploration would remain with Irwin for the rest of his life.

Born on St Patrick's Day, 17 March 1930, Irwin's boyhood was thus spent in the hills of Pittsburgh and later in the Wasatch Mountains, near Salt Lake City in Utah. It was from East High School in Salt Lake City that he graduated in 1947 and entered the United States Naval Academy, receiving a degree in naval science. Irwin opted to join the Air Force and received initial flight instruction at Hondo Air Base in Texas – "the most primitive living I had ever seen" – although his work was impaired by a bout of viral pneumonia which consigned him to the hospital for a spell. Thankfully, after he recovered an understanding instructor volunteered to teach him to fly at weekends.

Next came Reese Air Force Base in Texas, from which Irwin hoped he would get the opportunity to fly the multi-engined B-25 Mitchell bomber and perhaps become a civilian airline pilot at a later date; however, it soon became clear that the Air Force wanted interceptor pilots and immediately began training them for all-weather instrument-based flying. Despite having to fly the little T-6 Texan aircraft, one spot of good fortune came when Irwin's roommate introduced him to his future wife, Mary Etta Wehling. She was the daughter of Reese's chief maintenance officer and *he* harboured reservations about her forming a relationship with the young pilot. "He was a devout Catholic," Irwin wrote, "and she had been raised in the Catholic Church. This religious problem was something we couldn't escape because of the intense anti-Catholic sentiment in my family. My whole background had conditioned me to believe that there could be no satisfactory marriage with a Catholic girl."

In Irwin's mind, there was no way that a marriage between them could go ahead until they had solved the problem of their religious differences. Twenty years later, only months before Irwin flew Apollo 15, similar problems would almost break them apart. In the meantime, he moved to Yuma in Arizona and began training on the P-51 Mustang; from this stage onwards, he lived to fly. Elsewhere, his relationship with Mary blossomed and his self-described "headstrong" opinions eventually persuaded her to reject Catholicism and join his church. With this in mind, it is perhaps surprising that when they married, it was Mary's father who gave her away and Mary's brother who acted as Irwin's best man ... but Irwin's own father and brother were not present. "We got married at the [Reese] base chapel," he wrote. "My mother was the only member of my family who came for the wedding – I guess it was a little funny that Dad and Chuck weren't there."

By 1957, Irwin's career, too, had blossomed and the Air Force sent him to the University of Michigan for a master's degree in aeronautical engineering and instrumentation engineering. This, he wrote, was designed to prepare him for management work and research into the design of guided missiles; it was also the most mentally tiring work he had ever undertaken. Michigan, propitiously, would be a common theme in the histories of all three Apollo 15 astronauts: for Dave Scott had earned his undergraduate degree there and both Irwin and Al Worden had studied for master's credentials there, too.

Gradually, as with so many pilots who wound up joining NASA as astronauts, Irwin was steadily cementing his qualifications and expertise. Then, in 1961, barely a month after graduating from the Air Force's prestigious Experimental Test Pilot School, disaster struck. One day, he and a nervous, over-reactive student pilot named Sam Wyman took off on a practice flight. "I was in the back seat," Irwin wrote. "He pulled the plane up too abruptly, turned it too tightly onto the crosswind and we went into an uncontrollable flat spin." From his perspective, Irwin did not know if Wyman had overcontrolled or somehow 'frozen' the controls, but as the instructor he was powerless to recover from the spin. They crashed hard; Wyman threw up his hand to cover his face and his head rammed straight into the front panel and caused it to cave in. Irwin hit the back of Wyman's seat, which then collapsed onto his feet. "I was wearing tennis shoes," Irwin recounted, "and when we hit ... the seat came right down above my ankle, giving me a compound fracture, with the bones sticking out through the flesh. I had two broken legs, a broken jaw and a head injury. I was pinned in the wreckage. Fortunately, the plane did not catch on fire."

The control tower had seen the crash and immediately dispatched a fire truck and an ambulance. At first, it was assumed that Wyman's condition was fatal – he was even given the last rites – but, miraculously, *both men* survived. Both also suffered severe amnesia; Irwin's memory was a blank for the 24 hours prior to the accident and that of Wyman did not return for five years, together with all recollections of his wife and the memory of his young daughter. For a time, it was feared that Irwin might require the amputation of his right foot, due to fears of the onset of gangrene, although one of the Air Force's orthopedic surgeons, John Forrest, was instrumental in preventing this.

Recovery was slow and hard, but it is astonishing that within five months of the accident, Irwin was flying again. Graduation from the Aerospace Research Pilots School came in 1963 and he later headed the Advanced Requirements Branch at the Headquarters of the Air Defense Command. When he was selected as an astronaut, alongside Al Worden in April 1966, Irwin became the first Air Defense Command officer ever chosen by NASA. He took an interest almost immediately in the workings of the Apollo lunar module – hoping that it "increased your chance of getting on the surface of the Moon" – and was assigned by Neil Armstrong to have prime responsibility on Lunar Test Article (LTA)-8, putting the lander's systems through a rigorous series of thermal and vacuum tests in the summer of 1968.

"In order to approximate the conditions [the lander] would encounter," Irwin wrote, "we tested the vehicle in what amounted to a gigantic thermos bottle that could reproduce the pressure and temperature of space." Together with fellow selectee John Bull, he oversaw the implementation of new, fire-retardant materials, such as beta-cloth, after the Apollo 1 fire. Irwin's performance in this role led to Tom Stafford requesting him for support crew duties on Apollo 10; and after *that*, Dave Scott asked him to join the backup team for Pete Conrad's mission.

Like Worden, though, the incessant grind of training and frequent absences from home took its toll on Irwin's marriage and family life, to such an extent that by Christmas 1970 his wife Mary was on the brink of filing for divorce. "She had gone so far as to draw up two lists," wrote Irwin, "one giving reasons why she should stay with me and the other why she should go. There was a long list of reasons why she should leave and only two why she should stay." During Irwin's brief time at home, by his own admission, he was totally absorbed in The Mission, behaved "like an automaton" and family time was filled with long episodes of cold, stony silence, punctuated by open talk of divorce, in front of their young children. "It was a terrible mistake," Irwin concluded. "They felt threatened, because Mary and I were not getting along ... these long periods of quiet would bring the children close to tears. Then the air would clear and we would come back together."

In a sense, going to the Moon would bring Irwin full-circle from the hills of Pittsburgh to the mountains of Utah and Colorado and the desolation of California to the high peaks of the lunar Apennines. It is hardly surprising that Irwin was attracted to Lee Silver's geology classes like a duck to water. During one expedition to Brooks Lake in Katmai National Park in Alaska, studying the ash and debris from the United States' then-most-recent active volcano, Irwin took time out for a spot of salmon fishing. His attention was suddenly drawn across the stream ... to behold an enormous Kodiak bear – "one of the largest carnivores in the world" – staring at him. As it began to cross the stream, Irwin took to his heels and fled the whole distance, perhaps a couple of kilometres, back to base camp. "I *really* was glad that I had done a lot of jogging," he wrote. "When I got there, I discovered that I was still clutching my fishing rod!"

Irwin's gladness was shared by Dave Scott in having this quick-witted, quiet, deeply religious, though constantly-aware pilot on his crew. "There is no one I would rather have spent time on the Moon with than Jim Irwin," Scott wrote. "He always had a pleasant demeanour, came up with good suggestions and was very easy to work with." Together with the independent Al Worden, who would be spending a sizeable portion of his mission alone in lunar orbit, they were a perfect team – a 'Golden Crew'. The Golden Mission of Apollo 15 that they were about to undertake would test them all, but the rewards would be truly astounding.

JOURNEY TO HADLEY

The countdown to the launch of the fourth manned lunar landing mission at 9:34 am EST on 26 July 1971 was near-perfect. In fact, Launch Director Walter Kapryan described it as "the most nominal countdown that we have ever had". The astronauts were awakened early that morning, breakfasted on steak and eggs, caught a brief nap as they were being suited-up and Guenter Wendt and Vance Brand helped them into their couches aboard Endeavour at around 7:00 am. The clang of the hatch shutting them in startled Irwin. "I think that is when the reality of the situation hit me," he later wrote. "I realised I was cut off from the world. This was the moment I had been waiting for. It wouldn't be long now."

From his couch on the right-hand side of the spacecraft, Irwin had little to do and had some brief respite to reflect on his life, consider the enormity of the mission ahead of him and, more than anything, give himself over to an air of anticipation and expectancy as he waited for the Saturn V to boost them toward the Moon. Fifteen minutes before launch, they had felt and heard the unearthly clanking of the access arm moving away from the spacecraft, then beheld the stunning blaze of sunlight through the command module's only uncovered porthole. As the countdown entered its final seconds, the glare of the Sun was so intense that Scott had to shield his eyes, just to read the instrument panel in front of him.

Precisely on time, the five F-1 engines came to life with a muffled roar which flowed through every fibre of their bodies. "You just *hang* there," Irwin wrote. "Then you sense a little motion, a little vibration and you start to move. Once you realise you are moving, there is a complete release of tensions. Slowly, slowly, then faster and faster; you *feel* all that power underneath you."

For Dave Scott, the rise from Earth seemed nowhere near as violent as his previous launch on Apollo 9. There was, he admitted, more lateral shaking and more vibration during the first few seconds than he had anticipated, but after clearing the tower of Pad 39A the ride rapidly smoothed out. From his seat on the left side of the cabin, his attention was mainly riveted on three data streams: firstly, the 8-ball attitude indicator, then secondly a 'Q-meter', which monitored aerodynamic pressure, and finally the engine-pressure readout. Lastly, but perhaps even more importantly, Scott's eyes were always on the big red Abort light, his left hand firmly clutched around the T-shaped abort handle and his ears keenly aware of each status report from Mission Control, from Jim Irwin or from Al Worden. If he were to initiate an abort, Scott wrote, it would require his attention to have been drawn to *two* independent cues of a major problem. "The mechanics of initiating an abort took only a 45-degree anti-clockwise twist of the T-handle," he recounted in his autobiography. "I could not afford to make a mistake, because any error in moving this device could have catastrophic consequences for the mission. Yet the timeframe which an abort could be activated measured only fractions of a second."

For Apollo 15, the Saturn V's trajectory had been modified to take into account the larger payload requirements of the SIMbay and the Rover and the long-duration reserves aboard the lander. The rocket was despatched in a more efficient azimuth of 80–100 degrees, essentially due east, and its initial Earth-parking orbit was lowered to 166 km. This enabled an additional 500 kg to be boosted to the Moon. Moreover, the propellant reserves were reduced and the four outboard engines of the S-IC first stage, together with the centre engine, were burned for longer before shutting down.

Shortly after the separation of the first stage, its instrumentation went abruptly dead. The cause would later be identified as being due to the elimination of solid-fuelled ullage rockets in the interstage and a reduction in the number of motors in the S-IC tail from eight to four. These were normally employed to settle the fuel and oxidiser in the second stage and had been removed as part of a weight-saving measure to improve the Saturn's performance. The S-IC's engines did not shut down cleanly – they actually took over four seconds to drop to zero – and this left the first and second stages uncomfortably close together. When the S-II ignited, the exhaust from its five J-2 engines effectively disabled the telemetry package on the first stage. Subsequent investigation revealed that a failure of any one of the solids could have caused the S-IC and the S-II to collide. From Apollo 16 onwards, NASA returned to the original configuration of charges.

Eleven minutes after leaving Florida, the crew was satisfactorily inserted into a parking orbit of 171 x 169 km ... and after *that*, satisfactorily performed their translunar injection (TLI) burn to set a course for the Moon. This was quite

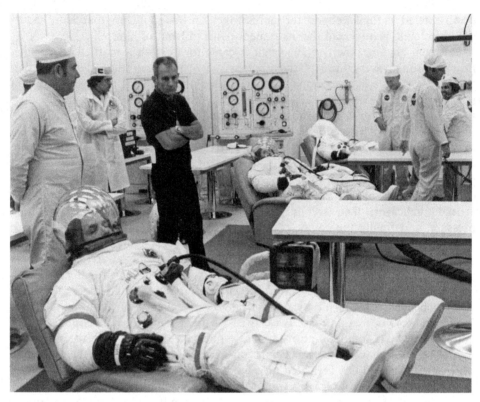

After suiting up, the Apollo 15 astronauts spent several hours pre-breathing pure oxygen in their suits. Front to back are Dave Scott, Al Worden and Jim Irwin. In his autobiography, *To Rule the Night*, Irwin would recall asking for a towel to be placed over his bubble-like helmet to enable him to take a nap. Deke Slayton, wearing a dark polo shirt, looks on.

disappointing, since they had so much work to do to prepare their craft and little time to gaze at the breathtaking view 'beneath' them. The TLI was followed by a nominal transposition and docking with the lander, Falcon. Not until three or four hours into the mission did the crew manage another chance to look at the Home Planet ... and realised, suddenly, that Earth had begun to shrink! "We could see the *full Earth*," Irwin wrote, his astonishment almost jumping from the words of his autobiography, "North and South America, Europe, Africa. You could see the blues and greens, the tans of the deserts and the whites of a few clouds and there was black all around the Earth. You couldn't see any band of atmosphere, no blue at all. It was the full Earth with the Sun shining right on it, fully illuminated, against the blackness of space." Irwin found it electrifying.

Getting out of their bulky suits, he wrote, was akin to three people dressing for a formal dinner in a space the size of a broom cupboard. They had to help each other, lest they risk damaging equipment or the $60,000 suits themselves, and it must have been a comical sight to behold arms and legs flailing in Endeavour's cabin. The novelty of weightlessness, though, did have its advantages and the interior of the command module seemed far larger than it had done on Earth. Irwin found that he could easily get from one side of the cabin to the other with nothing more than a push from his fingers. Mealtimes took some getting used to – the three men's food packages were colour-coded; red for Scott, white for Worden, blue for Irwin – and to keep track of things they resorted to Velcro-sticking their cornflake-box-sized rations to any suitable surface. Water pistols allowed them to reconstitute the dried foods into a kind of gruel-like mush in a squeeze-bag fitted with an exit nozzle.

Worse still was the dreaded process which Americans love to refer to, rather euphemistically, as 'going to the bathroom'. For Irwin, eating low-residue foods for the last few days before launch made absolutely no difference; on the morning of 27 July, the need arose. "It meant taking off all your clothes," he wrote, "and going down to the lower equipment bay with this plastic bag that has an especially designed top with a round receiver and a flat rim. You peel off a circle of tape with a sticky surface that you put right on your bottom." For a man like Irwin, who prided himself on his cleanliness, it was aggravating and bordering on intolerable. Only three sets of underwear for almost two weeks, together with an inability to clean oneself properly, left all three men grungy for almost the entire mission. Irwin *had* packed a bar of soap and from time to time the three men would take it from its container, just to let its sweet scent fill the cabin. "It *almost* made us feel clean," he wrote.

Elsewhere, the relative humdrum of a journey to the Moon continued. Only a couple of minor midcourse correction burns were required during the translunar coast and on 27 July Scott and Irwin opened the hatch to Falcon and began routinely checking switch settings and inspecting its electrical, environmental and communications systems. A glass cover on one of the tape meters had been shattered by the vibration of launch and the men discovered flecks of glass in the cabin. Although the meter itself was undamaged, it raised the very real risk that the men might inhale minute fragments of glass ... or even inadvertently eat it if it entered their food supply. "But the problem was relatively easily solved," Scott wrote. "Most of the loose debris was sucked into the filters of the LM's environmental control system, which we then cleaned off using sticky tape." They were also asked to switch on the vacuum cleaner and leave it running in order to remove any other debris.

Another potentially serious obstacle arose when the astronauts noticed globs of water floating around Endeavour's cabin. Irwin was in the lower equipment bay dechlorinating the water supply when he first noted it. He called Scott and at first they wondered if a line had cracked. "It was a big problem," wrote Scott, "and would be a major safety hazard should water penetrate into the sealed system of electrical wiring. Most of our systems were cooled with water-glycol, and if our water supplies ran low this would seriously compromise the running of the spacecraft." As the astronauts soaked up the globs of water with towels, Mission Control struggled to identify the source. At length, capcom Karl Henize told them to tighten a seal on the chlorination system and *that* seemed to do the trick ... though the tunnel between the command module and the lander took the form of a laundry room for a while, as the crew hung out their water-sodden towels to dry.

In Irwin's mind, the episode highlighted that lunar missions were *still* not routine and that success or failure sometimes hung by a thread. "If the fitting had broken," he explained, "or if we had been unable to stop the leak, we could never have landed on the Moon." Yet it also represented another example of the teamwork that had become NASA's trademark. "A technician at the Cape was driving home from work late that night when he heard on the radio that we were having this problem," Scott wrote. "He had pulled over, got to a telephone and called Mission Control to say he had detected a leak in one of the chlorination valves before launch and had worked on a procedure for stopping this from happening. It was *his* procedure that was transmitted to us in space." In Scott's mind, it was a perfect illustration of the total awareness and utter dedication of every individual in their contributions to the lunar effort. Other activities during the coast to the Moon included jettisoning the door for the SIMbay to expose its scientific payload. It is interesting that, only a few weeks after the Soyuz 11 disaster, the risk of depressurisation was at the forefront of everyone's mind, and so Scott, Worden and Irwin were instructed to don their space suits during the jettison process.

The remainder of the journey was largely uneventful and Endeavour and Falcon slipped perfectly into lunar orbit at 4:06 pm EST on 29 July. The reward at the end of a long voyage was dramatic. "As we passed the terminator," Scott explained, "on the [lunar] far side, we broke into sunlight and caught our first sight of the surface of that normally-hidden portion of the Moon, close up and fully lit. The far side is quite different from the near side. It has fewer giant dark craters and a more rounded topography. It looked spectacular." Irwin, too, was astounded, as they came from darkness into sunlight and the forbidding bulk of the Moon sprang at them in all its glory: a dark mass of gun-metal greyness, looking uncannily like clay, utterly surreal in appearance.

"The back side of the Moon," Irwin wrote, "is distinctly different from the front; it has no basins, none of the flat surfaces that you see on the front of the Moon with basins, like Mare Serenitatis and Mare Imbrium. Flying over the front you see craters of every size and description; you see mountain ranges, canyons or sinuous rilles – you could gaze at it hour after hour, particularly when you get to the terminator. There you pick up the sharp relief because of the long shadows; it makes the mountains stand up higher. The line between darkness and light sort of jags around, depending on the topography."

Early the following day, Scott and Irwin transferred to the lander and began powering up its systems and updating the guidance computer in readiness for their descent to the surface. Separation from Endeavour was scheduled to occur at the end of their 11th orbit, but was postponed when it became apparent that Falcon *wouldn't undock*! "Dave and I looked at each other," explained Irwin. "We couldn't figure it out." At length, Scott radioed Worden to get into the tunnel to check the umbilical connections and Houston gave them 40 minutes' grace before they would need to revise the plan. At length, Worden found the problem: an improperly-attached umbilical, used to signal the docking probe to disengage. "Al evidently had not had the connections firmly mated," Irwin continued. "He pushed the connectors in solidly, tried it again and we undocked. Of course, we were late undocking, but as it turned out it worked perfectly."

They were actually about an hour – approximately half a lunar orbit – behind schedule, but still on track to make their descent towards Hadley without any compromise of lighting and tracking constraints. Worden then performed an SPS burn to raise the low point of his orbit and circularise it. Should Scott and Irwin need to perform an abort during the final stages of their descent, their rendezvous would be easier with him in a circular orbit than in an elliptical one.

Like the other Apollo commanders, Dave Scott had flown extensively in the notorious Lunar Landing Training Vehicle (LLTV) and he and Dick Gordon had both undergone weeks of helicopter tuition at the naval school in Pensacola, Florida. "We had a great time," Scott wrote, "concentrating just on flying. To maintain our proficiency after that, we flew helicopters at Ellington [Field in Houston] and I often spent Saturday mornings practicing landings and flying in the chopper in the area around Clear Lake." Like Neil Armstrong before him, Scott felt that the LLTV was the best training analogue to the real thing – and he instinctively knew that guiding a *real* lunar module down towards the surface would be no less forgiving. "Flying the LLTV," he wrote, "was the best way of learning to fly in the co-ordinated mode, using both hands, that was needed to control the forward-aft, up-down and left-right motion of the lunar module. The pilot of an airplane has to contend with only four degrees of freedom: an airplane only flies forwards, though it can roll, pitch and yaw. A helicopter has, like the LLTV, six degrees of freedom – forward-aft, up-down, left-right; but it is still easier to fly than the LLTV, since it does not have the variable-thrust rocket engine of the Flying Bedstead." Scott was, however, in agreement with other Apollo commanders that in the humid Houston air the hydrogen peroxide thrusters tended to produce quite dense clouds, which meant astronauts flew virtually 'blind' for short periods. Still, as launch drew closer, Scott braved the LLTV two or three times per month.

Landing on the Moon, of course, was only one facet of the challenge; actually *living* and *operating* there as an explorer for several days was quite another. More than a year before launch, in January 1970, Scott had been one of a team of observers from NASA to spend a week in Antarctica. The team, which also included Deke Slayton, were there to observe scientific research in a hostile environment and they instantly came face-to-face with some of the challenges that living on the Moon would present. It also revealed that any hopes of establishing, maintaining and supplying a long-term lunar base would be *much* more difficult than previously supposed. "The reflection of sunlight on ice, day and night, during the southern summer," Scott wrote, "would, in some ways, replicate the intense glare of the Sun on the surface of the Moon. The difficulty of moving any distance in the Antarctic bore some resemblance, too, to the restrictions we would experience in lunar mobility."

For now, spending just three days on the surface was the challenge ahead of Scott and Irwin. At 6:04 pm EST on 30 July 1971, Scott ignited Falcon's descent engine, burning it steadily at 10 percent thrust for the first half-minute, then ramping it up to full thrust. Descending in a steadily sweeping arc towards the Apennines, at an altitude of 2.7 km Scott spotted the peak of Hadley Delta off to his left and, shortly thereafter, began to discern the long, meandering channel of the rille. The terrain, he noted later, was much less sharply defined that he had anticipated on the basis of simulations. In fact, the best Lunar Orbiter images of Hadley-Apennine had a resolution of only 20 m, and had been contrast-enhanced in order to emphasise the topography, thereby giving the impression of roughness. Still, Scott was able to find four familiar craters: Matthew, Mark, Luke and Index, the latter of which they had used in landmark sightings from orbit. (The name 'Index' was chosen instead of 'John' in order to stave off complaints from the atheist Madalyn Murray O'Hair, whose criticism of overtly religious symbolic gestures on missions had scalded NASA in the past.)

Dropping through a gap in the lunar mountains, Scott suddenly had the surreal feeling that he was 'floating' with strange slowness towards his landing site. "No amount of simulation training," he wrote later, "had been able to replicate the view we saw out of our windows as we passed by the steep slopes of the majestic lunar Apennine Mountains." In the simulator, they 'flew' a television camera towards a small, relatively flat patch of plaster-of-Paris; now, doing it for real, they drifted between the astonishing 5 km peaks of the mountains to both their left and right as they threaded their way towards Hadley. "It made us feel," he added, "almost as if we should pull our feet up to prevent scraping them along the top of the range."

As they continued to descend, twice as steeply as previous missions, Falcon's computer was transitioned to the so-called 'Program 66', enabling Scott to fly the spidery craft with a measure of semi-manual control. Glancing through his window, he was surprised that he did not recognise all of the craters and geological formations which *should* have been there. "Dave didn't want me looking at the surface at all," Irwin wrote. "He wanted me to concentrate on the information on the computer and other instruments. He wanted to be certain that he had instant information relayed to him. He was going to pick out the landmarks. But Dave couldn't identify the landmarks; the features on the real surface didn't look like the ones we had trained with." Scott could see Hadley Rille, though, and used that long, dark, meandering gouge as his marker, but was worried that they might still land 'long', and far to the south of their intended spot. This fear was confirmed by capcom Ed Mitchell – they were, indeed, a kilometre or more south of track. Scott knew that, even with the Rover, this might impair the effectiveness of their geological traverses. During those final, tense moments, he carefully clicked his hand controller 18 times, forward and to the side, adjusting their trajectory to bring Falcon back onto its prescribed path.

Those last seconds were so unreal – the clarity of the scene, the weird behaviour of the outflying dust, the strange, almost-unpowered sense of drifting like a snowflake through the majestic lunar mountains – that Irwin mentally convinced himself that he was still sitting in the simulator back in Houston. If he *had* admitted to himself that this was for real, he felt that he would have been just *too* excited to do his job properly. Yet if this *was* a simulation, it was one of the smoothest that he had ever flown.

They were very close to the surface now, perhaps just a few tens of metres, and lunar dust, kicked up and sprayed outwards in perfect arcs by the engine, obscured the landing site entirely, like a thick fog. It was only Irwin's call that the blue Contact

Light had illuminated which finally convinced them that they had touched down. The time was 6:16 pm and, with a firm thud, the seventh and eighth men from Earth had reached the surface of the Moon. "Okay, Houston," radioed Scott triumphantly, "the Falcon is on the Plain at Hadley!" His reference to the landing site as a 'plain' paid due tribute to Scott's alma mater, the Military Academy at West Point, whose parade ground was also nicknamed 'the Plain'. Although they had actually touched down about 600 m north and 175 m west of their intended spot, they did not know this. In any case, mission controllers were not overly concerned, since the Rover should enable them to fulfil each of their traverses and perform each of their main geological stops.

What *did* cause some concern was that Falcon had come down on uneven ground and one of its rear footpads had planted itself inside a small crater. (In fact, Mission Control would later call their lander 'the Leaning Tower of Pisa', an epithet which Scott showed little sign of appreciating!) Irwin remembered the landing as the hardest he had ever been involved in; "a tremendous impact with a pitching and rolling motion. Everything rocked around and I thought all the gear was going to fall off. I was *sure* something was broken and we might have to go into one of those abort situations. If you pass 45 degrees and are still moving, you *have* to abort. We just froze in position as we waited for the ground to look at our systems. They had to tell us whether we had a STAY condition". With some relief, 77 seconds after touchdown, Mission Control radioed their approval for Scott and Irwin to stay.

"The excitement was overwhelming," Irwin wrote, "but now I could let myself believe it." They had set down in a beautiful little valley, with the high mountains of the Apennines on three sides and Hadley Rille a kilometre and a half to the west. In his mind, it conjured up memories of the mountains of Colorado, high above the timberline; yet there was something else about it, too. Irwin was certainly one of the more religious men in the astronaut corps and he would later make little secret of the fact that he acutely *sensed* the presence of a supreme being on the Moon. This was at its sharpest whenever he looked up at the colourful Earth in the black sky. "That beautiful, warm living object looked so fragile, so delicate, that if you touched it with a finger it would crumble and fall apart. Seeing this *has* to change a man, has to make a man appreciate the creation of God and the love of God." This profound experience would remain with Irwin and guide his steps for the rest of his life.

One of the skills that Scott learned from Lee Silver during their impromptu geological trips in the mountains of California was the need to gain a visual impression of the variety of the site that they were about to explore. With this in mind, he had requested mission planners to schedule a 'stand-up EVA' a couple of hours after touchdown, in which he would stand on the ascent engine cover, poke his helmeted head through Falcon's top hatch and photograph and describe his surroundings. Deke Slayton had initially opposed the idea, on the grounds that it would waste valuable oxygen, but Scott had fiercely argued his case and eventually won approval. To conduct this half-hour 'SEVA', Scott pulled a balaclava-like lunar excursion visor assembly over his clear bubble helmet, clambered onto the ascent engine cover and removed the top hatch. It was, he later wrote, "rather as if I was in the conning tower of a submarine or the turret of a tank".

Meanwhile, Irwin shaded the instrument panel from the sunlight and arranged Scott's oxygen hoses and communications cables to enable him to stand upright. "He offered me a chance to look out," Irwin wrote, "but my umbilicals weren't long enough and I didn't want to take the time to rearrange them."

In the weak lunar gravity, Scott found that he could easily support himself in the hatch on his elbows ... and beheld the stunning view of the brown-and-tan Apennines, tinged by the intense blaze of golden sunlight, against the ace-of-spades blackness of the sky. Irwin passed up a bearing indicator and a large orientation map, which Scott used to shoot a couple of dozen interconnected pictures of the landing site now officially known as 'Hadley Base'.

As his eyes became accustomed to the view, and his mind connected it with months spent examining Lunar Orbiter maps, Scott began reeling off the recognisable landmarks. There was Pluton and Icarus and Chain and Side – intriguing craters in an area known as 'the North Complex' – and on the lower slopes of Mount Hadley Delta was the vast, yawning pit of St George Crater. One particularly prominent, rocky landmark which they had dubbed 'Silver Spur' in honour of their professor, showed clear evidence of stratigraphy in its flanks. The tops of the mountains were smooth and rounded; nothing like the forbidding, Fifties-era paintings of the Moon which showed jagged, menacing peaks.

"The SEVA was a marvellous and useful experience, for a lot of reasons," Scott later explained to the *Apollo Lunar Surface Journal*. "One of our problems at Hadley was that the resolution of the Lunar Orbiter photography was only 20 m, so they couldn't prepare a detailed map. The maps we had were best guesses and we had the radar people tell us before the flight that there were boulder fields – *massive boulders* – all over the base of Hadley Delta, just boulders *everywhere*. So another reason for the stand-up EVA was to look and see if we could drive the Rover, because if there *were* boulder fields down there, and nobody could prove there were *no* boulder fields, it changed the whole picture."

The view set Scott's mind at ease; not just because it looked totally unhostile, but because it contradicted pre-flight fears. He could clearly see that despite the undulatory nature of the terrain, it was surprisingly gentle, with no rocks bigger than about 15–20 cm nearby. The trafficability, as he put it, would be excellent. The next three days of exploration, surely, would be charmed.

As he prepared to come back inside the cabin after 33 adrenaline-charged minutes, Scott was electrified. "Tell those geologists in the back room to get ready," he breathed, "because we've *really* got something for them!"

INTO THE MOUNTAINS

Back inside Falcon, acutely aware that they were the only inhabitants of Earth ever to visit this barren, beautiful place, the astronauts removed their suits and set about preparing their evening meal and getting ready for sleep. "Tomato soup was big on the menu, as I recall," Scott wrote. "There was no hot-water supply in the LM, as there was in the command module, so all our meals on the lunar surface were served cold and we soon discovered that there was not really enough to eat, either." In the coming weeks, they would recommend that more food be carried on Apollos 16 and 17, for Moonwalking would require both energy and stamina and prove to be hungry work.

Irwin, too, remembered Apollo 15's steady supply of soups. "Eating them required some acrobatics," he wrote. "They were ... in plastic bags, but they had a Teflon seal that you had to peel off. We added water to the soups, then very carefully pulled the tab to open them up. If you opened them slowly, invariably the soup would start coming out in bubbles or blobs that would float all over the place. The trick was to open the bag *fast*, so that the viscosity or capillary action would encourage the soup to adhere to the plastic. The object was to take advantage of whatever adhesiveness the soup had." When it had been thus 'contained', they could eat it quite normally, with a spoon, directing it approximately towards their mouths – if they were lucky, they might succeed; if not, the soup globs would separate and float past. On the Moon, of course, one-sixth gravity aided matters and the eating of soup was considerably more pleasant.

Sleeping in their long johns, without the bulky space suits, was much more comfortable in one-sixth of terrestrial gravity than it had been in rehearsals on Earth before launch; very much like a water bed, Irwin wrote, and they felt as light as feathers in the weak lunar gravity. They popped in earplugs, pulled down the blinds over the two triangular windows and, lulled a little by the soft hum of the lander's life-sustaining machinery, drifted into a fitful sleep. Scott arranged his hammock in a fore-to-aft direction above the ascent engine cover, whilst Irwin stretched "athwart ship" beneath Scott's feet.

During their rest period, mission controllers in Houston watched with some alarm as the pressure inside the descent stage's oxygen tanks began to steadily drop. It was not possible to accurately determine what was causing the problem, because the systems were run on a low-data-rate telemetry stream during the night to conserve electrical power. At length, Flight Director Pete Frank opted to wake the astronauts an hour earlier than scheduled and they were asked to switch back to high-data-rate telemetry. It subsequently became clear that the valve of the urine transfer device had been left open, even though its receptacle had been capped, and nearly 10 percent of the 43 kg of oxygen had been lost. This was particularly worrisome, since *half* of this total was necessary as an emergency reserve. (After the flight, the astronauts would point out in their debriefings that it would have been more beneficial if Mission Control had awakened them as soon as they became aware of the oxygen loss; a mission commander, said Scott, should be kept aware of the state of his vehicle.)

Despite having long since accepted *being here*, Scott still succumbed to the temptation to raise the blind and take a long look at the astonishing panorama beyond Falcon's windows. There was, however, little time to wonder and the strictness of the timeline forced them to begin preparations to put on their suits for the first of three Moonwalks. Irwin would subsequently relate, with a hint of humour, that he and Scott did more talking to one another during the donning of the suits than they had in the past several days. With all the added bulk of a PLSS backpack, oxygen and water hoses and electrical cabling, and with the suit fully pressurised, Scott found it surprising that he actually *fitted* through Falcon's small, square hatch when the time finally came to venture outside.

With the exception, perhaps, of Pete Conrad's comment at the bottom of the ladder, it had become something of a tradition by now for each commander to make a meaningful statement when he took his first steps on the Moon. Dave Scott's historic handful of words a few seconds past 9:29 am EST on 31 July were entirely appropriate for a man who had started out as a fighter pilot and been steadily, yet totally, won over by geology and science in general. "As I stand out here in the wonders of the unknown at Hadley," he said as he gazed at the Apennines, "I sort of realise there's a fundamental truth to our nature. Man *must* explore ... and *this* is exploration at its greatest!"

With a squeeze, and almost falling onto his backside in the lunar dust, Irwin quickly joined Scott and the two men set to work deploying the Rover from its berth in Falcon's descent stage. To do so, they tugged on a series of pulleys and braked reels and the effect required both of them, working in tandem. "One astronaut would climb the egress ladder on the LM and release the Rover," wrote Mark Wade on his website, www.astronautix.com, "which would then be slowly tilted out by the second astronaut on the ground. As the Rover was let down from the bay, most of the deployment was automatic. The rear wheels folded out and locked in place and when they touched the ground, the front of the Rover could be unfolded, the wheels deployed and the entire frame let down to the surface by pulleys." As it flopped into the lunar dust, the contraption – "a brilliant piece of engineering," Scott would later write, "with sealed electric motors in the hub of each wheel" – was secured with pins.

Scott clambered aboard to give it a test drive and found a problem: the front steering was inoperable, so they would have to rely on rear-wheel steering instead. After installing the colour television camera and loading up the geology tools, they buckled themselves aboard and set off. It must have been a peculiar sight for any onlooker to see this space-age dune buggy bouncing across the lunar surface; even at top speeds of just 8–10 km/h, it was a wild ride and if the Rover hit a rock, it literally *left the ground* for a couple of seconds. Irwin later likened it to a bucking bronco or an old rowing boat on a rough lake.

"I've never liked safety belts," he wrote, "but we couldn't have done without them on the Rover. You could easily get 'seasick' if you had any problem with motion." In fact, Irwin's seat belt turned out to be *too short* and before they could set off Scott had to come around to his side of the Rover to buckle him in properly. "We didn't realise," Irwin explained, "when we made the adjustments on Earth, that at one-sixth-G the suit would balloon more and it would be difficult to compress it enough to fasten the seat belt."

The 'real' Rover was also slightly different to drive than the one in which the men had trained on Earth. From his seat, Scott found that he had to concentrate all of his energies simply driving and keeping track of craters: the harsh glare of sunlight made the terrain appear deceptively smooth, literally 'washing-out' surface features, and hummocks and furrows appeared out of nowhere, at a split-second's notice. Its manoeuvrability was good – "it could turn on a dime," Scott recalled – but its wheels kicked up enormous rooster-tails of dust, which were thankfully deflected by its fenders. "When you go over a rock," Irwin wrote, "[the wire-mesh wheels] just bow up and absorb the impact and spring back again." As Scott drove, Irwin, who had been notoriously quiet during the geological training with Silver, demonstrated what he had learned by smoothly reeling off detailed descriptions of the setting. As the navigator, Irwin tried to plot their course on the map, but had difficulty identifying their route because they were uncertain of precisely where they had set Falcon down. However, Mount Hadley Delta was clear to see, with St George Crater – an enormous gouge the size of two dozen football fields – on the lowermost slopes and all they had to do was drive with it on their port quarter and they knew that eventually they would come upon the rille.

Cresting the top of a ridge, they were rewarded with their first unearthly glimpse of Hadley Rille and gained a clear awareness of its enormous size. Half an hour after leaving Falcon, they made their first scheduled halt at a place called 'Elbow Crater', right on the rim of the rille at the base of the mountain. From here, with the aid of a telephoto lens, Scott took a series of pictures of the far side of Hadley Rille, whose interior wall showed clear evidence of layering in outcrops not far below its rim. The two men took a few minutes to gather samples. As Scott wielded his geological hammer, Irwin toted the specimen bags. Using their well-rehearsed sampling procedure, the two men identified, photographed, described and bagged samples of basalts, breccias, olivines and plagioclase and dug a trench into the friable soil. Then they set off towards their second site, near the very rim of St George Crater. It had been expected that the area would be littered with large blocks of rock, but upon finding the flank of the mountain remarkably clean, Scott decided to halt short of the rim and sample an isolated boulder. It was more than a metre across and its 'half-in-half-out' nature, part-buried in the soft soil, intrigued him.

Simply *walking* was as alien as the very strangeness of the world upon which they were now operating. It felt, Irwin explained, very much like walking on the surface of a trampoline – it had the same sense of lightness and 'bounciness' – although the bulk of the space suit made it virtually impossible to move in a natural, Earthly gait. "When you don't have the weight of your legs available to push against the suit," he wrote, "you are constrained as to how far you can move. Consequently, you just use the ball of your foot to push off. That's why we looked like kangaroos when we walked. We flexed the boot and *that* pushed us forward."

"One of the Moon's most striking features," Scott related in his autobiography, "was its stillness. With no atmosphere and no wind, the only movements we could detect on the lunar surface, apart from our own, were the gradually shifting shadows cast to the side of rocks and the rims of craters by the Sun slowly rising higher in the sky." There was *absolutely* no trace of *anything* which exhibited either life or colour or movement and the only sound came from the gentle hum of life-sustaining machinery in their backpacks, the whoosh of the air flowing through their suits and the crackle of each other's voices or the voice of Houston in their earpieces.

The problem of judging distances had already been noted by previous crews. "There's *nothing* of scale which is familiar," Scott told the *Apollo Lunar Surface Journal*. "There are no trees, there are no cars, there are no houses … and, as an

Dave Scott inspects a boulder on the flank of Mount Hadley Delta. The Rover is clearly visible on the right.

example, we all know what size trees are in general. There *are no trees* and there's *nothing* in the landscape that has *any* familiarity. There's no hook. So when you look out there, you see boulders, but you can't really tell whether it's a large boulder at a great distance or a small boulder nearby. If it's *very* nearby, it's easy because you can run out along the ground and start calibrating your eyes. If you're looking close to the LM, you know what three or four inches are, but as you start going out, you start losing your perspective, because there's nothing to measure out there. It's a very interesting phenomenon that everybody gets fooled on these distances."

Having said this, Scott added that the tracks of the Rover lent some indication of distance. "Once you have some tracks," he recounted, "you can start seeing things. As an example, up on the side of Hadley Delta, looking back at the Lunar Module, boy, it was *small!*" In the absence of an atmosphere or the slightest trace of haze, Falcon appeared far closer and far smaller than it actually was. "But it gives you a scale of how far away it is," Scott concluded. "Five or six kilometres ... and when you're on the Moon, boy, that's a *long* way away." Even decades after the mission, Scott expressed frustration in his inability to properly describe how it *felt*: the ability of his eyes and how well they transmitted images to his brain was good on the Moon.

Yet, by his own admission, there was nothing from his own experience on Earth to compare with it. Words like 'crisp' and 'clear' and 'distinct' and 'definitive' sprang to mind, but for a man with a burgeoning scientific and geological vocabulary, even Dave Scott had reached the limit of his descriptive powers.

Heading back towards Falcon after a little more than two hours, the men could take great pride in their achievements so far. However, they still had a sizeable portion of work to do before returning inside. Of primary importance was the assembly of their ALSEP on the surface. Scott picked a spot about a hundred metres from the lander for the package and Irwin lugged it over, one pallet on each end of a carrying bar, not dissimilar to a giant dumbbell. On his cuff checklist, the LMP checked a small 'map' of where each component was supposed to go. First came the nuclear-powered central station, with its two-dozen bolts that had to be removed before it could spring upright. It was difficult work in a pressurised suit and bulky gloves and by the time he had completed the assembly he found that he was behind schedule.

Meanwhile, Scott was experiencing his own problems. One of the ALSEP's experiments was a heat-flow investigation. This had been assigned to Apollo 13, but never made it to the Moon. It was not carried over to Apollo 14, so this was its first outing. It required Scott to use a small, box-like drill to bore a couple of deep, three-metre holes into the surface and emplace a pair of temperature probes. He would then drill a *third* hole for a core sample. He made excellent progress on the first hole for about half a metre, then met a hard subsurface. Despite leaning on the drill to give it extra bite, he fell behind the timeline and was advised by Mission Control to accept what he had, which was only half of the intended depth, and insert the first set of probes. The second hole proved even more difficult and Mission Control, concerned not just because it was pushing them behind schedule ,but also that Scott's oxygen usage was much higher than expected, perhaps due to the demands of driving the Rover amongst the craters, called a halt with the drill only one metre into the ground. Capcom Joe Allen told Scott to take a breather, then help Irwin with deploying the retroreflector and a solar-wind experiment. They would have to complete the drilling later. Their first Moonwalk ended slightly earlier than planned, after six and a half hours.

Back inside Falcon, both men were exhausted. The stress of driving and the toughness of manhandling the drill for the heat-flow experiment had worn out Scott's hands and forearms. Irwin, too, described the pain in his fingers as excruciating. They took each other's gloves off to inspect the damage: perspiration poured from them, but there was no evidence of bleeding or bruising. Then they realised that their *fingernails*, which had grown during the last five days, had been immersed in sweat for the last seven hours. To aid movement, their gloves had been designed to fit tightly against the tips of their fingers; the *pressure* and the *pain* was on the ends of the nails. Irwin resolved to cut his nails and advised his commander to do the same, but for some reason – perhaps fearful that it might compromise his dexterity on the surface – Scott declined.

Irwin was also uncomfortable. A problem with his drinking water bag had left him absolutely parched for more than seven hours. "There was a nozzle that you'd bend down to open a valve so you could suck the water out and drink it within the protection of the space suit," he explained, "but I could never get my drink bag to work and I never got a single drink of water during the whole time I was out on the surface of the Moon." He *did*, however, manage to gobble down a fruit stick inside his helmet ... and *that* helped him to keep going when the time came to assemble the ALSEP. Now, having doffed his suit, Irwin guzzled water like a jogger, then settled down with Scott for their second night on the Moon. 'Settled' probably was not an appropriate word, for conditions inside Falcon cannot have been pleasant: with the presence of all the rocks and soil specimens, the *smell* of the Moon – a strong, gunpowder-like aroma – pervaded the air and dust covered everything. They stashed their suits at the back of the cabin, making sure that the gloves were fitted, so as not to impair their seals, then debriefed to Houston and bedded down for the night.

With the first excursion over, the geologists were more than satisfied with Scott and Irwin's performance on the Moon. That evening, Dale Jackson of the US Geological Survey had dinner with capcom Joe Allen (who was also the mission scientist), backup commander Dick Gordon and several friends at a restaurant called 'Eric's Crown and Anchor'. Jackson could hardly contain his enthusiasm for the day's events ... and his booming voice quickly ensured that *everyone else* in the restaurant could do little but listen in on the conversation. At length, with everyone's attention riveted on him, Jackson exulted that Scott and Irwin had done everything expected of them: they had identified appropriate rocks, taken specimens of the soil around them, chipped shards from them and sampled the material *underneath* them. Everyone in the restaurant finally heard him proclaim loudly: "Why, they did everything but *fuck* that rock!"

Next morning, capcom Gordon Fullerton had some unwelcome news: the lander had lost a sizeable amount of water and because it was tilted with its rear leg in a crater he asked them to check behind the ascent engine cover to see if the water had gathered there. Fullerton was accurate and Scott and Irwin quickly scooped it up into a spent lithium hydroxide canister, more than a little shocked to observe that it had collected so close to some critical electrical connectors. They doubtless expressed quiet thanks to the Grumman designers, who had waterproofed all of the lander's electronics.

Scott and Irwin's second Moonwalk would feature a slightly shortened traverse, which had been redesigned in such a way that it would offer more exploration, combined with less travelling time between geological stops. One relatively low-priority activity had been eliminated and a greater measure of freedom was given to the efforts of the astronauts themselves; indeed, Joe Allen told them that Mission Control and the geologists in the Houston back room would depend heavily upon their descriptions and observations and it would be Scott and Irwin's choice on exactly where they chose for their major sampling. "We're looking now, primarily, for a wide variety of rock samples from the [Apennine] Front," Allen told them. "You've seen the breccias already. We think there may very well be some large crystal[line] igneous [rocks] and we'd like samples of those and whatever variety of rocks which you're able to find for us – but primarily a large number of *documented* samples and fragment samples." Scott was in full agreement; Allen was talking *their* language and after two years of geological field training, he felt ready and confident to explore.

A few minutes before 9:00 am on 1 August, safely buckled aboard the Rover, the astronauts set off due south, heading for Mount Hadley Delta, upon whose slopes they would concentrate their energies. On starting out, Scott had found, to his delight, that the Rover's front-wheel steering had mysteriously come online, giving him full four-wheel control.

It was a scenic trip, Irwin recalled. Ahead of them, and all around them, the terrain was literally splattered with craters, right up the slopes of Hadley Delta, and the *height* of the mountain meant that it rivalled the tallest peaks of the Rockies. After passing the vast cavity of Dune Crater, whose rim was littered with large blocks, they started up the mountain. On the plain, the going had been rough, but on the slope the surface smoothed out markedly. Reaching a point just below Spur Crater, they swung left and drove cross-slope. Looking down the slope, they were astonished to realise how far they had come. The lander was a tiny speck on the undulating plain, with the astronauts now at an elevation of about a hundred metres, and the view, completely unimpaired by atmosphere or the slightest hint of haze, almost knocked Dave Scott's socks off. Their first task was to find a small 'drill hole' crater that could have excavated material from the mountain, but the flank was remarkably clean. Scott curtailed the planned drive and they sampled a small crater and then an isolated boulder which was coated in greenish material. The green hue captivated Jim Irwin, whose Irish descent and birthday on St Patrick's Day – and the fact that he stowed some shamrocks in the lunar module – made this a special find. At first, the two men wondered if their eyes or Sun visors were playing tricks on them, but when it was unpacked a few weeks later in the Lunar Receiving Laboratory, their initial impressions would be confirmed: it *was* green, made entirely of minuscule spheres of glass, tiny droplets of magma spewed from a fissure by a 'fire fountain'. In time, it and other samples would contribute to making Apollo 15 one of the greatest voyages of discovery ever undertaken in human history.

Finally, they headed for Spur Crater, which proved to be a *real* gold mine of geological treasure. "As soon as we got there," Irwin described, "we could look over and see some of this white rock. Immediately, I saw white, I saw light green and I saw brown. But there was one piece of white rock that looked different from any of the others. We didn't rush over to it; we went about our job the usual way. First I took down-Sun shots and a locator shot about 45 degrees from the Sun-line and Dave took a couple of cross-Sun shots." Scott and Irwin slowly threaded their way around the rim to the strange white rock. "It was lifted up on a pedestal," Irwin wrote. "The base was a dirty old rock covered with lots of dust that sat there by itself, almost like an outstretched hand. Sitting on top of it was a white rock almost free of dust. From four feet away I could see unique long crystals with parallel lines, forming striations." Scott used a pair of tongs to pick it up from its small pedestal and held it up, close to his visor, to inspect it. The rock was about the same size as his fist and even as he lifted it, some of its dusty coating crumbled away ... and he saw large, white crystals.

The Genesis Rock, atop its light-grey 'pedestal', shortly before collection. The astronauts' bootprints and a tripod-like gnomon provide a measure of scale.

"Aaaahh!" he exulted.

"Oh, man!" added Irwin.

The rock was almost entirely plagioclase – an important tectosilicate feldspar mineral used by petrologists on Earth to help determine the composition, origin and evolution of igneous rocks – and from their expeditions into the hills of the San Gabriels, Scott instantly recognised it as a specimen of 'anorthosite', which is the purest form of plagioclase. For some time, lunar geologists had suspected that anorthosite formed the Moon's original, primordial crust; indeed, data from the unmanned Surveyor 7 lander had suggested its presence in the ejecta of the crater Tycho and tiny fragments had actually been found in samples from both Tranquillity Base and the Ocean of Storms.

"Explaining why most of the Moon's crust should be composed of anorthosite," wrote Andrew Chaikin, "led some geologists to an extraordinary scenario. Within the infant satellite, they proposed, there was so much heat that the entire outer shell became an ocean of molten rock. As this 'magma ocean' cooled, minerals crystallised. The heavier species, including the iron- and magnesium-rich crystals, sank to the bottom. The lighter crystals, specifically, the [aluminium-rich] mineral plagioclase floated to the top."

Recognising the find as probably a piece of the Moon's primordial crust, Scott could hardly contain his enthusiasm.

"*Guess* what we just found!" he radioed. "I think we just found what we came for!"

"Crystalline rock, huh?" said Irwin.

"Yes, sir," replied Scott, which Joe Allen echoed in his excited, Iowan drawl. After briefly describing the rock's appearance, Scott placed it into a sample bag by itself. It would be labelled as sample number 15415, but a keen journalist, inspired by the term petrogenesis, the study of the origin of igneous rocks, would later offer it a far more lofty title: 'the Genesis Rock', a sample of the *original* lunar crust, coming from one of the earliest epochs of the Moon's history, some 4.1 billion years ago. This date was reached by geologists at the University of New York at Stony Brook and proved to be almost a billion and a half years *older* than the *oldest* rocks then known on Earth. If the Moon was any older than *that*, noted Andrew Chaikin, it wasn't *much* older; the Solar System itself was thought to have formed only a few hundred million years earlier.

Dick Gordon, the backup commander of the mission, had long since described the passage of time on a space flight as remorseless and Scott and Irwin had little time to linger at Spur Crater. Joe Allen passed on a request from the geologists to collect a series of walnut-sized fragments and whilst Irwin set to work with the lunar rake, Scott stole a minute or two to inspect a large boulder right on the rim of the crater.

Back in the vicinity of Falcon, shortly before two in the afternoon and five hours into their second Moonwalk, Scott and Irwin had other chores to finish; first, there was the need to complete drilling the heat-flow hole which had hit resistant soil the previous day. Scott had already noticed inside the lander that his injured fingers were starting to turn black and had to summon as much strength as he could muster – bringing his hands close to his chest just to squeeze the drill's trigger – to complete the task. He could physically stand only about a minute of the pressure on his fingernails, before breaking off for a breather. At length, both sensor packages were in place to a depth of about one and a half metres. However, when Scott attempted to extract the core sample which, at about 2.4 m long, was the deepest such sample yet attempted on the Moon, he managed to lift it about 20 cm, after which it refused to budge any further. Allen told him to leave it until tomorrow.

Meanwhile, Irwin had dug a trench, photographed it and used a device known as a penetrometer to test the bearing strength of its walls and floor. "If you think digging a ditch is dog's work on Earth," he wrote, "try digging a ditch on the Moon. The big limitation is the suit and the fact that you are clumsy at one-sixth-G. I had practiced on Earth and come up with a technique that most dogs use. You spread your legs and push the dirt between them. I solved a dog's job with a dog's technique. This method worked perfectly on the Moon." He easily dug through a fine grey material which he likened to talcum powder or graphite, then a coarser, darker soil, but had to give up on reaching a very resistant layer which, although it *looked* moist, had all the consistency of hardpan.

Scott and Irwin wrapped up the second Moonwalk by planting the American flag in the soil and loading that day's rock box aboard Falcon. Back inside after seven hours and 12 minutes, they concluded what was already being lauded by scientists and flight controllers alike as the most productive expedition on the Moon yet undertaken. Not only had most of the equipment operated flawlessly, but the live – and *colour* – images provided by the Earth-operated television camera on the Rover had proven a far cry from the crude black-and-white pictures of Armstrong and Aldrin. Furthermore, Scott and Irwin had truly done their mentor, Lee Silver, proud through their geological descriptions and discoveries. "I'm told," Joe Allen radioed that evening, "that we checked off the 100-percent science completion square some time during EVA-1 or maybe even shortly into EVA-2. From here on out, it's gravy all the way!"

The gravy of the third excursion would be tempered by the fact that it would also be the shortest, scheduled to last barely four and a half hours. It started with a check of the ALSEP and recovery of the core sample. For a few moments, their efforts to extract the tube from the ground were fruitless and Scott was almost ready to give up. However, with Irwin's persuasion, both men hooked an arm under each handle of the drill and after several firm tugs, the tube sprang from the ground. Precious minutes were wasted, though, when the vise carried on the Rover to dismantle the tube into storable sections proved to have been fitted *backwards*; Irwin broke out a wrench and used that, instead, but Scott's frustration was evident. He knew that for every minute wasted before the drive started, they would lose at least another two minutes of geological exploration due to the way in which the walk-back constraint operated. Some senior NASA managers in Mission Control, keen to press on with the timeline, wanted to abandon the core entirely. However, the astronauts and Joe Allen had an ally in Flight Director Gerry Griffin, who had shared several of their expeditions into the Californian mountains and knew how important the science was ... and how important the deep core sample was to the success of this mission. It was *he* who persuaded the managers *not* to abandon the core tube work. After they had partially disassembled the tube, it was decided that they should leave the remainder of the task until later. When the core was opened on Earth, it proved to contain several dozen layers which documented some 400 million years' worth of lunar history ...

At length, Scott and Irwin buckled into the Rover and headed west-northwest for a good look at Hadley Rille. After that, if time permitted, they hoped to grab an opportunity to inspect the mysterious North Complex of craters, which some geologists thought might be a cluster of ancient volcanoes. During the journey, they passed a small crater whose specimens would turn out to be a scant few million years old, making it the youngest ever visited by men on the Moon. Their arrival at Hadley Rille was truly breathtaking. Its far wall, bathed in the harsh, direct sunlight of the late lunar morning, showed distinct layers of rock pushing through a

As Dave Scott resumed work with the drill, Jim Irwin focused his energies on digging a trench to measure the bearing strength of its walls and floor, as well as the general structure of the lunar soil.

mantle of dust, lending credence to theories that Mare Imbrium had been built up as a succession of ancient lava flows.

One theory was that the rille was a fracture where the mare surface 'opened' like a cooling joint. "But since the scientists have studied the pictures," Irwin later wrote, "the most popular theory is that Hadley Rille was probably a lava tube that collapsed." All around the two men were enormous slabs of basalt and the geologists in the back room in Houston quickly began pressing for them to move further downslope, though Joe Allen was becoming nervous. Looking at televised pictures, it seemed to him as though they were right on the edge of a precipice. In fact, the rille had no dramatic, precipitous 'drop' at its rim; rather, it resembled the gentle shoulder of a hill and they were able to walk several metres downslope without difficulty. "In fact," Scott wrote, "the slope down which we descended was only about 5–10 degrees and the maximum slope of the rille was only 25 degrees – not steep for such a canyon-like formation." It was steep enough, however, that from their vantage point they were unable to see the floor. Owing to the thinness of the regolith in amongst the boulders, the footing was much firmer than the flank of Mount Hadley Delta the day before. "The rille's rim and upper slopes," Scott added, "were covered in hard-packed regolith and small rocks, compared with the much softer and unconsolidated lunar soil we had encountered at the base of the mountains." For two hours, he and Irwin laboured on collecting and documenting samples and describing their glorious setting.

Time, however, was escaping them and any chance to explore the North Complex very quickly disappeared; *that* would have to await another generation of lunar prospectors. Both astronauts found this bitterly disappointing: in Irwin's mind, it left their excursion only half-complete, whilst Scott would wonder for years afterward if the unique data from the deep core was really *worth* abandoning the chance to visit the North Complex. Equally, they both appreciated the need to get back to Falcon with enough time in hand to prepare for liftoff later that day.

Back at the lander, with the minutes of the final Moonwalk rapidly winding down, Scott had one last opportunity to give a scientific demonstration to an audience of millions back home. It stemmed from a suggestion by Joe Allen, a physicist by profession, and was inspired by the experimental work of the great Italian scientist Galileo Galilei. More than three centuries earlier, Galileo had stood atop the Leaning Tower of Pisa and dropped two weights of different sizes, proving that gravity acted equally on them, regardless of mass. Now, in front of his own Leaning Tower – the slightly-tilted Falcon – Scott performed his version of the experiment. "In my left hand, I have a feather," he told his audience, "in my right hand, a hammer. I guess one of the reasons we got here today was because of a gentleman named Galileo a long time ago, who made a rather significant discovery about falling objects in gravity fields. The feather happens to be, appropriately, a falcon's feather, for our Falcon, and I'll drop the two of them here and hopefully they'll hit the ground at the same time." They did ... and applause echoed through Mission Control. "How about that?" Scott concluded triumphantly. "Mr Galileo was correct in his findings!" He originally planned to try it first, to check that it would *work*, but was worried that it might get stuck to his glove. He decided to 'wing it' and, thankfully, it worked. In his autobiography, Irwin would relate that Scott actually carried *two* feathers on Apollo 15, one from the falcon mascot at the Air Force Academy. Much to Scott's irritation, however, Irwin accidentally stepped on it. They searched for the feather, but could only find his big bootprints everywhere. "I'm wondering," wrote Irwin, "if hundreds of years from now somebody will find a falcon's feather under a layer of dust on the surface of the Moon and speculate on what strange creature blew it there."

Shortly after 9:00 am on 2 August, a little more than four hours since setting foot on the surface for EVA-3, Scott drove the Rover, alone, to a spot a hundred metres east of the lunar module. From this place, Mission Control would be able to remotely operate its television camera to record the liftoff of the ascent stage. On the next mission, this final parking place would be nicknamed the VIP Site, because it would give the whole world a VIP-class ticket to observe the launch. For now, Scott pulled out a small red Bible and placed it atop the control panel of the Rover, in order to show, he explained, those who followed in the future *why* they had come. Then he climbed off the machine which had extended their exploration several times beyond any previous landing crew and strode a few metres toward a small crater. Here, he dug a small hollow and dropped a tiny aluminium figurine of a fallen astronaut onto the lunar soil. Months earlier, he and Deke Slayton had privately hoped that no more explorers would die in the conduct of exploring space; yet already, Soviet cosmonauts Georgi Dobrovolski, Vladislav Volkov and Viktor Patsayev had lost their lives when their Soyuz 11 spacecraft depressurised during re-entry.

The figurine, about 8.5 cm tall, had been arranged by Al Worden; and Irwin himself had been responsible for organising a small plaque, planted alongside, which listed the names of 14 astronauts and cosmonauts known to have died doing their duty. In addition to Dobrovolski, Volkov and Patsayev, the list included the tragic Apollo 1 crew of Gus Grissom, Ed White and Roger Chaffee, who died in a launch pad test in January 1967, and cosmonaut Vladimir Komarov, who plummeted to his death when Soyuz 1's parachute failed to deploy three months later. Also listed were unflown NASA astronauts Ted Freeman, C.C. Williams – who might have walked on the Moon himself as a member of the Apollo 12 crew – and Ed Givens. Yuri Gagarin, the first man in space, lost his life in an aircraft crash in 1968; his name, naturally, was there. Finally, there were the names of Elliot See and Charlie Bassett, whose Gemini mission was so cruelly snatched from them when their aircraft crashed in February 1966, and veteran cosmonaut Pavel Belyayev, who died from complications arising from an operation for a stomach ulcer.

A certain sense of brotherhood swept over Scott in those private moments, despite the fact that some of the cosmonauts' names had only come to his attention through reading the newspapers and scanning through their obituaries. The Americans, of course, were different; he had known them as friends and daily work colleagues and was acutely aware that they had all played their part in bringing him to this place. Scott knew that only four more astronauts were destined to walk on the Moon – Apollo 16's John Young and Charlie Duke in a few months' time and then the as-yet-unnamed crew of Apollo 17 in December 1972 – and he also knew that he would never again return here. "All I knew in those moments," he wrote, "was that I had come to feel a great affection for this distant and strangely beautiful celestial body – in effect a small planet – constantly circling our own. It had provided me with a peaceful, if temporary, home. But it was time to return to my own home back on Earth."

OPTIMISM AND PESSIMISM

Within the confines of Falcon, Scott and Irwin had little time to gaze out at the grandeur of Hadley; only a few hours remained before their scheduled 1:11 pm EST liftoff, bound for a rendezvous with Al Worden in Endeavour. It marked the first occasion on which a crew would complete a Moonwalk and perform the liftoff and rendezvous *without* a rest period in between; certainly, having been out for less than five hours on the surface, Irwin wished Mission Control could have postponed the inevitable by several hours to have enabled them to drive home by way of the North Complex and gather a few samples, but it was not to be.

Precisely on time, Scott punched the Abort Stage button and a television audience back on Earth had the chance to actually see a landing crew depart from the Moon. Falcon's ascent stage literally 'popped' away from the descent stage and shot directly upwards with all the speed and accuracy of an express elevator, spraying fragments of insulation outwards for a second or two. One journalist would later compare its shower-of-sparks departure as like something left over from a Fourth of July celebration. Watching from Mission Control, Chris Kraft would gape at the *speed* of the departure. "I had no idea it went so *fast*," he wrote. "We'd been told it was like that by the other Moon crews, but seeing it for real was a thrilling shock."

Ten seconds later, a strangely familiar sound came into Scott and Irwin's earpieces; 'The Air Force Song', courtesy of Worden, which they had intended to play to Houston only, but which somehow ended up being routed to Falcon, as well. "This was kind of surprising," wrote Irwin, "because Dave had briefed Al to turn on that music at one minute *after* liftoff (that first minute is rather critical), but here it came at ten seconds. It really caused some consternation in Houston. First, they thought somebody was playing a trick in Mission Control, so they conducted a big search. They asked for radio silence – it was a tense situation. Finally, they realised that probably *we* had turned it on." It has been suggested that Scott was not happy with this, but in his autobiography he noted that "it felt good to start our return journey to the strains of 'Off we go into the wild blue yonder' ... although in this case the yonder was black!"

The climb back up into lunar orbit was uneventful and both Scott and Irwin would admit to hearing a faint 'whistling', almost like wind; it was the sound of the ascent engine, boosting them aloft. Docking with Worden was perfect and after transferring their stash of equipment, camera films, rock boxes and core tubes over to the command module, they set about sealing the hatches in order to jettison Falcon. At 11:04 pm EST on 2 August, seven and a half days since leaving Florida, Worden pushed the LM JETT button and the last remnant of the lander drifted serenely away, destined to impact the surface and add another quake to the steadily growing amount of data from the Apollo seismometers.

The jettison actually came ten minutes later than planned, due to a glitch with the seals on the astronauts' suits, and caused another problem when Mission Control fed them some confusing and incorrect data immediately afterwards. "Had we performed the CSM-LM separation manoeuvre as they originally calculated," Scott wrote, "we would very likely have made physical contact again with the LM, which we could see out of the window right ahead of us. Mission Control recalculated the manoeuvre and we proceeded safely. But, as Jim and I had gone for nearly 20 hours without sleep by this stage, and had not eaten for eight hours, Mission Control seemed to think that *we*, not *they*, had got confused!" One final request from Houston was that Scott and Irwin each take a Seconal tablet; a barbiturate with sedative qualities. Scott countered that they were sufficiently tired *not* to require sleeping pills to rest.

Mission Control – in the person of Deke Slayton, no less – repeated this request the next night. At first, Irwin thought they were joking. "What we were not told," Scott would write, "was that both Jim and I had shown slight irregularities in our heart rhythms while on the lunar surface. The irregularities, technically known as premature ventricular contraction, were a result, it was discovered later, of a potassium deficiency caused by the rigours of our training regime prior to the flight. Subsequent crews stocked up on liquids fortified with potassium." Scott was later vocal in expressing his concern that, by *knowing* this when it mattered, whilst on the surface, could have allowed him to make better decisions as the mission commander. He was even more concerned when it became clear that Irwin showed evidence of a condition called 'bigeminy', an irregular pattern in which his heart first skipped a beat and then beat twice in rapid succession. It was later revealed that Irwin experienced this abnormality several times on the Moon's surface *and* just after the rendezvous with Endeavour in lunar orbit. When he got some well-earned rest on the return journey to Earth, Irwin's heart returned to normal ... but Scott and others would wonder years later if this had a detrimental effect on Irwin's subsequent health.

In Mission Control, physician Chuck Berry was taking it extremely seriously. If Irwin were here on Earth, he told Chris Kraft, he would need to be committed to an intensive care unit. The two men got the capcom to ask Irwin how he was feeling. Fine, the astronaut replied, except a little tired and with some strange heart sensations. "He was 240,000 miles and a high-G re-entry from the nearest hospital," wrote Kraft. "My *own* heart skipped a beat!" But then Kraft and Berry realised that Irwin *was* already in an intensive care unit ... he was *already* getting 100 percent oxygen, was *already* being medically monitored near-continuously and, crucially, was in *zero-gravity*. "Whatever strain his heart is under," Kraft said to Berry, "we can't do better than zero-G." Irwin's heart rate steadied and remained normal for the rest of the mission. None of the flight surgeons took their eyes off his EKG traces until Apollo 15 was back on Earth. Post-landing tests showed no problems ... but then, a few months later, in 1972, the astronaut suffered his first heart attack. He retired from NASA that same year to found a Baptist ministry known as 'High Flight' and would later lead an expedition to Mount Ararat in Turkey in search of Noah's Ark. Another heart attack would follow a few years later and then a final one, early in August 1991, would claim Irwin's life. More than a decade later, Dave Scott would wonder if, had he known of this condition, might he have been able to simplify some of his friend's tasks at Hadley Base? He even pushed NASA for some years to give an adequate explanation of the effect of Irwin's heart condition on his health later in life.

For now, though, aboard the homeward-bound Endeavour, the crew was able to derive great satisfaction from a job well done. During his three days of solo activity in lunar orbit, Worden and the SIMbay had undertaken a huge amount of scientific research and, despite some problems with a sensor for the panoramic camera, his work had gone exceptionally well. Nor was he considered as little more than an operator; his own geological training made him an integral part in the proceedings and he was regarded as very much a 'co-investigator' in many of the tasks. In fact, several scientific papers, published after the mission, include the name of A.M. Worden in their list of authors. "One very interesting application of the SIMbay on the way back from the Moon," wrote Scott, "was to study in detail the temporal behaviour of pulsating X-ray sources, later to [become] known as black holes. In co-ordination with an Earth-based observatory in the Soviet Union scanning the same part of the galaxy at the same time as we were, we took the first photographs in space of a series of suspected black holes, about which much less was understood at that time." One of these, Cygnus X-1, had only been discovered a few years before Apollo 15 set off and is now believed to represent a close binary system in which a normal star sheds material which falls into the black hole.

Also notable was the deployment of the subsatellite, which was ejected from the bowels of the SIMbay on the afternoon of 4 August. It was released into an initial orbit of 102 x 139 km, inclined 28.5 degrees to the Moon's equator, and studied the interactions of the solar, terrestrial and lunar magnetic fields until two successive electronics failures in February 1972 caused it to lose most of its data channels. Ground support for the subsatellite was finally terminated in January of the following year.

Since the service module could not survive re-entry, it was Worden's task to venture out and retrieve the film from the SIMbay's cameras. This would be humanity's first 'transearth EVA', the first spacewalk conducted in the cislunar gulf, some 300,000 km from home. Yet, as he recounted in a May 2000 oral history interview for NASA, there were originally 'other methods' studied for getting these films inside the command module before the agency went ahead and approved a deep-space EVA. "There had already been some preliminary work on how to get this film out of the SIMbay," he pointed out. "Of course, it had been in the pipeline for several years and there were a lot of schemes to get the film from . . . the back of the Scientific Instrument Module all the way up into the command module, [which was] a distance of about 30 feet. How do you get out there safely so that you don't lose it, so that you don't hurt something? One of the schemes was . . . an 'arm' on a hinge that would go out and pick up the film and . . . bring it back by the hatch, where you could pick it up." Other suggestions included an 'endless clothesline', onto which the film canisters could be hooked and reeled down the length of the service module to Endeavour's crew hatch. "I objected to all of those, once we [were] assigned to the flight," Worden recalled. "None of them were very practical. We actually *proved* it with the clothesline. It's nice to think about something like an endless clothesline, but the truth was, when you're in space and [there is no atmosphere or gravity], if that canister started to bounce around, there [would be] nothing to stop it." During a test in the parabolic aircraft, it proved exceptionally difficult and potentially damaging. In the end, it was decided that an EVA was the most effective option. By the time he set out to perform it on 5 August, Worden had practiced the task more than 300 times aboard the parabolic aircraft.

"The EVA itself was kind of unique," Worden said of the relatively brief, 39-minute excursion, "sort of a unique perspective. I did have a chance to stand up on the outside [of the service module] and look. I could see the Moon and the Earth at the same time; and if you're on Earth, you can't do that, and if you're on the Moon, you can't do that! It's a very unique place to be. I guess our biggest concern was that we had everything tied down so that when we opened the hatch, we didn't have something go wandering off into space! But outside of that, it was pretty easy."

Obviously, since the cabin was reduced to vacuum for the spacewalk, Scott and Irwin also had to don their suits. Suddenly, as soon as the hatch opened, everything that had not been secured began drifting around. "When we opened the hatch," wrote Irwin, "it was just like a vacuum cleaner pulling all the loose stuff from the inside out into space. My toothbrush floated by; it had been in hiding. A camera came by; one of us grabbed it. We were all leaping around, trying to catch the important stuff." It was Irwin's job to move slightly outside after Worden in order to televise his EVA, but he had 'goofed' when hooking up his suit's umbilicals by wrapping them the wrong way around a strut, which limited his range of movement. "I had to force my hand out to reach the movie camera," he wrote, "attached to a boom, to turn it on. I saw the green light and thought the camera was on, but it wasn't working."

Since they were so far from Earth, the experiences of both men were quite distinct from 'traditional' spacewalks in orbit around the Home Planet. They were surrounded by the pitchest blackness and, with Earth some three hundred thousand kilometres away, one of the few sources of illumination was sunlight reflecting off the surfaces of the service module. Irwin considered it strange and eerie from his perch just inside Endeavour's hatch, in silhouette against the full disk of the Moon. "The *National Geographic* did a painting of me," he wrote. "It almost looks like a photo." It is a pity that *real* photographs of what must have been an absolutely stunning event were never returned.

Two days later, after shedding the service module, the command module plunged into the atmosphere, heading for a splashdown point a few hundred kilometres to the north of Hawaii. Like so many astronauts before him, Irwin was particularly struck by the fireball effect and the loss of communication with the ground. "You look out the window and there is just this beautiful orange-yellow glow," he wrote. "You also have the glow of ionised particles that are streaming out behind the spacecraft." As Scott and Worden monitored Endeavour's systems, Irwin positioned cameras in the windows to shoot images of the re-entry. As they descended further, he began to discern the deep blue of the Pacific Ocean, the whites of clouds and, at length, his first sight of land: the snow-tipped mountains of New Zealand. To a man brought up with an innate love of mountains, the mission of Apollo 15 brought Irwin full-circle: from the mountains of the Moon back to the mountains of Earth.

Several metaphorical 'mountains' had still to be overcome. Shortly before splashdown, one of Endeavour's three parachutes failed to deploy properly, threatening to give them a hard landing. "The drogue chutes came out at 25,000 feet," Irwin recalled, "and that really slowed us down, just like a drag chute or speed brakes in an airplane. It doesn't jerk you, it brakes you – then you start oscillating wildly underneath the drogues. Then, just before you get to 10,000 feet, the drogues are released. You can *see* up there; you see them go, and then you *feel* them go. You free-fall for a few seconds, wondering whether the main chutes are going to come out. At 10,000 feet the main chutes are out and they slow you almost completely. We saw three chutes; then we saw that one had failed. It just collapsed. Well, Al saw it and I saw it and at the same time the helicopter crews … told us we had lost a chute." Splashing down under two chutes was hard; harder, Scott thought, than his return from Apollo 9, but both Worden and Irwin considered it completely nominal.

Seven weeks later, in a 28 September press release, NASA revealed that the most likely cause of the failure was either a dump of monomethyl hydrazine from the reaction control system or the links which connected suspension lines to the parachutes' risers. Certainly, the release noted, "tests have shown that MMH being dumped through a hot engine can result in tongues of flame from the thrusters which could affect parachute lines". Getting rid of any residual traces of this highly toxic fuel had always been done just prior to splashdown, thereby reducing the risk of the command module turning into "a nice bonfire", to quote Worden, if the lines were ruptured by the shock of splashdown and the fuel mixed with the oxidiser and ignited hypergolically. NASA decided to eliminate the MMH dump from future missions and the spacecraft would land with residual propellants aboard.

In Worden's mind, the dumping of the MMH *had* indeed been to blame for the loss of the parachute, but they were also unlucky in that they landed on a very calm day. "As we released this [fuel]," he recounted, "it went right up into that chute and the chute just *dissolved*. I think it was probably the first time that it happened in the programme, because I think on every other flight there had been some surface wind that [moved] the spacecraft [during descent]. When they [vented the] rocket fuel, it went off in [the] back of them and they never had a problem. It just got lost in the wind, but with us, coming straight down, this stuff went straight up into the parachute … and *there* we had a problem. It probably would have gotten into the second chute if we'd been much longer … "

Dave Scott and his team were probably thankful that, unlike the Armstrong and Conrad and Shepard crews, they would not be confined in the LRL for several weeks of quarantine. In April 1971, George Low had discontinued this practice, based on a recommendation from the Interagency Committee on Back Contamination. Having said that, Scott would admit to being exhausted; all three men had lost weight and their immune systems were low after operating in such a carefully-controlled environment for so long. The immediate onset of technical and scientific debriefings and press conferences was something that they could have done without. It did, however, give Scott his own platform to press the case for reinstatement of the cancelled Apollo 18 and 19 missions. It seemed highly unlikely that sufficient public interest would be drummed up, even by the triumph at Hadley, to sway President Nixon's judgement, but Scott considered it worth the try.

On 13 August, the prime crew of Apollo 17 was announced and Scott's crew were assigned as their backups. That assignment would be terminated just a few months later as Dave Scott, Al Worden and Jim Irwin became embroiled in a particularly ugly affair which would tarnish their individual reputations as well as that of the

Descending beneath two of its parachutes, the command module Endeavour returns safely to Earth on 7 August 1971.

astronaut corps itself for several years to come. On the chopper, heading back to the recovery ship *Okinawa*, Scott told them to all salute in unison when they stepped aboard the deck; he disliked the practice of several previous crews to salute out of synch with one another. However, when Scott stepped out, he did not wait for his two crewmates and saluted straightaway, followed quickly, but not immediately, by Worden and then Irwin. They received a heroes' welcome when the ship arrived at Hickam Air Force Base in Hawaii. Then, during the flight back to Texas, Scott pulled out a stack of first-day philatelic covers for them to sign. He had carried them all the way to the Moon, as part of an arrangement with a stamp dealer and the intention was to sell them after splashdown and split the proceeds to set up trust funds for their children. This deal would return to haunt them in the spring of 1972. The key problem was that the covers – four hundred in total, a hundred for each of the astronauts and the remainder for a German stamp dealer named Walter Eiermann – had not been authorised by NASA. The deal with Eiermann called for the covers to be sold exclusively to collectors, privately and with no publicity, after the Apollo programme was over, with Scott, Worden and Irwin each expecting to receive around $8,000 from the proceeds. Eiermann, however, had other ideas and began to sell the covers within weeks of Endeavour's splashdown. The furious astronauts contacted him in October 1971 and tried to cancel their agreement, but it was too late. The story had leaked into the European press and by late autumn some of the covers were fetching as much as $1,500 apiece. None of the three astronauts accepted any money.

When Deke Slayton found out, he doubted at first that *his* astronauts, and particularly straight-arrow Dave Scott, could possibly have been involved. Jim Irwin finally admitted the truth and Slayton and Chris Kraft were furious. Neither of them could be sure if the crew's actions were *wrong* or *illegal* … or *not* … but they instinctively knew that it did not smell good and passed the information to George Low in Washington. "He was a stickler," Kraft wrote. "He immediately got the NASA inspector-general and the lawyers involved. In a matter of days, we had a full-scale internal scandal on our hands." All three astronauts were removed from active flight status and lost their backup duty on Apollo 17. Irwin was planning to retire anyway – in his autobiography, he found nothing attractive in the prospect of going through the training grind again – and Scott and Worden were both assigned to administrative posts elsewhere in NASA.

Previous astronauts, it is true, had profited in less dramatic ways from their missions, perhaps by selling autographs, but on *this* occasion there was a *lot* of scope for personal gain, or at least their children's gain … and the affair turned into a public scandal. Yet as the inspector-general and Slayton and Kraft dug deeper, it was clear that *other* astronauts had signed covers for Eiermann, too, though many had devoted their proceeds to charity. Tom Stafford, for example, who served as chief astronaut in 1969–71, proved that *his* payment went directly to charity and never touched his hands. Elsewhere, the Apollo 14 crew carried some silver medallions, which were to be melted down after the flight, mixed with other commemorative medallions by the Franklin Mint and sold to the public. The Franklin Mint had actually *advertised* the deal before Shepard, Roosa and Mitchell blasted off. The deal was never consummated and nothing was printed in the media, but it prompted a few influential congressmen to grumble. One other astronaut, Apollo 13's Jack Swigert, ended up revealing more money in *his* bank account records than he theoretically should. His response was one of defiance and self-defence: "It's *none* of your damn business!" He took legal advice and challenged NASA to say which law he had broken, but *none could be found*. Chris Kraft was in a quandary, with some colleagues telling him to drop the case, balanced against continued notoriety since the covers were still in circulation. Ultimately, the remaining covers were confiscated by the inspector-general, sometimes under duress, and turned over to the Department of Justice. "Our own NASA lawyers didn't expect indictments," Kraft wrote, "but those Justice hotshots knew the law better than anyone else. If there was a crime here, they'd find it. They didn't."

Congress demanded an investigation into what it perceived as 'improper conduct' and many of Scott, Worden and Irwin's colleagues in the astronaut corps were mixed in their reactions; some felt a court-martial was in order, others that it was simply a dumb mistake. Years later, Scott would blame the pressure of getting ready for Apollo 15 and admit that he made a bad call. As for poor Jack Swigert, he seemed to have been a leading contender to serve as command module pilot for the joint Apollo-Soyuz Test Project with the Soviets – a mission to be discussed in the next volume in this series – but was formally suspended from astronaut status. He took it very badly, Kraft recounted, and left NASA shortly afterwards. Ironically, he ended up becoming a politician.

Other astronauts were already training for flights to the Skylab space station in 1973 and *they* made their apologies, took their brief suspensions, lived through it and remained on duty. For Kraft, Slayton and Low, the final indignity came when they were called before a secret session of a Senate subcommittee ... and Scott, Worden and Irwin turned up outside. "[Senator] Cannon [of Nevada] called a recess and brought them in," Kraft wrote. "The senators treated them like gods, shaking their hands, patting them on the back, falling all over themselves in adulation." Then Cannon reconvened the session and Senator Margaret Chase Smith of Maine verbally roasted Kraft for 'daring' to place "these fine young men" in a position where they could be so tempted. As Kraft silently fumed and clenched his fists, Deke Slayton stood up and accepted that, as Scott, Worden and Irwin's immediate superior, *he* was to blame. From then on, *everything* – absolutely *everything* – in an astronaut's personal pack would be listed, sealed and signed-off, first by Slayton, then by Kraft ... and *then* by the NASA Administrator himself. Kraft also forced astronauts to formally agree *not* to sell any items from their personal packs.

From Scott's perspective, it was up to senior management to have kept them on the straight and narrow. Slayton, after all, had *introduced them* to Eiermann in the first place and *he* was responsible for checking and approving the list of all personal items that they carried aboard the spacecraft. "But before the flight," wrote Scott, "for some reason, he neither asked us personally for each of our lists, as was customary, nor signed off on the list personally. He said the flight crew support team had already logged everything ... Somehow, the support team had missed them when they prepared ... the manifest."

Years later, Scott would be extremely vocal in his irritation over NASA's treatment of his crew; the agency did nothing to dispel (untrue) rumours that the men had been fired as astronauts and even told them to get their own legal representation. "It was turning into a witch-hunt," Scott wrote. "NASA, we were advised, expected us to keep quiet and take the Fifth Amendment at [the Senate] hearing. We did not. We told it like it was. We had nothing to hide." In December 1978, a Memorandum Opinion for the Assistant Attorney-General declared that the space agency had no claim to the covers taken from the astronauts, that they were never intended for sale, that there had been *no attempt at concealment* by Scott's crew and, finally, that the covers *would* have been approved for carriage aboard Apollo 15 if a formal request had been made. In Scott's mind, NASA had hung the crew out to dry and their bosses in flight operations had abrogated their responsibilities, leaving them out in the cold. His attempts to obtain full disclosure from the space agency on the stamp incident came to nothing, but he did learn that *ten* other astronauts had been involved in stamp-dealing with Eiermann. However, "the wave reached the shore on Apollo 15," Scott wrote, "and *we* were the ones who bore the brunt of the blame". Even the Air Force, eager to distance itself from any negative publicity, issued letters of reprimand to all three men. In June 1972, Irwin announced his intention to leave the service and a few weeks later, just before he was to appear before the Senate Space Committee to defend his position on the covers, his retirement papers were rushed over to him. "Take them and run," he was told, though not in so many words, by a sergeant, "before the Air Force changes its mind."

In a sense, therefore, 1972 began with a mixture of optimism and pessimism for the future of America's space programme. The astronaut corps was tarnished by the ugly truth of the stamp incident – and, for a time, the prospect of some of their number being hauled into court was banded around by journalists – and the Moon landing effort was drawing inexorably to a close. Dave Scott's suggestion to reinstate Apollos 18 and 19 had gone nowhere and it was evident that by the end of the year the final bootprints of the century would have been imprinted in the lunar dust. Yet there was reason for excited optimism, too. The last two missions, commanded respectively by John Young and Gene Cernan, carried all the promise of Scott's flight; for they, too, would land in stunning highland settings, they too would feature three days on the surface, three long-duration EVAs, a SIMbay packed with scientific gear and dramatic transearth spacewalks on the way home. If Apollo was ending, in 1972, it would certainly end in style.

MUNICH

When John Young, Ken Mattingly and Charlie Duke flew to the Moon in April 1972, they carried amongst their personal belongings a large Olympic flag. It measured 1.2 m high by 1.8 m across and was stowed in a fireproof container aboard the command module, Casper. Although weight constraints had already decreed that it was impossible to carry this large flag down to the lunar surface, a smaller version – just 1.2 cm by 1.8 cm – *was* carried aboard the Orion lander for humanity's fifth touchdown on alien soil. It was meant to honour that year's Summer Olympics – the latest in the time-honoured international multi-sport extravaganza – scheduled to begin in late August in the then-West German city of Munich. It is shockingly ironic, therefore, that a gathering intended to foster international peace and harmonious goodwill should have been ravaged by the evil of terrorism. Even today, the bloody events which tore out the very heart of the Munich Olympics, all those years ago, continue to inspire fear, resentment, inflamed emotions and a burning desire for vengeance.

The event began on 26 August, almost exactly four months to the day since Apollo 16 had returned safely to the waters of the Pacific. The West German government, keen to show itself as a reinvigorated, more optimistic and more 'open' member of the democratic world, presented them as 'the Happy Games' and displayed the bright Sun and the dachshund Waldi as its mascots. Then, in the early hours of 5 September, a little under a week before the games were due to conclude, a group of eight Palestinian terrorists from the notorious Black September movement broke into the Olympic Village and took eleven Israeli athletes, coaches and officials hostage. When two of the hostages resisted the initial onslaught, they were slain. The next 18 hours degenerated into an uneasy stand-off with Munich police, with the terrorists demanding the release and safe passage to Egypt of over 200 Palestinian captives held in Israeli jails, together with a couple of radicals from the German Red Army Faction. The response of Israeli Prime Minister Golda Meir and her government, not surprisingly, was a flat refusal; they would *not* negotiate. American marathon runner Frank Shorter, whose lodgings were nearby, summed up the terror of those 18 hours: "Imagine those poor guys over there. Every five minutes, a psycho with a machine gun says 'Let's kill 'em now' and someone else says 'No, let's wait a while'. How long could you stand that?"

It was hugely embarrassing for the Munich authorities because the hostages were Jewish and the Games, of course, were being held in a nation which had bloodied its hands and disgraced its reputation in the Nazi concentration camps a generation earlier. Police negotiators offered unlimited funds to release the Israelis, but the terrorists retorted that they cared nothing for their own lives and, by extension, nothing for the lives of others. At length, they issued an order to be flown to Cairo. Feigning agreement, the authorities drove them by bus to a pair of military helicopters, which would ferry them to a NATO airbase at Fürstenfeldbruck. The intention was that a group of police snipers would then make a full-scale assault to rescue the hostages. Tragically, the plan was botched from the start: the German authorities had only guessed the number of terrorists, and had used that incorrect assumption to guide their planning (there were *twice* as many as expected) and the police 'snipers' had been selected not on basis of their skills as sharpshooters or their counter-terrorism training ... but because they shot competitively on weekends! Remarkably, the snipers managed to kill all but three of the terrorists, but the poor organisation of the assault meant that *all* of the hostages were murdered; a handful being machine-gunned, the others incinerated when grenades were thrown inside one of the helicopters.

The story of what happened next has become almost as famous as the event itself: an ultra-secret Mossad plot, initiated under Golda Meir's auspices, to assassinate those responsible turned into the stuff of legend and, more recently, into two highly-successful movies.

So it was that, with a flag to commemorate this soon-to-become-infamous event tucked aboard, the Apollo 16 lunar module Orion, riding the fire of its descent engine, swept over the central highlands of the Moon on the afternoon of 20 April 1972 and set down on a patch of bright, possibly volcanic ground called the Cayley Plains, not far from the crater Descartes. During their three days on the surface,

John Young and Charlie Duke would change the image of another of the Moon's mysterious and unknown wildernesses.

OF MOON-MEN AND ORANGE JUICE

With four successful landings in the bag, by the spring of 1972 astronauts and planners were becoming more confident in the trajectories and flying skills needed to bring spidery lunar modules onto the shores of waterless alien seas, into hummocky ejecta and amongst forbidding mountains. The Armstrong and Conrad crews had both landed on relatively flat maria and returned with hauls of lunar basalt that spoke of ancient volcanism. Then, Shepard and Mitchell had sampled materials from the Fra Mauro Formation and the vicinity of Cone Crater, allowing them to tentatively date the impact which formed the Imbrium basin at some 3.85 billion years ago. Scott and Irwin's exploration of Hadley-Apennine refined the dating of the Imbrium cataclysm and, thanks to their discovery of the Genesis Rock and the remote-sensing of their SIMbay experiments, had revealed that anorthosite seemed widespread in the lunar highlands.

Directly sampling those highlands was the task of Apollo 16. The relatively fresh crater Tycho – an enormous, 80 km-wide basin in the Moon's southern hemisphere, easily visible from Earth at 'full' Moon with the naked eye – was the geologists' favourite, particularly when the unmanned Lunar Orbiter spacecraft revealed the presence of giant boulders, thrown up from deep within the crustal interior. However, in spite of the geological goldmine that Tycho offered, it was a particularly difficult place to reach; hundreds of kilometres from the equator, it would have required a lunar module to fly across a hellishly-rugged patch of terrain and land in a boulder-strewn wasteland. Tycho, barely 108 million years old, is a sharply defined crater, with a bright interior and a distinctive system of rays emanating like spokes as far as 1,500 km. Its central peak rises as high as 1.6 km. Smooth areas on its floor and in the terraces of its interior wall are probably solidified pools of rock melt. The ramparts beyond its rim possess considerably darker material, which geologists theorised may have formed from minerals excavated during the impact which made the crater. The unmanned Surveyor 7 probe touched down just beyond Tycho's rim in January 1968 and photographed a hilly landscape dominated by jumbles of forbidding boulders.

Former astronaut Jim McDivitt, by now the Apollo spacecraft programme manager, was adamant: Apollo would land at Tycho ... *over his dead body*! In McDivitt's mind, the fact that it was *too far* from the lunar equator meant that it would be more difficult for the crew to re-establish themselves from a hybrid trajectory and back onto a free-return, should an Apollo 13-type incident befall them. Moreover, setting down so far south would demand additional energy expenditure ... and *that* would cut down on the amount of payload the lander could transport to the surface and in orbit aboard the SIMbay. The prospects of landing at Tycho were already on shaky ground before Apollo 13; following the miraculous return of Jim Lovell's crew to Earth in April 1970, its chances dimmed even further.

If Jim McDivitt was decidedly unhappy about the prospect of touching down at Tycho, he was even less pleased by informal discussions about bringing lunar modules down in even more exotic places ... on the Moon's *far side*. Yet that is precisely what a hard-core team of geologists, championed by their own astronaut, Jack Schmitt, proposed, planned and developed over a period of months. Their target was the crater Tsiolkovski, in the southern hemisphere of the far side, unseen from Earth and first discovered by the Soviet Luna 3 probe in October 1959. Its high, terraced walls and well-formed central peak, together with an unusual, mare-like coating of dark material on its floor whetted many geologists' appetites. Of particular interest was the fact that this mare covering was *not ubiquitous* ... it was present in the eastern and southern parts of the crater and even reached the outermost wall in the north-west, but was absent elsewhere.

Schmitt was well aware that some kind of communications satellite 'relay' would be have to be placed in lunar orbit to communicate with a crew that landed at Tsiolkovski. Many of his mission controller friends doubted NASA would ever entertain such an audacious expedition – "out of sight, out of mind," they wisecracked – but in the pages of Gene Kranz' autobiography, the renowned flight director revealed that even *he* went to impromptu meetings in Schmitt's Houston apartment to see what they had to offer and was surprised to find not only geologists but also flight controllers and trajectory specialists participating in the discussion. "I had the impression," Kranz wrote, "they were a bunch of Boy Scouts setting up tents and starting campfires." *That* misconception quickly faded. The group listed all the pros and cons – scientific, technical and engineering – for landings on the farside and sketched out detailed mission and trajectory plans. In the spring and summer of 1970, with the budgetary axe hanging over Apollos 18 and 19, the group felt that a mission to Tsiolkovski would be so spectacular as to reignite interest in a fickle public and persuade the Nixon administration to continue lunar exploration.

Of course, a manned landing at Tsiolkovski did not happen and Kranz was not alone in lamenting this apparent loss of American spirit: "The crew would be on their own in a virtually uncharted world and like the early explorers, living by courage and ingenuity alone. We would not even know whether they had landed or crashed until the CSM relayed the status a half hour later. These would be explorers like Byrd, Scott, Peary and Cook ... "

With Tycho and Tsiolkovski out of the picture, the Site Selection Board had convened in May 1971 to select the landing site for Apollo 16 and opted for a patch of the Caylcy Plain adjacent to some hills about 50 km south of the heavily-worn crater Descartes and just to the west of Mare Nectaris (Sea of Nectar). This site was expected to yield volcanic rocks which would provide a window into the history of the Moon's interior. Although volcanic basalts had been picked up by previous missions, the mare regions from which they came were all formed during a relatively narrow epoch of lunar history, perhaps as brief as 350 million years. By sampling volcanic material from the more ancient highlands, at elevations several thousand metres above the maria, it was hoped to discover more about the history of the Moon both *before* and after the maria formed.

"There were two major volcanic-type rocks," Duke explained to a NASA oral historian in 1999 in recounting what they expected to find, "a very viscous rock that bulged up and created the Stone Mountain topographical relief, and then down in the Cayley Plain, the valley was another kind of less viscous rock that flowed out. We were looking at a contact between those two geologic features to see if there was any." The hilly terrain was formally known as the Descartes Formation and the undulating light-toned plain as the Cayley Formation. The geologists were so convinced that they were *both* of volcanic origin that almost all of the crew's field training was at volcanic sites on Earth.

When John Young, Ken Mattingly and Charlie Duke were assigned as the Apollo 16 crew in March 1971, some prophets of doom may have anticipated that illness might get in the way. A year earlier, in his position as a member of the Apollo 13 backup team, Duke had come down with German measles and *that* had led to the last-minute decision to ground Mattingly and replace him with Jack Swigert. Years later, Duke did not remember any problem between himself and Mattingly after the disappointment of the measles episode, although Fred Haise certainly had some fun on the morning of Apollo 16's launch. "We were climbing into the command module on the launch pad," Duke recalled, "and Guenter Wendt and the team were up there. John gets in and I'm the next in on the right side. As I start to climb in, I reach in and look over and taped to the back of my seat was a big tag that said 'Typhoid Mary'!"

In fact, Duke had entered the headlines *again* through illness. Early in January 1972, he had been admitted to Patrick Air Force Base's hospital with bacterial pneumonia. Although three days later NASA announced that Apollo 16 would be postponed from its original 17 March launch date to 16 April, it had nothing to do with Duke's ailment; it was a result of a combination of factors: the need to replace the command module's three main parachutes, the need to replace a burst Teflon bladder in the reaction control system and the need to change one of the explosive cords which would separate the lander's ascent stage from the command module prior to transearth injection. This required a rollback of the Saturn V to the Vehicle Assembly Building in late January 1972 – the first time this had happened in the Apollo programme. The need to make modifications to the space suits was another minor headache. "Duke's current illness," read the NASA news release, "is not expected to impact the crew's requisite training ... However, the delay will ensure more than enough time for Duke to fully recover before reinitiating his training schedule". In fact, in his autobiography, Deke Slayton noted that the option of replacing Duke was never raised; *that*, he pointed out, was *one* lesson learned from Apollo 13.

Charles Moss Duke Jr was born in Charlotte, North Carolina, on 3 October 1935. He attended high school in South Carolina and the Admiral Farragut Academy in St Petersburg, Florida, before studying for a bachelor's degree in naval science from the Naval Academy. By this time, he had already developed a love of aviation: at the academy, he managed to get occasional rides in an open-cockpit, bi-wing seaplane called the Yellow Peril ... and was instantly hooked. "I got seasick," he laughed, "but I *never* got airsick!" Duke credited this as one of the reasons why he ended up opting for the Air Force, as opposed to the Navy, as his career service.

He received his degree in 1957 and was commissioned into the Air Force, undergoing initial flight instruction at Spence Air Base in Georgia and Webb Air Force Base in Texas, then moved to Moody Air Force Base in Georgia for qualification as an F-86 Sabre pilot. Graduation from training led to three years as a fighter interceptor pilot, based in Germany, followed by a master's degree in aeronautics from MIT in 1964 and completion of test pilot school at Edwards Air Force Base the following summer. Duke remained in California, teaching control systems and flying in the F-101 Voodoo, F-104 Starfighter and T-33 Shooting Star.

When one looks at a career pattern like this, it has much in common with the 'standard' route needed to become an astronaut. Yet Duke would later admit that he followed this path for his career and *not* – at first – specifically with space in mind. "It was a *goal* in my career," he admitted to NASA's oral historian, "but back in those days, the Air Force was really seeking advanced education as a prerequisite for promotion and so I just knew that was really what I needed to do, but I didn't want to leave the cockpit, because I loved to fly." Only after finishing his master's studies in Massachusetts in 1964 did Duke realise that engineering was great, but he missed the flying. The next logical step was test pilot school and by the end of *that*, in September 1965, he happened to read an advert in the newspaper: NASA was going to recruit a new class of astronauts. Duke approached his immediate superior, Buck Buchanan, deputy commandant of the test pilot school, to ask what he should do. Buchanan advised that Duke *could* apply to NASA and also for the Air Force's proposed Manned Orbiting Laboratory project. "But if you apply for *both*," Buchanan cautioned, "we're going to pick you for MOL and not let NASA have y'all." Duke knew that the Air Force's association with manned space flight was not good and, indeed, the MOL effort was cancelled in 1969.

Less than seven months after completing test pilot school, Duke was selected by NASA as an astronaut candidate. His first couple of years with the agency focused on spacecraft systems and engineering assignments. For a time, he and Stu Roosa worked on automatic and manual guidance systems for the Saturn V booster. Later, Duke served as the Astronaut Office's oversight representative for the qualification of the lunar module's descent engine. By his own admission, though, he felt that his assignment as capcom for the landing of Apollo 11 in July 1969 was the high point of his career ... and, in some people's minds, he is *more* famous for *this* than for making his own arrival on lunar soil three years later!

Duke and Ken Mattingly were named to their first crew positions at the same time, but could hardly have imagined that they would actually end up flying together. In August 1969, just a couple of weeks after the triumphant return of Neil Armstrong's crew to Earth, the 33-year-old Mattingly was named as the command module pilot for Apollo 13, joining Jim Lovell and Fred Haise for what was to be the third lunar landing mission. Duke was assigned with John Young and Jack Swigert to the backup crew. When Swigert took Mattingly's place in early April 1970, only days before launch, the crew-cut bachelor saw his chances of going into space evaporate before his eyes. Having said that, in the months leading up to the first Moon landing, Mattingly had no expectation of receiving a flight assignment *at all* ... and went so far as to meet Al Shepard, at that time chief astronaut, with a view to

Ken Mattingly rehearses his transearth EVA in a command module simulator. Since the cabin would be depressurised, all three Apollo 16 astronauts needed to be fully-suited: John Young can be seen in profile on the left of this image and Charlie Duke is partially visible on the right.

resigning from the corps! Since his selection, he had worked solidly through his technical assignments and in the early summer of 1969 was supporting Apollo 12 preparations at Cape Kennedy when he received a call to get back to Houston. It was explained to him that Bill Anders, previously the backup CMP for Armstrong's mission, was planning to retire and Mattingly had been picked to stand in for some simulator sessions. "They wanted *me* to do that," Mattingly recalled in a 2001 oral history interview, "but I had no illusions: if the backup crew flies, it's *not likely to be you*!" Soon afterwards, he received a letter from the family of a friend who had been shot down over Vietnam. How could he sit around in NASA, Mattingly thought, whilst his old Navy buddies were fighting and dying in a *real* conflict? He went directly to Shepard's office, but Big Al told him to consider it for a couple of weeks. By the time those weeks were up, Mattingly had been named as the prime command module pilot for Apollo 13. Years later, he guessed that Shepard probably knew all along, but had kept quiet.

Thomas Kenneth Mattingly II was born in Chicago, Illinois, on 17 March 1936 and received much of his schooling in Florida. Aviation, he would later recount, was in his blood: his father worked for Eastern Airlines and his earliest memories were of toy planes and model aircraft. "I built every model I could find," he explained, "ate every box of cereal that had a cut-out paper airplane on the back, all that sort of stuff." If the young Mattingly stayed out of trouble for long enough, his father would take him aboard an aircraft and fly to the end of a route and back. Weekends would then be spent at the airport, watching planes take off and land.

He studied aeronautical engineering at Auburn University, graduating in 1958 and securing membership of the Delta Tau Delta fraternity. Whilst there, as part of the Navy's reserve officer training programme, Mattingly had the chance to fly a propeller-driven attack aircraft and his choice of aeronautical engineering did little to hide an ultimate ambition to become a test pilot. Later in 1958, he enlisted in the Navy as an ensign and was vocal in his desire for flight training. When his gunnery officer, Lieutenant-Commander Glenwood Clark, asked him if he *really* wanted to go through flight training, Mattingly, naturally, replied in the affirmative. "You're the dumbest ensign I've ever met," Clark announced. "Out!" More than a quarter of a century later, having flown to the Moon and commanded two Shuttle missions, Mattingly returned to active duty in the Naval Space and Warfare Systems Command ... and was introduced to his new commanding officer, one Vice-Admiral *Glenwood Clark*! Mattingly could hardly believe it. After briefing his new boss, the two senior officers sat down together.

"Admiral," Mattingly began, somewhat tentatively. "Do you remember me?"

Clark looked him straight in the eye.

"I sure do. You were the *dumbest* ensign I ever knew!" The two men laughed. They became firm friends after that.

So it was that the Navy's dumbest ensign completed initial flight instruction and received his wings in 1960. He went on to fly the A-1H Skyraider for three years and the A-3B Skywarrior for another two years, then his options included postgraduate school for a master's degree in aerospace engineering – which he was not too keen on doing – or test pilot school at Edwards Air Force Base. Mattingly picked the latter. In March 1965, witnessing the first manned Gemini launch planted the germ of an idea to someday become an astronaut. He knew that the Air Force was selecting candidates from the Edwards school for its MOL project, so he reckoned he might get a shot at *that*. Another Navy pilot, Ed Mitchell, had the same idea and the pair applied for MOL ... and were rejected. Although they could not ordinarily apply to NASA, a sympathetic senior officer at Edwards, Lieutenant-Colonel John Prodan, gave them a chance and submitted their names to the civilian space agency. In April 1966, they were both selected. Mitchell, of course, had already completed test pilot school, but Mattingly would fly back to California in May for his graduation ceremony.

Three years later, when he was assigned to Apollo 13, Mattingly had gained a reputation for himself as an expert at command module systems. Had he flown as planned, he would have been the first rookie CMP to fly solo in lunar orbit; quite an honour when one considers that he was just 34 at the time and *all* previous lunar CMPs had been picked from a pool of experienced astronauts. Eleven months would elapse after the unlucky voyage of Lovell, Swigert and Haise before Mattingly was finally named to a new lunar crew. However, it is interesting that he was one of few astronauts to have actually been given the *option* to decide which crew position he might take. He told the oral historian that in the autumn of 1970 Deke Slayton offered him two choices: he could either be the CMP of Apollo 16 or, if he wanted an actual landing, wait and become the LMP of Apollo 18.

"Well," Mattingly said, "I'd sure like to go down to the surface."

"I'll give you the choice," replied Slayton, "but I would always take a bird in the hand."

Then Mattingly was quiet for a moment. "You know, there won't be a chance to go back as a commander if I go on 16."

"It's your call. Just think about a bird in the hand."

Perhaps Slayton felt a pang of conscience for Mattingly's loss of Apollo 13 and his commitment to the subsequent rescue; certainly, such an offer was not common, particularly for a rookie astronaut. Years later, Mattingly would have mixed feelings about the choice he made. He *really* wanted to go down to the Moon – the journey would be incomplete, he felt, without actually touching its surface – but with the likelihood of impending mission cancellations it would be better to at least go *near it* than *not go at all*. Mattingly picked Apollo 16 and in March 1971 the formal NASA announcement was made. Ultimately, it proved the right call.

By the time that announcement was made, Apollo had changed markedly and it was recognised that the mission to be flown by Young, Mattingly and Duke would be the penultimate voyage to the Moon for many years to come. Had the remaining missions been kept on the manifest, it is likely that Fred Haise, Bill Pogue and Gerry Carr would have continued to back up Young's crew and then flown a lunar mission of their own. By the spring of 1971, the backup positions were essentially 'dead-ended', with no chance of a lunar flight down the line. It is therefore hardly surprising that Pogue and Carr – both rookies from the 1966 class – accepted reassignment to the forthcoming Skylab space station in order to bring their own dream of a mission to reality. Haise, however, was committed to Apollo. He had been one of the most promising geology students amongst the pilots and almost certainly would have performed admirably on the Moon with Jim Lovell. Now, Deke Slayton had nothing but admiration for him; in spite of an essentially zero chance of actually flying, Haise opted to stay as Apollo 16's backup commander. Replacing Pogue and Carr as his crewmates on the backup crew were Apollo 14 veterans Stu Roosa and Ed Mitchell.

However, fate *almost* had one more card to play for Ken Mattingly. A couple of weeks before launch, the doctors found an elevated level of bilirubin in one of his blood tests. "Bilirubin?" he snorted increduously. "Who's *he* and what's *that*?" It turned out to be an indicator of liver function and possibly indicated that he might be coming down with hepatitis. Mattingly was irritated and not just because it threatened his one remaining chance to get to the Moon; it also seemed that they were focusing on blood test after blood test, but little attention was devoted to actually averting whatever medical problem was looming. The relationship between the CMP and the doctors in those final days, Mattingly said, "was about as antagonistic ... as you could create". Thankfully, he was cleared as fit to fly.

Apollo 16 set off at 12:54 pm EST on 16 April 1972, entering a parking orbit around Earth and executing a perfect translunar injection burn a couple of hours later. From Mattingly's perspective, in the centre seat, the *feeling* of launching aboard the most powerful rocket ever brought to operational status was *exactly* as it sounded. "You feel the same staccato shake and rattle and bang," he told the NASA oral historian. Weightless was as 'natural' to him as it possibly could be, in view of the fact that it was totally *unnatural* to his normal Earthly state, but Mattingly's worries about space sickness proved unfounded. Following the translunar burn, he stole a glance back towards Earth and was impressed by the colours of the oceans and continents. It would be to his intense regret that Apollo 16 remained in Earth orbit for only one and a half circuits. There was simply *not* enough time to look at the grandeur and beauty of the Home Planet. Moreover, he was *worried*: he was seeing so many new sights that were, quite literally, *out of this world*. "I'm afraid to look again," Mattingly recounted, "because I feel like I have an erasable memory and if I see one more thing, it's going to write over something I just saw and I'll forget it. I know that's preposterous, but I had this very palpable fear that if I saw too much, I couldn't remember. It was just *so* impressive. These things kept coming for the next ten days. They *never* stopped."

The journey to the Moon was largely uneventful and the command and service modules, named 'Casper', and lunar module 'Orion' entered orbit late on the evening of 19 April. The only real problem of note had been an annoying flecking of thermal paint from the lander's exterior, which manifested itself in a cloud of light-coloured particles. Soon after midday on 20 April, Casper and Orion undocked and Young and Duke began preparations for their Powered Descent to the surface. Then, things began to take an ugly turn. Ken Mattingly, now alone in the spacious command module, was getting ready to fire the SPS engine and adjust his orbit from an elliptical one into a near-circular path. He noted each step on the checklist: electrical power was routed to the engine, the gyros activated normally ... and then the time came to test the secondary control system. As soon as he touched the yaw thumbwheel for the gimbal motors, the entire ship noticeably shuddered. He tried again; and the same thing happened. It *felt* for all the world like he was aboard a train on a *very* bad track.

"Hey, Orion?"

"Go ahead, Ken," radioed John Young.

"I have an unstable yaw gimbal number two."

"Oh, boy." Even Young, who knew the command module like the back of his hand, having guided it solo around the Moon on Apollo 10, could think of no quick ideas to resolve the problem. Charlie Duke suggested that Mattingly stop it and start it again; no, came the response, he had already done that twice. If the control mechanism to the SPS was damaged in some way, all three men knew that their mission was over ... and that, in all likelihood, Apollo 17 would never fly. From his position, Mattingly *knew* that he could move the nozzle, but had no idea if he could make a successful circularisation burn with it. The rules, one of which had been changed just a week ago, required *all four* thruster-vector-control circuits – both primary *and* secondary – to be operative in order for the circularisation burn to go ahead. Young and Duke's only option was to re-rendezvous and fly in formation while awaiting a decision from Houston over how to proceed. Redocking and rescheduling the landing for the following day was out of the question, because by then the Sun would have risen another 12 degrees and the lighting conditions would be unfavourable for a touchdown at Descartes. Whatever the problem might be, they simply *had* to solve it rapidly or cancel the landing.

As engineers pored over the data, one hour became two ... and then three. In Downey, California, the data had been fed into an SPS mockup and a tentative conclusion had been reached: if Mattingly *had* to use the secondary system for the burn, the engine might shake, but it *would* be controllable. Watching from Mission Control in Houston was Jim McDivitt, the Apollo spacecraft programme manager. On Apollo 9 in March 1969, he and his crew had deliberately made their SPS shake in a test-firing – part of a what-if engineering demonstration – and it had responded well. In his mind, that personal experience provided the second assurance he needed and he advised Chris Kraft, now the director of the Manned Spacecraft Center, and the other managers that it was safe to proceed. Capcom Jim Irwin radioed up the happy news shortly before 7:00 pm EST, about four and a half hours after the trouble was first noted. However, what McDivitt did *not* know was that *all* of the control signals to the SPS – primary and secondary – went through the *same* cable; Mattingly had learned of this during his work on the engine's systems several months earlier, but had thought little of it. After the mission, McDivitt would inform Mattingly with a grin: "We *didn't* know they went through the same cable – you're the only one who did! You're right; we *wouldn't* have let you land!"

Less than three hours later, at 9:23 pm EST, Young and Duke brought Orion low over the lunar mountains and set down on an undulatory plain between a pair of bright-rayed craters, unsurprisingly named North Ray and South Ray. They were only a couple of hundred metres north-west of their target point. At the instant of touchdown, Duke – whose Carolina drawl had endeared him to terrestrial audiences during the flight to the Moon – could hardly contain his infectious enthusiasm. "Wowwww," he exulted, like a child at the fair. "Whoa, man! Old Orion is finally here, Houston. Fan-*tas*-tic!"

By now, on America's fifth lunar landing, managers, planners and astronauts had become, as Deke Slayton described it, somewhat blasé about operating on the Moon. The plan had been to conduct the first excursion immediately, but the six-hour delay in landing prompted mission managers to decide that Young and Duke should get some sleep first, even though it would require adjustments to the checklists and timelines. To have gone ahead and performed EVA-1 would have subjected the astronauts to a 27-hour day. Flight Director Gerry Griffin was *not* ready to put their endurance to the test. The crew, of course, felt quite differently. "We agreed," Charlie Duke told the *Apollo Lunar Surface Journal*, "but it turned out I wish we hadn't. I wish we had gone on out and stuck to our timeline, because it required a lot of changes in the procedures once we landed. This was a *big* mess!"

Sleeping in Orion was undoubtedly more comfortable for Apollo 16's explorers than it had been for their H-series predecessors. They could, at least, remove their space suits, which actually 'stood up', partially, at the back of the cabin, in the weak gravity. Conditions, though, were cramped. "You're tight," Duke admitted, "but I wouldn't call it jammed in. Once we got off the suits, there was a place between us, over the ascent engine cover, where we could drape the suits and they were out of the way. For me, I could lean back and sort of semi-sit on the environmental control unit." They arranged their hammocks and Duke was surprised that Young drifted straight to sleep. "My mind's just racing like crazy," Duke said later, "and even though we were tired, I couldn't get to sleep. I asked Mission Control if I could take a sleeping pill, which I did, and then I drifted off to sleep."

Fourteen hours after touchdown, a fully-suited Young hopped down the ladder and became the ninth man to set foot on another celestial body. "There you are, our mysterious and unknown Descartes highland plains," he breathed at 11:59 am EST. "Apollo 16 is gonna *change your image!*" All around him were rocks – he and Duke would not have far to go in order to indulge in some prospecting – and over the next three days they put their field geology training to good use. Young had become hooked on geology as Jim Lovell's backup, when he visited the Orocopia Mountains with Lee Silver and Jack Schmitt, followed by other sites in readiness for Apollo 16. Early in their training programme, in the summer of 1970, he and Duke and their *original* backups, Haise and Carr, had ventured into New Mexico's San Juan Mountains, Medicine Hat in Alberta and the Colorado Plateau, and Young had lapped it all up. Andrew Chaikin has characterised Young's appearance and personality as something which outsiders often underestimated: he would arrive for simulator sessions in a rumpled shirt and jeans, sit through lectures without saying a word and his Florida country-boy drawl made some engineers think he was slow-witted. They only *ever* made this mistake once, Chaikin wrote, for Young was one of the most quick-witted, responsible, intelligent and focused astronauts on the team. When Al Shepard retired in 1974, Young took his place as chief of the corps ... and held onto it for longer than any other astronaut.

Despite finally reaching the lunar surface, Young and Duke's mission was still far from secure because, in view of the late landing, some managers were considering scrubbing their third Moonwalk. In the science support room at Mission Control, Bill Muehlberger, a professor from the University of Texas who was serving as chief of the surface geology team and Young and Duke's scientific mentor, had spent much of their first night on the Moon convincing the NASA brass to keep EVA-3. Also fighting the astronauts' corner was Flight Director Gene Kranz, who was furious ... but for slightly different reasons. He remembered that, on Apollo 15, Scott and Irwin had overtaxed themselves to the point of exhaustion when their time on the surface had been shortened – it was typical of astronauts and their insatiable drive to achieve *every* objective on the flight plan to push themselves *beyond* the limit – and Kranz knew that Young and Duke would do the same. Eventually, he succeeded in persuading Chris Kraft that such a move would increase the risk on a mission which, thus far, had proceeded well.

The Descartes region, and the lunar highlands in general, were too valuable to waste this opportunity. In truth, Muehlberger's team knew little about the site – the best-resolution images of it had been acquired by Stu Roosa during his orbital passes on Apollo 14. If the Hycon camera had worked, it would have provided very high quality pictures. All they had were some telephoto shots in which the smallest object discernible was fully 20 m in size. In the months preceding Apollo 16's launch, a pair of scientists from the US Geological Survey, named Don Elston and Gene Boudette, subjected Roosa's pictures to the magnifying power of stereographic plotters and somehow managed to resolve boulders two or three metres in diameter. Given the almost universal belief that the site was of volcanic origin, their final map, upon which Young and Duke would rely, was dotted with cinder cones, lava flows and explosion craters.

With the two astronauts on the surface, the radio link to Earth became a kind of vaudeville with Duke's excited chatter balanced by a series of mumbled, dry one-liners from Young. In a sense, it seemed more fun and more comical than even Pete Conrad and Al Bean's capers on the Ocean of Storms. Already, the lateness of their landing had lost them the chance to take telephoto images of a landmark known as Stone Mountain in the early lunar morning. It had been hoped that Stone might display patterns of linear grooves similar to those on Mount Hadley Delta. Nevertheless, the men had much to do. As Duke busied himself with maps and charts, Young set to work erecting an automatic telescopic camera to photograph ultraviolet radiation from stellar sources, Earth and the lunar horizon; working in one-sixth gravity, he found, was delightful, and the instrument, which was heavy and cumbersome on Earth, was easily carried on his shoulder. Next came the Rover, unfolded from its berth in Orion's descent stage, then the Stars and Stripes was planted in the soil ... and Duke could not resist taking a photograph of his commander, jumping a metre or more off the ground, whilst saluting the flag.

Assembly of the ALSEP came next and for one scientist in particular, it was a chance to finally see his experiment laid on the Moon. Marcus Langseth was a geophysicist from Columbia University. His heat-flow investigation had originally been aboard the ill-fated Apollo 13. Some data had been acquired by Apollo 15, but Dave Scott had been unable to drill to the required depth and the heat flow rates were twice as high as anticipated. Langseth hoped the data from Apollo 16 would allow him to properly interpret the earlier data. Specifically, he wished to know how the heat flow from the lunar interior – whether this be heat left over from the Moon's formation or from radioactive decay – differed between a mare plain and the highlands. With the aid of a redesigned and more hardy drill, Duke set to work boring into the surface and found it a surprisingly easy task – within ten minutes, the first hole was finished to the required depth and the thermometers inserted. Duke, Langseth and capcom Tony England – himself a geophysicist – were all delighted, but their joy was not to last.

At the time, Young was finishing off his own duties at the ALSEP's central station – the boxy unit which provided command and power utilities to each of the four experiments connected to it by lengths of coloured ribbon-cable. In addition to Langseth's investigation, there were passive and active seismometers and a surface magnetometer. Unfortunately, as Young moved away from the central station, his boot caught the cable for Langseth's heat-flow experiment. Everyone watching the live television in Mission Control saw this, but before a warning could be issued, the cable had been yanked out of the central station. Young could only sheepishly apologise to Langseth. To be fair, with a large control unit mounted on their chests, the Moonwalkers could not see their feet very well and both men had warned engineers before launch that the cables refused to lie flat after deployment. Duke abandoned drilling the second hole. As engineers set to work determining whether or not the heat-flow experiment could be repaired, the astronauts had no option but to press on with the remainder of their seven-hour EVA.

Based on an initial survey of the televised view from the Rover, the site bore a similarity to Hadley-Apennine: an undulating grey-brown plain, backdropped by round-topped mountains. The expected volcanics were proving elusive. The area was littered with rocks, but many of them had the light-and-dark appearance of breccias, a type of rock composed of a mixture of fragments 'welded' together by the enormous temperatures and pressures of a major impact. As the two men drove westwards toward their first geology stop at Flag Crater, they found only more breccias. Indeed, at one stage, a puzzled Bill Muehlberger sent a message via capcom Tony England asking if they had seen *any* rocks which definitely *weren't* breccias. Duke glumly replied in the negative.

As has already been noted, the lunar crews all received extensive geological training, so Muehlberger was convinced that Young and Duke were relaying reliable information and accurate observations to Mission Control. Each rock would typically be photographed – using a tripod-like 'gnomon' and colour chart to measure the local vertical, Sun-angle and colour of the surface – lifted and placed in an individual Teflon-coated bag which measured approximately 19 x 20 cm in size and was prenumbered. The bags came in a 20-bag dispenser which could either be mounted on a bracket on the Hasselblad camera or on a hand-tool carrier. Young and Duke also had slightly larger sample collection bags at their disposal, which featured interior 'pockets' for holding drive tubes and cap dispensers and extra bags for additional samples. These bags would typically be stashed in the back of the Rover during traverses.

Driving to Flag Crater took its toll on Young, particularly since the Sun was 'behind' him and washed-out the landscape into a featureless blanket. In fact, Elston and Boudette had predicted scarps four to five metres high and, try as they might, the astronauts could see next to nothing, not even the tiniest craters, until they were almost on top of them. Before launch, Young had cautioned against visiting Flag on these grounds and now, on the Moon, he was nervous, pushing the Rover no faster than four or five kilometres per hour, lest he drive off a scarp. Thankfully, no such precipitous drops materialised. "We never encountered any of these features on the geology map," he reported during his post-mission technical debrief. "They were mapped at scarps of steep features that said we were going to have to drive around or over; we just never ran into those. I think that's because those photo-analysis guys were reaching for and pulling out features that weren't there. I mean, I *looked* for these things and, sure enough, if you really imagined it, you *could* see something there, but with that scarping we had [on the map] they were reaching for it. They sure weren't there in the real world and if they *were* there, they looked like every other slope that was around there."

At length, the 350 m Flag Crater gave them a breather and a chance for a few minutes of geological prospecting, followed by a short trip over to nearby Plum Crater. It was hoped that Plum might harbour rocks from the ejecta thrown from whatever impact had created Flag. Shortly before 5:00 pm EST, a little under five hours into the excursion, Young and Duke were trekking around the rim of Plum, picking up breccia after breccia ... but *no* evidence of anything volcanic. Then, Young spotted another boulder, chipped a piece away with his geology hammer and wondered aloud to Mission Control if it was a welded ashflow tuff: a rock composed of ash fragments 'welded' together in a matrix of volcanic glass. "Welded ashflow tuffs," explained Andrew Chaikin, "can look just like breccias and they are rich in silica, just as the Cayley rocks were thought to be. Some of Muehlberger's team had predicted there would be welded ashflow tuffs at Descartes." At length, after humming about it for a few minutes, Young pronounced it to be a breccia.

Before leaving Plum, Duke was directed to pick up another rock, lying on the eastern rim of the crater, which looked like it might be igneous. Young was not sure if they should bag it: it seemed as if it must weigh about 10 percent of the 90 kg limit that Orion could carry. It was so large, in fact, that Duke had to get down in the dust, roll the rock onto his knee and then try to stand up without losing his

Charlie Duke at Plum Crater. The Rover can be seen in the background.

balance. It was certainly a biggie, weighing-in at 12 kg, and would be nicknamed 'Big Muley' in honour of Muehlberger. "I never heard anyone call Bill M. 'Muley'," wrote Tony England in a 1996 letter to Eric Jones, published in the *Apollo Lunar Surface Journal*, "but I was aware that it referred to him. Charlie played with words a lot, so he probably invented the association with the rock on the spot. We would have all understood it. Bill is not only big, but he is a little stubborn. The 'Muley' played on his name, his size and, just a little, on his being stubborn. Of course, I'd never say that when I was within Bill's reach!" Muehlberger himself would get in on the act – "I was 210 pounds and 6 feet 1" – and would actually later admit that his nickname at college happened to have been 'Mully'.

Heading in the direction of the lander, they made a slightly abbreviated stop at Buster Crater, where Young took readings with a portable magnetometer, but the Moonwalk had already been trimmed when it became clear that Duke was using more cooling water in his suit than expected. By the time they returned to the vicinity of the lander, around six in the evening EST, some scientists were beginning to wonder whether Cayley was really volcanic. Some suggested that it was, but was masked by a surface layer of impact rocks.

Back aboard Orion, Young and Duke could feel justifiably proud, in spite of losing Langseth's heat-flow experiment. They had spent more than seven hours outside and had brought back over 20 kg of samples. The environment inside the cabin was decidedly grimy, with the concentrations of lunar dust, and both men would echo previous crews by comparing the aroma to spent gunpowder. In Duke's case, he noticed a peculiar *taste* to it, as well. "It had a greasy [feel]," he told Eric Jones for the *Apollo Lunar Surface Journal*. "It really magnified your skin oils ... but what I think it was, it was picking up your skin oils, because there is no moisture in the stuff, and it *did* have that taste: to me, gunpowder."

The flight surgeons were interested in something else: the men's potassium intake. In the summer of the previous year, Dave Scott and Jim Irwin had displayed unusual heart irregularities and physicians suspected that this was probably due to a lack of potassium. Consequently, Young, Mattingly and Duke flew to the Moon with an overstocked supply of potassium-laced orange juice. "The concoction did not taste quite like nectar," wrote Gene Kranz, "and John Young was quick to inform us that it made his crew gassy and nauseated, *not* a good state for a confined cockpit in zero-G." In preparation for one of his press conferences, Kranz tried some and found it to be thick, heavy and almost 'metallic'. His judgement: "It tastes like *crap*!" After four days of torture, Young finally complained to Tony England; the effect on his intestinal tract was no longer a joke.

That night, getting ready to sleep, Young accidentally left his microphone button in the 'on' position and his audience was treated to a full and frank conversation between two acid-stomached astronauts on another world.

"I got the farts again, Charlie. I don't know what gives 'em to me, I really don't. I think it's acid in the stomach, I really do."

"Prob'ly is," replied Duke.

"I mean," Young pressed the issue, "I haven't eaten *this* much citrus fruit in 20 years, but I'll tell you one thing: in another 12 fuckin' days, I ain't *never* eatin' any more ... and if they offer to serve me potassium with my breakfast, I'm gonna throw up! I like an occasional orange – really do – but I'll be damned if I'm gonna be *buried in oranges!*" At length, Tony England, who had been squelching his own microphone as a subtle cue to the two men, came on and told them that they had been transmitting. Not surprisingly, Young's graphic descriptions of his uncomfortable intestinal problems lessened after that.

UNTOUCHED SERENITY

When Young and Duke awoke after their second night's sleep on the surface, there was bad news for Marcus Langseth's heat-flow experiment. Thanks to a change in the design of its electrical connector, which had a sharp edge, the power cable had probably been sheared. Repair options were studied and a tiger team suggested bringing the heat-flow electronics box and the cable back inside Orion's cabin at the end of EVA-2 to allow the astronauts to work on it without the bulk of their space suits, but eventually such plans were called off. Even though backup commander Fred Haise tested the whole procedure – and it *worked* – the senior managers were worried that since it had four dozen individual wires, it would consume a large chunk of time. By the time Young and Duke ventured outside for their second Moonwalk, the decision had been made.

The astronauts' first destination was Stone Mountain, to the south-east of the lander, which Young had clearly seen through his window during the descent, and which now formed a stunning backdrop to the landing site. They parked the Rover in a small crater, partway up its flank. From an elevation of almost two hundred metres, they were higher than any other Apollo astronauts before them and their reward was a spectacular view of South Ray Crater – a yawning bowl, five times the size of a football pitch, radiating rays of boulders, some dark and others bright, in all directions. Before launch, Young had pushed very hard to add a fourth Moonwalk to Apollo 16 and mount an expedition into the crater itself, but radar soundings by terrestrial astronomers suggested that there would be too many rocks for the Rover to safely negotiate. From his vantage point partway up Stone Mountain, Young realised that such predictions had been overestimated.

The *evidence*, though, of whatever impact had carved South Ray was all around them, which was unfortunate, because the purpose of visiting Stone Mountain was to obtain samples of Descartes material. With the domical hill 'polluted' by Cayley material ejected by South Ray, it would be difficult to unambiguously identify the material they sought. The plan called for them to drive up Stone Mountain to a cluster of five craters that they had nicknamed 'the Cincos'. After this, they were to drive back down, pausing to sample at a number of positions that spanned the 'contact' between the two geological formations. The ordering of the stations was dictated by the walk-back constraint. Although the Cincos were clearly visible as Young and Duke drove across the plain, when they started up the hill they found it difficult to ascertain precisely *where* they were. When they eventually stopped in a large field of boulders, they were only a few tens of metres from the largest of the

Charlie Duke works with the Rover on the boulder-strewn slopes of Stone Mountain. In the foreground is a sample container.

quintet, but it was out of sight beyond a nearby ridge. The boulders were ejecta and the manner in which they were strewn on one side of the crater was an indication that they were debris of the impactor, and thus not representative of Stone Mountain, so the astronauts ignored them. Instead, Young scraped an exploratory trench in the rim of the crater, seeking small fragments of the hill beneath, but found nothing of interest.

Disappointed, the astronauts drove back down the hillside to another crater, some 20 m wide, where Young set to work with a rake on the inner wall that was shielded from South Ray's ejecta. "His difficult harvest," wrote Andrew Chaikin, "was not what he expected; what looked like rocks turned out to be clods of dirt that fell apart in the sample bag."

As they worked their way downslope, Young spotted what he thought might be a crystalline rock, glinting in the sunlight. Like the Genesis Rock plucked from Hadley-Apennine by Scott and Irwin, it *looked* almost entirely plagioclase, although its crystals were minuscule, very much like grains of sugar. It seemed to be anorthosite, but certainly did not appear to be volcanic.

Back on the plain, the astronauts made several stops to sample a point within one of the bright rays from South Ray and another area between the rays. By the time they returned to Orion, they had been out for more than seven hours.

So far, their explorations had produced few 'eureka' moments in the style of Apollo 15 with its Genesis Rock, but Muehlberger's team were fully satisfied with Young and Duke's performance.

Aside from the scientific demands of their mission, and their light-hearted demeanour, neither astronaut was immune to the sheer wonder of *where* they were and the knowledge that they were mankind's *only* representatives on this barren world. Four or five months earlier, during a geological trip to Hawaii, Charlie Duke had a strange dream – a haunting, though not nightmarish, dream which he would recount two decades later for an Al Reinert film called *For All Mankind*. Duke had come down with flu and ran a high fever; at times, he was almost delirious. In his dream, he and Young were driving the Rover up to the north. "You didn't feel like you were ... *out there*," he remembered wistfully of the peculiar, unreal clarity of the lunar setting. "It was untouched. The *serenity* of it ... had a kind of a pristine purity about it." They continued to bounce across this endless landscape of nothingness until, startled, they came upon a set of tracks and asked Houston if they could follow them. "After an hour or so, we came across this vehicle. It looked just like the Rover, [with] two people in; they looked like me and John. They'd been here for thousands of years ... " When pieces of the alien craft were examined back on Earth, the startling truth was revealed: it was *one hundred thousand* years old! In telling this tale, Duke has always maintained that the dream was *far* from being a nightmare situation; in fact, during his three days on the Moon, he never once felt afraid and often it was easy to lose sight of the chilling reality: beyond his few layers of space suit material was a near-total *vacuum* ... "and if you spring a leak in that suit, you're gonna be *dead*!" Anything remotely hostile about the lunar landscape faded when he looked at its sheer, stunning beauty: the hills, as already noted, were rounded, completely at odds with the sharp, angular, malevolent-looking mountains often depicted in science-fiction stories. For a test pilot, unused to the art of a wordsmith, Duke admitted that he found it hard to describe his feeling as he stood there, in the glare of the lunar morning, looking across this dramatic but absolutely lifeless world. "We just felt like we were *supposed* to be there," he recounted on the website www.charliedukestory.com. "We did not feel like we were intruders in this foreign, foreign land."

Next day, as the two explorers headed for North Ray Crater, the dream about the tracks undoubtedly entered Duke's mind. Naturally, he saw no tracks on the *real* Moon, but the sense of strange clarity was present, nonetheless. As Young drove, it was Duke's responsibility to take photographs and describe the terrain for the geologists back home. As navigator, he kept tabs on the maps and the geology stops and the headings that they were following. "I remember as John started off," Duke recalled to the NASA oral historian, "I said 'Okay, John, steer 120 degrees for 1.2 km and then turn left to 090 degrees and go another 2 km' or whatever it was. That's the way we navigated. The Rover had a little directional gyro. There's no magnetic field on the Moon, so a magnetic compass wouldn't work. We had a little gyroscope that was mounted in the instrument panel and so we pointed it down-Sun and it was the old Navy lubber's line: You had a bar came down across it, cast a shadow on the gyroscope compass card. We assumed that the shadow was west and so we just turned the card till 270 was up underneath that shadow … and *that* was our direction. We had a little odometer on the wheel that counted out in kilometres and that was our distance." Generally, though, they were never overly concerned about losing their way; if they were really unsure, it would be fairly straightforward to turn around and follow their tracks home.

At 11:30 am EST on 23 April, they reached the south-eastern rim of North Ray. The crater was over a kilometre across, and even the television camera on the Rover, remotely controlled from Earth by engineer Ed Fendell, could not fit its vast expanse within its field-of-view. As the camera panned across it, mission controllers gaped; awed not only by its size but also by the steepness of its walls, which plunged 200 m from the rim to the rock-strewn floor. From Young and Duke's perspective, it was quite different. Like Scott and Irwin's adventure on the slopes of Hadley Rille, they were far from the edge. Before launch, Young had toyed with the idea of carrying a 30 m tether, so that one man could descend partway into the gaping maw of North Ray, anchored by his colleague. However, weight constraints and increasingly nervous managers nixed the idea.

Keeping well back from the rim, the astronauts set about taking panoramic images, and collecting and documenting samples. Then, the scientists viewing through the Rover's camera saw a huge, dark boulder on the eastern rim; they reasoned that it might represent the deepest point of excavation by North Ray, which had punched through bright rock into a dark material and which might be a buried volcanic unit. They suggested that the astronauts take a sample of the boulder.

With the minutes rapidly running out, Young told Duke to head over to the boulder, then set off after his colleague.

"It may be further away than we think," Young warned.

"Nah, it's not very far," replied Duke.

"Theoretically, huh?" deadpanned Young. "Like everything else around here: 'A couple of weeks later ... '" His humour belied a clear realisation of astronauts, managers and geologists alike that depth perception on the Moon was difficult; distances were often misleading, but there was still some surprise when Mission Control saw Duke steadily get smaller and smaller on their television screen. By this point, the geologists realised that the 'boulder' was considerably larger than they had anticipated. On account of its tremendous size, the astronauts named it 'House Rock'.

"We kept jogging and jogging," Duke recounted, "and the rock kept getting bigger and bigger and we were going slightly downhill, that we didn't sense it at first and so we got down to ... House Rock. You know, it must've been 90 feet across and 45 feet tall. It was *humongous*! We walked around to the east side, which was in the sunlight, and it was *towering* over us."

A view from the Rover as Young and Duke headed back towards Orion.

After Young had chipped off a few samples using his geology hammer, the two men ran back uphill towards the Rover.

Despite the hope that the dark material excavated by North Ray might be a volcanic unit, the samples established House Rock to be yet another breccia. Although the geologists would have liked the astronauts to keep searching for the elusive volcanics, their time had run out. On hearing that his colleagues on the surface were finding only breccias, Ken Mattingly had laughed. "Well, it's back to the drawing boards," he said, "or wherever geologists go."

Walking on the Moon and operating for long periods in a bulky suit – the three excursions had lasted, respectively, seven hours and 11 minutes, seven hours and 23 minutes and five hours and 40 minutes – required more strength and stamina than Duke had anticipated. Although the J-series suits had been modified to allow the wearer to bend more easily at the waist, it took a lot of getting used to, and both time and practice were essential in making the ensemble 'work' for them. "When you got back inside and you took everything off, you were exhausted," said Duke. "It was *hard work*. You're squeezing that glove for seven to eight hours – if you can imagine squeezing a [hard rubber] ball – and doing the curls and stuff in the suit . . . it was *real* work." More than one astronaut would compare it to a 'light workout' . . . for *eight* continuous hours. On a couple of occasions, the flight surgeons noted their hearts racing and told them to slow down.

More problematic was the risk of puncturing the suit. At one stage, during a few free minutes, Young and Duke planned to run their own lunar Olympics – again, to commemorate that year's event in Munich – to rival Al Shepard's golfing caper. Time was running away from them, so they decided against it, but not before Duke tried a lunar high jump . . . and fell over backwards, directly onto his life-sustaining backpack, in full view of the Rover's camera. "I was in trouble," he recounted. "You could watch me scrambling like that, trying to get my balance. I ended up landing on my right side and bouncing onto my back. My heart was pounding. The backpack is very fragile; I thought the suit would hold, but the backpack, with the plumbing and connections and all, if *that* broke, it was just like having a puncture in the suit." Young quickly came over to help, but for a few anxious moments, all of Duke's childlike excitement was gone. "I got *real* quiet," he would recall, "and you could hear the pumps running in the backpack. I checked my pressure; it was okay. This fear began to subside."

A little more than a day later, having performed a smooth ascent to lunar orbit and redocking with Mattingly, the crew of Apollo 16 were on their way home. In view of the problem with the SPS, the plan to spend another day collecting data using the SIMbay was cancelled. The astronauts would openly wonder if Mission Control was just too nervous about the engine to allow them to remain any longer. Later, Mattingly thought that someone probably told Jim McDivitt about the primary and secondary SPS signals going through the *same* cable and *he* had ordered them home early. Either way, the crew was miffed to have their mission curtailed.

Nonetheless, Ken Mattingly had been hard at work during his three solo days, operating an extensive battery of SIMbay instrumentation. Eight investigations determined the distribution of X-ray fluorescence, gamma radiation and alpha particles, provided laser altimetry and mass spectrometric data and imaged the lunar terrain with panoramic and mapping cameras. Of particular note were the mapping camera, with its 20 m imaging capability, and the panoramic camera, which could take stereoscopic images with a resolution of just a couple of metres. Before heading for home, a second subsatellite was deployed to measure particle abundances and magnetic fields, but the orbit into which it was released was far from ideal and it did not last as long as intended.

By the time he flew Apollo 16, Mattingly had been training on orbital geology with Farouk el-Baz for nearly three years – counting his stint on Jim Lovell's crew – and in his oral history for NASA he would admit to some scepticism at first. He had no innate love of geology, but wanted to be *more* than just a truck driver. When he was told that someone called 'Farouk el-Baz' wanted to see him, Mattingly's first response was that someone was pulling his chain; that it was another astronaut 'gotcha'. At first, when asked to say what he saw in terrain photographs and at geological training sites, all Mattingly could come up with was something along the lines of "dirt ... and a few rocks", but the efforts of el-Baz had sharpened his eyes. Now, in orbit around the Moon, the training and preparation had paid off and Mattingly found himself identifying features that abounded with detail. He would routinely stay up beyond his scheduled sleep period simply to ensure that he did not miss anything interesting. He worked to the background music of his favourite classical pieces – Berlioz' *Symphonie Fantastique* and Holst's *The Planets*, among others – which set an appropriate mood for the amazing vista 'beneath' him. Moreover, Mattingly later joked, it made a pleasant change from the country music cassettes that Charlie Duke had played all the way to the Moon ...

As well as monitoring the SIMbay experiments, Mattingly was tasked with a number of other investigations. One of these was the Gegenschein from Lunar Orbit, which required him to take long exposures on high-speed black and white film to gather evidence of a faint light source – the 'Gegenschein' – known to cover a 20-degree field of view centred along the Sun-Earth line on the anti-solar axis. It was suggested that the Gegenschein was sunlight reflecting from particles of matter, trapped at the theoretical 'Moulton point'. From his perch, Mattingly would have traded nothing for the experience he savoured during those three days circling the Moon ... except, perhaps, for one thing: the chance to go *down there* and plant his own bootprints in lunar soil. "If you were going to devise a programme for personal enjoyment," he said, "the only thing you'd change in the way things worked out for *me* is I would have had *another flight* to go land on the Moon! *Nothing* can take the place of being there ... *but* you wouldn't want to skip the lunar [orbit] part to go to the surface. You need *both*, because the lunar [orbit] piece, especially solo, was probably more sense of exhilaration. I can't explain it, but it was really, *really* something!"

During the transearth coast, Mattingly's exhilaration reached its zenith on the afternoon of 25 April with the opportunity to venture outside, in a space suit, to retrieve film cassettes from the SIMbay. "The space suits were the same for command module pilots as for others," he recounted, "except we didn't have so many fittings and we didn't have a need for the gold-plated visor, because we're not supposed to be out in the Sun." Of course, like Worden, Mattingly *would* be going outside and consequently he did so wearing John Young's red-striped helmet cover with its gold visor. His primary task during the 84-minute excursion was to collect the cassettes, but a secondary objective was to set up a package of microbial specimens for ten minutes in the command module's hatch and retrieve it before he came back inside. At some point, Mattingly – who had married a couple of years earlier – misplaced his wedding ring somewhere in Casper's cabin; he presumed it lost forever.

As Mattingly moved outside, Charlie Duke passed him a few trash containers to discard, then held onto his suited legs as the CMP proceeded to clamber, hand over hand, to the rear end of the service module. Occasionally, he took a few seconds to look around at the ethereal blackness: the apparent absence of stars startled him and the view of the crescent Earth and the Moon, now some 80,000 km away, captivated him. Establishing himself securely in a pair of foot restraints at the end of the service module, Mattingly felt for the first time that he was *really* away – he was *really* in space; an incredibly powerful experience. Heading back towards Casper's open hatch, Duke suddenly broke in: "Look at *that*!" Mattingly looked . . . and *there* was his missing wedding ring, just floating outside. "I grabbed it," he said, "and we put it in the pocket. We had the chances of a gazillion-to-one." Doubtless, his wife Elizabeth was relieved.

GEOLOGIST WANTED: EXPERIENCE NECESSARY

In the weeks after the command module Casper splashed down into the western Pacific, concluding a spectacular 11-day mission, the unexpected absence of volcanic rocks at the Descartes and Cayley Formations forced a radical rethink of their origin. "While the Apollo 16 rocks could not rule out that lava or cinders had erupted elsewhere in the highlands," Andrew Chaikin wrote, "they all but proved that none had flowed at Descartes. It was not the first time the geologists had misinterpreted photographs of the Moon; it would not be the last. But that did not mean there was anything inherently wrong with photogeology. The error, one geologist would write years later, was that they had neglected to define more than one working hypothesis." The problem, Chaikin explained, was that too many geologists had paid too much attention to the close physical resemblance of lunar highland formations to terrestrial volcanic features. Nevertheless, the findings of Young, Mattingly and Duke were welcome, because they demonstrated a key maxim: that science *advances* most when one of its predictions is proved *wrong*.

The pilots had proven themselves to be exceptionally acute field geologists and their transmissions from the Moon showed a confident understanding of the scientific terminology and an awareness of specific rock types. Yet by the time Apollo 16 splashed down, the call from the scientific community for a *real*, professional geologist to walk on the lunar surface had finally been answered. In August 1971, a week after the return of Dave Scott's crew, the astronauts for Apollo 17 were announced by NASA. Commander of the mission would be Gene Cernan, whose performance on the Apollo 14 backup crew and the Apollo 10 prime crew assured him the support of Al Shepard and Tom Stafford, two extremely influential members of the corps. Interestingly, Stafford had been chief astronaut since the summer of 1969 and Shepard had resumed his duties in charge of the office in June 1971, just weeks before the Apollo 17 announcement. With the backing of these two powerful men, it might not seem surprising that Cernan should have received command of the last landing mission.

Having said this, the story of Cernan's helicopter crash a few months earlier had made it into the newspapers – to such an extent that even Vice-President Ted Agnew had telephoned the Cape Kennedy crew quarters to check on his health – and an accident report, published in October 1971, would blame 'pilot error'. It has already been noted that the three-flight crew rotation system from backup to prime was *not* a rule, and even Cernan's exemplary performance as backup commander of Apollo 14 did not guarantee him Apollo 17. Chomping at his heels was the Apollo 15 backup crew of Dick Gordon, Vance Brand and Jack Schmitt; Gordon, certainly, made no secret of the fact that he dearly wished to descend those last hundred kilometres from lunar orbit and leave his own bootprints on the Moon. If Cernan had Shepard and Stafford's backing, then Gordon could bank on the support of the equally influential Pete Conrad, Dave Scott and Jim McDivitt ... *and* he also had a trump card in his back pocket.

That card was Harrison Hagan 'Jack' Schmitt, the first professional geologist ever to enter the ranks of the astronaut corps. He had been born in Santa Rita, New Mexico, on 3 July 1935 and his fascination with the Earth sciences began at an early age: his father was a respected mining geologist for the US Defense Minerals Production Agency and the young Schmitt spent weekends during his teenage years working in the field. When he entered the California Institute of Technology in 1953, his initial focus was upon physics, but by the time he graduated with a bachelor's degree in science four years later, he had firmly committed himself to geology. He won a Fulbright scholarship and spent a year at the University of Oslo and, since his budget offered precious little to pay a field assistant, he often climbed alone in the fjords of western Norway, and went on to receive his PhD from Harvard in 1964.

Schmitt's next steps were into the new field of 'astrogeology' and he joined the eminent scientist Gene Shoemaker at the US Geological Survey in Flagstaff, Arizona, helping to develop exploratory procedures, sampling techniques and photographic methods to be used by astronauts on the Moon. In the autumn of 1964, he learned of NASA's intention to hire scientist-astronauts and submitted his application to the National Academy of Sciences. The academy screened a large number of applicants and forwarded a shortlist to the space agency; many did not even pass NASA's physical. After much deliberation, in June 1965 Schmitt joined physicists Owen Garriott and Ed Gibson, physicians Joe Kerwin and Duane Graveline and astronomer Curt Michel as the first group of scientist-astronauts. Many of the pilots, wrote Andrew Chaikin, strongly felt that science was little more than excess baggage on an exclusively engineering endeavour like Apollo, and so the newcomers were treated as virtual outsiders from the moment they arrived in Houston.

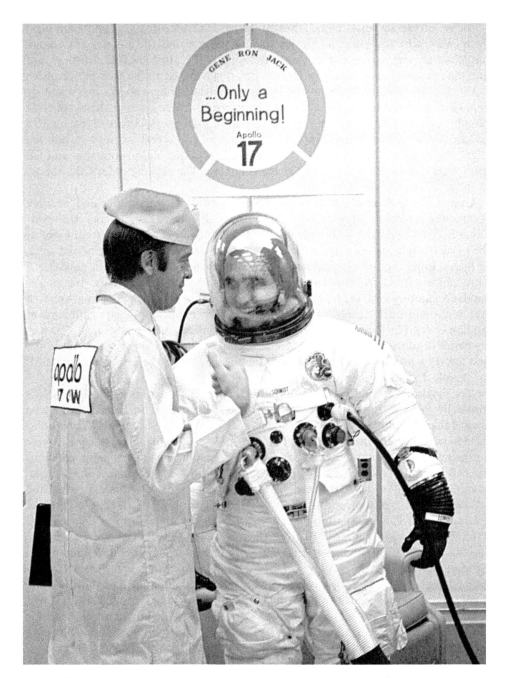

Jack Schmitt – the first and so far only geologist to practice his profession on another world – engages in a spot of banter with Al Shepard on the evening of 6 December 1972. Only hours after this picture was taken, Schmitt would be on his way to the Moon.

A few of the scientists could already fly jets and the others, Schmitt amongst them, were sent to Williams Air Force Base in Arizona for a year of training. Initially, they were wrung out in the modest Cessna 172, then the T-37, which was at that time the service's primary jet trainer, before advancing to the high-performance T-38 used by NASA. Competing against younger flying cadets, the scientists graduated near the top of the class, but their return to Houston in the summer of 1966 brought a rude awakening: *another* group of astronauts, chosen a few months earlier, was *already* scooping support crew assignments on real missions ... and the scientists, supposedly more senior, were ignored. "No one had to come out and tell them what was suddenly obvious," wrote Chaikin. "The pecking order didn't apply to them. They were never in line."

Unperturbed, Schmitt set to work on his technical assignments and in 1967 became the astronaut representative on the ALSEP surface experiments. However, under his own initiative, he broadened his scope to cover the entire descent stage of the lunar module and its cargo of scientific instruments and tools. During this time, he hosted impromptu geological meetings in his apartment with scientific colleagues, with senior managers, with engineers and with trajectory specialists. He single-handedly came up with the lunar science programme for Bill Anders during Apollo 8 and went on to work closely with Lee Silver on the geological training of subsequent landing crews. His focus on Apollo was almost fanatical, continued Chaikin, as was his desire to someday walk on the Moon. When Al Shepard approached him in the first few days of January 1970 to tell him that he was being assigned as backup LMP for Apollo 15, Schmitt was delighted.

Yet Shepard's decision, and that of Deke Slayton, had been forced down NASA's throat by the scientific community. Some had even pushed for a geologist to be aboard the *very first* landing mission and as each new team of test pilots journeyed to the Moon, it became increasingly difficult for NASA to explain why scientists were not being seriously considered. When Schmitt's announcement was made public at the end of March 1970, his backup slot made him a likely contender for a prime position on Apollo 18. Certainly, his commander, Dick Gordon, felt that the presence of the geologist on his crew gave them something of an edge over Gene Cernan's team. The cancellation of Apollos 18 and 19 in September 1970 hit them all hard, but the ongoing pressure to fly a geologist ignited fresh hopes that maybe Slayton would break the crew rotation system and assign the last landing to Gordon's crew. That hope grew brighter early the following year, when Cernan crashed his helicopter into the Indian River; Gordon thought he had it made.

When the announcement of the Apollo 17 crew was made in August 1971, it was good news for Cernan, Schmitt and CMP Ron Evans, but extremely *bad* news for Gordon, Vance Brand ... and Joe Engle, the hotshot test pilot who was now being pushed aside in favour of the geologist. In his autobiography, Cernan admitted that Engle – chosen as an astronaut by NASA in April 1966 – was "a magnificent aviator", who had reached the fringe of space on three occasions flying the X-15 rocket-plane. On the other hand, Cernan wrote, Engle did not understand the lunar module's quirky systems well enough to serve as an LMP. In Cernan's mind, it did not matter; his experience from Apollo 10 meant that *he* could carry both of them.

As commander of a crew – and the fact that it was a backup crew notwithstanding – Cernan considered himself responsible for Engle's performance and in the run up to Apollo 14 he took it upon himself to offer his LMP additional one-to-one tuition. Slayton's decision to drop Engle was probably eased by criticism of his performance by Tom Stafford, standing in as chief astronaut.

Having said this, Slayton remained impressed by Engle's flying abilities and submitted his name, along with those of Cernan and Evans, to NASA Headquarters accordingly. It was rejected. The top brass were not concerned about Cernan and Evans being aboard, they told Slayton, but the one person who absolutely *had* to be on the Apollo 17 list was Jack Schmitt, the geologist. In fact, high-level discussions had been ongoing within NASA for some time and on 1 March 1971 the agency's Associate Administrator for Manned Space Flight, Dale Myers, had written to Charlie Townes, chair of the Space Science Board of the National Academy of Sciences, lending his support. Still, from the perspective of the astronauts, nothing was set in stone. As late as 3 August 1971, in an interview with Stuart Auerbach of the *Washington Post*, Jack Schmitt considered it unlikely that he would make it to the Moon. Ten days later, he was formally announced as a member of the Apollo 17 crew ...

Cernan and Evans and their wives were holidaying in Acapulco when Slayton called to inform them that they had been picked to embark on humanity's final Apollo mission to the Moon. As Cernan received the happy news, he flashed a thumbs-up to Evans to signify that *he* too would be going with him. Then, when Slayton told him that he needed them back in Houston to 'discuss' the rest of his crew, Cernan was not sure whether to laugh or cry. "The four of us adjourned to the bar for a few rounds of rum and Coke," he wrote, "elated because we had gotten what we wanted, but disappointed, too. Even without being told, we knew that Deke probably had been forced to shuffle the crew, that Jack was going to fly and Joe was going to stay home."

For Engle, it was a devastating, crushing blow and he would later tell a journalist that one of the most difficult things he had ever done was going home and telling his children that he *wasn't* going to walk on the Moon. Jack Schmitt tried to sympathise, but the man he had unwittingly edged out of a lunar seat was visibly upset and bitter, yet handled himself with surprising dignity and grace. A month after the announcement, on 8 September, Engle told Jim Maloney of the *Houston Post* that "when something like this happens, you can do one of two things: You can lay on the bed and cry about it ... or you can get behind the mission and make it the best in the world". When one considers these words and this admirable attitude, Engle should rightfully be lauded for rededicating himself to helping Schmitt perfect his skills as Cernan's LMP.

For his own part, Cernan, too, was upset to lose Engle. The crew with whom he had trained and moulded into such a close-knit unit had been broken; the months spent tutoring Engle and becoming a team in the simulator was gone. Privately, Cernan was not sure if he and Evans could develop the same rapport with Schmitt as they had with their former crewmate. Their wives, Barbara and Jan, were even less sure. Schmitt's intelligence – Cernan often felt that he could *hear* the thinking wheels turning inside the geologist's skull – was balanced by a penchant for terrible puns, a brashness not normally associated with a scientist and a tone as abrasive as the very soils he would someday pluck from the lunar surface. Barbara Cernan and Jan Evans had also become good friends with Joe Engle's wife, Mary, during their two years of backup training for Apollo 14 and they were heartbroken. Nonetheless, Schmitt, a bachelor, was more than capable of performing the LMP's role and, like it or not, he was now on the crew for Apollo 17. It was Cernan's task to mould him into the team.

This proved tricky at first, particularly since Schmitt had apparently little regard for the chain of command to which the military pilot astronauts had become accustomed. His attempts to persuade senior managers to agree to a landing on the floor of the crater Tsiolkovski on the lunar farside had raised many eyebrows and Cernan regularly found himself in the firing line: as Apollo 17's commander, *he* was responsible for keeping Schmitt in line. "We knocked heads for a while," wrote Cernan, "until he realised that although NASA was a civilian agency, its mission leaders came primarily from the military services, where a commander's word was final. We could discuss differences and problems, but the old Supreme-Being argument *had* to apply. As commander, I wouldn't have time to debate when a critical decision was needed. I finally made it clear that he was going to work *through* me, like it or not. Period, end of story, sit down. Jack got the message, subdued his independent streak and became a solid crew member instead of a rebel." In time, the two men would develop such respect for one another that Cernan gave Schmitt the lead role in planning their three days on the lunar surface. He was keen to use the geologist's expertise in the interests of the mission.

Of course, flying Schmitt on Apollo 17 *because* he happened to be a geologist was not enough; and Slayton and even the more senior managers within NASA would have fought the scientists tooth and nail if they had the slightest notion that the man nicknamed 'Dr Rock' was not up to the task. Others wondered to what extent Schmitt could really *use* his geological expertise, encased within a bulky space suit and controlled by the limits of the timeline on the Moon, but in truth it was already clear that this would be humanity's last journey to its closest celestial neighbour for many years to come. It also had to be borne in mind that, even after five previous landings, Apollo 17 was by no means a walk in the park; it would be just as dangerous as the others and Chris Kraft, for one, was well aware of that fact. One day in the spring of 1972, he took Cernan aside and told him, point-blank, about his concerns: "Geno, put away that fighter pilot's silk scarf and just bring your crew home alive. If you run into something you don't like out there and decide not to land, I'll back you one hundred percent."

Cernan felt quite differently. The very fact that Apollo 17 *was* to be the final mission of the programme made him even more bold and he impressed on the journalists, the workers building and checking out their spacecraft and the politicians that this was *not* the 'end', but the 'end of the beginning'. However, it *was* the end. For two years, NASA's workforce had been on the decline and Richard Nixon's budget cuts had made this immensely painful for the dedicated men and women who had made possible the greatest achievement in human history. Although

Cernan meant it with all sincerity, his words grew to become a popular joke: a cartoonist drew a pair of workers at the top of a scaffold, one holding a notice informing him that he had just been fired. The other man was on the telephone, saying "Can we get Gene Cernan up here to give Smith that 'it's not the end, it's the beginning' speech again?"

Four decades later, and particularly since President Barack Obama's February 2010 speech that effectively scrapped plans for another Moon landing, Cernan's words come hauntingly to mind. One hundred years from now, when our species has hopefully returned to the Moon, it may well be seen differently, but for the last three generations, pessimistic space enthusiasts, myself included, have seen little promise for the future of lunar or planetary exploration. We remain chained to low-Earth orbit, with, it seems, neither the technology nor the political will to venture outwards. Abandoning our first footprints on the Moon, even for budgetary reasons, *has* been a backward step; as backward, indeed, as abandoning the New World or Australia might have been after just a handful of explorers had been there. "History," Stu Roosa once said, "will *not* be kind to us, because we were *stupid*."

Roosa's involvement with Apollo 17 was much more than a mere spectator, for in July 1972 he had been named, along with John Young and Charlie Duke, as a member of the backup team for Cernan, Evans and Schmitt. Previously, Deke Slayton had tried to offer positions to rookie astronauts, but with no more missions on the horizon, there seemed little need to train first-timers for 'dead-end' backup slots. Despite a one-in-a-million chance of flying Apollo 17, Roosa had accepted the backup CMP job, but Duke, newly returned from his own landing at Descartes, instinctively *knew* that he would *never* fly in place of Schmitt. Even if the geologist broke his leg, NASA would probably postpone the launch. As for John Young, he would stick around for whatever missions he could possibly get ... and in the autumn of 1972, he came close to gaining the Apollo 17 command for himself. During this time, a degree of camaraderie developed between Young, Roosa and Duke: they called themselves the only all-Southern Apollo crew and even jokingly grew moustaches.

The process of finding a landing site for Apollo 17 began in November 1971. "The flight planners," wrote Eric Jones in the *Apollo Lunar Surface Journal*, "wanted a site far enough from the eastern limb that they would have at least 12 minutes of flying time – and preferably 15 minutes – between Acquisition of Signal and the start of the final descent." This requirement restricted the list of possible sites to no farther east than a lunar longitude of 34 degrees. In a memo on 23 November, Jim McDivitt wrote that only in the case of extraordinary scientific merit should a site between 34 degrees and 43 degrees be considered. The final decision in February 1972 of a valley in the Taurus Mountains on the edge of the Serenitatis basin, just south of the crater Littrow, was heavily influenced by images taken by Al Worden.

In his autobiography, Cernan remarked that the chosen landing site was so far off the 'usual' track that it did not have a name and it was later dubbed 'Taurus-Littrow'. The steep-sided mountains were expected to yield samples from the pre-

Imbrium era, because the Serenitatis basin predates that of Imbrium. Yet one of the main mysteries was the floor of the valley, which was covered with some of the darkest material ever seen on the Moon and looked strangely out of place amidst the light-coloured highlands. "The targeted landing point itself," read a NASA news release of 16 February 1972, "will be on the other prime sampling objective, which is the very dark, non-mare material filling the valleys between the mountains. On occasion, the dark material is found in small troughs on the mountainsides, indicating that it once thinly covered the mountains, but has eroded off the steep slopes."

On Earth, such geological enigmas can arise when pockets of volcanic gas at high pressure find release at the surface and spray droplets of lava into the sky as a veritable 'fire fountain'; if a similar situation had occurred on the Moon, it could have blanketed a wide area with ash. During his observations the previous August, Worden had noted the possible presence of small volcanoes – which he termed

Jack Schmitt at work with the Rover in the Taurus-Littrow valley, close to Shorty Crater.

'cinder cones' – in this region and actually suggested over the space-to-ground radio link that a future landing there would be highly desirable. If it *was* volcanic in origin, scientists theorised, its relatively low density of craters and covering of dark material was strongly suggestive of it being one of the Moon's youngest such areas. "The explosive nature of the volcanism," continued the news release, "indicates a relatively high content of volatiles or gases, both of which have been exceedingly rare in all lunar samples seen thus far. If the Moon, as the preferred models indicate, has indeed cooled from the outside in, these youngest lunar volcanics should be derived from the greatest depths and may give the first good samples of the deep lunar interior."

At the same time, the North and South Massifs that bounded the valley were expected to provide ancient lunar rocks. The scientists were intrigued by what appeared to be landslides which would offer Cernan and Schmitt a means of sampling from an elevation that would not be directly accessible to them. Further east, the domical Sculptured Hills looked like they might be a result of the much-sought highland volcanism. All in all, Taurus-Littrow promised to be the most complex geological site yet visited on the Moon ... and as such was the perfect prospecting ground for Jack Schmitt.

So it was that at 7:08 pm EST on 11 December 1972, four days into the Apollo 17 mission, the first professional geologist *ever* to practice his art on a different world clambered down the ladder of the lunar module Challenger and planted his boots into the soft soil of Taurus-Littrow. His first words, however, were hardly geological: "Hey, who's been tracking up *my* lunar surface?" The person responsible, Gene Cernan, was not far away, learning to maintain his balance on a surface which felt like walking on a bowl of Jell-O. Despite his test-pilot instincts, though, Cernan was by no means immune to wonderment – and some would speculate over the years that NASA considered *him* a better ambassador than Dick Gordon to hold the mantle of 'last man on the Moon' – and his sense of amazement at standing in a place that no one had *ever* visited was profound. "The soil that was firmly supporting me," he wrote, "was *not* the dirt of Earth, but of a *different* celestial body, and it glittered in the bright Sun as if studded with millions of tiny diamonds." Low in the morning sky, the blazing Sun cast a long shadow beyond the lander.

In fact, it was the Sun which had helped dictate that Apollo 17 would be the only Apollo lunar landing mission to launch from Earth at night. "It would be impossible to land on the Moon in darkness," Cernan wrote, "but if the Sun was too high, its furious light would wash out the lunar surface below during our approach. Therefore it was determined that Apollo 17 must leave Earth after dark in order to arrive over the landing area, days later, with the lunar morning Sun at just the proper angle to cast the stark shadows we needed to outline the surface details."

If Cernan did not have enough on his plate with the planning and preparation for the mission, two other situations outside of his control cropped up: one was an injury which almost bumped him from the crew ... and the other came as a direct consequence of the massacre of Israeli athletes at the Munich Olympics. Shortly after that blood-letting, intelligence sources picked up hints that Black September militants may have been targeting the crew and families of Apollo 17. As summer turned to autumn and launch drew ever closer, Cernan arrived at Cape Kennedy to find Charley Buckley, the head of security, fitting a new, bullet-proof door to the crew quarters. When Buckley and Deke Slayton told him that Black September were no longer threatening the astronauts or their spacecraft, but a considerably more valuable target – their *children* – Cernan was livid. So too was Ron Evans, but both accepted the police protection that was offered. In the final weeks before launch, police minders sat in unmarked cars, twenty-four hours a day, monitoring the astronauts' homes. During the day, well-dressed, polite, but armed, federal agents sat in their children's classrooms at school.

As if this pressure and stress was not bad enough, in late September, Cernan underwent a routine physical examination and a prostate infection was found. Chuck La Pinta, the flight surgeon, kept it quiet and treated him accordingly. Years later, Cernan would remember La Pinta with fondness as one of the few flight surgeons *not* to have gone out and rung alarm bells; he had dealt with it quietly and discreetly. Then, something far more serious reared its head. One day in October, the crew was playing softball at the Cape, when Cernan felt something 'snap' in his right leg. Fearing a ruptured tendon, he had to be carried from the diamond by Evans and Schmitt. La Pinta's prognosis was not good: if the tendon *was* indeed ruptured, then it would take months to heal, effectively eliminating Cernan from Apollo 17, but if not it would still require a lengthy spell of bedrest and time on crutches. Thankfully, when the X-ray results came back, they showed no rupture, but La Pinta advised Cernan not to overly tax himself; if he did, the tendon might tear and he would lose Apollo 17 for good. At the same time, La Pinta kept Cernan shielded from the managers, including Deke Slayton, who feared the worst. At length, the flight surgeon's treatment and advice worked. La Pinta, wrote Cernan, "was a great doctor, a *terrific* liar and an even better friend".

INTO THE VALLEY AT TAURUS-LITTROW

By the evening of 6 December 1972, Cernan was long-since back to full health and readiness; in fact, he had flown the Lunar Landing Training Vehicle above Ellington Field just a few weeks earlier. The usual rituals of Lew Hartzell's steak-and-eggs 'breakfast' were followed, then the laborious process of applying biomedical sensors, suiting-up and transferring to the launch pad. At length, all three men were ensconced in their seats in the command module, which they had named 'America'. Launch was scheduled for 9:53 pm EST, at the start of a four-hour 'window', and the countdown was proceeding nicely. Then came a glitch. The automated launch sequencer on the ground failed to properly command the oxygen tank of the Saturn V's third stage to pressurise. The launch controllers issued the command manually, but the sequencer knew that it had not sent the command and refused to proceed. With only minutes remaining in the countdown, hearts missed beats as the very real question arose: Would Apollo 17 fly?

The problem did not necessarily eliminate the chance of launching that night and, at length, a work-around was devised and the countdown clock resumed ticking a

On its last mission with a human crew, the Saturn V provided Florida – and most of the eastern seaboard – with a midnight sunrise as it carried Apollo 17 towards the Moon.

few minutes before midnight. Finally, at 12:33 am on 7 December – a little more than three decades since one of the most infamous days in United States history, the attack on Pearl Harbour – the Saturn took flight, stunning the assembled crowds and most of America's East Coast with a spectacle rivalling sunrise. "It's lighting up the sky," announced public affairs commentator Jack King. "It's just like daylight here at the Kennedy Space Center." From his left-hand couch, Cernan could clearly see the fiery glow reflecting off the clouds and the shuddering cabin seemed 'painted' with a fearsome reddish hue. Ron Evans and Jack Schmitt, experiencing their first ride into space, were jubilant.

One of Schmitt's areas of interest was meteorology and as soon as Apollo 17 reached Earth orbit, he began chattering a storm about what he could see: the snow-covered white expanse of Antarctica figuratively and literally took his breath away, as did cloud patterns over Australia and Zanzibar and a nasty storm brewing over southern California. At one stage, the capcom likened him to a human weather satellite. Problems during humanity's last journey to the Moon were more irksome than serious; in fact, Ron Evans' loss of his scissors intended for opening a package of food was the only obstacle which stood in their way.

At 12:20 pm on 11 December, the command ship America released the lunar module Challenger in orbit around the Moon and Cernan and Schmitt readied themselves for their Powered Descent towards Taurus-Littrow. By this time, Cernan was an old hand at such matters and hardly noticed when they passed beneath the 15.4 km 'floor' that he had reached with Tom Stafford on Apollo 10. His eyes and brain were too busy, literally riveted to every nuance and idiosyncratic response from the spidery machine that was guiding him to his home for the next three days. "Thanks to the simulators back on Earth, with their computer-enhanced photos of the approach to the landing site," Cernan wrote, "I knew this place better than I knew my own palm and there were no surprises as we zoomed toward the jagged highlands that separate the Sea of Tranquillity from the Sea of Serenity. I called out the passing landmarks that verified we were on track to the narrow entrance to the Valley of Taurus-Littrow."

Dropping closer now, both men's eyes remained on their instruments ... until Cernan spotted something that he wanted Schmitt to take a look at. Halfway through the 12-minute burn of Challenger's descent engine, he told Schmitt to look at "something spectacular" outside his window. Expecting to see some unexpected example of lunar geology, Schmitt looked; but he couldn't see a thing, he said, except for the blue and white globe of Earth. "*That's* what I'm telling you to look at!" chuckled Cernan. It amazed him that, even after more than a year training with Schmitt, scientists continued to perplex him. Three kilometres above the surface, now plunging like a fast elevator, Earth hung in the black sky, directly in front of their windows. "Down we flew toward crop-duster altitudes," Cernan wrote, "scooted over the dome-like Sculptured Hills, some of which were more than a mile high, and roared into the eastern entrance of a crater-pocked lunar valley deeper than the Grand Canyon, surrounded by mountains whose crests were *above* us."

As Schmitt called out altitudes and rates of descent, Cernan confidently set Challenger down on a smooth spot at 2:54 pm EST. At last, less than five days since leaving Florida, they were here. Paradoxically, the cacophony of roaring and groaning and screaming and shuddering which the Saturn V had unleashed as it climbed for the heavens were gone ... to be replaced by the absolute silence, ethereal stillness, perfect serenity and utter lifelessness of the Moon's Taurus-Littrow valley. To their right, the North Massif stood taller than eight Eiffel Towers, and to their left the "wretched slab" of the South Massif equalled the height of half a dozen Empire State Buildings, stacked one atop the other.

Four hours later, at 6:54 pm, Cernan shuffled his way down Challenger's ladder and planted his boots into the soft lunar dust. "Getting down the ladder in the suit wasn't any particular problem," he told the *Apollo Lunar Surface Journal*. "It would have been a lot easier *without* a suit, but ... they performed tremendously. There had been a tremendous amount of technological development in the suit between my EVA on Gemini IX and Apollo 17. Walking on the surface with the mobility of a Gemini EVA suit would have been damn near impossible. It was just an order of magnitude difference in technological development. Many people don't realise the importance and significance of the suit. It was *everything*. It was radio communications, it was cooling, it was breathing, it was pressurisation, it was protection from the Sun, it was protection from abrasiveness. It had to provide mobility, dexterity, safety and reliability." It was an oasis of life on a barren and hostile world. Cernan was, in essence, a small mobile spacecraft in his own right.

Shortly afterwards, he was joined by a second mobile spacecraft named Jack Schmitt and the two men set about the first of their tasks. Moving around was tricky to start with: only ten other humans had gone from terrestrial gravity to weightlessness to *one-sixth gravity* and Cernan and Schmitt bobbed around like rubber ducks in a bathtub. At length, the professional geologist in Jack Schmitt overcame the awestruck voyager: the rocks, he radioed, looked "like a vesicular, very light-coloured porphyry of some kind; it's about 10 or 15 percent vesicles". A porphyry is an extremely finely grained basaltic matrix that contains mineral crystals.

Walking on the Moon was also extremely *dirty*; the dust clung to their suits, to their visors and to their gloves, with the result that within minutes Cernan looked like he had been outside for a week. Trying to brush it off his suit only made matters worse. At one point, reaching to pick up a rock, Schmitt slipped and, laughing, tumbled into the dust: his pure-white suit ended up charcoal grey *from the knees down*!

"You adapt very, very quickly," Cernan later told the *Apollo Lunar Surface Journal*. "You very quickly realise – probably in the first couple of minutes – that you don't need to take baby steps or regular steps to get anywhere. Somehow, your brain and your body co-ordinate your movements and if you're going to go any distance, you start skipping or hop-skipping to get where you're going. It's not like you start running. It's just that you move with such ease. Later on, when you start moving at faster paces than we were doing here, if you decide to turn or change directions, you have to think about your high centre-of-mass and plan how you're going to handle that ... or you're going to go tail-over-teakettle ... but you adapt very readily, physiologically and psychologically. You're conscious, as soon as you're on the surface, that you're in this one-sixth-G environment and that you can move around so much more easily. The human being is a very unique, very adaptable creature."

Their first major task was to unpack the Rover from Challenger's descent stage; for the next three days, they would rely upon its capabilities. Unreeling lanyards and watching the framework fold into place was, wrote Cernan, like assembling a Christmas bike for his daughter, Tracy. "Hallelujah, Houston," yelled Cernan as he took it for a test drive. "Challenger's baby is on the roll!" As they loaded their geological tools aboard the Rover, Cernan kept looking up at the Earth, hanging like a decoration in the sky above the South Massif. At one stage, he even told Schmitt to take a look for half a minute; the geologist owed himself *that* much. Schmitt, a man who had spent his life exploring the rocks and soils of that blue-and-white world, feigned disgust.

"What? The *Earth*?"

"Just *look* up there!"

"Aaaahhh," drawled Schmitt. "You seen one Earth, you've seen them all!"

By now Cernan was familiar with Schmitt's dry humour and good-natured sarcasm, but in his autobiography he expressed disappointment at how off-handedly the geologist dismissed this awe-inspiring sight. Before launch, Cernan had urged Schmitt and Evans to embrace each and every experience from this mission since it was sure to be the most remarkable and breathtaking and exhilarating adventure of their lives and he also *knew* that *none* of them would *ever* come this way again. It was perhaps with this in mind that, after erecting the Stars and Stripes, during which Cernan managed to capture a now iconic image of his colleague with Earth in the background, they set to work assembling their own package of experiments, the ALSEP, which would relay data from this strange place.

Many of the experiments on Apollo 17 were new ones. The Lunar Ejecta and Meteorites investigation measured the physical parameters of the minuscule particles impacting the surface. A seismic profiling sensor featured four geophones arranged in a 90 m equilateral triangle – one at each point and one in the centre – which measured the detonation of small explosive charges which would be fired after the astronauts had left the Moon, in order to better understand the structure of the valley floor. Elsewhere, another instrument monitored the tenuous lunar 'atmosphere', detecting gases from hydrogen and helium at the 'low end' of the atomic mass scale up to krypton. A gravimeter would be carried on the Rover to investigate the deep substructure of the valley floor. A second gravimeter, of a completely different type, was to be set up at the ALSEP site. This was a speculative venture. Although Einstein's theory of relativity predicted the existence of gravity waves, they were proving difficult to confirm. Working with a counterpart on Earth, any signal detected by just one instrument could not be a gravity wave, but a signal sensed by both could *only* be a gravity wave passing through the Solar System. In effect, this instrument was a highly sensitive seismometer.

Marcus Langseth, sitting in Mission Control in Houston, was undoubtedly pleased to see Cernan boring *three* holes into the surface for his heat-flow investigation, though even with the benefit of an improved and hardier drill, it was tough work. "I had to grip it tightly," wrote Cernan, "and force my whole weight on it, but progress was no better than haphazard. The drill would find easy access for a few inches, then clunk against rock and kick back. My heart rate went up to 150 beats per minute, my hands hurt from squeezing the handle and dust swirled in a sticky haze." Cernan's increased heart rate alarmed flight surgeons and, indeed, he ate seriously into his precious oxygen supply. After a while, Flight Director Gerry Griffin told capcom Bob Parker to ask Schmitt to help Cernan extract the deep core sample from the ground.

For his own part, the geologist had his own troubles with the gravimeter – it needed to be perfectly level and, at length, a slightly exasperated Schmitt was forced to give it a good smack with one of his tools to 'adjust' it. In a sense, his efforts were in vain, because when the instrument was commanded to 'uncage' its sensor, a flaw in the mechanism prevented it attaining the sensitivity needed for its intended role. Nevertheless, the instrument proved able to detect normal seismic activity.

Called over to help Cernan, Schmitt tried to throw his weight on the jack that was being used to extract the core tube ... and abruptly lost his balance and fell flat on his face. This prompted a few chuckles from Houston, but Cernan was worried: had his partner damaged the precious machinery in his backpack? He need not have been overly concerned, for Schmitt was quickly back on his feet. By the time they finished with the core sample, they were 40 minutes behind schedule. Their first geological traverse would have to be shortened. "Instead of the mile-and-a-half trip south to [the crater] Emory," wrote Cernan, "we would stop halfway, in a boulder field near the crater Steno."

Not surprisingly, Schmitt was unhappy at losing part of their traverse – in fact, at their single geology stop somewhere near Steno, they could do little more than grab a few rocks and a bagful of pebbles and dust with the lunar rake.

Back in the vicinity of Challenger, Schmitt spontaneously broke into song:

"I was strolling on the Moon one day ... " he began.

Cernan joined the duet: " ... in the merry, merry month of December ... no, May!"

"May!" Schmitt corrected himself.

"May's the month!" confirmed Bob Parker.

Schmitt continued: " ... When much to my surprise, a pair of bonny eyes ... "

"Sorry, guys," Parker interjected, "but today *may* be December!"

Returning inside the lander after seven hours and 12 minutes, the two men were exhausted: their forearms ached and, after removing his gloves, Cernan noticed blood under his fingernails, undoubtedly from too much time spent struggling with the drill. In his autobiography, he also recalled the unusual sensation of *repressurising* Challenger's thin-walled cabin: it was almost as though an oil can had suddenly been filled with air. A loud 'bloop' noise was followed by the pressure forcing the hatch to visibly bulge outwards. It reminded Cernan of his visits to Grumman and made him realise how *fragile* this machine really was.

During their time inside the lander, the astronauts stored their suits at the back of the tiny cabin. However, they were sodden with sweat and in order to dry them in time for tomorrow's EVA, Cernan and Schmitt attached the helmets and gloves and hooked up the oxygen hoses to circulate air through them. "That was like inflating a pair of big balloons," wrote Cernan, "and it seemed as if two more guys had just crawled into our lunar pup tent." Their massive backpacks, meanwhile, were hung on the walls. They had a quick dinner, then debriefed over the radio with Mission Control and, for a few moments, rolled a couple of the rock samples over in their hands. Cernan was amazed. These pieces of dun-toned regolith had lain undisturbed for maybe three billion years and had been exposed to fearsome solar and cosmic radiation ... yet they did not look totally different to samples he had seen on his geology trips back on Earth.

By the time Cernan and Schmitt dozed off to sleep for their first night on an alien world, they had already been awake for the better part of 24 hours. Like previous crews, they strung their hammocks – Schmitt near the floor, Cernan above – and even without their pressurised suits it could hardly be described as comfortable. Initially, Cernan was too keyed-up to sleep; his mind raced with plans for tomorrow's excursion, which would take them to the South Massif. Every so often, he heard Schmitt breathing steadily, and sneezing occasionally in the midst of so much dust, but otherwise the lander and everything around it were eerily still and silent. There was no hushed breeze or patter of raindrops, Cernan wrote, or the slightest hint of anything else alive, save the two of them. He *was* physically and mentally exhausted, yet the irony of *sleeping* when they only had about 60 hours left on the surface seemed too much; at times, he lifted the window blind and gazed outside at the motionless flag and the Earth slowly rotating in its fixed position above the South Massif.

Wagner's rousing *Ride of the Valkyries* shocked them from their slumbers a few hours later and reminded Jack Schmitt of his days at Harvard University, almost a decade earlier. Every morning of the final exams, the students in each dormitory tied their stereos together and directed the speakers, at full-blast, into the outside courtyard ... and *anyone* in the vicinity literally *levitated* out of their beds! Fortunately, Mission Control did not play it quite that loud, but Cernan and Schmitt were more than ready for their second day's exploration on the surface. Schmitt found, to his pleasure, that the pain in his forearms had disappeared overnight; after the mission, he would guess that his cardiovascular system was so much more efficient in one-sixth gravity that it literally 'cleansed' the muscles of lactic acid and other waste products before they could cause any further damage.

There was also good news when they set foot on the surface. In preparing the Rover on the first day, Cernan had snagged the pull-out extension of one of the

Although their rest periods aboard Challenger were considerably more comfortable than earlier crews, conditions for Gene Cernan and Jack Schmitt cannot have been entirely pleasant ... or clean. In this image, an exhausted Cernan – clad in his water-cooled long underwear and dishevelled communications hat – smiles for Schmitt's camera, whilst their two space suits stand upright at the back of the tiny cabin. Note the lunar grime on the suits and on Cernan himself.

orange fenders with the handle of the geological hammer in his shin pocket. He had repaired it using duct tape, but during the traverse it became detached and rained dust down on the vehicle. John Young and the backup crew had crafted a repair during the night. Together with the engineers, they had folded four of the crew's geology maps into a rectangular shape, about the same size as a child's Halloween mask, and had taped them all together. Cernan and Schmitt were to do the same and then affix it to the remains of the fender using a pair of clamps normally used to hold lamps in the cabin. This improvisation *worked*, but it put them a full 80 minutes behind schedule. Nevertheless, Cernan would remember the hour-long drive to the base of the South Massif as one of the most exciting experiences from the entire mission, dodging craters and jerking the T-bar to negotiate each ridge and furrow.

On arrival at a broad trough-like depression at the base of the South Massif, they spent an hour sampling boulders which had tumbled down the flank of the 2 km mountain. "In fact," wrote Cernan, "we had tapped such a geological goldfield that Houston stretched our time there to the maximum and it was still frustrating to leave such a promising area." By now, Cernan was far more than an aviator – if Schmitt had learned to fly a lunar lander, then *he* had become an exceptional field geologist – and they would find common ground in that there was *never* enough time to explore properly. On the Moon, the demands of the clock were forever their enemy. In fact, the trough at the base of the massif was a fairly large crater – called Nansen in honour of the Norwegian explorer and later statesman Fridtjof Nansen – which had been partially filled in by material that had slumped off the mountain. For Schmitt, the landscape felt strangely familiar, reminding him of the Alpine valleys he had studied during his days at the University of Oslo.

In the minds of Schmitt's colleagues back on Earth, including Lee Silver and Bill Muehlberger, Apollo represented classic 'government science'. By this, they meant that the pace of missions was so fast that it was impossible to use the insights gained from one mission to inform the planning of the next. In fact, many findings from Dave Scott's and John Young's missions were still being discussed when Apollo 17 was on the Moon. "When this mission was over," wrote Andrew Chaikin of Apollo 17, "the great enterprise called Apollo was shutting down, with so much of the Moon left unexplored – and *that* was also classic government science."

Government science or not, the clock continued to tick remorselessly; and the sites that the astronauts could explore was dictated by the walk-back limit. Like Scott and Irwin and Young and Duke before them, Cernan and Schmitt were not allowed to *drive* further from Challenger than they would be able to *walk home* if the Rover conked out. This walk-back limit was extremely conservative, taking into account the possibility of damaged equipment, excessive oxygen usage and a multitude of other worries, to such an extent that Cernan and Schmitt had tried to push it as far as they could. Here they hit a solid wall: the walk-back limit would *not* be compromised. As Charlie Duke had once observed, amidst the grandeur and beauty and serenity of the Moon, it was all too easy to forget the cruel fact that there was a *vacuum* just millimetres away and the slightest leak in the suit could spell instantaneous death.

For now, though, frustration was taking its toll. The men would certainly have benefited from longer at Nansen, but they had the craters Lara – named for the character from *Doctor Zhivago* – together with Shorty and Camelot to explore. Approaching Lara, Schmitt described what he could see, partly for himself and capcom Bob Parker, but chiefly for his colleagues in the science support room. While Cernan took a core sample, Schmitt did some solo-sampling and occasionally toppled over into the soft lunar dust. After a while, Parker nicknamed him 'Twinkletoes' and radioed that the switchboard at Mission Control was lighting up with calls from the Houston Ballet Foundation, requesting Schmitt's services, prompting the scientists to start referring to this crater as 'Ballet'.

It was their next stop, at Shorty Crater, that provided one of the real surprises of the mission. More than a hundred metres wide, it was, radioed Schmitt, "a darker-rimmed crater ... the inner wall is quite blocky ... and the impression I have of the mounds in the bottom is that they look like slump masses that may have come off the side". Orbital images acquired by Al Worden during his solo mission on Apollo 15 had shown clear evidence of a dark halo around Shorty, which contrasted with the lighter surroundings. It was suggestive of a volcanic explosion crater and perhaps the source of the dark deposits elsewhere in the Taurus-Littrow region. Cernan and Schmitt had only half an hour scheduled at this site. Soon after they began sampling, the geologist's boots scuffed away at the dust and, there before his very eyes, was the ubiquitous grey ... and a slight tinge of *orange*.

Schmitt thought he was imagining it. Was his gold-tinted visor playing a trick on his eyes? He partly lifted the visor; it was still there. *Orange soil.* He called Cernan, who came bounding over. They confirmed it and jointly agreed that it looked like it had been oxidised, like the rust-coloured soil they had often seen in the desert during their geological expeditions with Lee Silver. In the science support room, Silver himself was excited: this *had* to prove that Shorty *was* a volcanic vent. Cernan was excited, too: it was, he wrote, *unexpected treasure* – like a Spanish conquistador finding jungle gold. Meanwhile, Schmitt set to work digging an exploratory trench into the orange deposit in order to trace its extent, and found that it spread along an ellipse-shaped area which ran parallel to the rim of the crater. In minutes, his opinion of Shorty had changed. When they arrived, he was convinced that it was an impact crater, but now – "if ever I saw a classic alteration halo around a volcanic crater, *this* is it!"

There was, however, no time to explore further. They had already spent ten minutes longer than planned at Nansen and the walk-back limit precluded an extension at Shorty. With only minutes available, Cernan drew his hammer and pounded a core tube into the orange soil at the base of the trench and both men were surprised that when they pulled it out, its surface was red for part of its length, but about a metre down it was a sort of purplish-grey, almost black. Subsequent analysis would show that the orange material was composed of tiny beads of glass which had once been molten lava droplets spewed into the lunar sky in a fire fountain, but Shorty was not the volcanic vent.

"The origin of the fountain," wrote Andrew Chaikin, "was a form of lava that contained dissolved volcanic gases. As it ascended from deep within the Moon to the surface, the effect was that of shaking up a bottle of soda and then uncapping it: the gas rapidly came out of solution, propelling molten rock high into the lunar sky. Just as water pressure in a decorative fountain causes the liquid to break up into droplets, this so-called 'fire fountain' was composed of an intensely hot spray. In the weak gravity, the droplets arced hundreds or perhaps thousands of feet through the vacuum. During their flight, they cooled into tiny glass spheres, which rained down on the valley of Taurus-Littrow." The beads derived their colour from the specific chemical composition of the lava – as indeed did the green beads recovered by Apollo 15. The dark soils of Taurus-Littrow would prove to be chemically identical to the orange stuff; the difference being that if the lava cooled rapidly it formed glass and if it cooled more slowly it produced dark crystals. In conclusion, wrote Chaikin, both the orange soil and the dark stuff on the valley floor were evidence of volcanism, but the fire fountains had not occurred recently: in fact, they were around three and a half *billion* years old. The impact some 19 million years ago that made Shorty had excavated the buried materials to the surface, ready for an astronaut nicknamed Twinkletoes to someday sample them.

Only one more period on the Moon's surface awaited them. On the afternoon of 13 December, a little under seven days since leaving Florida, Cernan and Schmitt were outside for the third time. Before launch, Schmitt – who had examined almost every aspect of the J-series space suits and their capabilities – had lobbied hard to get a fourth excursion tacked onto the mission, but to no avail: the conservative managers were aware that any emergency might leave them dangerously close to their water and battery reserves. Gene Cernan, too, was convinced that it *could* be done, but ultimately bowed to the judgement of programme manager Owen Morris.

The two men drove north past Sherlock Crater, swung right at Turning Point Rock and then headed across the lower flank of the North Massif. Orbital photographs had shown a large, dark-hued boulder, trailed by a five-hundred-metre furrow down the hillside. As they neared the boulder, Cernan and Schmitt could now see that it had broken into *five* fragments as it came to rest. Schmitt was in his element, making a clear and decisive field study of the boulder in an effort to piece together its history, whilst Cernan huffed and puffed upslope to take a series of panoramic images of the geologist at work. One of these images forms the front cover of this book.

By the time that Cernan and Schmitt had completed their final sampling stop at the Sculptured Hills, they had effectively explored Taurus-Littrow from one end of the valley to the other and with three extravehicular sessions in excess of seven hours apiece, they had easily amassed more time on the surface than any other crew. They had driven 30 km in the Rover – whose makeshift fender finally snapped off during the ride back towards Challenger – and their 75 hours on the Moon were drawing inexorably to a close. Both were exhausted, grimy and sore in their arms and hands, but their *suits* – those remarkable miniature spacecraft, upon which their lives depended – had come through with flying colours.

The glory of Apollo was ending and Cernan had known for some time that *he* would be the last man on the Moon for *many* years to come. Further missions to the Moon were not even a prospect for the Shuttle era – the longer-term plans for permanent space stations and a base on our closest celestial neighbour had seemingly evaporated. Notwithstanding all the rationale about the expense and the oft-cited need to 'solve problems on the ground' – many of which, even in 2010, remain *unsolved* – it is left to our generation to explain to our children and *their* children why we took such bold steps into the unknown ... and then *stopped*. A few years ago, during his election campaign, President Barack Obama asked "why should we send people into space when we have kids in the US who can't read?" In many minds, his words echo another politician's dreary excuse for failing to adequately fund space exploration ... and yet continue to more than adequately fund ongoing, seemingly unwinnable conflicts abroad. By playing on the urgency of solving gritty problems on Earth, many of which have never been solved, with or without space funding, presidents have always had a ready get-out-of-jail-free card and an assurance of popular electoral support. The comments made in 1969 by John Furst and George

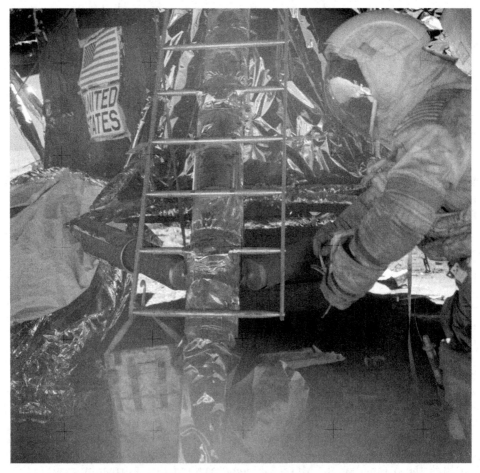

"As we leave the Moon ... ": Gene Cernan's words as he prepared to climb Challenger's ladder still haunt us in 2010, as yet another dream to return to the Moon is shattered by yet another short-sighted president. The question remains: *When* will we go back?

Romney and Edward Koch and others are both honourable and commendable, but today, as back then, there are *still* rats in apartments and there are *still* "problems on Earth". One must wonder if, four decades from now, there will also *still* be "kids in the US" who remain illiterate. Is this the fault of the space programme and is it our future space dreams which must suffer as a consequence? Or has the space programme actually done a great deal – and *promises* to do a great deal more – to inspire and to educate and to advance our younger generations? Will future generations understand and forgive us? Will historians speak kindly of the presidents – of Nixon and Ford and Carter and Reagan and Bush and Clinton and Bush and Obama – whose undoubtedly well-intentioned policies on Earth often proved sadly shortsighted in space? Will the 20th century be remembered as the time in which the greatest scientific, technical and engineering endeavour was triumphantly accomplished ... or as the time when our species won a stupendous victory and then lost its nerve to go further?

Such thoughts were clearly on Cernan's mind when, shortly after 12:30 am EST on 14 December 1972, he took his final steps. He turned for one last look at the stark landscape – the Sculptured Hills, the North and South Massifs, the thousands of craters, the dark sky and the Earth hanging silently above – and suddenly found the words that he wanted to say. More than three years earlier, Neil Armstrong's first words had been uttered in triumph; now Cernan's *last words* were uttered quietly and with undisguised angst. "Bob," he radioed to the ever-present capcom Bob Parker in Mission Control, "this is Gene. As I take these last steps from the surface, back home for some time to come, but we believe not too long into the future, I believe history will record that America's *challenge* of today has forged man's *destiny* of tomorrow ... And as we leave the Moon at Taurus-Littrow, we leave as we came and, God willing, as we shall return, with peace and hope for all mankind. Godspeed the crew of Apollo 17."

CONVERGENT PATHS

A few hours later, Cernan and Schmitt were reunited with a joyous Ron Evans, who would also enter the history books as the last man to orbit the Moon solo. Like Cernan, he was a naval aviator and had spent the last three days operating a SIMbay packed with scientific gear ... and taking time to simply *marvel* at a sight only a handful of his kind had ever seen. Like those few enlightened ones who came before him, Evans knew that he would never come this way again and he had been determined to make the most of it. One event that would prove particularly memorable would be the final transearth EVA, when he spent an hour outside recovering the precious SIMbay film cassettes.

Ronald Ellwin Evans had been born on 10 November 1933 in St Francis, Kansas. He received a degree in electrical engineering from the University of Kansas and joined the Navy as an ensign through the reserve officer training corps. After initial training, he served as a combat pilot and later as a flight instructor. He had been one of the aviators whose names were forwarded to NASA in the summer of 1963, along

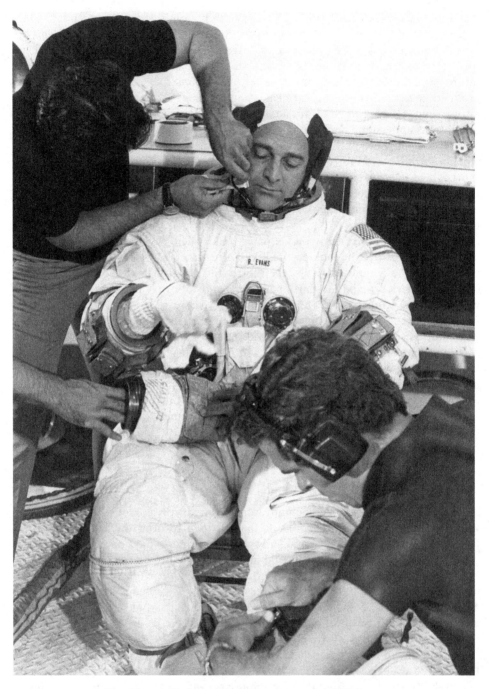

Ron Evans, the last man to fly solo around the Moon.

with those of Cernan and Dick Gordon, but on this occasion Evans was not selected. When Cernan *was* picked that October, Evans was deeply disappointed. However, his chance would finally come three years hence. Still, wrote Cernan, "no fortune teller could have predicted that, in not too many years, Ron and I would be together in a rocket ship, flying to the Moon".

After missing the 1963 NASA selection, Evans was assigned to the Naval Postgraduate School in Monterey, California, from where he earned a master's degree in aeronautical engineering. At length, just a few weeks before Gene Cernan made his first space mission aboard Gemini IX-A, Evans was bringing his F-8 Crusader in to land on the deck of the aircraft carrier *Ticonderoga*, on station in the Gulf of Tonkin. He had just completed another bombing run over North Vietnam – one of more than a hundred on his tally – and was asked to report, sweaty and tired, to the ready room. Evans flopped into a chair and must have nearly *floated out* again when the ship's captain entered the room and read a teletype message for him: Lieutenant-Commander Ronald E. Evans had been picked as one of NASA's newest astronaut candidates. One of the strangest ironies was that the ship aboard which Evans was serving at that time would be the very same one that would pluck the command module America from the Pacific Ocean on 19 December 1972.

In the same manner as Evans' career had come full-circle, so America's space programme had ridden a wave from John Kennedy's rousing speech in 1961 to the completion of humanity's initial exploration of the Moon a little over a decade later. From a historical perspective, Apollo was, is and will remain an enigma – a phenomenon, locked in time, whose genesis occurred in an impossibly short period, took place at astonishing speed ... and ended in a heartbeat. Fifty years or more did not elapse after the first sighting of the New World before another footprint was planted there; nor did Captain James Cook's successors wait so long before landing again in Australia or New Zealand or Tahiti or Hawaii. These voyages, though they took place on a planet capable of supporting life, were no less hazardous for those involved; Ferdinand Magellan, for example, lost his life making the first circumnavigation of the world's oceans. When Kennedy spoke of the *choice* to go the Moon, he did not speak in words which implied that it would be a singular occurrence – his implication was that the journey was *not* merely a race with the Soviet Union, but was demonstrative of a new-found ability to explore the heavens, coupled with an eagerness to share its meaning with others. Kennedy's words may well have been sincere, but half a century on, the footprints of Armstrong and Aldrin, Conrad and Bean, Shepard and Mitchell, Scott and Irwin, Young and Duke and Cernan and Schmitt are an achievement that we have yet to surpass. Maybe it *was* just a cynical political ploy to beat the Russians, after all. If this was the case, then perhaps Kennedy's promise to commit America to the Moon and share its meaning with others has failed. Perhaps it represents our species' greatest failure of the 20th century. From my perspective, born four years after Cernan and Schmitt left their bootprints at Taurus-Littrow, and now in my 34th year, having just heard news that future Moon landings are again off NASA's agenda, it is difficult not to be pessimistic. The Moon today seems further away than ever; both it and Mars remain the exclusive preserve of our robotic emissaries. Yet, reading between the lines of

Nicknamed 'The Blue Marble', this stunning view of Earth was acquired by the Apollo 17 astronauts. As the Sixties wore into the Seventies, the remarkable adaptability of our species made it possible for us to leave the cradle which has nurtured us for millions of years and learn to live elsewhere. Through triumph and tragedy, the United States and the Soviet Union pursued their space ambitions purely for political and technological prestige ... but, in doing so, established a firm foothold in the heavens and opened our eyes to the Universe around us.

Barack Obama's 2010 redirection of NASA, I have to wonder if the Moon *will* be revisited in my lifetime – perhaps private entrepreneurs *do* hold the key to the future. Many have already spoken cogently and realistically of $100 million fare-paying circumlunar trips as soon as 2020. When one reads of these plans, it is hard *not* to be encouraged. We can only hope that the next generation of lunar explorers represent something different from the exclusively 'government science' of Apollo.

Throughout the years covered by this book, two quite different paths in space were followed by the Soviet Union and the United States. The former, despite its lunar ambitions, sought to establish a more gradual and increasingly permanent human presence in low-Earth orbit, through long-duration flights and ultimately the first civilian and military Salyut stations. Whilst not 'spectacular', in terms of popular appeal, these missions had, by the end of the Seventies, firmly established the Soviets as record-breakers in terms of space endurance and made them the first to truly set up a home in orbit. America, on the other hand, would bask in the glory of Apollo and a series of triumphant landings on the Moon and then operate its own space station, Skylab, while developing the Space Shuttle, which was intended to be a 'cheaper' means of getting people into orbit.

The race between the communist East and the capitalist West during the late Sixties and into the early Seventies seesawed back and forth: first came the shock of Yuri Gagarin's pioneering flight, then Valentina Tereshkova and the spacewalk of Alexei Leonov, rapidly followed by the successes of Gemini, the journey of Apollo 8 to the Moon and the triumph of Neil Armstrong and Buzz Aldrin. Tragedy, too, had halted both nations in their tracks: the Apollo fire, the horror of Soyuz 1, the near-disaster which befell Apollo 13 and the tragic loss of Georgi Dobrovolski, Vladislav Volkov and Viktor Patsayev aboard Soyuz 11.

It is ironic, in a way, that one of the fundamental reasons why space spending declined so much – in *both* nations – as one decade wore into the next, was the steady thawing of their mutual relationship. The need for a cheaper means of placing men into space was acute: there were no real missile gaps between the superpowers and neither nation could afford to continue signing blank cheques to fund a series of celestial spectaculars simply to out-do one another. Nor could they afford to sustain continued hostility; America was being increasingly bloodied in Vietnam and Soviet relations with China were tense. A series of US-Soviet meetings in Helsinki and Vienna, known as the Strategic Arms Limitation Talks, hammered out the details of a chain of agreements to limit the stocks of ballistic missiles on both sides. One of the later deals, the Anti-Ballistic Missile Treaty, signed by Richard Nixon and Leonid Brezhnev in May 1972, allowed each nation only *two* sites on which they could base major defensive systems. Two years after the treaty, it was further decided to limit this number to one site each: in America, based at Grand Forks Air Force Base in North Dakota, and in the Soviet Union, centred on Moscow. Although the United States unilaterally withdrew from the Anti-Ballistic Missile Treaty in 2001, it was – for nearly three decades – considered one of the landmark events in the limitation of arms of mass destruction.

When the Prevention of Nuclear War Agreement was reached in June 1973, and both nations agreed to implement policies to restrain hostility, it was quite possibly the high-watermark in Soviet-American détente. In fact, discussions of co-operation in space matters had been underway for several years, with the sharing of biomedical data and exchanges of lunar soil samples from manned and unmanned missions: Apollos 14 and 15 and the Soviet Luna 16. Talks to stage a joint manned mission were also underway; in 1970–71, several high-level NASA managers travelled to Russia and their Soviet counterparts visited the United States. Such plans were set in stone when the Anti-Ballistic Missile Treaty was signed: Nixon and Brezhnev agreed that a three-man Apollo would dock with a two-man Soyuz in the summer of 1975. The ambitious mission, it was hoped, would bring a sense of closure to the embittered competition which had previously dominated the Sixties. It would also offer Deke Slayton his only chance of ever riding a rocket into the heavens.

The mission, later called 'Apollo-Soyuz', would also bring closure to the first episode of the American manned exploration of space; for it would be the last time their astronauts ventured aloft until the arrival of the Shuttle. A glum hiatus of more than half a decade would pass before the vehicle would spread its wings. As the Seventies wore into the Eighties, the United States introduced a totally new era of exploration. The Shuttle, it was promised, would fly fortnightly or even weekly, trucking crews of up to *seven* astronauts into orbit at a time. They would launch and repair satellites, operate scientific and medical experiments and perhaps build and sustain a permanent space station. The Soviets, for their part, had already operated several space stations, and by 1989 *their* cosmonauts would have flown missions lasting a full year. Whilst there would be no more bootprints on the Moon and no chance of *any* on Mars, by the end of the Eighties we would have begun to learn the skills needed to *live* in space for the first time. The early Seventies provided a taste of what it was like to operate off the planet for long periods of time and to undertake serious scientific exploration of another world. By the close of the Eighties, we would have taken still longer strides and would have begun to establish a real *home* in space.

Bibliography

'A Bit of Fear.' *Time*, 25 February 1966.

'Poetry and Perfection.' *Time*, 3 January 1969.

'The Russians' Turn.' *Time*, 24 January 1969.

'Post-Lunar Quarantine.' NASA News Release, Manned Spacecraft Center, Houston, Texas, 24 January 1969.

'The Spider and the Gumdrop.' *Time*, 14 February 1969.

'Apollo 9 Press Kit.' NASA Headquarters, Washington, DC, 23 February 1969.

'Apollo's Unsung Hero.' *Time*, 28 February 1969.

'Apollo 9 Launched.' *Flight International*, 6 March 1969.

'A Spectacular Step Toward Lunar Landing.' *Time*, 14 March 1969.

'Apollo 9: A Total Success.' *Flight International*, 20 March 1969.

'Rousing End to a Relaxed Flight.' *Time*, 21 March 1969.

'Photography at New Heights.' *Time*, 28 March 1969.

'Apollo 10 Press Kit.' NASA Headquarters, Washington, DC, 7 May 1969.

'Is the Moon the Limit for the US?' *Time*, 9 May 1969.

'Dress Rehearsal.' *Time*, 16 May 1969.

'Nine Miles From the Goal.' *Time*, 30 May 1969.

'An Uncluttered Path to the Moon.' *Time*, 6 June 1969.

'Apollo 11 Lunar Landing Mission Press Kit.' NASA Headquarters, Washington, DC, 6 July 1969.

'The Crew: Men Apart.' *Time*, 18 July 1969.

'Scoopy, Snoopy or Sour Grapes?' *Time*, 25 July 1969.

'A Giant Leap for Mankind.' *Time*, 25 July 1969.

'Awe, Hope and Scepticism on Planet Earth.' *Time*, 25 July 1969.

'The flights of Soyuz 6, 7 and 8.' *Flight International*, 23 October 1969.

'Back to the Moon.' *Time*, 24 October 1969.

'Orbital Troika.' *Time*, 24 October 1969.

'Off to the Moon Again.' *Time*, 14 November 1969.

'Blithe Spirits in Space.' *Time*, 21 November 1969.

'Towards the Ocean of Storms.' *Time*, 21 November 1969.

'Bull's Eye for the Intrepid Travellers.' *Time*, 28 November 1969.

'The Moon – Through the Looking Glass.' *Time*, 28 November 1969.

'A New View of the Ocean of Storms.' *Time*, 5 December 1969.

'Aims and Costs of the Soviet Space Station Program.' Central Intelligence Agency Directorate of Intelligence, January 1970.

'Rescheduling of Apollo 13 to April 11.' NASA News Release, Manned Spacecraft Center, Houston, Texas, 8 January 1970.

'Naming of Apollo 15 Crew.' NASA News Release, Manned Spacecraft Center, Houston, Texas, 26 March 1970.

'Apollo 13 Press Kit.' NASA Headquarters, Washington, DC, 2 April 1970.

'Apollo 14 Landing Site & Launch Slip.' NASA News Release, Manned Spacecraft Center, Houston, Texas, 7 May 1970.

'Discovery of 4.6 Billion Year Old Moon Rock from Apollo 12.' NASA News Release, Manned Spacecraft Center, Houston, Texas, 26 May 1970.

'Back in Orbit.' *Time*, 15 June 1970.

'Success for Soyuz.' *Time*, 29 June 1970.

'Site Selection for Apollo 15.' NASA News Release, Manned Spacecraft Center, Houston, Texas, 1 October 1970.

'Apollo 14 Nears Jan. 31 Launch.' *The Astrogram* XIII, No. 4, 10 December 1970.

'Apollo 14 Press Kit.' NASA Headquarters, Washington, DC, 21 January 1971.

'Accident Board established for Cernan's helicopter accident.' NASA News Release, Manned Spacecraft Center, Houston, Texas, 25 January 1971.

'To Fra Mauro and Beyond.' *Time*, 1 February 1971.

'Announcement of Apollo 16 Crew.' NASA News Release, Manned Spacecraft Center, Houston, Texas, 3 March 1971.

'Apollo 14 Rock Report.' *Flight International*, 11 March 1971.

'Quarantine Discontinued.' NASA News Release, Manned Spacecraft Center, Houston, Texas, 28 April 1971.

'A Salyut for Russia.' *Time*, 3 May 1971.

'Earth Week and Beyond.' *Time*, 3 May 1971.

'A Troubled Salyut.' *Time*, 10 May 1971.

'Apollo 16 Site Selection.' NASA News Release, Manned Spacecraft Center, Houston, Texas, 17 June 1971.

'A Russian Success.' *Time*, 21 June 1971.

'Duration Record for Soyuz 11?' *Flight International*, 24 June 1971.

'Apollo 14 Soil Results.' NASA News Release, Manned Spacecraft Center, Houston, Texas, 25 June 1971.

'Triumph and Tragedy of Soyuz 11.' *Time*, 12 July 1971.

'Wear Suits During LM Jettison.' NASA News Release, Manned Spacecraft Center, Houston, Texas, 19 July 1971.

'Apollo 17 Crew Announcement.' NASA News Release, Manned Spacecraft Center, Houston, Texas, 12 August 1971.

'Age Dating of Genesis Rock.' NASA News Release, Manned Spacecraft Center, Houston, Texas, 17 September 1971.

'Apollo 15 Parachute Failure.' NASA News Release, Manned Spacecraft Center, Houston, Texas, 28 September 1971.

'Cernan Accident Report.' NASA News Release, Manned Spacecraft Center, Houston, Texas, 18 October 1971.

'LM Pilot Charles M. Duke Hospitalised.' NASA News Release, Manned Spacecraft Center, Houston, Texas, 4 January 1972.

'Apollo 16 Mission Rescheduled.' NASA News Release, Manned Spacecraft Center, Houston, Texas, 7 January 1972.

'Apollo 17 Site Selection.' NASA News Release, Manned Spacecraft Center, Houston, Texas, 16 February 1972.

'Apollo 15 Stamps.' NASA News Release, Manned Spacecraft Center, Houston, Texas, 11 July 1972.

'Apollo 17 Press Kit.' NASA Headquarters, Washington, DC, 26 November 1972.

'Salyut 2 Mission in Difficulties?' *Flight International*, 19 April 1973.

'Soyuz 14 Crew Back.' *Flight International*, 25 July 1974.

'Soyuz/Salyut In Trouble Again?' *Flight International*, 5 September 1974.

'Soyuz Setback.' *Time*, 9 September 1974.

'Cosmonauts Train at JSC; US Reps Meet in Moscow.' *Roundup*, NASA Lyndon B. Johnson Space Center, 13 September 1974.

'Soyuz 18 Cosmonauts Aboard Salyut 4.' *Flight International*, 5 June 1975.

'Space Spectacular.' *Time*, 16 January 1978.

'Soyuz 5's flaming return.' *Flight Journal*, June 2002.

'A cosmonaut's once-secret story.' *St Petersburg Times*, 8 June 2006.

Aldrin, Buzz and McConnell, Malcolm (1989) *Men From Earth*. New York: Bantam.

Andrews, James T. 'In Search of a Red Cosmos.' In *Societal Impact of Spaceflight* by Steven J. Dick and Roger D. Launius (eds.) Washington, DC: NASA History Division, Office of External Relations, 2007.

Bilstein, Roger E. (1989) *Orders of Magnitude: A History of the NACA and NASA, 1915-1990*. Office of Management: Scientific and Technical Information Division, NASA Headquarters, Washington, DC.

Brooks, Courtney, Grimwood, James M. and Swenson, Loyd S. (1979) *Chariots of Apollo: A History of Manned Lunar Spacecraft*. NASA Headquarters, Washington, DC.

Cernan, Eugene and Davis, Don (1999) *The Last Man on the Moon*. New York: St Martin's Press.

Chaikin, Andrew (1994) *A Man On The Moon*. London: Penguin.

Clark, Phillip (1988) *The Soviet Manned Space Programme*. London: Salamander.

Coleman, Fred (1997) *The Decline and Fall of the Soviet Empire: Forty Years That Shook the World, From Stalin to Yeltsin*. New York: St Martin's Press.

Collins, Michael (1974) *Carrying The Fire*. New York: Farrar, Straus and Giroux.

Conrad, Nancy and Klausner, Howard (2005) *Rocketman*. New York: New American Library.

Day, Dwayne A. (2006) 'Murdering Apollo: John F. Kennedy and the retreat from the lunar goal.' www.thespacereview.com/article/735/1. Retrieved September 2009.

Dick, Steven J. and Launius, Roger D. (eds.) *Societal Impact of Spaceflight*. Washington, DC: NASA History Division, Office of External Relations, 2007.

Doran, Jamie and Bizony, Piers (1998) *Starman*. London: Bloomsbury.

Evans, Ben (2009) *Escaping the Bonds of Earth*. Chichester: Praxis.

Hall, Rex D. and Shayler, David J. (2003) *Soyuz: A Universal Spacecraft*. Chichester: Praxis.

Hansen, James R. (2005) *First Man: The Life of Neil Armstrong*. London: Simon & Schuster.

Irwin, James B. and Emerson, William A., Jr. (1973) *To Rule the Night*. Philadelphia and New York: A.J. Holman Company.

Ivanovich, Grujica S. (2008) *Salyut: The First Space Station – Triumph and Tragedy*. Chichester: Praxis.

Jones, Eric M. and Glover, Ken (eds.) (2008) *Apollo Lunar Surface Journal*. Washington, DC: NASA History Division.

Kranz, Gene (2001) *Failure Is Not An Option*. New York: Berkley.

Kraft, Chris (2002) *Flight: My Life in Mission Control*. New York: Plume.

Mandrovsky, Boris. 'The Soyuz 6, Soyuz 7 and Soyuz 8 Mission'. NASA Headquarters: Office of Advanced Research and Technology, Biotechnology and Human Research Division, January 1970.

Portree, David S.F. and Treviño, Robert C. (1997) *Walking To Olympus: An EVA Chronology*. Washington, DC: History Office, NASA Headquarters.

Reinert, Patty 'Conversation with John Young.' *Houston Chronicle*, 17 December 2004.

Schweickart, Russell 'No Frames, No Boundaries.' In *Earth's Answer* by Michael Katz, William P. Marsh and Gail Gordon Thompson (eds.) Lindisfarne Books/ Harper and Row 1977.

Scott, David and Leonov, Alexei (2004) *Two Sides of the Moon*. London: Simon & Schuster.

Shayler, David J. (2002) *Apollo: The Lost and Forgotten Missions*. Chichester: Praxis.

Shayler, David J. and Burgess, Colin (2007) *NASA's Scientist-Astronauts*. Chichester: Praxis.

Shepard, Alan and Slayton, Deke (1994) *Moon Shot*. London: Virgin.

Siddiqi, Asif A. (2000) *Challenge to Apollo*. Washington, DC: NASA History Division, Office of Policy and Plans.

Simmons, Gene (1972) *On the Moon with Apollo 16: A Guidebook to the Descartes Region*. Washington, DC: NASA History Division.

Slayton, Donald K. and Cassutt, Michael (1994) *Deke*. New York: Forge.

Spudis, Paul D. 'The Moon.' In *The New Solar System* (4th edn.) by J. Kelly Beatty, Carolyn Collins Petersen and Andrew Chaikin (eds.) Cambridge, Massachusetts: Sky Publishing Corporation 1999.

Stafford, Thomas P. and Cassutt, Michael (2002) *We Have Capture*. Washington, DC: Smithsonian Books.

Swanson, Glen E. (ed.) *Before This Decade Is Out: Personal Reflections on the Apollo Program*. Gainesville, Florida: University Press of Florida 2002.

Thompson, Neal (2004) *Light This Candle*. New York: Three Rivers Press.

Wheeler, Robin (2009) 'Apollo lunar landing launch window: the controlling factors and constraints.' In *Apollo Flight Journal* by David Woods *et. al*. Washington, DC: NASA History Division.

Index